Second Edition

Handbook for

CRITICAL CLEANING

CLEANING AGENTS AND SYSTEMS

T0139006

Second Edition

Handbook for
CRITICAL
CLEANING
CLEANING AGENTS
AND SYSTEMS

Edited by Barbara Kanegsberg
Edward Kanegsberg

CRC Press
Taylor & Francis Group
Boca Raton London New York

CRC Press is an imprint of the
Taylor & Francis Group, an **informa** business

Cover design by David Kanegsberg.

CRC Press
Taylor & Francis Group
6000 Broken Sound Parkway NW, Suite 300
Boca Raton, FL 33487-2742

First issued in paperback 2017

© 2011 by Taylor and Francis Group, LLC
CRC Press is an imprint of Taylor & Francis Group, an Informa business

No claim to original U.S. Government works

ISBN 13: 978-1-4398-2827-4 (hbk)
ISBN 13: 978-1-138-07456-9 (pbk)

Visit the Taylor & Francis Web site at
http://www.taylorandfrancis.com

and the CRC Press Web site at
http://www.crcpress.com

To a beautiful, safe, productive world for the next generation

Noa Raeli Kanegsberg

To our most valuable collaborative efforts

Deborah Joan Kanegsberg and David Jule Kanegsberg

And to the memory and positive influence of

Israel Feinsilber

Jule Kanegsberg

Murray Steigman

Dr. Jacob J. Berman

Contents

Preface to the Second Edition

Why a Second Edition?

In the last few years, challenges to the manufacturing community have increased and so have performance expectations. With the ever-decreasing size of components, these expectations are becoming more difficult to meet. The second edition of the *Handbook for Critical Cleaning* hopes to help you meet these expectations and produce high-quality products in a cost-effective manner. Although cleaning is a process and not a chemical, increased awareness of the consequences of chemical use to workers, to the general public, and to the environment has prompted more stringent regulatory measures worldwide. Environmental and worker safety regulations are imperative to maintaining a decent quality of life on this planet and, perhaps, to our very survival. However, the goal of manufacturing is not to jump through regulatory hoops, but to produce efficient products. Compromising on the efficacy of cleaning and thereby producing a suboptimal product can affect public safety and can compromise on the quality of life. Manufacturers face the challenge of doing it all.

What Is Critical Cleaning?

Identify, then qualify/validate and monitor the critical cleaning steps. The terms "critical cleaning" and "precision cleaning" are often used interchangeably. However, we prefer the former term. Precision cleaning suggests cleaning in a highly restricted clean room, where each individual component is perhaps cleaned separately by a highly trained technician, where there are perhaps wet benches with automated product handling, and where there may be a multichamber-automated spray system that feeds directly into the clean room. This is a limited view of the important cleaning step. Perhaps the best way to define a critical cleaning step is to consider the negative consequences that arise if that step is not performed or is performed inadequately. Based on our experience, the important cleaning step, the *critical cleaning step*, may occur in a machine shop or in a job shop (e.g., a coating facility) in what appears, at first glance, like an automotive repair facility. If the soil (matter out of place) is not adequately removed at that step, subsequent processing and cleaning steps may not resolve the problem but may actually exacerbate contamination by inadvertent chemical reaction of the soil, drying of the soil, or by embedding the soil on the surface of the product. Contamination happens long before the product enters the clean room. A clean room can only minimize recontamination, and even the most sophisticated clean room or controlled environment cannot correct a contaminated product (Kanegsberg and Kanegsberg, 2010).

Lean Cleaning and Supply Chains

Economic pressures have led to the implementation of such concepts as lean cleaning and six-sigma; thus, we have to clean smarter. In fact, it is imperative that we clean smarter. Cleaning must be value added. Assess your own processes and understand the role that cleaning plays in those processes. Sometimes the value added is only appreciated by factoring in the costs of not cleaning or under-cleaning at a particular step. Critical cleaning is not just about *what* is done or *how* it is done. It is also about *when* it is done. This becomes even more important when we realize that most products are not built from scratch within one facility. There usually is a complex supply chain of autonomous or semiautonomous facilities, which may sometimes be separate divisions or departments of the same company. Regardless of whether the supply chain involves inter- or intracompany processes, it is crucial that communication take place and that process understanding and process integration occur. The most critical cleaning step may be one that needs to take place at a supplier, before the part reaches your facility.

Critical Cleaning and Surfaces

Cleaning is removing undesired materials from surfaces without changing the surface in an unaccept-able manner. As products get smaller, the surface becomes a greater percentage of the product. When products are at the nanoscale, it can be said that the product *is* the surface.

New and Useful

Cleaning Is a Process

The economic and regulatory hurdles involved in introducing new cleaning agents have increased considerably (see "A balancing act" in book 2). Therefore, chemicals that have been developed for markets other than cleaning but have been adapted for the cleaning sector and complex blends have become increasingly prevalent. Therefore, the newer cleaning products are covered extensively in the second edition. Cleaning equipment has also evolved during the past decade, and meshing the appropriate cleaning agent with the right equipment requires a working understanding of chemis-try, physics, and engineering. We have added discussion of ultrasonic techniques and monitoring. Partially spurred by regulatory pressures, an increased use of so-called nonchemical approaches is included.

Process Implementation

All of the knowledge in the world about cleaning agents, cleaning equipment, and process flow is of no use if you do not improve the cleaning process. We provide guidance to actually do something: to select, validate, implement, and monitor the cleaning process. We also cover new approaches to defini-tive, lean, analytical testing and provide discussions related to clean rooms, including construction and working in a clean room.

Applications

The application portion of the *Handbook for Critical Cleaning* has been expanded to include critical cleaning processes for high-value products such as for medical devices, pharmaceutical, food process-ing, aerospace, and military. Electronics cleaning, which had been considered to be "solved" a decade ago, has resurfaced as a critical issue due to such developments as miniaturization, increased component density, and replacement of lead solders with lead-free, higher temperature solders. Conservation of fine art may not immediately be thought of along with manufacturing, but this involves critical cleaning and the requirements are in some ways similar. Two art conservators outline the thought processes and trial-and-error determinations to match cleaning agents to the soil when cleaning or restoring paintings.

Safety/Environmental Considerations

Safety and environmental considerations are not only global issues but are also important concerns at the national and local levels, and they do not always coordinate or mesh well. You cannot ignore them, and you should not ignore them. We have not attempted to outline all regulations. Dealing with such a moving target would be frustrating and futile, and most engineers would develop glazed eyeballs. Instead, we have attempted to discuss a few topics that are important to the critical cleaning community and to provide strategies for working constructively with the regulatory world. Some of this guidance comes from members of the regulatory community.

Resource conservation is becoming an important topic in the twenty-first century. Efforts to minimize or recycle water, chemicals, and energy will increasingly become a factor in keeping process costs competitive. Green cleaning, which considers both safety and environmental impact, is discussed throughout the two books. The definition of green is not set in stone; it will continue to evolve.

Web-Based Material

Some of the authors have submitted non-print media (color illustrations, animations, film clips, etc.) to augment their chapters. These can be accessed via the "Downloads & Updates" tab on the web pages for these books at CRCPress.com.

The Lady in the Saffron Sari

Barbara Kanegsberg

"You must run, you must flee," implored the earnest gentleman as he ran toward us.

Puzzled and slightly alarmed, our daughter Deborah and I peered down a corridor of immense, multicolored marble slabs while balancing a finished wood cabinet door, a celadon green tile, and some decorative hardware. It was a brilliant, Southern California morning. The silhouette of the plaster Disneyland Matterhorn broke through a cloudless blue sky. The only obvious danger was the trauma of remodeling the kitchen.

"Why do we need to run?" I asked.

"You must hide, my wife must not see you," he replied.

"Why can't your wife see us?" our daughter asked.

"Because, you see, I told her, first we will select the marble, then the cabinets, then the tile, then the door pulls. You are coordinating. If she sees you, she will want to coordinate."

At that very moment, an elegant woman wearing a luminous, saffron-yellow sari came gliding across the marble yard.

"You see, dear," she said, putting her arm around the gentleman and steering him purposefully toward the exit, "they are coordinating. Let's go, we must coordinate too."

Coordinate, Extrapolate, Optimize

The lady in the saffron sari had the right idea. You, too, must coordinate. Achieving a high-quality manufactured product in a cost-competitive manner requires coordination of critical cleaning and contamination control within your company and perhaps coordination with the efforts of a complex supply chain. If you are in charge of selecting cleaning equipment, please read over the chapters on cleaning agents and coordinate the two efforts (and vice versa). Coordinating cleaning efforts with regulatory requirements, including safety, environmental, and validation requirements, is also time well spent. Whether you are a job shop, an initial fabricator, a final assembler, or you have a repair facility, understanding the importance of critical cleaning is a must to achieve a cost-competitive advantage.

It is reasonably safe to say that your manufacturing situation is unique. We suggest that you consider, blend, and extrapolate from the information and advice provided in both books, even perusing those chapters that seem outside of your field. We often combine the approaches of what, at first glance, seem to be unrelated fields. As you read the chapters, think about how approaches might apply to your application and where cleaning is really necessary. Always clean critically.

Acknowledgments

We want to express our profound gratitude to all contributors. Many of you composed your chapters during a time of professional and/or personal challenges; we thank all of you for your wonderful, useful, practical chapters. The information, expertise, and guidance provided in these chapters are invaluable. We would also like to thank Cindy Carelli, Jessica Vakili, Jennifer Smith, and the staff at CRC Press for supporting us throughout the process. A special thanks to Dr. Vinithan Sedumadhavan, the production project manager, for careful attention to detail and to turning the manuscripts into printed pages.

Our thanks also to our children, Deborah Kanegsberg and David Kanegsberg, daughter-in-law, Sandra Hart, and parents, Ruth Feinsilber and Mimi Steigman, for standing by us during the writing and editing process. Our granddaughter, Noa Raeli Kanegsberg, was a very special inspiration for creating this second edition.

Reference

Kanegsberg, B. and Kanegsberg, E. Contamination detection basics, *Controlled Environments Magazine*, June 2010.

Ed Kanegsberg
Barbara Kanegsberg

Preface to the First Edition

Adapted from: What is critical cleaning?, First Edition, *Handbook for Critical Cleaning*.

Critical cleaning is required for the physical manifestation of technology.

We are in the information age, an age of thought, ideas, communication. However, this technology is based on physical objects, parts, or components. Many of these objects require precision cleaning or critical cleaning because they are either intrinsically valuable, or they become valuable in the overall system or process in which they are used. Some parts or components require critical cleaning not because of the inherent value of the part itself but instead due to their place in the overall system. For example, inadequate cleaning of a small inexpensive gasket can potentially lead to catastrophic failure in an aerospace system.

Nearly all companies which manufacture or fabricate high-value physical objects (components, parts, assemblies) perform critical cleaning at one or more stages. These range from the giants of the semiconductor, aerospace, and biomedical world to a host of small to medium to large companies producing a dizzying array of components.

Soil

The concepts of contamination, cleaning, and efficacy of cleaning are open to debate and are intertwined with the overall manufacturing process and with the ultimate end-use of the assembled product.

Contamination or soil can be thought of as matter out of place (Petrulio and Kanegsberg, 1998). During manufacture, parts or components inevitably become contaminated. Contamination can come from the environment (dust, smog, skin particles, bacteria), from materials used as part of fabrication (oils, fluxes, polishing compounds), as a by-product of manufacturing, and as from residue of cleaning agent ostensibly meant to clean the component.

Cleaning

Cleaning processes are performed because some sort of soil must be removed. In a general sense, we can consider cleaning to be the removal of sufficient amounts of soil to allow adequate performance of the product, to obtain acceptable visual appearance as required, and to achieve the desired surface properties. You may notice that surface properties are included because most cleaning operations probably result in at least a subtle modification of the surface. If a change in the cleaning process removes additional soil and if as a result the surface acquires some undesirable characteristic (e.g., oxidation), then the cleaning process is not acceptable. Therefore, surface preparation and surface quality can be an inherent part of cleaning.

Identifying the Cleaning Operation

Cleaning processes and the need for cleaning would seem to be trivial to identify. If you had a child who appeared in the doorway covered with mud, you would do a visual assessment of the need for cleaning, perform site-directed immersion or spray cleaning in an aqueous/saponifier mixture with hand-drying. However, people perform critical cleaning operations without knowing it. This lack of understanding can detrimental to process control and product improvement.

Recognizing a cleaning step when it occurs is probably one of the major challenges in the components manufacturing community. Cleaning is often enmeshed as a step in the overall process rather than being recognized as a concept in itself. It may be considered as something that occurs before or after another process, but not as a process to be optimized on its own (Dorothy Rosa, personal communication). A cleaning process often not called a cleaning process. For example, optics deblocking (removing pitches and waxes), defluxing, degreasing, photoresist stripping and edge bead removal in wafer fabrication, and surface preparation prior to adhesion, coating, or heat treatment can all be thought of in terms of soil removal (cleaning). Sometimes the cleaning process is identified only by the name of the engineer who first introduced it.

The sociological and psychological bases for this aversion to discussing cleaning are no doubt fascinating, but are beyond the scope of this book. The important thing is for you to recognize a cleaning process when you see it.

There are several reasons. One obvious reason is process control. A second is trouble-shooting or failure analysis. If the product fails and you need to fix the process, it is crucial to identify not only where soil might be introduced but also what steps are currently being taken in soil removal. If the chemical being used comes under regulatory scrutiny, identifying cleaning is even more important. If a supplier provides the component and a problem arises, it is important to be able to recognize where the cleaning steps occur. Finally, identifying the cleaning steps allows you to apply technologies developed in other industries to your own process.

Critical Cleaning

Defining critical cleaning or precision cleaning is a matter of ongoing debate among chemists, engineers, production managers, and those in the regulatory community. Certainly the perceived value or end-use of the product is a factor as are the consequences of remaining soil. The level of allowable soil remaining after cleaning is a consideration. Precision cleaning has been defined as the removal of soil from objects that already appear to be clean in the first place (Carole LeBlanc, personal communication). In some instances, however, high levels of adherent soil are involved in the processing of critical devices. Precision cleaning was once euphemistically said to be YOUR cleaning process for YOUR critical application, whereas everyone else's process could be considered as general cleaning (Kanegsberg, 1993). In one sense, there is some truth that the manufacturer is often the one best able to understand process criticality. At the same time, recognizing general cleaning and critical cleaning as parts of other operations can lead to overall industrial process improvement.

Why Should You Be Concerned about Critical Cleaning?

Critical cleaning issues are becoming increasingly important. Competitive pressure is increasing. Higher demands are being made of industry. A clean component produced efficiently and in an environmentally preferred manner (or at least in an environmentally acceptable manner) is a given in today's economy.

Performance, Reliability

Products are becoming smaller, with tighter tolerances and higher performance standards. Some products, such as implantable biomedical devices, are expected to perform for decades without a breakdown. Small amounts of soil and very tiny particles can irreparably damage the product.

To successfully remove the soils, you have to understand the various cleaning chemistries and cleaning equipment, and how they are combined and meshed with the overall build process.

Costs

Pressure to keep costs down increases constantly. The costs of the effective processes have tended to increase. Choosing the best option for the application can keep costs down.

Safety and Environmental Regulatory Requirements

The manufacturing community needs a wide selection of chemicals and processes to achieve better contamination control at lower costs. However, our understanding of health and the environment has led to restrictions on chemicals and processes. The manufacturer needs an understanding of atmospheric science and of the approaches used by regulatory agencies to foresee future trends.

Overview of This Book

Philosophy

In setting out to put together this comprehensive book on critical cleaning, I sought inputs from the experts in the field. Frequently these are people associated with vendors of cleaning equipment and/or cleaning agents. Naturally, each person's viewpoint is somewhat colored by their own portion of the market. However, on the whole, I was impressed with the scope and fairness of the material submitted. An attempt has been made to minimize use of brand names. In some cases, there are several contributors in a similar area. In general, my philosophy has been to include all but the most blatant material; by having a large number of contributors, a wide range of products and viewpoints are presented; the reader is expected to be intelligent enough to weigh the advantages and/or disadvantages of each approach for his or her own application.

Conclusions

While each application is very site specific, contamination control problems cut across industry lines. At the same time, each industry still tends to work in a separate little world. It is hoped that this book will provide a synthesis of cleaning approaches.

A diverse assortment of components and assemblies require critical or precision cleaning. Some examples include

Accelerometers
Automotive parts
Biomedical/surgical/dental devices (e.g., pacemakers)
Bearings
Computer hardware (metal, plastic, other composites—the insides of your computer and printer)
Consumer hardware (telephones)
Digital cameras
Disk drives

Electronics components
Flat panel displays
Gaskets
Gyroscopes
Motion picture film
Optics
Space exploration hardware
Wafers/semiconductors/microelectronics
Weapons, defense systems (missiles)

Acknowledgments

This book is the result of a phenomenal level of effort by those involved in the worlds of critical cleaning, surface preparation, and environmental issues. The information, expertise, and guidance provided by the contributing authors is invaluable. Dr. Ed Kanegsberg, business associate and spouse, provided support, encouragement, and invaluable participation in the editing process. He also provided the viewpoint and experiences of a physicist and practicing engineer. Bob Stern and the staff at CRC Press provided excellent guidance throughout the process.

I would also like to thank family members Deborah Kanegsberg, David Kanegsberg, Ruth Feinsilber, and Mimi and Murray Steigman for their patience and encouragement.

Finally, I would like to thank Dr. Shelley Ventura-Cohen, a wise colleague and adviser. She tells the story of her aunt, who, on observing Shelley staring blankly at a cookbook while an inert, raw chicken sat on the counter, exclaimed: "look at the chicken, not the book." Dear reader, critical cleaning, surface preparation, and contamination control are complex subjects, but they are also intensely practical subjects which relate to a product—your product. My suggestion, therefore, is to look at this book, and at the same time look at the chicken.

References

Carole LeBlanc, Toxics Use Reduction Institute, Lowell, MA, personal communication.
Dorothy Rosa, *A²C² Magazine*, personal communication.
Kanegsberg, B. Options in the high-precision cleaning industry: Overview of Contamination Control Working Group XIII, in *International CFC & Halon Alternatives Conference*, Washington, DC, October 1993.
Petrulio, R. and Kanegsberg, B. Back to basics: The care and feeding of a vapor degreaser with new solvents, in presentation and proceedings, *Nepcon West '98*, Anaheim, CA, 1998.

Barbara Kanegsberg
BFK Solutions
Pacific Palisades, California

About the Second Edition

Philosophy

We want to help you clean critically, productively, and profitably; our goal was thus to make the second edition of the *Handbook for Critical Cleaning* even more comprehensive than the first edition. Contributors are experts in their field. We have included the viewpoints of manufacturers of parts/components of those who supply cleaning chemistries and cleaning systems, of people in regulatory agencies, and even of other consultants. We have minimized the use of brand names, but have included enough information to be unambiguous. Our philosophy is to include a range of viewpoints, some differing from our own. We urge you to make the optimal decision for your application.

Organization

Chapters in the Second Edition

The second edition of the *Handbook for Critical Cleaning* is substantially new. While we have reprinted a few classic chapters from the first edition, most chapters are new or have been substantially updated. We suggest that readers peruse not only the chapters related to their line of work and applications, but also look at what might at first glance appear to be unrelated applications. By providing a synthesis of cleaning approaches, we hope to help you make better decisions about your own cleaning processes.

We strive to achieve the impossible (or highly improbable)—a perfect balance of topics. After the publication of the first edition, we received comments that we did not devote enough space to aqueous processes, an approximately equal number of comments that we did not devote enough space to discussions on solvent processes, and assorted comments about a lack of attention to other advanced cleaning processes. We thank everyone for their comments; you are probably correct. Therefore, if you have a different viewpoint or unique cleaning application, let us know. This is how we keep learning and improving.

This series is divided into two books with five parts:

- Book 1: *Handbook for Critical Cleaning: Cleaning Agents and Systems*
  - Part I: Cleaning Agents
  - Part II: Cleaning Systems
- Book 2: *Handbook for Critical Cleaning: Applications, Processes, and Controls*
  - Part I: Process Implementation and Control
  - Part II: Applications
  - Part III: Safety and Regulations

Following is a capsule summary of each of the book chapters.

Book 1: *Handbook for Critical Cleaning: Cleaning Agents and Systems*

Part I: Cleaning Agents

An overview of cleaning agents is presented by the editor, Barbara Kanegsberg. In this expanded overview, Barbara attempts to capture the diversity of cleaning chemistry options and to put those options in perspective.

The part begins with a discussion on aqueous cleaning agents (see Chapter 1, Kanegsberg). Water is the most common cleaning agent. Michael Beeks and David Keller of Brulin & Company, a producer of aqueous cleaning equipment, expand and update their chapter from the first edition and give a comprehensive review of aqueous cleaning essentials (Chapter 2). Much of the information is also applicable to nonaqueous solvent cleaning.

Many of today's chemicals, both aqueous based and solvents, are blends. JoAnn Quitmeyer of Kyzen Corporation presents a new chapter that is a comprehensive review of cleaning agent chemistries, including single components and blends (Chapter 3).

John Burke of the Oakland Museum of Art, California, updates his particularly informative discourse on solubility and the techniques used to classify solvents (Chapter 4). It becomes clear from this chapter as to why certain solvents are applicable to removing certain types of soil.

John Owens of 3M updates his chapter on the hydrofluoroethers (HFEs), a class of solvents that have been introduced as replacements for the ozone-layer depleting chemicals (ODCs) (Chapter 5).

Joan Bartelt of DuPont updates the chapter by Abid Merchant (retired fom DuPont) that discusses the hydrofluorocarbons (HFCs), another class of ODC replacements (Chapter 6).

John Dingess and Richard Morford of EnviroTech International Inc. update the chapter by Ron Shubkin (retired from Albermarle Corporation and Poly Systems U.S.A. Inc.) on normal-propyl bromide (NPB), a substitute for the aggressive ODC solvent, 1-1-1-trichloroethane (Chapter 7).

Stephen P. Risotto, formerly of the Halogenated Solvents Industry Association (HSIA) and now with the American Chemistry Council, updates his contribution on the chlorinated solvents, a group of traditional solvents that are seeing a resurgence of use in certain applications (Chapter 8).

Ross Gustafson of Suncor Energy discusses critical cleaning applications of the bio-based D-limonene (Chapter 9).

Dan Skelly of Riverside Chemicals reviews benzotrifluorides, a group of VOC-exempt compounds (Chapter 10).

Part II: Cleaning Systems

This part reflects the wide range of process choices. The importance of drying is emphasized. Advanced and so-called nonchemical systems, such as CO_2 cleaning, steam cleaning, and plasma cleaning, are also covered. In these systems, the cleaning agent and the cleaning equipment are inseparable.

The part begins with an overview of cleaning systems contributed by the editors (Chapter 11). As with the overview for cleaning agents, this reviews processes that are treated in this book by other authors as well as those for which there are no additional chapters.

There are six chapters dealing with ultrasonics and the closely related megasonics technologies. The technology is widely used, and the diverse insights of the authors will be helpful to select equipment. John Fuchs, retired from Blackstone—Ney Ultrasonics, and Sami Awad of Ultrasonics Apps., LLC each give an overview of ultrasonics (Chapters 12 and 13). Sami Awad then teams up with K.R. Gopi, from Crest Ultrasonics Corp., to provide a new chapter on multiple frequency ultrasonics (Chapter 14). Mark Beck of Product Systems Inc. covers the basic technologies of megasonics (Chapter 15). The theory of cavitation has been absent from most discussions of critical cleaning geared to the manufacturing community. Further, the important yet elusive topic of ultrasonics metrics has seen much progress. Along these lines, we are pleased to present two new chapters. In the first chapter, Mark Hodnett of the National Physical Laboratories (U.K.) provides graphics covering theory and discusses a new technique for ultrasonics metrics along with case studies (Chapter 16). In the second chapter, Lawrence Azar of PPB Megasonics covers the principles and theoretical/mathematical basis of cavitation and discusses ultrasonics metrics (Chapter 17).

Edward Lamm of Branson Ultrasonics Corp. contributes a useful chapter, on optimizing the equipment design, covering solvent, aqueous, and semiaqueous cleaning equipment as well as rinsing, drying, automation, and other ancillary equipment (Chapter 18).

Ron Baldwin of Branson Ultrasonics Corp. contributes an important new chapter on equipment design for aqueous cleaning to help you during scale-up from laboratory cleaning to production cleaning (Chapter 19).

Dan Skelly of Riverside Chemicals contributes a chapter on equipment for cold cleaning, that is, where cleaning agents (notably solvents) are used below their boiling point (Chapter 20).

Richard Petrulio of B/E Aerospace provides a revised, expanded, and very readable chapter on the design of flushing systems (Chapter 21). This is one example where a company was able to design equipment for its own cleaning application. The chapter also provides very good guidance for the process of developing and testing a cleaning process.

Joe McChesney of Parts Cleaning Technologies updates and revises techniques for minimizing waste streams in solvent vapor degreasers, including methods for calculating the size or capacity of the required equipment (Chapter 22). Some recent case studies for the minimization of emissions have also been added.

Arthur Gillman of Unique Equipment Corporation contributes retrofitting vapor degreasers to allow the use of different cleaning chemicals or to meet newer emission control standards (Chapter 23). This option can obviate the need for new equipment.

John Durkee of precisioncleaning.com and Dr. Don Gray of the University of Rhode Island update their chapter on contained airless and airtight solvent systems, one approach for remaining in compliance with air regulations while using emissive chemicals (Chapter 24).

Wayne Mouser of Crest Ultrasonics Corp. contributes a new chapter on vapor phase organic solvent cleaning, a classic critical cleaning technique (Chapter 25).

In some cases, the cleaning agent and the cleaning equipment are inseparable. In particular, this is true for what are called "nonchemical" cleaning approaches. Several examples are provided in the next five chapters.

Ed Kanegsberg of BFK Solutions provides a new chapter, an overview to nonchemical cleaning, that addresses aspects covered in more detail by four of the authors and also reviews additional approaches, such as laser, UV/ozone, and fluidized dry bath cleaning (Chapter 26).

Jawn Swan of Crystal Mark, Inc. contributes a new chapter on micro-abrasive blasting, a technique with applications ranging from electronics and medical devices to architectural restoration (Chapter 27).

Robert Sherman of Applied Surface Technologies authors a new chapter on solid carbon dioxide cleaning, with applications for removing particles and small levels of soils from such critical surfaces as semiconductor wafers and precision optics (Chapter 28).

William Nelson of the U.S. EPA updates his chapter on supercritical and liquid CO_2 cleaning (Chapter 29).

William Moffat of Yield Environmental Systems (YES) teams with Kenneth Sautter, also of YES, to update the chapter on another approach to removing organics, plasma cleaning (Chapter 30).

Max Friedheim of PDQ Precision Inc. teams with his process engineer, Jose Gonzalez, to update and expand the chapter on the use of steam vapor cleaning for critical cleaning applications; additional case studies are included (Chapter 31).

John Russo of Separation Technologists has completely revised his chapter on selecting the best waste water treatment for aqueous operations (Chapter 32). This comprehensive chapter discusses pretreatment, posttreatment, and water recycling techniques.

Cleaning with liquids frequently means that drying is required. The final three chapters in Part II deal with this sometimes neglected process.

Barbara Kanegsberg of BFK Solutions provides an overview to drying (Chapter 33). Daniel VanderPyl of Sonic Air Systems updates his chapter on physical methods of drying (Chapter 34). Robert Polhamus of RLP Associates along with Phil Dale of Layton Technologies, Ltd. update their chapter on chemical displacement drying techniques (Chapter 35).

Book 2: *Handbook for Critical Cleaning: Applications, Processes, and Controls*

Part I: Process Implementation and Control

Part I integrates the topics of process selection and maintenance, contamination control, analytical techniques, and materials compatibility.

Barbara and Ed Kanegsberg lead off the part with a revised, expanded discussion of "How to Work with Vendors?" that applies to print, electronic, telephone, and face-to-face communication of information (Chapter 1). Barbara Kanegsberg continues with a new chapter, "The Balancing Act," discussing the technical, economic, political, and regulatory trade-offs and conflicts involved in developing and maintaining a process (Chapter 2).

Art Gillman of Unique Equipment Corporation contributes a new chapter, drawing on his many decades of experience as well as the experiences of his colleagues to present "Blunders, disasters, horror stories, and mistakes you can avoid," a compilation of cleaning lore that should be read and absorbed by all readers (Chapter 3).

Mike Callahan of Jacobs Engineering expands his chapter about optimizing and maintaining the process (Chapter 4). A number of topics such as fixturing, process monitoring, and process improvements are included.

Part I contains four new chapters related to clean room design, operation, and behavior. Controlling the cleaning environment improves the success of the cleaning process by minimizing product contamination. Scott Mackler of Cleanroom Consulting provides a comprehensive chapter on "Basis of design for life sciences cleanroom facilities" (Chapter 5). Kevina O'Donoghue of Specialised Sterile Environments brings a view from "across the pond" in Ireland with her chapter on "Validating and monitoring the cleanroom" (Chapter 6). Jan Eudy of Cintas Corporation provides a chapter on "Cleanroom management and gowning" (Chapter 7). Howard Siegerman of Siegerman & Assoc. and Karen Bonnell of Production Economics coauthor a chapter on the "Principles of wiping and cleaning validation" (Chapter 8). Ed Kanegsberg of BFK Solutions provides an overview of issues related to detection and measurement of contamination (Chapter 9). Finally, Ben Schiefelbein of RJ Lee Group provides a new, insightful chapter on the philosophy of and choices for analytical analysis with "Practical aspects of analyzing surfaces" (Chapter 10). Many of the techniques should be considered whether or not you operate in a clean room.

The next four chapters in this part address knowing when to clean, when the part is clean enough, and materials compatibility.

Mantosh Chawla of Photo Emission Tech. (PET), Inc. updates his chapter on the important topic of "How clean is clean? Measuring surface cleanliness and defining acceptable levels of cleanliness" (Chapter 11).

Two chapters are a must for those involved in process validation for implantable medical devices. Kierstan Andrascik of QVET Consulting brings her experience with analysis of medical devices in a new chapter on "Cleaning validations using extraction techniques" (Chapter 12). David Albert of NAMSA has expanded his chapter on "Testing methods for verifying medical device cleanliness" (Chapter 13).

Eric Eichinger of Boeing North America has expanded his chapter on the critical issue of materials compatibility both for metals and nonmetals (Chapter 14).

Part II: Applications

While each manufacturing situation may be thought of as unique, there are commonalities, and it can be helpful to explore common contamination problems in specific industrial sectors and to see how manufacturers in similar situations tackle cleaning problems. Therefore, the number of specific applications presented in this part has been expanded in the second edition.

Barbara Kanegsberg of BFK Solutions, with a contribution by Bev Christian of Research in Motion, provides "Clean critically: An overview of cleaning applications" (Chapter 15). Specific examples and case studies drawn from aerospace, electronics, and biomedicine are given.

A number of comprehensive new chapters are devoted to applications within the life sciences. John Broad of NAMSA and David Smith of Tissue Banks International provide a chapter on "Cleaning validation of reusable medical devices" (Chapter 16). Paul Lopolito of Steris Corporation provides insight into "Critical cleaning for pharmaceutical applications" (Chapter 17). Hein Timmerman of Diversey, Inc. in Belgium provides a chapter on "Cleaning in food processing" (Chapter 18).

Three new chapters reflect renewed importance of cleaning in electronic assembly applications. Mike Bixenman of Kyzen Corporation provides a comprehensive chapter on "Electronic assembly cleaning process considerations" (Chapter 19). Harald Wack of ZESTRON contributes a chapter on "Surfactant-free aqueous chemistries" (Chapter 20) and Helmut Schweigart, also of ZESTRON, writes about "Contamination-induced failure of electronic assemblies" (Chapter 21).

Ahmed Busnaina of Northeastern University treats the case of particle removal in his expanded chapter (Chapter 22). This is an effective treatment of surface physics presented in a readable and understandable manner.

Shawn Sahbari of Applied Chemical Laboratories, Mahmood Toofan of Semiconductor Analytical Services, and John Chu discuss the challenges faced in semiconducting wafer fabrication for aluminum interconnects (Chapter 23). In a new companion chapter, Shawn Sahbari and Mahmood Toofan discuss microelectronic cleaning with "Copper interconnect and particle cleaning" (Chapter 24).

The world of fine art is the subject of a new chapter, "The cleaning of paintings," by Chris Stavroudis, a paintings conservator in private practice, and Richard Wolbers of the University of Delaware (Chapter 25). Cleaning these valuable, critical surfaces involves careful formulation of cleaning chemistries as well as considerations involved in conserving surface qualities and attributes. If you are in manufacturing, peruse the chapter for ideas and approaches; there are more commonalities with the world of fine art than might be apparent on the surface.

Jason Marshall of Massachusetts Toxic Use Reduction Institute (TURI) provides a new "Road map for cleaning product selection for pollution prevention," outlining decisions that can be made to clean with less impact on workers and the environment (Chapter 26).

Bill Breault, Jay Soma, and Christine Fouts of Petroferm contribute a new focused case study on removing wax from aerospace build and assembly operations (Chapter 27). The approaches can be extrapolated to other operations.

Wayne Ziegler of the Army Research Laboratory and Tom Torres of the Naval Facilities Engineering Service Center (NFESC) team to provide a new chapter, "Implementation of environmentally preferable cleaning processes for military applications," that describes the efforts of the military Joint Services Solvent Substitution Working Group to implement processes across all military agencies with less use of hazardous or environmental degrading cleaning products while ensuring uncompromised performance (Chapter 28).

Part III: Safety and Regulations

Because cleaning almost always involves using materials or processes with environmental or safety "baggage," sucessfully navigating regulations and working with regulators has become part of the overall picture. This final part provides tools and approaches to achieving successful critical cleaning processes in a highly regulated world.

Barbara Kanegsberg begins with an expanded, frank overview of safety and environmental issues (Chapter 29).

Jim Unmack of Unmack Corporation contributes an updated and greatly expanded chapter that outlines health and safety aspects associated with cleaning processes (Chapter 30). Recognizing that manufacturing is a global issue, he provides insight on European as well as U.S. requirements.

Mohan Balagopalan of southern California's South Coast Air Quality Management District (SCAQMD) expands his thoughtful and frank discussion on working with regulators—from a regulator's viewpoint (Chapter 31).

Steve Andersen, recently retired from the U.S. EPA, teams with Margaret Sheppard of the U.S. EPA Significant New Alternatives Program (SNAP) to update and expand the discussion of how industry and government can work together, with a discussion on how lessons learned from the ODC phaseout can drive continuous environmental improvement (Chapter 32).

The book closes with an expansion of the chapter by Don Wuebbles of the University of Illinois-Urbana (Chapter 33). His chapter reviews "Screening techniques for environmental impact of cleaning agents."

Barbara Kanegsberg
Ed Kanegsberg
BFK Solutions
Pacific Palisades, California

Editors

Barbara Kanegsberg is the President of BFK Solutions LLC, Pacific Palisades, California, an independent consulting company established in 1994. BFK Solutions is now the industry leader in critical cleaning. As a recognized consultant in the areas of critical cleaning, contamination control, surface quality, and process validation, she helps companies optimize manufacturing cleaning processes, improve yield, resolve regulatory issues, and maintain trouble-free production. She has participated in projects that include aerospace/military equipment, electronics assembly, medical devices, engineered coatings, metals, pump repair, and optical and nanotechnological devices. She has also participated in a number of product development- and intellectual property-related projects for manufacturers of cleaning chemicals and industrial equipment. Prior to establishing BFK Solutions, she was involved in the substitution of ozone-depleting chemicals at Litton Industries. She also developed clinical diagnostic tests at BioScience Laboratories. She has a background in biology, biochemistry, and clinical chemistry.

Barbara is a recipient of the U.S. EPA Stratospheric Ozone Protection Award for her achievements in implementing effective, environment-friendly manufacturing processes. She has several publications to her credit in the areas of surface preparation, contamination control, critical cleaning, method validation, analytical techniques, and regulatory issues. She has organized and participated in numerous seminars, tutorials, and conferences, including programs at USC and UCLA. She regularly coauthors technical columns that appear in *Controlled Environments Magazine* and *Process Cleaning Magazine*. She also participates in standardization and guidance committees relating to aerospace, military, electronics, medical devices, and safety/regulatory issues. Examples

include the JS3, (interagency military group), the ASTM Medical Device Cleanliness Testing Task Force, and the IPC group revising the cleaning/defluxing handbook.

She has a BA in Biology from Bryn Mawr College and MS in biochemistry from Rutgers University, New Jersey.

Barbara, "the cleaning lady," can be reached at 310-459-3614 or Barbara@bfksolutions.com.

Ed Kanegsberg is the Vice President of BFK Solutions, Pacific Palisades, California; he is also a chemical physicist and engineer who troubleshoots and solves manufacturing production problems. Ed is a recognized advisor and consultant in the areas of industrial cleaning process design and process performance. He uses his four decades of practical experience along with his background in physics and engineering to help companies solve production problems and optimize their cleaning and contamination control processes. As a member of the technical staff at Litton Guidance and Control Systems Division, he was responsible for precision instrument development and technology transfer from prototype to the production facility.

Ed is an educator and believes firmly that most of rocket science can be easily understood by non-rocket scientists. He has authored several technical articles and coauthors technical columns with Barbara. He has delivered numerous presentations and particularly enjoys discussing the physics of cleaning and successful automation. He has nine patents in high-reliability instrumentation and multiple "company private" inventions.

He has a BS and PhD from Massachusetts Institute of Technology and Rutgers University, New Jersey, respectively.

Ed, "the rocket scientist," can be reached at (310)459-3614 or Ed@bfksolutions.com.

Contributors' Bios

Sami B. Awad, PhD is the head of Ultrasonic Apps., LLC, an advisory service that specializes in ultrasonic applications and helps current users in solving production issues and in process optimization. Dr. Awad helps companies in all aspects of a new project, from system and process design to production optimization, including selecting frequencies, equipment, chemistry and waste minimization. He has decades of experience in developing ultrasonic cleaning processes and developing innovative chemistries for general and ultrasonic precision cleaning and for surface treatements. Dr. Awad has a PhD degree in organic chemistry and has served as a professor of chemistry on the faculties of Drexel University, Philadelphia, PA, and Cairo University, Egypt. Dr. Awad was the VP of Technology and the Director of the Lab for 19 years at Crest Ultrasonic Corp.

> Contact: PO Box 180, Drexel Hill, PA 19026
> Phone: (610) 348-5895; e-mail: sawad@ultrasonicapps.com and sawad@rcn.com

Lawrence Azar is the president of PPB Megasonics, Lake Oswego, Oregon, established in 1996. He has a BS in engineering from the University of California at Berkeley, and an MS in engineering from Massachusetts Institute of Technology (MIT). His research at MIT focused on the nondestructive evaluation of materials and ultrasonic testing. He has published a number of articles and has two patents relating to ultrasonic cleaning. He is a member of the Acoustical Society of America, IEEE, and SEMI.

> Contact: 3 Monroe Parkway, Suite P438, Lake Oswego, Oregon 97035
> Phone: (503) 697-0828; e-mail: azar@megasonics.com

Ronald Baldwin is the engineering manager for precision cleaning at Branson Ultrasonics Corporation, Danbury, Connecticut. He is an expert in cleaning equipment and cleaning process design with 27 years of experience in the cleaning industry. Ronald is a graduate of Cornell University, Ithaca, New York and has a BS in mechanical engineering.

> Contact: 41 Eagle Rd., Danbury, Connecticut 06813-1961
> Phone: (203) 796-0471; email: ron.baldwin@emerson.com

Joan E. Bartelt joined DuPont in 1990 after receiving a PhD in Chemistry from Indiana University. She has worked predominately in the fluorinated materials groups of DuPont, in various technical roles in R&D, product development and technical service. She holds several patents for new fluorinated fluids and cleaning formulations.

> Contact: Chestnut Run Plaza Bldg 711, 4417 Lancaster Pike, Wilmington, DE 19805
> Phone: (302) 999-3625; email: joan.e.bartelt@usa.dupont.com

Mark Beck is the CEO of Product Systems Inc., Campbell, CA. He has over 30 years of experience in semiconductor processing and equipment design. Mark is a graduate of the University of California, Berkeley and has a BS in mechanical engineering.

    Contact: 1745 Dell Ave. Campbell, California 95008

    Phone: (408) 871-2500x104; e-mail: mbeck@prosysmeg.com

Michael Beeks is currently science advisor, R&D at Brulin & Company, Inc., Indianapolis, Indiana. He has forty plus years of experience in the cleaning industry as formulator and technical manager of household and industrial detergents. He has a bachelor's degree from Yankton College, Yankton, South Dakota.

    Contact: 2920 Dr. Andrew J. Brown Ave., Indianapolis, Indiana 46205

    Phone: (800) 776-7149x3630; e-mail: mbeeks@brulin.com

John Burke is the chief conservator for the Oakland Museum of California, Oakland, California. He has been active in the field of art conservation for over 38 years. He also serves as adjunct professor of conservation at John F. Kennedy University, Pleasant Hill, California, where he teaches a graduate course in preventive conservation, and is a director on the Board of the American Institute for Conservation.

    Contact: 1000 Oak Street, Oakland, California 94607

    Phone: (510) 238-3806; e-mail: jb@museumca.org

Phil Dale is the managing director at Layton Technologies, Ltd., Staffordshire, United Kingdom. He has 32 years of experience in the manufacturing business with extensive knowledge of the design and manufacture of complex precision industrial equipment.

    Contact: Unit 33 Parkhall Business Village, Parkhall Road, Weston Coyney, Staffordshire, UK ST3 5XA

    Phone: +44 (0) 1782 370400; e-mail: phildale@laytontechnologies.com

John Dingess has worked in the chemical industry for more than 30 years. As director of technology and chief chemist at Enviro Tech International (ETI) Inc., Melrose Park, Illinois (retired), he is an expert in n-propyl bromide chemistry for cleaning applications. Earlier affiliations include Safety-Kleen Corporation and Stepan Company. John is a longtime member of the American Chemical Society. He has a bachelor's degree in chemistry from the University of Illinois at Chicago Circle Campus.

    E-mail: jadingess@afo.net

John Durkee heads PrecisionCleaning and has over 30 years of experience in chemical and engineering businesses. He is a registered professional engineer in Texas. PrecisionCleaning is a global consulting and expert witness firm. Dr. Durkee works with both manufacturers and end users of cleaning equipment and cleaning fluids. As a consultant and industry leader, he provides guidance about the development and implementation of new technology including marketing advice. Dr. Durkee also works with legal teams both in an advisory or participatory capacity. As an experienced provider of support, he is familiar with the MPEP, creation of expert witness reports, the need for brevity in depositions, and the nature of court proceedings. Dr. Durkee's third book, on the science behind solvent cleaning, will be published by Elsevier in 2011.

    Contact: PO Box 847, Hunt, Texas 78024

    Phone: (830) 238-7610; e-mail: jdurkee@precisioncleaning.com

Jan Eudy is the corporate quality assurance manager for Cintas Corporation, Mason, Ohio. A registered microbiologist, Jan oversees research and development, directs the quality system and ISO registration at all cleanroom locations, and supports validation and sterile services. She is a certified quality auditor. She is also an active member of several professional organizations and is a fellow and IEST President Emeritus. Jan has a degree in medical technology from the University of Wisconsin, Madison, Wisconsin with graduate studies in medical microbiology at Creighton University, Omaha, Nebraska.

    Contact: 6800 Cintas Boulevard, Mason, Ohio 45040

    Phone: (513) 573-4165; e-mail: eudyj@cintas.com

Max Friedheim is the president of PDQ Precision Mini-Max, San Diego, California. He has been in the tool supply business since 1947. He is extensively involved in product development and is the holder of numerous patents. In 1997, on behalf of his company, he accepted the Clean Air Award for Technology from SCAQMD.

    Contact: PO Box 99838, San Diego, California 92169

    Phone: (858) 581-6370; e-mail: pdq@minimaxcleaner.com

F. John Fuchs has decades of experience with ultrasonics. He has been involved in significant developments in ultrasonic cleaning and related technologies. He has authored numerous educational articles and presented major papers on ultrasonics. Fuchs has a BS in industrial engineering from the University of Michigan. He is retired from Blackstone–Ney Ultrasonics.

Arthur Gillman is the president of Unique Equipment Corporation, Montrose, California, and has over half a century of practical and theoretical experience in an array of critical cleaning applications. He has developed equipment for both aqueous- and solvent-based processes. He has been an advisor on several SCAQMD committees. He has assisted numerous components and parts manufacturers in areas ranging from benchtop to very large airless applications.

    Contact: 2029 Verdugo Blvd. M/S 1005 Montrose, California 91020-1626

    Phone: (818) 409-8900; e-mail: agillman@uniqueequip.com

Jose Gonzalez is the general manager and specialist in technical support at PDQ Precision Mini-Max, San Diego, California. He has extensive experience in the development of industrial equipment.

    Contact: PO Box 99838, San Diego, California 92169

    Phone: (858) 581-6370; e-mail: ftcn1@aol.com

K.R. Gopi holds PhD in chemical engineering from Indian Institute of Technology, Madras, India and MTech in ceramic technology from Anna University, Chennai, India.

    Contact: Crest Ultrasonics, Advanced Ceramics Technology Sdn. BhD. Penang 14000, Malaysia

    Phone: +60(4)-507 0018; e-mail: kr_gopi@yahoo.com

Don Gray has been a full-time faculty member in the Department of Chemical Engineering at the University of Rhode Island, Kingston, Rhode Island, for 30 years. He has spent 20 years designing environmentally safe solvent processing equipment. Dr. Gray has over 10 patents or patent-pending designs including those related to pollution prevention methods and/or equipment for metal degreasing, carbon desorption, cavitation enhancement of mass transfer, and dry-cleaning techniques.

    Contact: Room 202, Crawford Hall, University of Rhode Island, Kingston, Rhode Island 02881

    Phone: (401) 874-2651; e-mail: gray@egr.uri.edu

Ross Gustafson is a technical specialist with the Asphalt Marketing Division of Suncor Energy, Calgary, Alberta. As technical director for Florida Chemical Company, Inc., Winter Haven, Florida, he worked for more than 12 years in developing citrus terpene formulations and applications for the cleaning industry. He has a BA in chemistry from Gustavus Adolphus College, Saint Peter, Minnesota and an MS in chemistry from the University of Colorado.

    Phone: (303) 506-1313; e-mail: rtgus@comcast.net

Mark Hodnett is a Senior Research Scientist at the National Physical Laboratory (NPL), Teddington, United Kingdom. He has 16 years' experience in developing measurement solutions for high-power ultrasonic fields and cavitation, and is responsible for NPL's; scientific research programme in this technically-challenging arena, helping to maintain NPL at the forefront of ultrasound metrology worldwide. He is also the manager of measurement services for low-power ultrasonic field characterization, applicable to medical devices. Mark has a BSc in Physics from the University of Surrey, Guildford, United Kingdom.
    Phone: (44) 20-8943-6365; e-mail: mark.hodnett@npl.co.uk

Barbara Kanegsberg, president of BFK Solutions, Pacific Palisades, California, is a recognized consultant in critical/industrial cleaning and contamination control. She helps companies optimize manufacturing cleaning processes, validate cleaning methods, improve yield, resolve regulatory issues, and maintain trouble-free production. She conducts dynamic workshops and training programs. Barbara received a U.S. EPA Stratospheric Ozone Protection Award. She has a BS in biology from Bryn Mawr College and an MS in biochemistry from Rutgers University.
    Contact: 16924 Livorno Dr., Pacific Palisades, California 90272
    Phone: (310) 459-3614; e-mail: Barbara@Bfksolutions.com

Ed Kanegsberg, vice president of BFK Solutions, Pacific Palisades, California, helps companies solve production problems and optimize their cleaning and contamination control processes. He has decades of experience in physics and engineering, including the transition of products from prototype to production. His writings emphasize the physics of cleaning and surface quality. He has a BS in physics from Massachusetts Institute of Technology and a PhD in physics from Rutgers University.
    Contact: 16924 Livorno Dr., Pacific Palisades, California 90272
    Phone: (310) 459-3614; e-mail: Ed@bfksolutions.com

David Keller has been an industrial/process chemist at Brulin & Company, Inc., Indianapolis, Indiana since 1997. Previously, David was employed at Wayne Chemical Corporation as a formulating chemist in the food sanitation industry. David has a BS in chemistry from Lawrence Technological University in Michigan and has performed graduate work in synthetic organic chemistry at Indiana University.
    Contact: 2920 Dr. Andrew J. Brown Ave., Indianapolis, Indiana 46205
    Phone: (800) 776-7149x3600; e-mail: dkeller@brulin.com

Edward W. Lamm is the worldwide technology manager for Precision Processing at Branson Ultrasonics Corporation, Danbury, Connecticut. After 15 years of experience in the petrochemical industry, he has been active since 1989 in both chemical and equipment manufacturing in the precision cleaning field. He holds a BA in chemistry and an MS in chemical engineering.
    Contact: 41 Eagle Rd., Danbury, Connecticut 06813
    Phone: (203) 796-0392; e-mail: edward.lamm@emerson.com

Joe McChesney is the director of sales and marketing at Parts Cleaning Technologies, Bowling Green, Kentucky. He has over 30 years of experience in engineering, design, and field applications of all types of cleaning equipment—both aqueous and solvent. Joe has an engineering degree from Western Kentucky University, Bowling Green, Kentucky. He has several published technical papers and participates in numerous organizations in regards to technological and environmental issues.
    Contact: 307 Emmett Ave., Bowling Green, Kentucky 42101
    Phone: (270) 746-0095 x 103; e-mail: jm@pct-1.com

Abid Merchant was with DuPont for over 30 years and was a lead technical person in the Cleaning Agent Group. He played a key role in developments of the HFC cleaning agents and holds many patents. He was a member of UNEP's Solvent Technical Options Committee and served as an expert to a task force to reconcile the Kyoto and Montreal Protocols. Abid received an M.S. degree in Chemical Engineering and an M.B.A degree. He is now retired.

William Moffat is the CEO and founder of Yield Engineering Systems (YES), Inc. He is a process engineer turned entrepreneur. He founded YES in 1980, introducing a vacuum bake/vapor prime system that innovated front-end processing in the semiconductor industry. Moffat received an HNC in electronics and an HNC in mechanics from Stockport Technical College. He incessantly whistles happy tunes.

Contact: 203-A Lawrence Drive, Livermore, California 94551-5152,
Phone: (925) 373-8353x 202; email: bmoffat@yieldengineering.com

Richard Morford is the current CEO of Enviro Tech International (ETI), Inc., Melrose Park, Illinois. He has served ETI as general counsel since 1996, leading the corporation's regulatory, and health and safety programs. Rich served as the executive director of the International Brominated Solvents Association, Ltd. from 2003 to 2007 and currently acts as president of a nonprofit youth sports program in the Chicago area. Richard is an alumni of DePaul University College of Law and DePaul University, Chicago, Illinois.

Contact: 2525 West LeMoyne Ave Melrose Park, Illinois 60160
Phone: (708) 343-6641x25; e-mail: rmorford@ensolv.com

Wayne L. Mouser, group vice president, Crest Ultrasonics Corporation, Trenton, New Jersey, has more than 42 years of experience in cleaning applications including development of aqueous, semi-aqueous, liquid/vapor, and low flash point solvent systems. Wayne holds multiple patents related to precision cleaning and has a BS from Western Kentucky University, Bowling Green, Kentucky.

Contact: 1886 Berkshire Lane N, Minneapolis, Minnesota 55441
Phone: (763) 559-1785; e-mail: wmouser@crestminneapolis.com

William M. Nelson is an environmental scientist at the U.S. EPA Office of Inspector General/OPE. He has been active in green chemistry for over 12 years and is a recognized expert in the area of green solvents. He received his PhD in organic chemistry from the Johns Hopkins University.

Contact: 77 W. Jackson Blvd., Chicago, Illinois 60604
Phone: (312) 886-6611; email: wmnelson@illinois.edu

John G. Owens is a lead research specialist with the Electronics Markets Materials Division Laboratory of the 3M Company. He has over 23 years of experience in the development of fluorinated compounds and their use in applications such as precision cleaning and holds 50 U.S. patents related to this work. He has bachelor's and master's degrees in chemical engineering from the University of Minnesota, Minneapolis, Minnesota, and the University of Virginia, Charlottesville, Virginia, respectively.

Contact: 3M Center Bldg., 236-3A-03, St. Paul, Minnesota 55144-1000
Phone: (651) 736-1309; email: jgowens@mmm.com

Richard Petrulio is the senior repair engineer at B/E Aerospace, Inc., Anaheim Facility. He has 24 years of experience in designing, manufacturing, testing and repairing of specialized compact refrigeration systems. He has extensive hands-on experience in cleaning process and equipment development. He holds a BS in mechanical engineering from California State Polytechnic University in Pomona, California and a MA in occupational studies from California State University in Long Beach. In addition, he holds a California State teaching credential in the areas of engineering, refrigeration and automotive mechanics.

Contact: 3355 East La Palma Avenue, Anaheim, California 92806
Phone: (949) 677-9863; email: richard_petrulio@beaerospace.com or rpetrulio@sbcglobal.net

Robert L. Polhamus is the principal for RLP Associates, Chester, New York, involved with sales of industrial cleaning equipment, and is a marketing consultant for industrial cleaning equipment. He has over 33 years of experience and has served on several industrial committees including ad hoc CFC elimination with the U.S. EPA. He has a BA in chemistry and an MBA.

   Contact: 25 Split Rock Rd, Chester, New York 10918

   Phone: (845) 469-6965; e-mail: rpolham@optimum.net

JoAnn Quitmeyer is the director of research and development at Kyzen Corporation, Nashville, Tennessee, a major chemical supplier of cleaning chemistries used in electronics, semiconductor, metal finishing and optics markets. She spent over 30 years formulating cleaners and lubricants for industrial applications including 9 years at Kyzen Corporation, 15 years at W.R. Grace, Columbia, Maryland, where she developed the Daraclean product line, and 13 years with the Magnus Division of Economics Laboratory. She was educated at the University of Minnesota, St. Paul, Minnesota.

   Contact: 430 Harding Industrial Drive, Nashville, Tennessee 37211

   Phone: (615) 831-0888; e-mail: joann_quitmeyer@kyzen.com

Stephen P. Risotto served as the Executive Director of the Halogenated Solvents Industry Alliance (HSIA), representing manufacturers and users of chlorinated solvents, for 11 years and has more than 25 years of experience understanding how these solvents are used in industrial and precision cleaning processes. He currently serves as a Senior Director for the American Chemistry Council in Washington, DC.

   Contact: Faye Graul, Halogenated Solvents Industry Alliance, Inc.1530 Wilson Boulevard, Suite 690, Arlington, VA 22209

   Phone: 703-875-0683; e-mail: srisotto@gmail.com (Risotto); fgraul@hsia.org (Graul)

John F. Russo is the president and founder of Separation Technologists, Methuen Massachusetts. He has worked in the fields high-purity water, filtration, and wastewater treatment for more than 35 years. He has authored more than 10 technical papers. He is a recipient of a U.S. EPA Stratospheric Ozone Protection Award for leadership in closed-loop water recycling. He holds an associate's degree in engineering and a BS in chemistry.

   Contact: Separation Technologists, Inc., 7A Raymond Ave., Unit A-7, Salem, NH 03079

   Phone: (603) 898-0020x101; e-mail: johnfrusso@separationtech.com

Kenneth Sautter, a senior process engineer at Yield Engineering Systems, Inc., Livermore, California since 2005, runs tests and troubleshoots processes for customers. He served as an officer in the U.S. Navy and worked for GCA, Solitec, and Vertical Circuits. He holds 16 patents related to equipment and processing. Sautter has a BS in chemistry from MIT and an MS in imaging and photographic science from RIT.

   Contact: 203-A Lawrence Drive, Livermore, California 94551-5152

   Phone: (925) 373-8353x204; e-mail: KSautter@yieldengineering.com

Dr. Robert Sherman is the president of Applied Surface Technologies, New Providence, New Jersey. He has a BS in physics from The Cooper Union, New York and a PhD in material science and engineering from the University of Illinois, Champaign, Illinois. His early work centered on surface analysis for practical material problems and process development. He has published over 40 papers, 3 patents, and numerous presentations. He started Applied Surface Technologies in 1991, where he focuses on new applications and methods with an emphasis on surface cleaning using CO_2 snow

   Contact: 15 Hawthorne Dr., New Providence, New Jersey 07974

   Phone: (908) 464-6675; e-mail: roberts@co2clean.com

Ronald L. Shubkin is retired. He was involved in a wide variety of projects in industrial research and development, including lubricant and solvent development. He holds 31 U.S. patents and has published extensively. He has a BS in chemistry from the University of North Carolina, and a PhD in inorganic chemistry (organometallics) from the University of Wisconsin.

P. Daniel Skelly is the technical service manager, Riverside Chemicals, North Tonawanda, New York. Previously, he was senior technical service specialist at Occidental Chemical Corporation, Dallas, Texas. He has worked extensively in cleaning applications with chlorinated solvents and the benzotrifluorides. He has a BS in chemical engineering from the University of Illinois and an MBA from Niagara University.

Phone: (716) 692-1350; e-mail: daniel.skelly@rivchem.com

Jawn Swan is the president of Crystal Mark, Inc., Glendale, California With over 47 years of experience he has been involved at all levels of micro sandblasting. He started out repairing SS White micro sandblasters in high school. He was responsible for maintaining various SS White micro sandblaster systems, including the LAT-100, MAT-800, and AT-701 thick-film resistor trimmer systems in the western United States. Over the years he has developed numerous micro sandblasting applications, including the design and building of automated systems.

Contact: 613 Justin Ave, Glendale, California 91201

Phone: (818) 240-7520x223; e-mail: jawn@crystalmarkinc.com

Daniel J. VanderPyl is the president of Sonic Air Systems, Inc., Brea, California, and during his 30 years in the blower and air knife industry, he has developed a wide array of product innovations and application engineering firsts. In addition to holding two blower and air knife patents, he has written numerous articles for a wide range of industry journals and is recognized as an authority in his field.

Contact: 1050 Beacon St., Brea, California 92821

Phone: (714) 255-0124; e-mail: dvanderpyl@sonicairsystems.com

I

Cleaning Agents

1

Cleaning Agents: Overview

Barbara Kanegsberg
BFK Solutions

Introduction—More than Molecules

Cleaning is a process, not simply a chemistry, so it is important to look at the cleaning agent and at the entire cleaning process, including the specific equipment selected. Having said this, matching the appropriate cleaning agent to the job at hand can be a traumatic experience, even to those with some background in chemistry. This chapter provides an overview of cleaning agents and highlights a few additional cleaning agents not discussed in other chapters; it is by no means exhaustive.

Cleaning agents are generally divided into aqueous and solvent. When most people involved in cleaning refer to the term solvent, they actually mean organic solvent. Organic solvents are not those processed from, say, free-range, herbicide-free lemons. Instead, organic refers to materials that have the element carbon in them. However, when you think about it, both organic- and aqueous-based cleaning agents are solvents. If you dissolve sugar in tea, the water is acting as the solvent; the sugar is the solute. Because many oils are carbon based, organic solvents have been classically used for very heavy

3

degreasing jobs. However, other mechanisms such as saponification (as described in Chapter 2, Beeks and Keller) have been successfully used to lift heavy grease off of parts. Certainly, water-based cleaning is widely used; most of us have successfully removed oils and greases from dishes using semiautomated aqueous cleaning, not a vapor degreaser.

While a segment of the cleaning industry has traditionally been carried out by aqueous cleaning, until the mid-1990s, most cleaning was conducted using classic chlorinated solvents or with ozone layer depleting compounds (ODCs). In fact, many products were designed specifically around the solvency properties of chlorofluorocarbon-113 (CFC-113) and 1,1,1-trichloroethane (TCA). The loss of ODCs and increasing regulatory constraints on classic solvents (notably chlorinated solvents) has lead to upheaval in the manufacturing world, and, because there are no true drop-in substitutes for ozone depleters, the market has fragmented (Kanegsberg 1996). This fragmentation has continued for a number of reasons, including lack of understanding as to how to use the new methods, dissatisfaction with selected new approaches, high cleaning costs, development of new solvent and aqueous blends, increasingly stringent and ever-changing regulatory conditions, differences in regulations in various geographical locations, increasingly stringent cleaning and performance requirements, new product design including micro and nanoscale designs, and the use of new process fluids including metalworking fluids.

To make matters even more complex, the line between cleaning agent and cleaning equipment or cleaning is often blurred. Sometimes the cleaning agent is generated in use, for example, CO_2 (solid, liquid, and supercritical) and plasma. Solid or abrasive media are, in a sense, both the cleaning agent and the cleaning process; these topics are discussed later in Chapter 26 (E. Kanegsberg), and Chapters 27, 28, and 29 (Swan, Sherman, Nelson, respectively).

No one product or class of products is likely to satisfy all cleaning requirements. The cleaning agent must be matched to the soil, the substrate (the component or part to be cleaned), the cleaning requirements, drying requirements, and other performance and environmental constraints. Inorganic soils are often referred to as hydrophilic; they dissolve effectively in water. Organic-based soils, often referred to as hydrophobic, tend to dissolve more effectively in organic solvents. The choice of cleaning agent should ideally be based on technical considerations. Unfortunately, chemists, engineers, production people, and those involved in regulatory agencies may themselves become hydrophilic or hydrophobic (Kanegsberg 1996). While we all have cleaning agent prejudices, irrationally ruling out one class of cleaning agents can result in inadequate cleaning and can be economically and ecologically detrimental.

Another reason to keep an open mind and to try to understand a range of approaches to cleaning is that the line between aqueous and organic cleaning tends to blur, so even if you and your firm are unalterably devoted to organic solvents, it will be helpful to learn about aqueous cleaning (and vice versa). Some inert organic cleaners are blended with small amounts of surfactants to improve removal of soils. Many aqueous cleaners contain significant amounts of organic additives. Some may be basically blends of water-soluble organic compounds. In fact, certain similar organic solvent blends may be classified as semi-aqueous or co-solvent only in that very small formulation differences allow for rinsing in water (for semi-aqueous) or another solvent (for co-solvent). Therefore, this overview emphasizes a number of specific organic compounds, not simply because they can be used for solvent cleaning per se but also because they are used in cleaning agent blends, including aqueous blends. We have not discussed ionic liquids; they may eventually have widespread adoption in cleaning applications.

An Introduction to CAS Numbers

In this chapter and in various places in *Handbook for Critical Cleaning*, periodic allusion is made to the Chemical Abstracts Service (CAS) number for a particular cleaning chemical. The CAS number or CAS Registry Number® is a registered trademark. While CAS numbers do not have chemical significance, they are valuable tools for everyone involved in critical/precision cleaning and manufacturing because they are internationally recognized, unique chemical identifiers of elements, inorganic compounds, and organic compounds. This means that even two optical isomers would each have a unique CAS number.

Because one chemical may have multiple names, in my experience, a CAS number is helpful in many aspects of precision cleaning, including process control, failure analysis, and safety/environmental regulatory compliance. For example, an engineer may refer to the cleaning agent as a nickname. "Trich" can refer to trichloroethylene or it may mean that, despite the phaseout of ozone-depleting chemicals, 1,1,1-TCA is being used. Freon® is a trade name, and it can refer to a number of chemicals.

We often find that it is difficult to identify the cleaning chemistry based on labeling information on drums, aerosol containers, and even material safety data sheets (MSDSs). Suppose you find that a cleaning process uses "FabuloKleen." You want to know the following: What is in the product? What are the technical/performance/compatibility characteristics? Are there regulatory issues? Someone hands you a bottle. You grab a magnifying glass, peer at the label, and find half a dozen chemicals, displayed in all of their incomprehensible, polysyllabic glory. Someone locates the MSDS; you find larger print, but the same problem. But wait! Next to each listed chemical, there is a CAS number. With that number and access to the Internet, it is typically possible to identify the chemicals in short order. This may set your mind at ease, or you may find a chemical that is overly reactive for your purposes or one that is under regulatory distress.

The CAS numbers provided in this and other chapters are believed to be from reliable sources. We consulted MSDS and Technical Data Sheets from established companies. In practice, we have encountered some inaccuracies, even in MSDSs, so it is best to cross-reference several sources. Additional clarification about CAS Registry numbers, including authoritative, definitive information (including issues such as intellectual property or Environmental Protection Agency [EPA] registration) is available online (CAS Registry 2010).

Please also be aware that the South Coast Air Quality Management District (SCAQMD), a regulatory agency in Southern California, has a CAS program; in this case, CAS is an acronym for Clean Air Solvent. The distinction should be apparent contextually.

Aqueous Cleaning Agents

Simple Additives

Water removes some soils. With appropriate cleaning action and constant rinsing to remove soils, water alone can clean. However, additives improve performance.

Some blends are relatively simple and are blended in-house. Such blends can be particularly desirable where residue is an issue either for the product or for disposal of the spent cleaning agent. Small amounts of peroxide (0.5%) have been added to water to clean and remove bacteria. Dilute hypochlorite (bleach) is often used to prevent biological contamination. Ammonia is often used for simple cleaning. Alcohol and acetone are sometimes added to boost cleaning power and promote rapid drying. Where low-flashpoint solvents are used, flammability must be considered in process development, even with aqueous cleaning agents. Sodium bicarbonate may be added. Acid washing and acid etching with Piranha, chromic acid, and other strong acids can be thought of as selective cleaning. While the solutions are simple, process control, process monitoring, employee safety, potential flammability, and environmental regulatory issues must all be considered.

Commercial Blends of Aqueous Cleaning Agents

Commercially available aqueous cleaning agents contain additive blends, often consisting of a dizzying array of organic and inorganic compounds. Some additive packages are totally inorganic; most are a mixture. While a few companies disclose the additive package, more typically, for competition-sensitive issues and other factors, the exact formulation is a closely held secret. A few examples of additives are provided in Table 1.1 (Cala 1996). Well-designed aqueous formulations are complex, sophisticated, and

TABLE 1.1   A Few Additives Used in Aqueous Formulations

Additive	Function	Description with Examples
Surfactants	Wettability Soil displacement/dispersion Solubilization	Single molecule with hydrophilic and hydrophobic portions May have long-chain organic portion (many carbons in a row) For example, alcohol ethoxylates
Defoamers	Control excessive foam Allow use in high-pressure spray applications, etc.	Poorly soluble in bath at operating temperature Impart slight "oil-like" quality Usually nonionic surfactants For example, nonionic block copolymers
Solvents, assorted	Decrease surface tension Adjust pH Improve solubility range	Typically soluble in water May be VOCs For example, butyl cellosolve, pyrrolidone, morpholine, glycol ethers, alcohols
Corrosion inhibitors, passivating	Prevent corrosion of metals	React with metal surface to reduce reactivity Typically oxidizing agents For example, chromates, nitrates, permanganates, chlorates Some reducing agents, e.g., Na sulfite
Corrosion inhibitors, non-passivating	Prevent corrosion of metals	Adsorption, formation of protective films For example, silicates (most common), pyrophosphates, carbonates, amines, gelatin, tannic acid, thiourea
Builders	Promote efficacy of cleaning by surfactants Sequester water hardness Maintain pH (acidity, alkalinity) Decrease metal, lead content of waste stream	Typically salts Chelating agents, for example, sodium tripolyphosphate; sodium hexametaphosphate, sodium citrate Precipitating builders (e.g., carbonates)
Hydrotropes	Promote solubility of organics in presence of high levels of inorganic salts	Important with inorganic surfactant packages For example, toluene sulfonates, short chain alcohols, benzoate salts
Oxidizers	Corrosion inhibitors Adsorption, dissolution in soils, oxygen release Better soil removal	For example, hydrogen peroxide

Source: Adapted from Cala, F.R. and Winston A.E., *Handbook of Aqueous Cleaning Technology for Electronic Assemblies*, Isle of Mann: Electrochemical Publications, 1996.

are specifically designed to remove certain soils. Also, be aware that even though aqueous cleaning agents are used in dilute form, they are not formulated from organic carrot juice. As with other cleaning agents, aqueous cleaning agents must be used with understanding and respect.

For general metals cleaning, aqueous formulations with relatively broad range of acceptability for substrate and soil have been found. However, for most high precision applications the aqueous cleaning agent must be specifically matched to the soil, the expected soil loading, the substrate, and the expected end use of the product. For example, in some applications, phosphate or silicate residue is not acceptable. In addition, in cleaning certain metals, notably aluminum, careful selection of the aqueous cleaning agent and process is required.

Water-Soluble Organics

Some cleaning agents, nominally referred to as aqueous cleaning agents, are primarily or significantly high in organic solvent blends including long-chain nonlinear alcohols or D-limonene. They can be rinsed with water or, in some cases, with either water or solvents. Providing both solvent and aqueous cleaning in the same process has advantages. However, recovery of the waste stream and carryover can become a problem.

Specific Blends

We will see more and more cleaning agent blends. As the number of manufacturing process fluids (e.g., metalworking fluids) proliferate, blends become important to fine-tune the performance aspects of the cleaning agents. Further, with the increasing proliferation of epoxies, potting compounds, elastomers, and composites, materials compatibility requirements must be met. Further, for a variety of reasons, including economic and regulatory constraints, the marketplace favors introduction of complex blends. There is a huge investment in developing individual molecules for cleaning. Individual molecules are readily identified by regulatory agencies and are therefore more likely to be restricted. Cleaning agent formulation combines science and art. This is true for aqueous, semi-aqueous, water-soluble organic, and organic solvent blends. Formulation is a bit like cooking. Even though the ingredients must be carefully defined both quantitatively and qualitatively, formulators have developed a style based on their education and experience. As you read the specific chapters dealing with formulated cleaning agents, you will develop an understanding of the basics and principles of formulation as well as the rationale and formulation style of the authors and of their companies. The goal is not to turn you into the next Julia Child of cleaning chemistries but rather to allow you to make rationale informed decisions as opposed to selecting cleaning agents based on Internet hyperbole.

Solvents and Solvent-Based Cleaning Agents

As in aqueous cleaning, there are mystery blends. However, it is often easier to identify the components of solvent-based cleaning agents. With the proliferation of new organic solvents, a number of by-products and natural products have been or are now used in cleaning processes. Cleaning agents based on orange, pine, cantaloupe, and grapes have been developed—an entire fruit basket of possibilities. Some solvents have been discussed in detail; others are alluded to in discussions or cleaning equipment or of specific cleaning agents.

A few additional solvents and solvent categories are worth noting. In general, one must consider that while many of these solvents have been used for years, long-term inhalation toxicity may not be available. In addition, blends and azeotropes of both new and well-established solvents may have their own solvency, compatibility, or toxicity properties. This holds true for organic and water-based cleaning agents. The best advice remains to test the solvent or blend in the application under consideration and to be conservative—minimize employee exposure and minimize loss of cleaning agent to the environment. In addition, compounds with higher boiling points and fairly complex blends with certain additives may leave undesirable residue, so rinsing is often required.

General Desirable Attributes of a Precision Cleaning Solvent

The following section is an excerpt from *Handbook for Critical Cleaning*, First Edition (Agapovich, 2000).

> For purposes of definition in this section, a solvent is defined as a cleaning agent which readily evaporates after use, when cleaning a component. No follow-up cleaning is required after the use of this class of material. These solvents will evaporate even after cold-cleaning. This will generally mean that the solvent has a vapor pressure of greater than 25 torr at ambient temperature. Another better quantifier of volatility is the evaporation rate compared to *n*-butyl acetate. *n*-Butyl acetate is set at unity. (ether and carbon tetrachloride are also used as reference materials) The majority of solvents discussed in this chapter all have an evaporation rate > *n*-butyl acetate or 1.

Another parameter related to volatility is the heat of vaporization, the heat required when a solvent transforms to the vapor phase from the liquid phase. When possible, it is best to choose solvents that have a low heat of vaporization as these materials will evaporate without absorbing heat (the part cools). Solvents with high heats of vaporization cool a part as more heat is absorbed from the environment. Often water condenses on a device that is cleaned, if the environment is humid. This is not desirable in precision cleaning.

Volatility and ease of vaporization has drawbacks. These include issues of containment, flammability, toxicity, and local regulations on emissions and cost. These issues will be discussed later in the chapter. Cleaners such as any aqueous based, terpenes and hydrocarbons in the combustible range (flash point >100°F) do not fall under the definition of a solvent and will not be discussed in this chapter.

Critical precision cleaning areas that use solvents under this definition include but are not limited to medical devices, directional devices (gyroscopes, accelerometers and including components within), computer components (disc drives), precision ball bearings, oxygen transport systems and circuit boards.

In addition to the need for volatility discussed previously, a precision cleaning solvent must meet other critical requirements. Obviously, the contaminant requiring removal must have a finite solubility in the cleaning solvent. This definition varies as factors such as time, temperature and agitation can be altered. Generally speaking, increasing the temperature of a solvent will improve cleaning effectiveness, as will increasing the exposure time. As an example, 3M defines "soluble" when a material dissolves in another in the range of 5 to 25 grams per 100 grams of solvent at room temperature (3M 1991). In another article, solubility has been defined as 50 grams per 100 ml of solvent (Agapovich 1997). There are many other range of definitions. The solubility and cleaning effectiveness required will vary depending on how contaminated the part is and also the users final requirements.

Other solubility parameters include a Kauri Butanol (KB) number. The KB numbers for CFC-113 and 1,1,1-trichloroethane are 31 and 124, respectively. Generally speaking, the highly chlorinated compounds have the higher reported KB values. The KB number reflects the ability of a solvent to dissolve heavy hydrocarbon greases. Specifically, it is a measure of ability to dissolve a solution of butanol and kauri resin. ASTM D 1133–90 describes the standard test method for determining the KB value of hydrocarbon solvents. This procedure has been extended to evaluate ODC replacements discussed in this chapter. It is not applicable to oxygen containing solvents.

However, the author [Agapovich] believes that as a first approximation, that the "like dissolves like" concept is very useful. This means that polar solvents dissolve polar contamination and non-polar solvents dissolve non-polar contaminants. Hydrocarbon solvents will dissolve hydrocarbon oils and fluorocarbon based solvents dissolve fluorocarbon oils and greases. To clean solder flux residues from printed circuit boards, polar oxygen containing solvents like alcohols or chlorinated solvents are required.

In precision cleaning, it is beneficial to have a solvent with a low viscosity and low surface tension. This property will allow solvents to enter very narrow gaps in a complicated device, to clean a contaminant. Particle removal is often a critical part of precision cleaning operations. A solvent with a low viscosity and low surface tension also facilitates particle removal. A high density solvent provides additional momentum to remove particles from surfaces.

Another critical requirement is chemical stability during use and also a solvent having a long or infinite shelf life. This was been a problem with the use 1,1,1-trichloroethane without stabilizers. Some azeotropes and mixtures discussed later require stabilizers.

Of even more importance is compatibility with the component one is cleaning. There cannot be a chemical reaction, physical change such as irreversible swelling or extraction of the materials of construction of the component being cleaned. CFC-113 and 1,1,1-trichloroethane were compatible with most materials. ODC alternative solvent manufacturers are very cognizant of the concerns of customers about solvent compatibility. Extensive solvent/materials compatibility tests are performed on a wide range of materials when a new solvent is introduced to the public. If one still has a question of the compatibility of material and solvent, it is best to have the solvent user perform the compatibility test in ones particular application.

Another critical solvent property is the non-volatile residue (NVR). It is critical when a solvent evaporates from a surface after cleaning, no residue is left behind. Solvent manufacturers typically have NVR specifications in the range of 1–10 parts per million (ppm) for precision cleaning. Very often the reported NVR of a given batch of solvent is well under the company set specification. The issue of NVR is also important when expensive solvents such as the PFCs, HFEs and HFCs are reclaimed and recycled. Any recycling process (such as distillation) must produce a product with the NVR meeting the original manufacturer (OEM) specifications.

Recycling is desirable for cost savings when using expensive solvents. The halogenated containing solvents (PFCs, HFCs and HCFCs) are particularly expensive. Recycling of used solvents is possible when solvents are used in cleaning and subsequently contaminated with particles or high boiling oils and greases. A simple strip or low theoretical plate distillation can be performed to purify the reclaimed solvent, to obtain NVR levels equal to or better than the OEM specifications.

Summary of Physical Properties of Primarily Unblended Organic Solvents

Table 1.2—consolidated, adapted, and where appropriate modified from the first edition (Agapovich 2000)—provides a comparison of some key physical characteristics of organic solvents. While a few are simple blends of two molecules, they can be thought of as simple cleaning agents. Almost all of them have relatively low boiling points and have a surface tension, density, and viscosity that are favorable for good wetting. This is important for cleaning components with miniature and microstructures.

We have not listed allowable inhalation levels or environmental regulatory information. Aside from the fact that such information does not represent physical characteristics, the safety and regulatory picture is a moving target that is geographically specific, nationally, regionally, and locally. The allowable inhalation level can vary by orders of magnitude, depending on the location. Many of aggressive cleaning solvents, and even some nonaggressive ones, have environmental and safety-related regulatory baggage. Very few chemicals have been banned; even highly restricted ones can be used safely and responsibly with the correct process equipment.

Old Reliable, Assorted Organic Solvents

Examples of classic organic solvents include toluene, hexane, heptane, benzene, and xylene. Flammability, worker exposure, air toxics, and company liability issues reduced the use of these solvents when ODCs were available. They have remained popular for specialized uses or as blends, particularly for specific, high-value applications. With decrease in availability of ODCs and an increase in low-flash-point and well-contained cleaning systems, these solvents have enjoyed a resurgence in popularity. They have a wide solvency range, but they are particularly "oil-like." If they are to be used, appropriate environmental and safety controls must be employed. Some specific organic solvents and categories of solvents are highlighted because they have found utility in critical cleaning applications.

Hydrocarbons (Mineral Spirits)

Hydrocarbon blends (mineral spirits or Stoddard solvent or kerosene) consisting of a petroleum cut of hydrocarbons with a range of molecular weights have been used in cleaning applications for many years. They are a mixture of molecules with closely related properties. Given the high boiling point and the potential for contaminants in some formulations, care must be taken in use and removal, and lot-to-lot variability may impact process control. Mineral spirits are not exempt as volatile organic

TABLE 1.2 Physical Properties of Solvents

Category	Solvent	CAS Number	Boiling Point (°C)	Vapor Pressure (Torr) at 20°C	Flash Point (TCC)°C (°F)	Density (g/cc) 20°C	Viscosity (cP) 20°C	Surface Tension (dyne/cm) 20°C
Hydrocarbon	n-Hexane	110-54-3	69	124	−26 (−15)	0.66	0.31	18 (25°C)
	n-Heptane	142-82-5	98	36	−4 (25)	0.68	0.41	20.3
	Isooctane	540-84-1	99	41	−12 (10)	0.69	0.50	18.8
	Cyclohexane	110-82-7	81	78	−27 (−17)	0.78	1	25
Ketone	Acetone	67-64-1	56	185	−17 (0)	0.79	0.36	23.3
	MEK	78-93-3	80	74	−1 (30)	0.80	0.43	24 (25°C)
Alcohol	Methyl	67-56-1	65	97	12 (54)	0.79[a]	0.55	22.6
	Ethyl (200 proof)	64-17-5	78	45	14 (58)	0.79[a]	1.1	22 (25°C)
	Isopropyl	67-63-0	82	32	12 (53)	0.78[a]	2.4	21.8 (15°C)
Chlorinated	Methylene chloride	75-9-2	40	350	None	1.33	0.44	28.1
	Trichloroethylene	79-01-6	87	47[b]	None	1.46	0.57	29.5
	Tetra-chloroethylene (perchloroethylene)	127-18-4	121	18[b]	None	1.62	Nl	Nl
	Trans-1,2-dichloroethylene	107-60-2	48	324[b]	6 (43)	1.26	Nl	Nl
Brominated	n-Propyl bromide	106-94-5	71	111	None	1.35	0.49[a] cS	26
Siloxane	Hexamethyl-disiloxane (OS-10)	107-46-0	100	42	−3 (27)	0.76[a]	0.65[a] cP	15.2
Benzotrifluoride	Para-chloro-benzotrifluoride (PCTBF)	98-56-6	139	7.9	43 (109)	1.34[a]	0.79[a] cP	25
HFC/HFE	HFE-7100 Methyl nonafluoroisobutyl ether, methyl nonafluorobutyl ether	163702-08-7 163702-07-6	61	202	None	1.52[a]	0.61[a]	13.6
	HFE-7200 Ethyl nonafluoroisobutyl ether, ethyl nonafluorobutyl ether	163702-06-5 163702-05-4	78	109	None	1.43[a]	0.61[a]	13.6
	HFC-XF 2,3-dihydroperfluoro-pentane	138495-42-8	55	226	None	1.58[a]	0.67[a]	14.1

HCFC Class II ODC	AK-225 (HCFC-255ca/cb)	422-56-0ca, 507-55-1cb	54	283	None	1.55[a]	0.59[a]	16.2 (25°C)
ODCs, phased out	CFC-113, 1,1,2-trichloro-1,2,2-trifluoroethane, Freon® 113	76-13-1	48	334	None	1.56[a]	0.68[a]	17.3 (25°C)
	1,1,1-TCA	71-56-6	74	121	None	1.32[a]	0.8[a]	25 (25°C)
	HCFC 141b, Dichlorofluoroethane	1717-00-6	32	572[b]	None	1.24[a]	0.43[a]	19.3
Perfluorinated	PF-5060, Perfluoro compounds, primarily with six carbons	86508-42-1	56	232	None	1.68[a]	0.4[a]	12.0
	PF-5070, Perfluoro compounds, primarily with seven carbons	86508-42-1	80	79	None	1.73[a]	0.6[a]	13.0
	PF-5052, Perfluoro compounds, primarily with five carbons	86508-42-1	50	274	None	1.70[a]	0.4[a]	13.0

Source: Adapted and expanded from Agapovich, J.W., Review of solvents for precession cleaning, in *Handbook for Critical Cleaning*, 1st edn., Boca Raton, FL, 2000.

Note: Nl, not listed in literature reviewed; TCC, tag closed cup.

[a] At 25°C.

[b] mmHg at 25°C.

compounds (VOCs). Some mineral spirit blends work effectively because they contain other chemicals that may show aggressive solvency but may also have safety and/or environmental regulatory restrictions. These additives are sometimes apparent from the MSDS, but not always. It is possible to obtain very pure mineral spirit blends with a narrow, defined range of hydrocarbon chain lengths. Such well-defined hydrocarbon blends are better suited to high-precision applications, both for cold cleaning and, in the appropriately designed system, even for vapor phase cleaning. The military has incorporated some of these well-defined mineral spirits into cleaning specifications.

Chlorinated Solvents

Methylene chloride, perchloroethylene (tetrachloroethylene), and trichloroethylene are used in precision cleaning applications (refer to Chapter 8, Risotto). The solvents are effective; they are also subject to many environmental and worker safety regulations. They can be used with appropriate safeguards. They tend to be good wetting agents, meaning that they are useful in removing soil that is trapped in tightly spaced components. The Hansen solubility parameters (refer to Chapter 4, Burke) are such that they are useful in removing a wide range of soils. However, soil blends that contain large amounts of inorganic, polar contaminants (such as salts) are generally more effectively used in aqueous cleaning.

Trans-1,2-Dichloroethylene

Trans-1,2-dichloroethylene (156-60-5) is a daughter compound of these chlorinated solvents, and it has many physical and solvency properties that are at least somewhat similar to the chlorinated solvents (refer Table 1.2). In the United States, *trans* has been adopted in industrial and precision cleaning applications, and it is increasingly used as a blending agent. Some desirable characteristics include aggressive solvency, rapid evaporation, high density, an odor that is similar to other chlorinated solvent, and reasonable solubility in a range of organic solvents; it can be used in aerosol formulations (Shaw 2010). *Trans* could also be used neat as a chlorinated liquid/vapor degreasing solvent with the proviso that it be used in an appropriately designed system for low-flash-point solvents. It should also be noted that the level of *trans* needed to provide appreciable solvency to an hydrofluorocarbon (HFC) or hydrofluoroether (HFE) azeotrope also contributes a significant level of VOCs. Therefore, while the base material may be noted as VOC exempt, the blend may or may not be a good candidate for cleaning in your location.

Trans-1,2-dichloroethylene is said to have a relatively favorable worker safety profile, particular as compared with other chlorinated solvents and the *cis* isomer of *trans*. The notable safety and environmental constraints are that *trans* is a VOC and that it has a low flash point. The inhalation limit is 200 ppm (U.S. Department of Labor, Occupational Safety and Health Administration n.d.). As this book goes to press, the U.S. EPA appears to be at the initial stages of a health risk assessment of *trans*-1,2-dichloroethylene, and my understanding is that there is external peer review; one might anticipate that results will be posted in late 2010. The status and progress can be assessed online (U.S. EPA 2010).

Perfluorinated Compounds

Perfluorinated compounds (PFCs) contain fluorine and carbon, but no chlorine or bromine. They are exceedingly mild, inert cleaners that can be used for removal of fluorinated lubricants and as rinsing and drying agents. PFCs are very effective for particulate removal. They are not ODCs. They are still sold alone and in various formulations. However, because of the global warming potential associated with a long atmospheric lifetime, often ranging in the thousands of years, industry has been under what might be termed strong regulatory "encouragement" to find substitutes. HFCs and HFEs can and has replaced PFCs in many, if not most, applications.

TABLE 1.3   Some Characteristics of HFC-365 mfc and HFC-245 fa

Characteristic	HFC-365 mfc	HFC-245 fa
Structure	$CF_3CH_2CF_2CH_3$	$CF_3CH_2CF_2H$
Molecular weight	148	134
Boiling point (°C)	40.2	15
Flashpoint (°C)	Below −27	None
UEL/LEL (% by volume)	3.8–13.3	None
ODP	Zero	Zero
Atmospheric lifetime	10.8 years	Low to moderate
VOC status	Exempt	Exempt

Hydrofluorocarbons and Hydrofluoroethers

Some HFCs and HFEs with significant use in cleaning processes are discussed in Chapter 5 (HFEs) (Owens) and Chapter 6 (HFCs) (Bartelt and Merchant). Other HFCs and HFEs are used as part of cleaning products. Some of them have low boiling points but are useful as blending agents to suppress flammability, moderate aggressive solvency, and lower the VOC content. They are attractive for blending in aerosols and for other single-use applications as opposed to process baths (Knopeck 2000). Both HFC-365 mfc (Solvay, Solkane® 365 mfc) and HFC-245 fa (Honeywell) are very mild solvents. Both have been accepted under the EPA SNAP program as substitutes for ozone depleters; both are indicated as having relatively favorable worker exposure profiles. They were originally developed as foam blowing agents (Zipfel 2000). It should be noted that HFC 365 has a very low flash point. In nonflammable blends, and even in azeotropes, there is the potential for flammable mixtures to develop in-use. Therefore, you as the end user should work with your advisors and with responsible cleaning agent suppliers to carefully evaluate your own production situation. It is interesting the HFC 365 is blended with *trans*-1,2-dichloroethylene; the product literature indicates that commercially available blends of these two flammable solvents as well as a blend containing ethanol do not have a flashpoint as tested by DIN/EN/ISO 13736 (Solvay Fluor GmbH 2009). HFC-365 mfc has a significantly higher boiling point than HFC-245 fa. The differences may not seem significant, but remember that as a rule of thumb chemical processes double in rate with every 10°C increase. For both materials, some plastics compatibility studies have been conducted. Because synergistic behavior can occur, you would be well-advised to confirm the compatibility of any proposed blend to the specific mix of materials and in the application at hand (time of exposure, temperature, force of cleaning action). With a 15°C boiling point, many end users will find that HFC-245 fa would not be suitable for use in classic, unmodified vapor degreasers; flushing systems or airless/airtight systems might be useful. Some properties of HFC-365 mfc and HFC-245 fa are summarized in Table 1.3. Additional products are reportedly under development, and detailed information is not available as the book goes to press. Based on early indications and on the current global regulatory environment, while we might be pleasantly surprised, one would predict that these will be relatively mild solvents.

Oxygenated Solvents

Examples of oxygenated solvents include the short-chained alcohols (methyl alcohol, ethyl alcohol, isopropyl alcohol [IPA]), methyl ethyl ketone (MEK), methyl isobutyl ketone (MIBK), and acetone. Compared with, say, hexane or heptane, the addition of oxygen makes these compounds more polar, that is, more like water. Relative to hydrocarbon blends, they are more suited to polar or inorganic soils. It should be noted that oxygenated solvents cannot systematically be used as drop-in replacements for chlorinated solvents; they have different physical, flammability, and solvency properties. Despite the low flash point, some oxygenated compounds, such as IPA, can be used in appropriately designed vapor degreasing systems.

Alcohols and Ketones

IPA is widely used in critical applications, including aerospace and medical device cleaning. Despite its widespread use, IPA is not a universal solvent. Because of the use in hospital environments, some people feel that it connotes cleaning. It evaporates rapidly, and it typically does not leave a residue. It should be noted that where a residue is found, particularly in benchtop cleaning, one culprit can be plasticizers. Residue can come from using the wrong plastic container or from a well-loved plastic bottle that may have lurked on the benchtop for years (even decades). Another culprit is the shipping/ storage container. A third is the purity of the IPA. Purchase high-quality IPA. For exacting applications, it may need to be stored in glass. Finally, for benchtop use, select the correct dispensing containers. If you use plastic dispensers, please do not allow you or your assemblers to become devoted to a single plastic dispensing bottle. Change those dispensing containers often. Please also remember that IPA has a low flash point (12°C). This means that filling an ultrasonic tank with IPA is not acceptable; it presents a potential fire hazard. This means that you have to evaluate all processes involving IPA as well as those near IPA uses.

Long chain alcohols such as tetrahydrofurfuryl alcohol (THFA) are used alone and in blends. The longer the carbon chain, the more they become fat-like (able to dissolve oil). However, the alcohol portion confers some water-like qualities. Often, these alcohols can be part of aqueous, semi-aqueous (rinse with water), or co-solvent (rinse with solvent) blends. Alcohols of varying chain length have been used in blends, particularly aerosol blends, to confer appropriate solvency properties.

Acetone is a ketone; there is a double-bonded oxygen in the molecule. Acetone has an exceedingly low flash point (–17°C), so whatever (Pourreau 2006) cautions are used with IPA need to go up by orders of magnitude with acetone. Acetone has very different solvency properties than hydrocarbons, halogenated solvents, or alcohols (refer to Chapter 4, Burke). Because it evaporates very rapidly, it has been used for decades and decades as a final quick rinse and drying agent. After being delisted as a VOC, there has been increased interest in acetone. Acetone has been appropriately adopted for some processes, wishfully or inappropriately for others. The extremely rapid evaporation rate, incompatibility with some plastics, and low flash point remain issues. With, and only with, appropriately designed equipment, acetone can be used in cleaning systems. The exposure times tend to be longer than with benchtop cleaning, heat may be applied, and the impact of water dissolved in the acetone must be considered. For example, in a heated acetone system, a visible white "bloom" on magnesium was observed; the problem was observed after brief benchtop cleaning (Kanegsberg 2009). There has been increased use of acetone in aerosol applications, because it is a relatively low-cost alternative to other VOC-exempt compounds such as HFCs and HFEs. It also replaces hydrochlorofluorocarbon (HCFC)-141b, an ozone-depleting chemical. Of course, the use of acetone-based aerosols near sources of ignition is not advisable. In addition, acetone can react with some plastics even after brief exposure.

N-methyl pyrrolidone (N-methyl-2-pyrrolidone, NMP) is a high boiling (295°C), high flash point (91°C, 196°F) ketone, which is used alone and in blends. Because it is water-soluble, it can be blended for removal of both rosin and organic acid flux, and it is used in photoresist systems. Other pyrrolidones are being developed, notably n-octyl pyrrolidone (NOP) and n-hydroxy ethyl pyrrolidone (HEP). Used alone and in blends, they may serve to extend the range of cleaning in degreasing. For example, NOP has a longer carbon chain and is therefore more oil-like, so it could be used in formulations for degreasing, paint stripping, and de-inking of paper. HEP shows promise in photoresist removal (Waldrop 2000). Dimethyl sulfoxide (DMSO) has also been developed in some photoresist applications as well as in other processes requiring a fairly aggressive solvent.

Esters

Various monobasic and dibasic esters and notably lactate esters are used alone and in blends in semi-aqueous and co-solvent applications. They have proven particularly effective in the removal of pitches, waxes, and other difficult, mixed soils.

Most esters have a fairly strong or distinctive odor. They are high boiling and must be rinsed in high-precision applications. Esters may also be used in coatings formulations. *t*-Butyl acetate (TBAC) and dimethyl carbonate (DMC) have relatively low tropospheric reactivity and may therefore be particularly useful where VOCs are an issue. Methyl soyate, a biobased cleaning agent, has been developed as a cleaning agent alone, in blends, and in sequential cleaning processes. While methyl soyate is not exempt as a VOC at the Federal level, there has been acceptance of some methyl soyate products as clean air solvents in SCAQMD (see discussion of biobased materials in this chapter).

Esters hydrolyze in the presence of water; they break down to form an acid and an alcohol. Therefore, when any ester is used in a process bath for an extended period of time and perhaps in the presence of heat, it is important to consider the extent to which water can be dissolved in the ester, to consider how rapidly it hydrolyzes, to consider the potential impact of hydrolysis products on efficacy of cleaning, and to determine any worker safety, flammability, and/or environmental impacts. There could be four potential cleaning agents in the cleaning tank: the ester, the water, the acid, and the alcohol.

t-Butyl Acetate—Regulatory Complexity

However, the situation with TBAC illustrates issues associated with the complexity of environmental regulations, as discussed in "The Cleaning Agent Balancing Act (*Handbook for Critical Cleaning: Applications, Processes, and Controls,* Chapter 2)". TBAC has shown promise of utility in aerospace and other precision cleaning applications, including military and aerospace (Foreman 1999), for brake cleaners, and as a component of coatings and inks (Pourreau 2006). In the precision cleaning of electronics assemblies, TBAC has been found to provide a good complement to acetone. TBAC has a higher boiling point. Unlike acetone, it does not dissolve nitrile gloves, and in one test evaluation, the assemblers found the odor to be acceptable (Elias n.d.).

At the Federal level, while TBAC was eventually accepted by the EPA as having negligible reactivity relative to ethane and therefore is VOC exempt, acceptance by the EPA and at the State level took much longer than would have been expected. At the State level, the California Air Resources Board (CARB) was concerned with the potential toxicity of TBAC, both in and of itself and because it can break down to *t*-butyl alcohol (California EPA Air Resources Board 2006). Because TBAC is an ester, it can break down to an acid and an alcohol. The wording of the EPA exemption for TBAC is unique among those compounds that are listed as VOC exempt (U.S. EPA 2009; definitions of VOC and ROG 2009). While TBAC is not a VOC and is therefore exempt from limitations on use or emissions, it is treated like a VOC. That is, recordkeeping and reporting for TBAC is like that for VOCs; you have to keep records of use and report emissions to the State along with other VOCs. One might expect this to discourage the use of the chemical as a cleaning agent or process fluid, and it might be expected to result in confusion. In fact, the EPA indicated that SCAQMD did not fully comply with the EPA definition of TBAC as a VOC (U.S. EPA 2009). (Also see Chapter 29, B. Kanegsberg, in *Handbook for Critical Cleaning: Applications, Processes, and Controls.*)

Dimethyl Carbonate

In the United States, DMC (carbonic acid dimethyl ester, CAS Number 616-38-6) is a potentially attractive cleaning agent in part because it has been determined to have negligible reactivity and is therefore VOC exempt at the Federal level (U.S. EPA 2009). Not all states have approved the VOC-exempt status. DMC has a very low flash point and a fairly high boiling point. DMC is fairly soluble in water and therefore may be subject to breakdown by hydrolysis. Some important physical and chemical properties are summarized in Table 1.4.

TABLE 1.4   Physical and Solvency Properties of DMC

Property	Level
Boiling point (°C)	90
Freezing point (°C)	2–4
Vapor pressure (torr, 20°C)	42 (55 at 25°C)
Flash point °C (°F) closed cup	(63°F)
Density (g/cc) (20°C)	1.07
Viscosity (cP 20°C)	0.625
Surface tension (dyne/cm 20°C)	3.19×10^{-2} N/m
Hansen solubility parameter dispersion	15.5
Hansen polar	3.9
Hansen hydrogen bonding	9.7
Solubility in water (g/100 g water)	13.9

We have not evaluated DMC. A status summary has been provided by one supplier (Smith 2010):

DMC is a moderately evaporating solvent with an evaporation rate of 3.22 (BuAc = 1.0), which puts DMC right between methyl ethyl ketone (MEK, (4.03) and toluene (2.0). The solubility profile is similar to a number of common glycol ethers such as cellosolve acetate, propylene glycol monobutyl ether (PNB) and propylene glycol monoethyl ether acetate (PE acetate). DMC remarkably freezes approximately 36°F, although mixtures with other solvents can depress the freezing point. The extent to which DMC will be used in the cleaning industry will probably be dependent on what co-solvents can used with DMC to overcome the low flashpoint and high freezing point. Azeotropes may be developed as DMC is further researched for cleaning applications. The odor profile may be characterized as an alcohol or light ester odor. DMC has toxicological properties that favor industrial, automotive, or outdoor use, but it would not be recommended for indoor consumer or institutional use. DMC has a similar toxicological profile to methanol, its primary metabolite. The industrial exposure level recommended by a U.S. supplier for an 8-hour day by inhalation is 100 ppm (based on hydrolysis to methanol). California's CARB (Air Resources Board) does not immediately plan to exempt DMC for most cleaning or consumer applications. CARB needs to obtain typical formulations and determine what solvents DMC would replace in key applications. CARB can then do a health and safety assessment.

Biobased Cleaning Agents

Biobased (biologically based) products are industrial or commercial products (not for food products) that are composed of plant- or animal-based materials. One advantage of biobased products is that they are a renewable resource. Plant-derived biobased cleaning agents have been commercialized. Some are derived from one source, others are blends of biobased products; still others may have additives such as surfactants. Biobased materials can also be used as blending agents. Biobased products may be water- or solvent based. Many are used in inks and/or coatings; the extent of development as cleaning agents by different suppliers is variable.

Most have very high boiling points. Therefore, for critical/precision cleaning applications where residue is an issue, they may require rinsing and drying. Both water-rinseable and solvent-rinseable formulations are available. Some physical properties of biobased solvents are summarized in Table 1.5. We have found some variability in the numbers reported for boiling point and flashpoint; factors in this variability may include methodology, differences in chemical composition, and perhaps the presence of other chemicals.

TABLE 1.5 Some Physical Properties of Biobased Solvents

	Methyl Soyate[a]	Ethyl Lactate	D-Limonene
Specific gravity	0.88	0.83	0.84
Vapor pressure (mm Hg)	<1.0	1.7	1.4
Boiling point °C (°F)	216 (421)	154 (309)	178 (310)
Flash point °C (°F)	>100 (>212, approx. 360)	59 (139)	46 (115)
Kauri-butanol number	58	NA	67

[a] United Soy Board (2010).

Like all chemicals, biobased products should be managed with appropriate diligence, caution, and respect. The term "biobased" is often erroneously assumed to be synonymous with safe. Centuries ago, the de' Medici's were infamous for creatively dispatching their enemies through the use of poisons. I suspect that, rather than consulting with the formulator of a large chemical conglomerate, de' Medici's drifted serenely through the palace garden with the resident compounder and harvested a local, sustainable, renewable resource. The dose makes the poison. Understand what you are using and how you are using it. After all, biobased cleaning agents are extracted from living things and are therefore used in a concentrated form. They may be used where heat and mists are generated; some may oxidize and become ignition sources. They should be disposed of appropriately. When people use any cleaning agent, from any source, including biobased products, it is important to understand potential environmental and worker safety issues. It is also important to understand the potential for any residue, including residue from a biobased product, to interact with a living host. So, for those involved in medical device applications, evaluate all cleaning agents, including biobased cleaning agents, used within your company and by your supply chain. In conclusion, many biobased cleaning agents have very desirable properties. Use them with understanding and intelligence.

Soy-Derived Cleaning Agents

Methyl soyate (CAS Registry Number 67784-80-9) is a soy-derived product that was developed as a biofuel and can be used for cleaning applications, and may prove to be a renewable resource alternative to hydrocarbon blends in general metal cleaning; it could also be used in blends or sequential solvent processes for critical applications (Wildes 2002). Methyl soyate is used alone and in co-solvent blends. It is produced by the esterification of soybean oil. Heated soy oil and methanol react in the presence of a catalyst to produce methyl ester and glycerin. The general reaction is indicated below. The ester and glycerin are separated, and the ester is purified using a water-wash process and filtration. Catalyst

$$C_3H_5(O_2CR)_3 + 3MeOH \xrightarrow{catalyst} 3MeO_2CR + C_3H_5(OH)_3$$

Triglyceride   Methyl alcohol   Methylsoyate ester   Glycerin (Soy-derived)

Neither OSHA nor ACGIH have established guidance for inhalation exposure to methyl soyate. Per a recent MSDS (AG Environmental Products, L.L.C. 2009), there is no OSHA PEL or ACGIH TLV. The manufacturer has indicated that exposure via inhalation is unlikely because of its low vapor pressure. However, exposure to aerosolized methyl soyate is possible and has not been well studied. While acute (very short-term) toxicity studies have been performed (United Soybean Board 2010), we have found no indication that long-term inhalation studies have been performed. It does have a high flashpoint, but care should be taken with heated baths to avoid potential problems with breakdown or oxidation products.

In terms of reactivity relative to ethane, methyl soyate is not VOC exempt according to the U.S. EPA, but it can be used in certain areas under certain state implementation plans. While the VOC content is under 50 g/L (as estimated by EPA Method 24; ASTM D3960), for many cleaning applications, the VOC-exempt status is the one that matters. More pertinent to cleaning, for certain products, methyl soyate meets the requirements of the SCAQMD "Clean Air Solvents" (CAS) program; this program is not to be confused with the CAS system for uniquely identifying chemicals. Some products meet the CAS requirement because a technicality of the specific gas chromatography/mass spectrometry technique (SCAQMD 2004) results in some components not being picked up on the column and therefore not detected. Because the SCAQMD method has been accepted by the EPA as part of the SIP, to the best of my understanding and interpretation of the regulations, SCAQMD CAS solvents are treated as non-VOCs, either in SCAQMD or where SCAQMD regulations may have been cloned or adapted and where they may have been accepted by the Federal EPA (see Chapter 29, B. Kanegsberg, in *Handbook for Critical Cleaning: Applications, Processes, and Controls*). Where cleaning performance and/or regulatory characteristics are desirable, such a solvent could be rinsed with another lower-boiling, low-residue VOC-exempt solvent. The bottom line is that methyl soyate provides a possible option, even in areas of poor air quality.

Ethyl Lactate

Ethyl lactate (CAS Number 97-64-3), an ester produced from the fermentation of corn-derived feedstock, has also been considered for use as a cleaning agent along and in combination with methyl soyate (Henneberry 2000). It is produced from lactic acid, which is derived through the fermentation of cornstarch. It has been used periodically in cleaning for several decades.

D-Limonene and Alpha Pinene Blends

D-Limonene (citrus-derived) and alpha pinene (pine tree-derived) cleaning agents have been developed. Both have noticeable odors; both can leave appreciable residue, depending on the application, and must be rinsed completely. Both are VOCs. In practice, D-limonene-based cleaners have found more utility in critical cleaning applications (refer to Chapter 9, Gustafson). As with many esters, long-term inhalation exposure studies have not been completed.

Parachlorobenzotrifluoride

Parachlorobenzotrifluoride (PCBTF, CAS number 98-45-6) is a fairly aggressive solvent that has been used in cleaning agents, alone and in blends, as well as in coatings. It was commercialized and marketed in the United States by Occidental Chemical during the 1990s. PCBTF is VOC exempt. After fairly extensive testing, PCBTF processes have been developed for military and aerospace applications. Shortly before the first edition of *Handbook for Critical Cleaning* went to press, the manufacturer of the product exited the market. However, PCBTF from imported sources continues to be available and used. For additional information, refer to Chapter 10 (Skelly).

Volatile Methyl Siloxanes

The following is an excerpt from Cull and Swanson (2000)

> A new class of fluid chemistry has been introduced to the precision- and industrial cleaning markets, based on linear and cyclic volatile methylsiloxane (VMS). The use of a silicone based cleaner may seem "counter-intuitive" to some people, since the low surface tension of silicone contamination has historically made it very difficult to remove. However, as a cleaning solvent the low surface tension becomes a definite asset, since it helps wet out and undercut soils. Drag-out is also reduced, because of the liquid's low viscosity. One of the keys to success with what appears at first to be an unlikely technology is the ability to manufacture VMS that dries with ultra-low nonvolatile residue (NVR) so that it will evaporate completely, leaving behind a clean surface.
>
> While VMS materials are new as cleaning solvents, they have been commercially available since the 1950s, primarily used as the building blocks for higher molecular weight, nonvolatile

silicone fluids and polymers. In addition, VMS fluids are widely used in the personal care industry, including many antiperspirant, hair care and skin care products. The majority of the personal care applications employ VMS materials with a *cyclic* structure, designated "cyclomethicones" by the Cosmetic Toiletry and Fragrance Association. In contrast, industrial- and precision cleaning fluids are made primarily with *linear* VMS fluids, which have a faster rate of evaporation and higher recommended exposure levels, but higher cost than the cyclics.

They are compatible with a variety of surfaces, and can be used on metals, glass, polycarbonate, acrylic and other plastics. With their low Kauri butanol (Kb) value, pure VMS fluids are not very aggressive cleaners, however, making them primarily effective on nonpolar contaminates like silicones, oils and light greases. Patented azeotropes have been developed that improve cleaning effectiveness on more difficult soils like rosin solder flux.

Linear, branched, and cyclic siloxanes are used in cleaning applications. Examples of linear and cyclic siloxanes are indicated in Figures 1.1 and 1.2. Volatile methyl siloxanes (VMS) are used in a variety of applications. Some physical properties are summarized in Table 1.6. As expected, on the basis of "like dissolves like," they can be used to remove silicone-based coatings. A patented blend has been used in electronics applications. They have been blended with parachlorobenzotrifluoride. In our experience, the siloxanes have greater solvency than might be expected based on the Kb value. Examples of unexpected solubility are illustrated in Chapter 7 (Dingess, Morford, and Shubkin). A recent NAVAIR study indicates that cyclic siloxanes may be useful as a replacement for mineral spirits in some military applications (Arafat 2009).

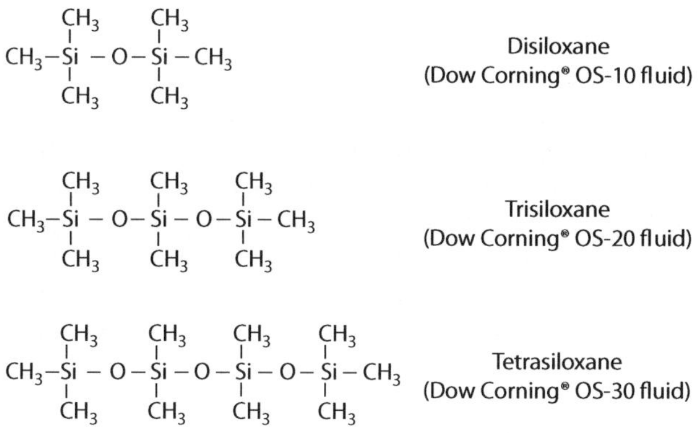

$$CH_3-\underset{\underset{CH_3}{|}}{\overset{\overset{CH_3}{|}}{Si}}-O-\underset{\underset{CH_3}{|}}{\overset{\overset{CH_3}{|}}{Si}}-CH_3$$

Disiloxane
(Dow Corning® OS-10 fluid)

$$CH_3-\underset{\underset{CH_3}{|}}{\overset{\overset{CH_3}{|}}{Si}}-O-\underset{\underset{CH_3}{|}}{\overset{\overset{CH_3}{|}}{Si}}-O-\underset{\underset{CH_3}{|}}{\overset{\overset{CH_3}{|}}{Si}}-CH_3$$

Trisiloxane
(Dow Corning® OS-20 fluid)

$$CH_3-\underset{\underset{CH_3}{|}}{\overset{\overset{CH_3}{|}}{Si}}-O-\underset{\underset{CH_3}{|}}{\overset{\overset{CH_3}{|}}{Si}}-O-\underset{\underset{CH_3}{|}}{\overset{\overset{CH_3}{|}}{Si}}-O-\underset{\underset{CH_3}{|}}{\overset{\overset{CH_3}{|}}{Si}}-CH_3$$

Tetrasiloxane
(Dow Corning® OS-30 fluid)

FIGURE 1.1 Chemical structure of linear VMS.

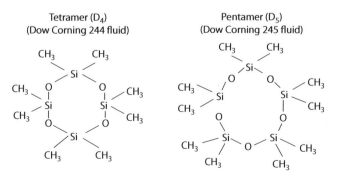

FIGURE 1.2 Cyclic VMS structure.

TABLE 1.6 Properties of Representative Methylsiloxanes

	Cyclotetra Siloxane	Cyclopenta Siloxane	Di-Siloxane	Tri-Siloxane	Tetra-Siloxane
Other names	Octamethyl-cyclotetra-siloxane, D4[a]	Decamethyl-cyclopenta-siloxane D5[a]	Hexa-methyldi-siloxane[b]	Octa-methyltri-siloxane[b]	Deca-methyl-tetra-siloxane[b]
CAS number	556-67-2	541-02-6	107-46-0	107-51-7	141-62-8
Flash point, closed cup (°C)	55	76	−3	34	57
Freezing point (°C)			−68	−82	−68
Boiling point (°C)	172	205	100	152	194
Evaporation rate (ASTM D 1901)	0.2		3.8	0.7	0.15
Viscosity, cSt at 25°C	2.5	4.2	0.65	1.0	1.5
Specific gravity at 25°C	0.953	0.956	0.76	0.82	0.85
Surface tension, dynes/cm at 25°C	17.8	18.0	15.2	16.5	17.3
Heat of vaporization, cal/g at 25°C	42	39	46 Estimated	44 Estimated	36 Estimated
Kauri-butanol value	14.5		16.6	15.1	13.4
VOC content, weight %	0	0	0	0	0

Source: Adapted from Cull, R.A. and Swanson, S.P., Volatile methylsiloxanes: Unexpected new solvent technology, in B. Kanegsberg and E. Kanegsberg (eds.), *Handbook for Critical Cleaning*, 1st edn., CRC Press, Boca Raton, FL, 2000.
  Dow Corning® designations OS-10, OS-20, OS-30.
  [a] Cyclic siloxanes.
  [b] Linear siloxanes.

VMS may be repackaged, blended, or sold under a commercial name. Therefore, it may not be readily apparent that the repackaged material is or contains a VMS. Determining where you have a VMS provides another illustration of the value of the CAS number. For example, in the NAVAIR study, three products were tested as a replacement for mineral spirits. Based on the MSDS, QSOL™ 300 is over 97% decamethylcyclopentasiloxane CAS number 541-02-6. Similarly, product literature indicates that Cyclo 147 F is primarily 541-01-6 (with another cyclic siloxane, CAS number 556-67-2). SB32 (apparently available through a VMS producer) is over 90% CAS number 541-02-6, with up to 5% mixed cyclosiloxanes (CAS number 69430-24-6).

Some of the VMS, particularly the cyclic VMS, have come under regulatory scrutiny, in part because they are marketed as alternatives for dry cleaning. Data have been provided by Dow Corning to the EPA (U.S. EPA 2009). While information received can be publicly accessed through EPA docket EPA-HQ-OPPT-2009-0180 and can be accessed online, the information is not presented in a form that is readily comprehensible to or digestible by me. Further, the EPA has not evaluated the data, and the timeframe for evaluating the data is not clear. Environment Canada has studied siloxanes extensively and has issued a summary report indicating that D4 and D5 do not pose a threat to human health; this appears to be based on the levels in personal care products. There is a proposal, not a regulation, that the levels of D4 and D5 in products be regulated to protect aquatic organisms (Environment Canada 2010).

Solvent Blends: Azeotropes, Co-Solvents

Aqueous additives are used for such purposes as improving wettability, solubilization and removal of soils, compensating for water quality, and forestalling corrosion. Solvents are blended for a number of reasons. Blends can blur the lines of demarcation among various categories of cleaning agents.

Stabilization

Water and acidity are the enemies of many halogenated solvents. Stabilizer packages are added to many chlorinated solvents (including TCA) and to *n*-propyl bromide (*n*PB) to prevent acid formation, breakdown, and reactivity with metals. Effective stabilization is important in degreasing (liquid/vapor cleaning). Stabilization becomes even more challenging in airtight and airless systems because the solvent is reused without replenishment over a much longer period than in traditional open-top degreasers. Stabilizers may be added to other solvents to prevent oxidation.

Extending the Solvency Range or Moderating the Solvency

Solvent blends can provide custom, fine-tuned cleaning options. Sometimes, blends provide surprises.

Azeotropes are strongly preferred over blends for vapor degreasing applications. An azeotrope is a constant-boiling mixture of two or more compounds. An extreme example of a non-azeotrope blend would be sugar in water. On heating, the water is boiled away, and the sugar remains. By contrast, IPA and cyclohexane form a constant boiling azeotrope. This means that the vapors contain both components in a constant proportion that does not change over the life of the blend. The cyclohexane/IPA azeotrope can be very useful in removing a range of ionic and nonionic soils from complex parts. Of course, it must be used in cleaning systems that are designed for low-flash-point solvents. Azeotropes have to be managed with care. Even azeotropes can vary in composition if they are used at a temperature not in the azeotropic range.

Blends that are not azeotropes will lose various components to evaporation at different rates. This means that the relative concentrations in the liquid and remaining blend will vary with time. Cleaning capability, compatibility, and flash point can all change. Blends that are not true azeotropes should be viewed cautiously, particularly in vapor degreasing applications.

In addition, azeotropes have been known to behave synergistically (non-additively) in terms of performance and compatibility. These properties are not necessarily predicted by solvency parameters. In other words, while two solvents may each show acceptable materials compatibility with a given component, the blend could produce component deformation. Therefore, even if you think you understand the solvency and compatibility issues of each component in an azeotrope, it is prudent to test the mixture.

Solvent blends are also used to modify or extend the solvency range in cold cleaning applications. An aggressive solvent can be toned down and a mild solvent made more aggressive. For example, HFE has been blended with *n*PB to tone down the aggressiveness of *n*PB. VMS may be blended with an alcohol or PCBTF to boost solvency.

Co-Solvents; Sequential Solvents, Bi-Solvent Process

The terminology of co-solvents is a bit confusing. Co-solvents have been thought of as two chemicals used in the same tank or used sequentially. Co-solvents are not necessarily miscible, so the cleaning process bath may require ongoing mixing or agitation. In one sense, any blended solvent could be thought of as a co-solvent system. Co-solvents can be blends that are primarily aqueous or primarily solvent. Supercritical or liquid CO_2 cleaning can also be accomplished with co-solvents; sometimes these are used in the same tank and at other times they are used sequentially. A patent for a process involving sequential solvents, termed a bi-solvent process, has also been described for cleaning precision components without VOCs (Mouser 2009).

Surfactants

Surfactants are used in solvent blends to provide some qualities similar to aqueous cleaners, to change emulsifying qualities and to allow the solvent to be readily rinsed in water (as in co-solvent processes). Some blended cleaning agents are offered as similar formulations, with or without the surfactant.

Emulsions: Macroemulsions, Structured Solvents, and Microemulsions

Oil and water do not mix, except in emulsions when they may coexist in a transient or relatively permanent form. An oil and vinegar salad dressing is typically a transient macroemulsion. Mayonnaise is a more permanent emulsion. Emulsifying qualities are used in aqueous blends, solvent blends, or both. In aqueous formulations, organic chemicals may be part of the mix only under certain conditions such as temperature. For example, the separation of the organic phase in an aqueous cleaner at the operating temperature may serve to defoam the blend, allowing for spray applications.

In the same way, immiscible organic chemicals may be used as emulsions, transient or permanent. Transient macroemulsions can be used to transfer the soil from one chemical to the other; the part is then rinsed in more of the chemical with very low solubility for the soil in question. Macroemulsions are typically cloudy. Sometimes one of the phases is aqueous; in other processes both are solvent.

Microemulsions, structured solvents, liquid crystals, or continuous phase emulsions have been introduced as cleaning agents. Structured solvents are stable mixtures of organic solvent, water, and coupling agents. The continuous phase may be solvent or water. They appear clear and may be primarily water or primarily solvent. Structured solvents can be made to separate during the application process. Such products are useful for mixed soils where both solvent and water-like characteristics are desirable (Shick 1996). Microemulsions are seen in consumer household products and in some single-use formulations for benchtop cleaning.

Mystery Mixes

Industry depends on mystery mixes. Blended solvents, particularly the high boiling blends involving hydrocarbons, esters, and nonlinear alcohols, greatly extend the specificity of cleaning that can be obtained. However, blended solvents can be particularly difficult for you, the components manufacturer, to evaluate. As with aqueous cleaning agents, many manufacturers consider the formulations to be highly proprietary and competition sensitive.

Many of the comments regarding mystery mixes apply to both aqueous and non-aqueous-based formulations. Improved performance and the desire on the part of formulators to maintain control of superior formulations have been in part responsible for the proliferation of mystery mixes. One additional observation is that a complex mixture where each chemical is used at a very low level could have the benefit to the cleaning agent producer of keeping the blend below the regulatory radar screen. Some areas of concern in using complex, mystery mixes include unexpected compatibility issues, regulatory constraints on one or more component, and unscheduled formulation changes. In such cases, it may be desirable to set up confidentiality agreements so that you understand the ingredients in detail. At the very least, particularly where the product is used in a process requiring high levels of validation and testing, it is prudent to obtain an agreement with the vendor that the product will not be changed.

In addition, supposed improvements in formulations can have unintended consequences. This author has observed many instances where a blended product was improved in such a manner as to adversely impact the process.

Ozone-Depleting Chemicals

CFC-113 and 1,1,1-Trichloroethane

CFC-113 and 1,1,1-TCA set the standards for cleaning for decades. CFC-113 has low to moderate solvency; TCA is an aggressive solvent. Both can be used for liquid/vapor phase degreasing. Both have high ozone depletion potentials; in the United States, they have been phased out of production. It is still possible to obtain recycled material, but at a high cost.

HCFC 141b

HCFC 141b (CAS no. 1717-00-6) has been phased out of production. Given the flashpoint-inerting properties, it has been widely used in aerosol cleaning agents, particularly where cleaning was needed near sources of ignition. While HCFC 141b is still available, we do not suggest using HCFC 141b blends; given worldwide distribution channels, we suggest that manufacturers be aware of what they are purchasing and that they test alternative cleaning agents.

Substitutes for HCFC 141b aerosols tend to be costly and/or to have other safety and/or environmental issues. Many manufacturers have switched to flammable blends for aerosol applications. Workers should be made aware of the chemical and flammability issues associated with such products.

HCFC 225

HCFC 225 is a versatile cleaning agent. The chemical was developed and marketed by Asahi Glass. It is typically sold as a mixture of two isomers, HCFC 225ca (3,3-dichloro-1,1,1,2,2-pentafluoropropane, CAS no. 422-56-0) and HCFC 225 cb (1,3-dichloro-1,1,2,2,3-pentafluoropropane, CAS no. 507-55-1). HCFC 225 has significant use in industrial cleaning, because of performance, overall environmental attributes, and worker exposure profile. HCFC 225 is the solvent with perhaps the closest properties to those of the late, lamented CFC-113 (Table 1.7).

One desirable attribute of HCFC 225 is that while it has some solvency for industrial soils, it can also be "fine-tuned" by blending to achieve the required solvency characteristics. Used neat, it is a mild to moderate solvent, much like CFC-113 (for those who recall manufacturing before the mid-1990s). HCFC 225 can also be blended with more aggressive organic solvents; stabilizers are also added. Examples of additives to increase solvency or performance include *trans*-1,2-dichloroethylene (*trans*), ethanol, and cyclohexane. Other blends include a proprietary surfactant, a proprietary fluorinated compound, and solvents at under 1%. Some but not all of the blends are azeotropes. Some blends are used as industrial process fluids as well as for critical cleaning.

TABLE 1.7 Comparative Attributes of HCFC-225 and CFC-113

	HCFC-225	CFC-113	CFC-113AES[a]
Boiling point (°C)	54	47.6	46.5
Freezing point (°C)	−131	−35	−41.8
Density (g/cm³)[b]	1.55	1.57	1.51[c]
Viscosity (cP)[b]	0.59	0.65	0.66
Surface tension (dyne/cm)[b]	16.2	17.3	18.5[c]
Latent heat of vaporization (cal/g,b.p.)	34.6	36.1	42.9
Relative evaporation rate (ether = 100)	90	123	120
Specific heat (cal/g,°C)[b]	0.24	0.229	0.272[c]
Solubility of water (wt%)[b]	0.031	0.109	0.25
Solubility in water (wt%)[b]	0.033	0.017	—
Flash point (°C)	None	None	None
KB value	31	31	39
ODP (CFC-11 = 1.0)	0.03	0.8	0.8
GWP (CO₂ = 1.0, 100 years)	370	5000	4800

Source: Adapted from Miki, T. et al., HCFC-225: Alternative precision and electronics cleaning technology, in *Handbook for Critical Cleaning*, 1st edn., CRC Press, Boca Raton, FL, 2000.

[a] Azeotrope of CFC-113 and ethanol.
[b] At 25°C.
[c] At 20°C.

The neat material and many of the blends do not have a flashpoint and have relatively favorable worker safety profiles. Asahi Glass Co. Ltd. has set an AEL (acceptable exposure level) of 100 ppm (8 h TWA) for the for the ca/cb mixture. The AELs for HCFC-225 ca and cb are 50 ppm and 400 ppm, respectively. Throughout the United States, HCFC 225 is exempt as a VOC, so it is favored for use in areas of poor air quality. Of course, the additives tend to be VOCs, and, since the blends may contain upward of 50% additive, it is important for manufacturers to evaluate blends in the context of allowable VOC emissions. It should be noted that a similar situation holds for many HFC and HFE blends.

HCFC 225 Phaseout

Unfortunately, because HCFC-225 depletes the stratospheric ozone layer, it is scheduled for global phaseout through the Montreal Protocol. In the United States, the Federal Clean Air Act calls for cessation of the use, sale, and production of HCFC-225 as of January 1, 2015 (EPA 2009). This might give pause to those using HCFC 225, particularly in the precision cleaning or critical cleaning of high value product.

This is an example of a situation where performance attributes must be balanced with the current and anticipated regulations. Cleaning protocols often require a substantial monetary and intellectual investment. With all of the pending regulations and discussions of regulations, in order to make informed, proactive decisions, it is important to understand exactly how manufacturers of critical components and products will be impacted.

It is helpful to clarify the meaning of "use" by the regulatory community (Kanegsberg 2010). Margaret Sheppard of the U.S. EPA's Stratospheric Protection Division explains that the EPA has interpreted the use ban to mean a ban on "use for manufacturing products." Sheppard explains that a solvent manufacturer or aerosol packager would no longer be able to use virgin material after December 31, 2014 to produce, for example, cleaning products. For more detailed information about EPA's interpretation of the regulations regarding HCFCs, the reader is welcome to peruse the 38-page document in the *Federal Register* (U.S. EPA 2009).

While users should be proactively looking at alternative processes, the door does not abruptly shut. The important point for those doing hard surface cleaning as part of manufacturing or repair is that end users could continue to use existing stocks of HCFC 225 that they have already purchased. Furthermore, the use of recycled HCFC 225 material is also allowed, either for end use or for use in the manufacturing of cleaning agents. This clarification should allow manufacturers to make informed, reasonable decisions about using HCFC 225 in current and impending projects. It also allows time to plan cleaning process changes.

Solvency and Physical Properties, Other Parameters

Kauri-Butanol Number

A number of solvency systems are described in Chapter 4 (Burke). In addition, other solvency systems are in use. One cloud-point test, the kauri-butanol (KB) number, is often alluded to. The KB number is determined by the volume of solvent required to produce a defined degree of turbidity when added to standard solutions of kauri resin in *n*-butyl alcohol. As a general rule, the higher the value, the stronger the solvent. The system was developed to indicate the relative solvent power of hydrocarbons, and it is not valid for oxygenated solvents. The KB number should be considered along with the boiling point, because, if the solvent can be heated to higher temperatures, more entropy is introduced into the system and better solvency occurs. Estimating solvency by mixing a cleaning agent with *t*-butyl alcohol and tree sap is a rather unsophisticated approach. However, the KB number remains widely used, and it is somewhat predictive of solvency (Kenyon 1995). Comparing the KB number with the Hansen system is somewhat analogous to comparing a black and white TV of the 1950s with a current, full-color, high-definition color broadcast. Table 1.8 lists the KB numbers and boiling points of several

TABLE 1.8 Kauri-Butanol (KB) Numbers and Boiling Points of Representative Cleaning Agents

Cleaning Agent	KB Number	BP (°C)
CFC-113	32	48
1,1,1-TCA	124	74
HCFC 141b	56	32
Methylene chloride	136	40
Trichloroethylene	129	87
n-Propyl bromide	125	71
D-Limonene	68	150
PCBTF	64	139
HCFC-225	31	54
HFC 43-10	9	55
HFC 43-10 blend, including *trans*-1,2-dichloroethylene	30	37
HFE 569 sf 2	10	76
VMS OS-10	17	100

representative cleaning agents. You will notice that many cleaning agents with high KB numbers are no longer produced or are under severe regulatory distress.

Wetting Index

The wetting index has been used as a guideline to the ability of a cleaning agent to penetrate closely spaced components. The wetting index is directly proportional to the density and inversely proportional to the surface tension and viscosity. The higher the wetting index, the more readily can a cleaning agent penetrate closely spaced components. Dr. W.G. Kenyon has discussed the wetting index for many years. As he emphasizes, it is most useful as a teaching tool. In general, many of the vapor degreasing solvents have a high wetting index than do water or hydrocarbon blends. As with other indications, however, wetting index alone does not determine the efficacy of cleaning. While hydrophobic chemists could use the wetting index as evidence that aqueous cleaning "won't work," it is more constructive and realistic to explain that aqueous cleaning depends less on wetting and more on cleaning force. Of course, where residue is of concern on product with blind holes, when water is used as a rinsing agent, cleaning force, time, temperature, and fixturing must be carefully designed into the process. The wetting indices of a few common cleaning agents are provided in Table 1.9.

Other Physical Properties and Regulatory Issues

Other physical properties such as boiling point, flash point, and evaporation rate must be considered in choosing a solvent, and the solvent must be considered in the particular regulatory microclimate where the process is being carried out. Tables 1.10 and 1.11 list some physical properties and a few regulatory considerations of currently used, developmental, and phased-out, longed-for solvents.

Many solvents, including those that are VOC exempt, can be used in vapor phase cleaning applications. Those with low flash points, however, must be used in specially designed equipment. Such equipment has a high initial capital cost. Many solvents do not have a flash point but do have an upper explosion level (UEL) and a lower explosion level (LEL); this must be considered in specialized operations and in selecting and maintaining emission control equipment.

The boiling point must be high enough to allow efficient cleaning, but not so high as to damage materials of construction or slow the build cycle. A very high boiling point may preclude the use of the

TABLE 1.9   Examples of Wetting Index

Cleaning Agent	Density (g/cm³)	Surface Tension (dynes/cm)	Viscosity (cp)	Wetting Index
Generally desirable	High	Low	Low	High
CFC-113	1.48	27.4	0.70	121
TCA	1.32	25.9	0.79	65
IPA	0.785	21.7	2.4	15
nPB	1.33	25.9	0.49	105
HCFC-225	1.40	16.8	0.61	145
HFE 449 sl	1.52	14	0.6	181
Hydrocarbon blend	0.84	27	2.8	11
Water	0.997	72.8	1.00	14
Saponifier solution, 6% aqueous	0.998	29.7	1.08	31

TABLE 1.10   Physical Properties, VOC, ODC Status

Cleaning Agent, Comments	Boiling Point °C (°F)	Flash Point	UEL/LEL %	Evap. Rate (Ref. for Evap. Rate)
1,1,1-TCA, ODC, phased out	74 (165)	None	15/7.0	5 (buac = 1)
CFC-113 ODC, phased out	48 (118)	None (TOC)	NA	0.45 (buac = 1)
HCFC-141b, ODC, phased out	32 (90)	None	17.7/7.6	>1 (ether = 1)
Stoddard Solvent, typical (hydrocarbon blend) (VOC)	152 (305)	40.6 (106)	6.1/1/1	Hydrocarbon blend, VOC
n-Propyl bromide (VOC), worker exposure profile	71 (160)	None	8/3	4.5 (buac = 1)
Methylene chloride VOC exempt Hazardous air pollutant	40 (104)	NA	19/12	NA
Per-chloroethylene VOC exempt Hazardous air pollutant	121 (250)?	None (TCC)	none	2.1 (buac = 1)
HCFC 225 VOC exempt, ODC, impending phaseout	54 (130)	None	None	0.9 (ether = 1)
HFE 569 sf2 VOC exempt	76 (169)	None (TCC, TOC)	NA	
HFE 449s1 VOC exempt	61 (142)	None (TCC, TOC)	NA	
HFC 43–10mee VOC exempt	55 (131)	None (TOC)	NA	
Water	100 (212)	None	None	

Note: These data were obtained from various standard publicly available references, primarily MSDS from the Cornell University Program Design Construction Web site (http://msds.pdc.cornell.edu/ISSEARCH/MSDSsrch.htm), University of Vermont Web site, with some confirmation by Lange's *Handbook of Chemistry*, 13th edn. (McGraw-Hill, New York), and *Dangerous Properties of Industrial Materials*, 3rd edn. (N. Irving Sax, Reinhold Book Corp., New York). They should be used as guidelines only—the evaporation rate data in particular are prone to inconsistency among references). Boiling points rounded to nearest integer. Please confirm all information with current MSDS (buac = butyl acetate).

solvent in a standard vapor phase degreasing operation. The evaporation rate must be sufficiently rapid to allow rapid drying, but not so rapid that the solvent is immediately lost. These considerations are all relative to the operation in question.

Costs

Costs are relative. Cleaning is not a chemistry, it is a process. Costs are more than the price of the cleaning agent and the initial investment in capital equipment. The manufacturing community would

TABLE 1.11 Physical Properties, VOC, ODC Status of Low-Flash-Point Solvents

Cleaning Agent Comments	Boiling Point °C (°F)	Flash Point	UEL/LEL %	Evap. Rate (Ref. for Evap. Rate)
TBAC proposed VOC exempt	98 (208)	15 (59)	NA[a]/1.5	VOC exempt, with provisos
Para-chlorobenzo trifluoride VOC exempt	139 (282)	43 (109)	10.5/0.9	
Di-siloxane VMS VOC exempt	100 (212)	−3 (27)	18.6/1.25	Dow VMS OS-10
		TCC		3.8
Tri-siloxane VOC exempt	152 (306)	34 (94)	13.8/0.9	Dow VMS OS-20
		TCC		0.7
Cyclo-tetrasiloxane VOC exempt	205 (401)	76 (170)		Dow VMS OS-245
		TCC		Not calc.
Acetone VOC exempt	56 (134)	−20 (−4)	13/2.5	6

Note: These data were obtained from various standard publicly available references, primarily MSDS from the Cornell University Program Design Construction Web site (http://msds.pdc.cornell.edu/ISSEARCH/MSDSsrch.htm), University of Vermont Web site, with some confirmation by Lange's *Handbook of Chemistry*, 13th edn. (McGraw-Hill, New York), and *Dangerous Properties of Industrial Materials*, 3rd edn. (N. Irving Sax, Reinhold Book Corp., New York). They should be used as guidelines only—the evaporation rate data in particular are prone to inconsistency among references). Boiling points rounded to nearest integer. Please confirm all information with current MSDS (buac = butyl acetate).

[a] Not available.

benefit from the availability of additional resources regarding cleaning costs. The efficiency of cleaning equipment and the impact on the manufacturing plant could be improved with better equipment insulation. In fact, environmental policy might benefit from a more holistic assessment of process impact. Extensive studies were performed a decade ago; the approaches hold true today (Kanegsberg 1999, 2000). Few solvents are inexpensive, particularly if total process costs are considered. To save money, invest in high-quality, well-designed cleaning processes, processes with excellent stewardship by your suppliers.

In terms of organic solvents, pound per pound, traditional solvents such as IPA, acetone, and the chlorinated solvents are relatively inexpensive. nPB and the VMS are moderately priced, and the engineered solvents (HCFC 225, HFEs, and HFCs) are the most costly.

Blended high boiling solvents can vary markedly in price. The costs may be perceived as high in applications where soil loading is a problem and frequent solvent change-out is required. With heavy soil loading, it may be more effective to perform initial cleaning in a relatively inexpensive product and then conduct subsequent steps in the more sophisticated cleaning agent.

Aqueous cleaning agents must be compared against each other in the intended application. Let us assume that two concentrates are under consideration and that one is twice as costly as the other. If the inexpensive cleaning concentrate must be used at a 1:4 dilution while the other provides equivalent performance at a 1:20 dilution, the picture changes (O'Neill 2000). Filtration markedly influences bath life and therefore modifies the overall cost of the cleaning agent.

How Not to Clean Critically with Household Products

This section was adapted from Kanegsberg (2007).

For many manufacturers, setting up a cleaning process, even for critical product, involves a trip to the grocery store or home improvement center to purchase household detergent or liquid dish soap (Kanegsberg 2007). Or, it may involve a trip to the hardware store for containers of kerosene or some other "fix-it" blend. What's wrong with that? Especially if the process is working, why not let sleeping dogs lie?

Aqueous cleaning processes are most efficient when the appropriate chemistry for the soil and substrate (the part being cleaning) is selected. Solvents designed for the home maintenance or for hobbyists do not have to meet the exacting standards of industry; these include products based on organic chemicals such as mineral spirits or bio-based chemicals. In our experience, there have been too many instances of problems with build processes that could be traced directly to the use of consumer or household products Productivity decreases; costs increase. Sometimes, the problem chemistry was used during final assembly; in other cases, the problem occurred at a sub-vendor/supplier.

Household products are not optimized to critical and industrial cleaning. Many household products contain perfumes, colorants and lotions to make the product more pleasant to use. However, such additives can leave a thin, overall surface residue that interferes with subsequent processes like coating or deposition. Even with rinsing, additives can become trapped in the nooks and crannies of the product, leading to assembly problems, yield issues or unexpected, catastrophic product failure. Appropriate additives are added to industrial aqueous formulations; time should be invested in selecting products with additives most beneficial to your cleaning application.

Household cleaning products are sometimes selected on a "temporary" basis for prototype products because they are effective in a dip tank. However, they may not adapt well to an upgraded process. Even if you are using an immersion process bath, there can be problems with bath life in that dish soaps tend to hold soil in suspension. The high soil loading capacity of a consumer cleaning agent may lead to a false sense of security and mask bath degradation. The capacity for a degree of soil loading is necessary for practical cleaning processes. However, there can be too much of a good thing. Because consumer products typically hold soils in suspension, there can be problems related to soil overloading and redeposition. Even if a consumer product appears to be economical and the decision is made to use a dip-tank process bath, the costs of changing out the process bath, including labor time and costs to dispose of the spent cleaning agent.

In contrast, many industrial cleaning agents are designed as "oil-splitting" chemistries. That is, the chemistries are designed so that the oil pops up to the surface so that it can be removed leaving a clean process bath. Of course, if you currently clean in a simple immersion tank, when you switch to an oil-splitting chemistry, you may also need to modify the process so that you are not re-soiling product by dragging it through a layer of oil. The most efficient way to remove the oil involves a sparger or weir; this involves some investment in process equipment. If a high pressure spray, the appropriate defoamers must be present or unpleasant soapsuds will result. In addition some cleaning agents for critical applications are designed specifically for ultrasonics applications.

In choosing the cleaning for a given application, it is crucial to optimize relative to the material(s) being cleaned and the metalworking fluids and polishing compounds being used. Consumer cleaning agents are designed to remove typical consumer soils from typical household items. They are designed for cleaning dishes or perhaps for quickly wiping surfaces. Even if the product appears to remove soils effectively, it is important to evaluate the potential for substrate damage by testing the product with all of the materials of construction that will be exposed to the cleaning chemistry.

For critical cleaning, the cleaning agent must be well-defined and it must be consistent over time. For years, a dishwashing detergent was successfully used to clean critical aerospace subassemblies. Then, the word "improved" appeared on the label, a citrus scent was apparent and the cleaning process stopped working. In this actual "legend of aerospace," the contamination source was apparent. Sometimes, unannounced formulation changes are not so quickly identified; valuable time and product can be lost before a cause of the process failure is found. What if this happened to you? You might find another process; but if you had qualified the household cleaner with your customer such an emergency change might not be greeted kindly.

Try asking the supplier of a household, consumer cleaning agent for product support. We did, in the course of helping clients where household cleaners were enmeshed in in-house and supply chain processes. We asked a number manufacturers of household cleaners used in industrial and critical cleaning applications and asked to speak with someone in the applications lab. For those products designed exclusively for household use, we were not put in contact with a chemist or product support specialist. After repeated conversations with company representatives who had not the foggiest notion about industrial and manufacturing processes, copious e-mail requests, an interminable amount of time spent on hold, and receipt of generous supplies of store coupons, some suppliers of household cleaning product either verbally or in a brief e-mail indicated that they declined to support the products for industrial assembly processes. There are a few "crossover" products. A few household cleaning agent manufacturers that also claim that their products can be used in industrial applications provide process and laboratory support. Occasionally, there is a genuine chemist to provide support. If you like a household product, they may be able to suggest a similar formulation that is better suited to industrial or critical cleaning applications. The above exercise is important for all cleaning chemistries, because the quality of support for products sold for critical and industrial applications varies.

Unfortunately, use of household products in critical cleaning applications is likely to be an ongoing problem for a number of reasons. The first is familiarity. If the dish soap removed last night's burnt-on dinner, it might seem reasonable to try it for burnt on lubricants. If it appears to work, assemblers continue to use it. There is the convenience factor. Particularly for small-scale processes, it may be easier to pick up a case of discount industrial cleaner at the "big box" store than to place an order with a distributor. Engineers may plan to worry about cleaning "later." Suddenly, it is time to validate or qualify the process; and the only tests have been run with non-industrial cleaners. We live in an age of advertising and most of us are probably responsive to publicity. Adverting appeals to our egos. In at least one instance, a cleaning agent was selected because it had the same name as the supervisor. We observed the power of the infomercial in the course of a comparison study of aqueous-based products for point of use or hand-wipe cleaning. One client was using a water-based spray cleaner that the operator had found on a late-night infomercial. We compared the product that supposedly worked with a number of other formulations and found it to perform at best marginally, even against the soils supplied by that client. The study was presented to the Joint Services Solvent Substitution Working Group (JS3WG) and is posted on their website (http://js3.ctc.com).

Other products have names that convey the concept of clean, or safe or ecologically-friendly. It is important to determine if the name is backed up by actual favorable safety or environmental attributes and if it performs acceptably in your application. In industrial applications there may be air and/or water regulations that impact use of the product. One individual commented that a consumer-oriented cleaning agent had grit to remove heavy soils from his fingernails, but left his hands soft. However, because industrial cleaning processes often involve heat, force and time, a consumer-oriented cleaning agent used in an industrial process cannot be considered immune to worker safety issues. Even if the product itself is biodegradable and kind to the earth, your process bath contains environmentally-unfriendly items like soils, polishing compounds and metal fines. In many cases, you cannot simply dispose of a process bath as if it were a dishpan. Finally, even if the dish soap does not damage your dishes and leaves your hands silky-smooth, even if it is totally organic and contains no hazardous ingredients, it may damage the product. You may not see the damage, but your customer could see an increase in defects.

A household product may be added when other parts of the build process change. All cleaning processes, particularly aqueous processes, have to be optimized to the total build process. If any factors change, for example a lubricant, the cleaning process may have to be modified. If the process is not working properly, and if the response from management is not rapid enough, it is very tempting to try a "patchwork" quick-fix from the local hardware store. If it seems to improve matters (at least visually), the product becomes an informal part of the process, but probably a poorly-controlled part of the process.

Do not let sleeping dogs lie, but wake them gently. Do *not* wait for a problem to arise. Review the written documentation. We have observed consumer products that are immortalized in aerospace documentation for critical assemblies—the documents often date back a generation or so. Next, and this is even more important, actually tour the fabrication areas. If you are a final assembler and use sub-vendor suppliers for part of the assembly and/or cleaning, contact them and perform a site visit there, too. If you find consumer-oriented cleaning products, flag those products, noting the brand, supplier, source and contact information. Then, contact the manufacturer or supplier and determine the level of technical support. If product support is not available, make plans to change the process. However, it is generally counterproductive to simply mandate immediate changes unless there is an urgent process or yield problem or a compliance issue. It is more effective to work with the technicians, the operators and the assemblers to determine why the consumer product was introduced. Rather than placing blame (which makes people avoid fixing the problem), this approach can provide invaluable information as to what is actually needed to achieve critical cleaning and contamination control. Only then should you evaluate cleaning agents designed for industrial and critical cleaning and find more supportive suppliers.

How to Choose a Cleaning Agent

One of the problems in developing a manufacturing process is the rather daunting list of considerations and provisos. To cope with the problem, there is the tendency to think linearly and to attempt to find the perfect cleaning agent. There is no perfect cleaning agent. However, we persist in our search for unattainable perfection.

All too often, when a cleaning process is being developed, a cleaning agent selection committee is established to screen out all undesirable applicants. The safety/environmental group is likely to rule out any environmentally challenged cleaning agent, even if it could be used in a non-emissive manner. Company management and sometimes the customer may submit a series of "don't" lists. Whole classes of cleaning agents may be ruled out as being unacceptable on general environmental principles. The materials and process chemists may insist that for any cleaning agent to be considered, it should be able to be in contact with all materials of construction at some elevated temperature for, let's say, 24 h. The purchasing department may insist that only one or two cleaning agents be selected - period. The manufacturing engineers may insist on an extremely rapid process time, instant drying, and aqueous cleaning. What is left? Sometimes nothing; sometimes a class of cleaning agents which is totally unsuited to the cleaning application at hand.

Perfection aside, for nearly every cleaning application, there are several workable solutions. Some of the considerations in cleaning agent selection are indicated in Table 1.12. The cleaning agent has to be considered in the context of the cleaning process and, indeed, in the context of the overall manufacturing process. The factors indicated in Table 1.12 are meant to provide a starting point. It usually becomes very apparent which factors are the most important in a given manufacturing situation.

It is more productive to proceed with a nonlinear approach that considers performance, costs, cleaning agent, cleaning equipment, suitability to the workforce, worker safety, and the local regulatory microclimate.

Cleaning is a process, not a chemical. The landscape of available cleaning agents and processes varies as advances in technology, requirements of build processes, and regulatory drivers change the perceived appeal of various options, because the assortment of products will inevitably change. To achieve ongoing, successful critical cleaning, our best advice is to develop an appreciation not just of the specific cleaning agents, but of the underlying, commonsense approaches to successful cleaning and contamination control.

TABLE 1.12 Overall Considerations in Choosing Cleaning Agents

Factor	Process Consideration
Cleaning properties	Cleaning requirements, your process (how clean is clean enough)
	Performance under actual process conditions
Cleaning performance	Solubility characteristics relative to soil of interest
	Wetting ability
	Boiling point
	Evaporation rate
	Soil loading capacity
	Ability to be filtered
	Ability to be redistilled
Materials compatibility	Compatibility under actual process conditions (temperature, time of exposure)
	Product deformation at cleaning, rinsing, drying temperatures
Residue	Nonvolatile residue (NVR) level
	Rinsing requirement
	Process time impact
Cycle time	Cleaning
	Rinsing
	Drying
	Product cool-down
	Component fixturing
	Loading and unloading equipment
	Product rework
Cleaning equipment	Suitability with current cleaning equipment
	Ability, costs of retrofit
	Costs of new cleaning equipment
	Auxiliary equipment required
	Maintenance, repair
	Automation, component handling
	Footprint (length, width, height)
	Equipment weight
	Component fixturing
Flash point	Choice of cleaning equipment
	Process control
	Control of proximal processes, activities
	Choice of auxiliary equipment
	Choice of emissions control
Toxicity	Acute
	Long term
	Anticipated exposure under process conditions (including sprays, mists)
	Employee monitoring
	Inhalation
	Skin adsorption
Worker acceptance	Method of application
	Drying speed
	Similarity to current process
	Automation
	Computer skills
	Perceived loss of control of process
	Odor
Cleaning agent management	Water preparation
	In-process filtration (water, organic, or aqueous cleaning agent)
	Waste water disposal
	On-board redistillation

(continued)

TABLE 1.12 (continued)   Overall Considerations in Choosing Cleaning Agents

Factor	Process Consideration
Regulatory, air	Global, national, local (ODC, VOC, HAPs, GWP)
	Neighborhood concerns
	Environmental justice issues
	Production phase-out
	Usage bans
	Disposal of waste stream
Regulatory, water	Global, national, local
	Disposal of waste streams
Company, customer, product performance requirements	Contractual requirements, restrictions
	Company policy
	Testing, acceptance qualification required
	In-house safety, environmental policy
	Insurance company issues
Costs	Cleaning agent
	Cleaning agent preparation and disposal
	Costs as used (dilution)
	Capital equipment
	Disposables
	Sample handling
	Total process time
	Rework
	Insurance
	Regulatory permitting
	Process qualification
	Employee education, training
	Process monitoring
Supplier stewardship, cleaning equipment supplier	Responsive distribution, supplier
	Supportive technical staff
	Clear, understandable MSDS
	Provides required technical information
	Provides required regulatory information
	Supports process development

References

3M Fluorinert™ Liquids Product Manual, 1991.

Agopovich, J.W., PFC alternative analyses, Precision Cleaning, March 1997.

Agopovich, J.W. Review of solvents for precision cleaning, in B. Kanegsberg and E. Kanegsberg (eds.), in *Handbook for Critical Cleaning*, 1st edn., CRC Press, Boca Raton, FL, 2000.

AG Environmental Products, L.L.C. Soygold 1000 MSDS. 2009.

Arafat, El Sayed. NAVAIR finds alternatives for petroleum-based solvents. *Currents*, Fall 2009:75–77.

Cala, F.R. and A.E. Winston. *Handbook of Aqueous Cleaning Technology for Electronic Assemblies*. Isle of Mann: Electrochemical Publications, 1996.

California Environmental Protection Agency Air Resources Board. Environmental impact assessment of tertiary-butyl acetate, Staff Report. January 2006.

CAS Registry. CAS Registry and CAS Registry Numbers. February 11, 2010. http://www.cas.org/expertise/cascontent/registry/regsys.html (accessed March 12, 2010).

Cull, R.A. and Swanson, S.P., Volatile methylsiloxanes: Unexpected new solvent technology, in B. Kanegsberg and E. Kanegsberg (eds.), *Handbook for Critical Cleaning*, 1st edn., CRC Press, Boca Raton, FL, 2000.

Definitions of VOC and ROG. www.arb.ca.gov/ei/speciate/voc_rog_dfn_1_09.pdf, January 2009.

Elias, W.G. Real-life applications with environmentally compliant solvents for electronics. *Proceedings of Nepcon West 2000*, Anaheim, CA, 2000.

Environment Canada. Background on Siloxanes D4, D5 and D6. February 15, 2010 (accessed March 16, 2010).

EPA. More Information on HCFC's. December 8, 2009. http://www.epa.gov/Ozone/title6/phaseout/hcf cuses.html (accessed February 2010).

Foreman, J.E. Tertiary butyl acetate: A potential VOC exempt solvent for hand wipe cleaning and coatings applications. *Presentation at the Tenth Annual International Workshop on Solvent Substitution and the Elimination of Toxic Substances and Emissions*. Scottsdale, 1999.

Henneberry, M. and Opre. J. Agrochemical based biodegradable solvent performance ethyl lactate & soy methyl ester. *CleanTech 2000 Proceedings*. Flemington, NJ: Witter Publishing, 2000. pp. 466–468.

Kanegsberg, B. Precision cleaning without ozone depleting chemicals. *Chemistry and Industry*, 20, 1996:787–791.

Kanegsberg, B. The cost of process conversion part 1. *CleanTech '99*, May 1999.

Kanegsberg, B. Costs of cleaning part 2. *CleanTech 2001*, Rosemont, IL, May 2001.

Kanegsberg, B. The joyful dawn of a new era. *Process Cleaning Magazine*, 2(3), May/June 2007.

Kanegsberg, B. Cost-effective cleaning for quality thermal spray coating. *International Thermal Spray Association, Presentation at Annual Meeting*. Orlando FL, April 2009.

Kanegsberg, B. and E. Kanegsberg. Industry watch column, HCFC 225—Ban or phaseout? *Process Cleaning Magazine*, March/April 2010.

Kenyon, W.G. and B. Kanegsberg. Accelerating the change to environmentally-preferred, cost-effective cleaning processes. *Precision Cleaning '95*, Chicago IL, Tutorial, May 1995.

Knopeck, G. Pentafluoropropane: An HFC solvent for aerosols. SATA (Southern Aerosol Technology Association) *Spring Meeting, Presentation*, April 2000.

Miki, T. et al. HCFC-225: Alternative precision and electronics cleaning technology, in B. Kanegsberg and E. Kanegsberg (eds.), *Handbook for Critical Cleaning*, 1st edn., CRC Press, Boca Raton, FL, 2001.

Mouser, W., R. Manchester, W. Barrett, and F. Bergman. Method, apparatus, and system for bi-solvent based cleaning of precision components. U.S. Patent 7604702. October 20, 2009.

O'Neill, E., A. Minemadi, A. Guzman, R. Romo, M. Shub, and B. Kanegsberg. Simplifying aqueous cleaning, the value of practical experience. *Products Finishing Magazine*, August 2000.

Pourreau, D. Tert-butyl acetate (TBAC) a technical overview and regulatory update on the latest voc-exempt solvent. *Plating and Coatings Industry*, January 2006.

SCAQMD. SCAQMD Method 313: Determination of volatile organic compounds (Voc) by gas chromatography/mass spectrometry (GCMS). October 26, 2004. http://www.aqmd.gov/rules/cas/app1.html (accessed March 8, 2010).

Shaw, D. 1,2 Transdichloroethylene the last chlorinated solvent? *Spray Technology and Marketing*, January 2010:22–23.

Shick, R.A. Formulating cleaners with structured solvents. *Proceedings of Precision Cleaning '96*, 1996. pp. 285–289.

Smith, M. Status of DMC. Kowa American Corp. March 2010.

Solvay Fluor GmbH. Solvokane Product Literature. November 2009.

U.S. Environmental Protection Agency. Product Stewardship Program for Six Siloxanes Conducted Under a memorandum of Understanding (MOU) Signed by EPA and the Dow Corning Corporation; Notice of Receipt and Availability of the MOU Data, July 2009. http://www.epa.gov/EPA-TOX/2009/July/Day-30/t18195.htm (accessed March 16, 2010).

U.S. Environmental Protection Agency. Revisions to the California State Implementation Plan, South Coast Air Quality Managment District and Ventura County Air Pollution Control District. Federal Register, CFR Citation 40 CFR Part 52 Approval and Promulgation of Proposed Rules, December 2009.

U.S. EPA. Definition of volatile organic compound (VOC). March 31, 2009. http://www.epa.gov/ttn/naaqs/ozone/ozonetech/def_voc.htm (accessed March 8, 2010).

U.S. EPA. IRIS toxicological review of cis-and trans-1,2-dchloroethylene. January 25, 2010. http://cfpub.epa.gov/ncea/cfm/recordisplay.cfm?deid=190184 (accessed March 19, 2010).

United Soy Board. Methyl soyate: Eco-friendly with performance potential. 2010. http://soynewuses.org/Opportunities/Default.aspx (accessed March 8, 2010).

United Soybean Board. Methyl soyate: Eco-friendly with performance potential. 2010. http://www.soynewuses.org/Opportunities/Default.aspx (accessed March 30, 2010).

United States Department of Labor, Occupational Safety and Health Aministration. 1,2-Dichloroethylene. http://www.osha.gov/SLTC/healthguidelines/1_2-dichloroethylene/recognition.html (accessed March 19, 2010).

United States Environmental Protection Agency. 40 CFR Part 82 Protection of Stratospheric Ozone: Adjustments to the allowance System for Controlling HCFC Production, Import and Export. Federal Register Doc. E9-29569, 74 FR 66412, December 15, 2009:66412–66418.

United States EPA. Definition of volatile organic compounds. March 31, 2009. http://www.epa.gov/ttn/naaqs/ozone/ozonetech/def_voc.htm (accessed March 2010).

Waldrop, M.W., Personal communication and technical summaries, BASF. January 2000.

Wildes, S. Methyl soyate: A new green alternative solvent. *Chemical Health and Safety*, 9(3), May–June 2002:24–26.

Zipfel, L. and P. Dournel. HFC-365mfc, the key for high performance rigid polyurethane foams. *UTECH 2000, Presentation*, April 2000.

2

Aqueous Cleaning Essentials

Michael Beeks
Brulin & Company, Inc.

David Keller
Brulin & Company, Inc.

Introduction

While aqueous cleaning is almost as old as man, industrial parts manufacturers and cleaners have long held the view that certain soils could only be cleaned adequately by non-aqueous methods. Starting in the 1970s, many environmental initiatives forced the industry to look at aqueous cleaning as an ecologically more responsible alternative to vapor degreasing and other solvent cleaning methods. As a result many, cleaning applications that were once strictly the province of non-aqueous cleaning methods are now being done quite successfully with aqueous processes. The following discussion serves as a brief primer on the many aspects and considerations needed to successfully understand the art of aqueous cleaning in the manufacturing environment.

Cleaning Overview

With few exceptions, there are certain principles treated generally here that apply to all types of cleaning.[4] Cleaning processes combine mechanical, thermal, and chemical energy sources to remove a soil from a substrate. The total energy needed is the sum of these energy sources over a given period. Within these parameters, the following general guidelines apply:

1. Cleaning efficacy and rate improve as temperature increases.
2. Agitation improves the rate and efficacy of soil removal. Agitation provides mechanical energy to physically remove soils and ensures that fresh cleaner will continuously contact the soil.
3. Cleaner solutions generally have a performance-versus-concentration curve. A minimum level of cleaner is generally necessary for effective cleaning. Cleaning improves with incremental increases in concentration up to some point after which further increases result in little or no further improvement in performance.
4. Removing soil requires a finite amount of time to apply the necessary energy to accomplish the task. There are four primary factors that govern the total energy that can be applied in a cleaning process: (1) cleaner concentration, (2) mechanical agitation, (3) temperature, and (4) time. The general expression for how these factors work together is

    Cleaning energy = Cleaner concentration × mechanical agitation × temperature × time

    If one of the factors decreases, one or more of the others have to increase in order to maintain performance. Certain limitations may apply; for example, high cleaner concentration can lead to diminishing returns and increased costs. Cleaning performance response is not linear as a function of concentration and excessive cleaner concentration can create problems in the following rinse steps. Mechanical agitation is usually a fixed property that cannot be changed on most cleaning equipment. Temperature can be raised up to a point but too much temperature may damage the parts and cause excessive water consumption due to the increased evaporation rate. The ability to increase the time parameter is limited by the substrate's resistance to corrosion under the cleaning conditions and the required parts throughput rate for the system. For most cleaning applications, it is relatively easy to meet the necessary energy input to obtain the required cleanliness, but there are also some very challenging applications that are difficult or impossible to clean. In some cases, solvent cleaning is still the more appropriate cleaning method to use.
5. Rinsing is necessary to remove any cleaner or soil residue remaining on the parts after washing:
    a. Rinse type and quality is dependent on the cleanliness requirements of the application.
    b. Multiple small rinses are generally more efficient and cost effective than one large rinse.
    c. An agitated rinse is more effective than a still rinse.
    d. Final part cleanliness or conversely residue on the part is limited by rinse quality.
6. Soil must be prevented from redepositing on parts. The most obvious answer is to stop the soils from contacting the substrate after initial removal. Soil removal may be accomplished by using cleaners that include one or more of these methods:
    a. Emulsification
    b. Emulsification followed by demulsification
    c. Deflocculation
    d. Displacement
    e. Saponification
    f. Sequestration
    g. Wetting
    Additionally, redeposition can be controlled by
    a. Microfiltration
    b. Oil skimming and coalescing

    c. Using cleaning tanks of sufficient size to disperse the soil and slow the rate of increase of contamination concentration

7. The cleaning method or solution should not harm the item (substrate) being cleaned.
8. Precleaning to remove bulk soils may be an economical and common sense way to increase overall cleaner life.
9. Cleaning systems should be designed as a unit. That is, the cleaner and the cleaning equipment should be chosen to work together and address the particular cleaning application. Typical concerns that should be addressed include

    a. Cleaning temperature and its effect on the cleaner and the substrates. Some considerations include method of heating, insulation, and evaporation.

    b. Equipment design should include an evaluation of cleaner and part compatibility with regard to materials of construction, economy of operation, electrochemistry, OSHA and other regulatory guidelines, and ease of service.

    c. The suitability of the mechanical energy input must be addressed in terms of effectiveness of removing soil from the substrate, controlling foaming tendencies of the cleaner, avoiding mechanical damage to the parts, and avoiding degradation of the cleaner.

Aqueous cleaners can generally be categorized as being acidic, alkaline, or pH neutral. Alkaline cleaners are by far the most predominant of the three and are used in all commercial/industrial cleaning applications. Thus, this discussion will mainly cover alkaline cleaning but the principles will be applicable to all types of cleaners.

Agitation techniques represent the greatest variation in cleaning methods. The most important factor is that it costs money in equipment and/or labor to provide high levels of agitation. The equipment must be designed to meet the objective of providing adequate-to-superior agitation for soil removal at the lowest cost. The major limitations for providing adequate agitation are equipment costs, equipment size (i.e., how big of a "footprint" the equipment makes), excessive foam generation, excessive generation of mist/spray, toxic vapors, and the creation of flammable/combustible gases.

Now that we have taken a broad look at some cleaning principles, let us look a little more closely at each.

Cleaning Parameters

Temperature

The majority of industrial cleaning is carried out at 140°F–180°F. The effect of temperature depends on the type of soil being removed and the specific cleaner. The first consideration is what type of soil needs to be removed. Temperature is very important in speeding the removal of fats, greases, oils, and waxes. Increased temperature reduces the viscosity of oils and greases, making them more mobile and therefore easier to displace from the substrate. Fats and waxes are often solids at room temperature. It is critical to melt these fats and waxes in order to remove them by aqueous methods. If the melt range of the fat/wax is above the boiling point of the cleaner, aqueous cleaning will not be effective on these soils; solvents must be used instead.

There is a well-established principle that the rate of a chemical reaction is doubled for each 10°C (18°F) increase in temperature. If the cleaning process works by reaction between a fatty acid/oil and alkali, by a paint coating undergoing chemical decomposition, or by an acid chemically removing rust and scale, then this reaction rate relationship is applicable. It is possible to remove solid fats/waxes if they can chemically react with the cleaner without melting them first but the rate of removal is often too slow for manufacturing applications.

On the other hand, excessively high temperatures could "set" proteinaceous soils or may cause an undesirable reaction between the soil and the substrate, resulting in the soil becoming more difficult to remove. Just as increasing temperature will increase the rate of cleaning, it will increase the rate of undesired reactions as well. Most corrosion inhibitors work by forming a loose barrier on the clean metal surface. Excessively high temperatures can disrupt this barrier and result in chemical attack, usually seen as discoloration and etching.

Agitation

As has been previously stated, agitation techniques represent the greatest variation in aqueous cleaning techniques. It is usually possible to find an aqueous cleaner to remove a soil from a substrate. One of the biggest problems users run into is inadequate cleaning performance due to inadequate or improper choice of agitation. The method of agitation should be matched to the size and shape of the part. For example, while spray washing may be very effective for cleaning large relatively flat parts, it is usually not suitable for parts with blind holes where direct impingement is problematic. Relatively flat objects and components that do not have hidden areas can be cleaned by immersion or spray wash. Parts that are very large or very small often cannot be cleaned effectively by spray wash. An exception on small parts is possible if specialty mounting racks can be built. Spray wash cleaning is also limited on the chemical side because the formula must not create excessive foam. Eliminating foam restricts the choices of raw materials a chemist can use in formulating a spray wash cleaner.

Parts that can be damaged from spray impingement should instead be cleaned by immersion. Virtually all parts can be cleaned by immersion. Ultrasonic cleaning is the most effective method of agitation for immersion cleaning. Ultrasonic cleaning is restricted by cost (comparatively expensive equipment) and size. (Ultrasonics are not as effective on tanks above approximately 1000 gal capacity.)

Spray under immersion, oscillating lifts, and turbulation are the next most effective methods of immersion agitation. Combining an oscillating lift with ultrasonics provides superior agitation and works well on removing highly viscous soils such as wheel bearing greases found in automotive repair applications.

Chamber-type cleaning units, commonly called "cellular" or "cell systems," allow the use of a combination of agitation techniques. In cell systems, parts are cleaned in a basket that resides in a chamber. The chamber is first subjected to spray washing until the basket is completely immersed. The chamber is then agitated ultrasonically and then drained. The parts are then rinsed using spray followed by ultrasonics agitation. Cell systems may even incorporate basket rotation in the wash and rinse cycles. The three agitation methods can sometimes create excessive foam if care is not taken in choosing a cleaner.

While air-sparging can be very effective at agitating an aqueous cleaner, there are several provisos:

1. The cleaner must be a low-to-non-foaming product. Air-sparging will cause foam to overflow the tank. Only cleaners designed for continuous spray washing should be used with this form of agitation.
2. The popping of the bubbles formed by air-sparging generates mist and spray that may create an exposure problem for workers. Aerosol mists generated from using air-sparging on alkaline derusters are very corrosive and must not be inhaled or get in the worker's eyes! Some approaches to addressing worker exposure to mists and sprays include covering the tank, installing aggressive exhaust ventilation to remove the mist and spray, or placing the tank within a cabinet to contain the mist/spray. However, it is usually impractical to put a cover on most immersion tanks due to the mechanics of opening and closing the cover during parts transfer in the cleaning cycle. Adding aggressive exhaust ventilation increases the cost of the cleaning system and may affect temperature stability in the work environment. Enclosing the system in a cabinet increases the cost of the cleaning system and will likely increase the "footprint."
3. Air-sparging can shorten the life of alkaline cleaners by neutralization with carbon dioxide (CO_2). Carbon dioxide is a weak acid. Acids and bases neutralize each other when combined. Even though CO_2 makes up only a fraction of a percent of the atmosphere (0.035% measured at Mona Loa Observatory, 1990, as reported in *Handbook of Chemistry and Physics*, 1996)[20], over time, the large volume of air passing through the tank results in exposure of the cleaner

to a significant amount of CO_2. The CO_2 neutralizes the alkaline builders, especially sodium or potassium hydroxide. The CO_2 reacts with free hydroxide ions (OH^-, the cause of alkaline pH) to form bicarbonate ions:

$$CO_2 + OH^- \rightarrow HCO_3^-$$

The bicarbonate then goes on to react with more hydroxide ions to form a carbonate ion and water:

$$HCO_3^- + OH^- \rightarrow CO_3^{2-} + H_2O$$

As the hydroxide ions (OH^-) are consumed, the pH of the cleaner decreases. Heavy-duty caustic cleaners are especially prone to this problem whereas mildly alkaline cleaners are much less sensitive. Acidic and pH-neutral cleaners are not affected by this problem.

General agitation from pump circulation can be adequate for noncritical cleaning applications, but it is not usually suitable for precision cleaning. Finally, soaking the substrate in a stagnant tank is unacceptable, even for noncritical cleaning applications.

Concentration

Concentration, also called use-dilution, can affect multiple attributes of the cleaning process. In many cases, minimum or maximum cleaner concentrations impact corrosion characteristics, chemical etching, or the deposition of protective barriers as well as cleaning efficacy. The required cleaner concentration depends on the type of agitation and temperature. As an example, in the absence of foaming problems, it may be possible to obtain similar cleaning performance from the same alkaline cleaner at 5%–10% by immersion, 3%–5% by spray, 1%–3% by steam cleaning or high-pressure hot spray, or 2%–4% in a high-pressure room temperature spray. For each of these applications, increasing the cleaner concentration provides better cleaning performance up to a certain point and then levels off. The leveling-off point is dependent on the specific chemistry used in the cleaner, the soil being removed, the agitation technique, and the temperature.

Time Required for Cleaning

It is important to emphasize that cleaning is not instantaneous; some time is required for the cleaner to perform its work on the soil. As with the leveling-off point, the cleaning time depends on the concentration of the cleaner, the specific chemistry used in the cleaner, the soil being removed, the agitation technique, and temperature. In a stagnant bath, cleaning may take between 5 min to over an hour to occur, if cleaning occurs at all.

In agitated systems, most immersion cleaning times do not exceed 10 min, although numerous exceptions can be found. Spray washes typically take no more than 5 min.

One general rule of thumb does exist for ultrasonic cleaning; if it takes more than 5 min to clean, either there is something deficient in the cleaner or the process itself is not suited for the application.

One notable exception to the 5 min rule for ultrasonics is the cleaning of used automobile cylinder heads and other major engine components. Many automobile repair facilities have adopted ultrasonic processes as an alternative to heavy-duty caustic or solvent cleaning tanks. Heavy-duty caustic tanks have fallen out of favor because they cannot be used on aluminum parts; nearly all major engine parts are now made of aluminum. Corrosion hazards and waste disposal considerations also have contributed to the decline of the caustic stripping tank. Solvents have been on the decline mainly due to environmental regulations that govern volatile organic content (VOCs), ozone-depleting substances (ODS), and hazardous air pollutants (HAPs). The few solvents (chlorinated solvents and cresylic acids) that are effective on baked-on soils also pose serious health risks. Ultrasonic cleaning with highly concentrated, moderately alkaline cleaners has been found to be effective at removing most of the baked-on soils. The drawbacks include long processing times (often 15–45 min) and some cavitational erosion. Cavitational erosion occurs with softer metals (mainly aluminum) and typically appears as a "star- or Y-shaped" pattern on the surface.

It must be stressed that the time available for cleaning is very closely related to the economics of the cleaning operation. The increased cost in equipment, energy, and chemicals to reduce time must be weighed against the economic gains in increased production rates. Additionally, consideration must be given to effects on reject rates and customer satisfaction.

Rinsing

No matter what cleaning method has been employed, the surface of the freshly cleaned part contains some soil and cleaner residue. In some situations, the residue may present no problem, but in many others, the residue must be removed to yield acceptable parts.

The value of pressure sprays and mechanical action in rinsing is often neglected in general parts' cleaning applications. In our experience, direct spraying is far more effective in flushing away the loosened soil than just soaking the part in a stagnant immersion tank. Static or slow-moving rinses in which there is improper/inadequate flow usually results in parts that must be recleaned or discarded. The use of a short spray rinse followed by an agitated immersion rinse is very effective at reducing residue.

The number of rinses performed is an important factor affecting final parts' cleanliness; two rinse steps are more effective than one, three are better than two, etc. Multiple rinses can be of shorter individual duration and still be more effective than a single rinse due to the exponential dilution of contaminants as the parts proceed from one rinse to the next.

Another important consideration is the quality of water used in the rinse step. The quality of rinsing can only be as good as the quality of water used in the last rinse. Unsoftened water obtained from a municipal source or well contains varying levels of hard water ions, carbonates, phosphates, sulfates, and organic by-products from treatment processes. The water hardness can often be extremely high, leading to hard water deposits and soap scum residue. Softening the water to remove hard water ions (calcium, iron, magnesium, and manganese) will eliminate those hard water salts but may still leave other impurities and therefore may not be adequate. Using deionized, distilled, or reverse osmosis (RO) purity water gives the best rinsing performance. As expected, as the quality of water required for the application goes up, so does the cost. The level of water quality must meet the requirements of the application. The application requirements must be evaluated for each system.

It is possible to equip the multistage rinse with a set of deionizing resin beds and an activated carbon filter. Closing the rinsing loop by deionizing the overflow water can reduce water consumption, replacing only the water lost to evaporation. Users with significant wastewater disposal costs should consider a closed loop setup.[3,4]

Redeposition

The design of the tank and cleaner is an important factor in reducing/eliminating redeposition of soils. The tank must be of sufficient capacity to provide room for the soils to move away from the parts. The size is also important to moderate the rate at which soil loading increases; too small of a tank can result in the cleaner becoming saturated in a matter of hours or days. Bag filters can be used to remove gross particulate matter. The cleaner may incorporate phosphates, silicates, specialty surfactants, and synthetic polymers which remove and suspend soils in solution. The use of aqueous cleaners that can "split out" oils instead of emulsifying them in combination with an oil coalescer/skimmer slows down and possibly even prevents the soil from reaching a saturated condition in the cleaner. Oil splitting followed by physical removal results in longer bath life and minimizes soil redeposition. Self-emulsifying oils cannot be removed by oil skimmers/coalescers; they can only be removed by microfiltration.

A good rinse must always follow the cleaning step to prevent soil redeposition. Waxes typically require a hot rinse to prevent resolidification on the clean surface.

Protecting the Substrate

The cleaner must be compatible with the substrate. The cleaner must not discolor, etch, or otherwise damage the substrate, unless these side effects are desirable in a specific application. Thus, the cleaner must contain additives to protect the substrate from these effects. Good rinsing, careful drying, or use of inhibitors may be necessary to avoid tarnishing. Transfer time between cleaning and rinsing tanks should be minimized to avoid drying of cleaner residue on the substrate, especially with silicate-based cleaners. If silicates are allowed to dry on a part, the residue is very difficult to remove without resorting to use of acid fluoride cleaners or scraping/sanding. The dried silicate residue can turn from a corrosion inhibitor to a corrosion promoter on some metals if it is allowed to remain on the part for too long.

Controlling the Cleaning Line and Monitoring the Cleaner

It is necessary to determine when the cleaner is nearly exhausted so that fresh cleaner can be prepared or the old cleaner can be rejuvenated; this is not always easy. Measuring properties such as alkalinity, conductivity, and pH are useful in determining the state of the cleaner but it is not uncommon for tanks to fail even when the above-mentioned test results are within normal tolerance. The properties that should be measured are those that are critical to the specific process. For example, silicate-based aluminum cleaners should be monitored mainly for pH; silicate testing should also be done if affordable. Silicates protect aluminum from corrosion and they participate in soil anti-redeposition. The problem with silicates is that their solubility in water decreases as the pH drops in the cleaner. At some point, the silicate level will drop below the minimum level necessary to protect aluminum. When this happens, spotting or etching may occur.

Heavy-duty caustic cleaners are best monitored by active and total alkalinity titration or by conductivity. Acid cleaners are best monitored by total acidity titration or by conductivity.

Cleaners that undergo microfiltration to remove emulsified oils present the greatest difficulty in monitoring. The pH may need to be monitored if they contain silicates. Alkalinity titration is useful but it does not detect many of the surfactants that are stripped out by the microfiltration process. The emulsified oils may contain ingredients that severely impact the pH and skew the alkalinity titration. Monitoring the refractive index, how much the solution bends light, can help in tracking loss of cleaner due to stripping by the microfilter. Unfortunately, refractive index requires that the solution be relatively clear; cloudy solutions are difficult or even impossible to measure.

Some tanks are so difficult to maintain that the user must enlist the assistance of a chemical management firm or change some of the manufacturing processes that precede the cleaning step.

Improving Bath Life

Great strides have been made that allow some aqueous baths to last for several years with periodic replenishment. However, cooperation between chemical and equipment manufacturers as well as industrial end users is critical to achieving this result.

One way of improving bath life is to have two cleaning tanks in series. Most of the soil is removed in the first tank. The parts going into the second tank are relatively clean. After the first tank becomes heavily contaminated, it can be discarded and the cleaner in the second tank pumped over to the first tank. In this manner, the cleaners can be used for long periods. With careful management, it is possible to continue operating the first bath with very high contamination levels that would require replacing the bath if only one wash tank was employed.

The contamination of the second tank can be reduced by employing a short rinse after the first cleaning operation; the rinsing water either goes to sewer or back to the first tank. Besides better performance and economy, this procedure may be necessitated or modified by governmental regulation of effluent that prohibits or limits the discarding of cleaning baths.

The single biggest problem in obtaining long tank life is understanding that cleaning is part of the overall manufacturing process and must be evaluated as part of the whole. The design of the cleaning process must consider the soils and their effects not only on processing but also on disposal. The choice of cleaning equipment and cleaner should be made as a coordinated effort that takes into account soil removal from the part and from the bath. These choices must also take into consideration the ultimate costs of operation, including the costs of disposal. It is in this area that chemical and equipment manufacturers and their customers most often fail.

The best example of failure to design a process from beginning to end is using self-emulsifying oils in a machining step and then expecting long tank life from the aqueous cleaner. Self-emulsifying oils are easy to remove from the parts but usually cannot be removed from the cleaning tank by oil coalescers/skimmers. Self-emulsifying oils contain their own detergents that form very stable emulsions. The emulsions formed by self-emulsifying oils cannot be broken by the aqueous cleaner without destroying the cleaner at the same time. Thus, the emulsions build up in the tank and eventually lead to soil redeposition problems.

Microfiltration in the 0.1–0.5 μm pore size range can remove self-emulsified oils; the equipment is expensive; it is not a simple "turnkey" process; and some chemical add-back is needed to replenish cleaner that is removed in the filtration process.

While the microfilter will remove nearly all of the emulsion, typically >90% reduction per turnover, the cleaner must be designed to pass through the filter. Only immersion cleaners that are operating at temperatures significantly below their cloud points can be processed through a microfilter in an economical manner. The *cloud point* is the temperature at which the surfactants become insoluble and the cleaner turns from clear to cloudy/turbid. Most spray wash cleaners cannot be microfiltered because their surfactants purposely become partially insoluble in water to control the foam. When a spray wash cleaner reaches its cloud point, the surfactants form aggregates that are too big to pass through the filter. In microfiltration of spray wash cleaners, surfactant stripping rates of >90% per turnover are common. Therefore, the microfiltered spray wash cleaner is no longer able to clean and control foam. Generally, it is not cost-effective to use chemical additives to maintain a spray wash cleaner in a microfiltration operation.

Microfiltration has been successfully used with immersion cleaners; but there will still be some stripping of surfactants, typically 5%–30% per turnover on properly formulated cleaners. Monitoring and adjusting a microfiltered immersion tank can be difficult and may not be an economically viable option.

Similarly, ultrafiltration (pore sizes below 0.1 μm) is not recommended because the stripping rate of cleaner components becomes too aggressive regardless of the cloud point. It is not cost effective to offset additive losses through add-back packages.

The ideal manufacturing process uses straight oils (non-emulsifying) which can be split out of solution by the cleaner and then isolated through coalescing/skimming. Unfortunately, many machining processes require the use of self-emulsifying oils to extend tool life due to the excellent heat dissipating property of water. If the overall process cannot sacrifice the use of self-emulsifying lubricants and the soil loading rate of the cleaning tank is high, then tank life will be short (possibly only several hours to 1 week at most). Very short tank life increases overhead due to the increased down time for recharging the tank and increased disposal costs. Incorporation of a crude precleaning tank/spray wash to remove the bulk of self-emulsified oil residue will significantly extend the life of the main wash tank. Manufacturers should review their overall process and find ways to minimize the use of self-emulsifying oils when possible.

Another good example of failure in designing a process for long bath life is choosing to use hard water to charge and replenish the wash tank. Most aqueous cleaners are formulated with some hard water tolerance. However, over time, water must be added to the tank to replenish losses due to evaporation. The problem is that while the water evaporates off, the hard water ions remain in the tank. As more hard water is added to the tank, the hard water ions will build up and eventually overwhelm the built-in tolerance of the cleaner. Once this occurs, problems such as soap scum formation and deposits on the parts and in the tank arise. The cleaning solution must be discarded when this occurs. If the problem is allowed to continue, scale formation will build up. Scale results in increased costs to heat the tank (the scale is an insulator) and eventual burnout of the electrical heating elements.

Water

For most aqueous cleaners, water comprises 80%–99% of the cleaning solution and is used in practically all rinsing steps. Although most people do not think of it in this way, water is actually a solvent in aqueous cleaners. A major key to understanding the efficacy of aqueous cleaners lies in the role played by water, its natural properties, and impurities.

Water has been vital to man and nature since the beginning of time. The basic cycle by which water evaporates, condenses, and flows along the surface of the earth governs all animal and plant life. Approximately 61.8% of the human body is water.[1] Almost 70% of the Earth's surface is covered with water, most of it in oceans with the balance found in lakes, rivers, the atmosphere, and absorbed into soil and rocks. Water is never absolutely pure in nature and its impurities are factors of concern in industrial applications. Man has contributed to impurities found in water sources. The disposal of spent cleaning solutions into surface waters has been one source of man-made impurities; but this is on the decline due to the rise of environmental protection regulations since the 1970s.

When an aqueous cleaner is used to remove contaminants from a surface, the water is basically the solvent in which the cleaning takes place. The importance of water's function cannot be overstated. As the solvent, water is able to dissolve and disperse the soils being removed. Additives such as acids, alkalis, chelants, and detergents significantly augment the cleaning process. These additives are not nearly as effective by themselves unless they are dissolved in a solvent, i.e., water. The combination of these additives with water yields the powerful, synergistic effects that we exploit today.

Physical Properties of Water

Pure water is colorless, odorless, and tasteless. Its chemical formula is H_2O, which shows that it is made from the two elements, hydrogen and oxygen, in a ratio of two to one, respectively. These two specific elements combined in that ratio yield physical properties unmatched by any other molecule. These properties are:

1. Very small size.
2. Not flammable or combustible.
3. Very high boiling point for its size.
4. The two elements that make it up are so different that they impart a high polarity to the molecule.
5. The high polarity of water accounts for the high boiling point. It also accounts for
   a. The high level of thermal energy that it can absorb per degree of temperature increase (heat capacity) and the level of thermal energy needed to get it to boil once it has reached the boiling point (heat of vaporization). The high heat of vaporization is what makes water so effective in steam boiler heat exchanging systems.
   b. The ability to dissolve numerous substances, especially minerals and other polar substances.
   c. The inability to dissolve nonpolar substances like fats, greases, and oils.

The very high boiling point gives aqueous cleaning processes the flexibility of a range of operating temperature options. The temperature of choice can be fine-tuned to the properties of the soil and substrate. This property also minimizes loss due to evaporation. However, water loss due to evaporation may become a concern, especially at temperatures above 150°F. The very high boiling point, 212°F, is beneficial in that most aqueous cleaning operations do not exceed 180°F so outright boiling is not a problem. Many substrates cannot tolerate the extreme heat of boiling water without suffering from discoloration, etching, or mechanical deformation.

The high heat capacity of water makes it very effective in heating metal parts up to the cleaning temperature of the bath while having minimal impact on the bath temperature. Because metals have a low heat capacity, significantly less energy is expended in raising them to the bath temperature. In contrast, traditional organic solvents have low heat capacities like metals, so they are more prone to temperature fluctuations when used as heated immersion cleaners.

The high polarity of water can be viewed as a double-edged sword. The high polarity makes it possible for water to dissolve many inorganic compounds such as caustic soda, caustic potash, borates, carbonates, phosphates, and silicates. Water is also an effective solvent for many surfactants used in formulating aqueous cleaners. Unfortunately, this polarity also results in water being contaminated by numerous impurities both from the earth's crust and man-made pollution. Some of the impurities in the starting water are identical to cleaning ingredients, i.e., carbonates and phosphates. The key is which impurities are beneficial and which are detriments. Contaminant levels in water used to make aqueous cleaners are usually low, so we have to focus on which impurities are detrimental. The elimination/suppression of these impurities is essential to prevent problems, including reduced cleaner performance, longevity, corrosion, contaminated surfaces, and water spotting.

Chemical manufacturers and industrial users spend millions of dollars annually on water conditioning equipment to reduce or remove the impurities as part of preventive maintenance. Many users remain uninformed about their water quality needs. Undesirable consequences of the impact that poor water quality has on the cleaning process include increased cost of recleaning and rejects. See the section "Water Pretreatment" for more details on water purification methods.

An often forgotten property of water is its ability to dissolve oxygen gas. Oxygen gas dissolved in water can be corrosive and attack metals during immersion. High chrome and mild carbon steels as well as cast iron are exceptionally prone to rusting in these situations. When such parts are damp and left exposed to the air, flash rusting will occur. The boiler water treatment industry knows all too well how detrimental dissolved oxygen is in boiler systems. They have to treat boiler systems with what are known as "oxygen scavengers" to chemically remove the oxygen gas from water. Oxygen scavengers are not normally used in aqueous cleaning but corrosion inhibitors may have to be used to combat the corrosive effects of oxygen and other materials dissolved in water.

Impurities

If water were H_2O and nothing else or if all waters carried the same impurities, the use of water for industrial applications would be simple and straightforward. However, natural waters, even rain, snow, sleet, and hail, as well as all treated municipal supplies contain some impurities. The type and amount of contaminants in natural waters depend largely on the source. Well and spring waters are classified as ground waters whereas rivers and lakes are designated as surface waters. Ground water picks up impurities as it seeps through the rock strata, dissolving some part of almost everything it contacts. On the other hand, the natural filtering effect of rock and sand usually keeps the water free and clear of suspended matter. Surface waters often contain organic matter, such as leaf mold, and insoluble matter, such as sand and silt. Pollution from industrial waste and sewage is also frequently present. Stream velocity, amount of rainfall, and where this rain occurs on the watershed can rapidly change the character of surface water. All these forms of water contain inorganic salt impurities with the most common being in the following descending order: (1) bicarbonates, (2) sulfates, (3) chlorides, and (4) nitrates. Below is a list of the more common and troublesome water impurities and properties:

Turbidity: Turbid water is characterized by suspended insoluble matter, including coarse particles (sediment), that settles rapidly on standing. Amounts range from zero in most ground waters to over 6% or 60,000 parts per million (ppm) in surface sources such as muddy and turbulent rivers.

Hardness: The level of soluble calcium and magnesium salts is called "hardness" and is expressed as calcium carbonate equivalents in gpg (grains per gallon) or ppm (parts per million). 1 gpg = 17.1 ppm. Calcium salts are typically present at about twice the concentration of magnesium salts. Hardness ions are undesirable because they become less soluble and drop out of solution as the water is heated. Upon drying, they produce hard, stony water spots that can be difficult to remove.

Iron: The most common soluble form of iron in ground water is ferrous bicarbonate. Although some water is clear and colorless when drawn, upon exposure to air, ferrous bicarbonate can cause water to become cloudy and deposit yellowish or reddish-brown stains on everything it contacts. Iron can also shorten the life of a water softener by contaminating the ion exchange resin. Although the majority of iron-bearing waters have less than 5 ppm, as little as 0.3 ppm can cause trouble.

Manganese: Although rarer than iron in water, manganese occurs in similar forms. Manganese can form deposits in pipelines and tanks very rapidly at levels as low as 0.2 ppm. Manganese deposits are dark gray to black in color. The Delaware River area of the United States is well known for having elevated manganese content in the water.

Silica: Most natural waters contain silica at levels ranging from 1 ppm to over 100 ppm. Silica spotting can be very difficult to remove. The only way to remove silicate scale is mechanically (scraping and sanding) or by dissolving in acidic ammonium bifluoride or hydrofluoric acid solutions. Acidic fluoride cleaners are highly corrosive and poisonous and may pose acute worker safety hazards on inhalation or adsorption!

Mineral acidity: Surface waters contaminated with mine drainage or trade wastes may contain sulfuric acid plus ferrous, aluminum, and manganese sulfates. These contaminants are corrosive; and therefore, waters contaminated with mineral acidity are unfit to use without a pretreatment system.

Carbon dioxide: Free carbon dioxide is found in most natural water supplies. Surface waters tend to have the lowest levels of CO_2, although some rivers contain up to 50 ppm. In ground waters (wells), the concentration varies from zero to levels so high that carbon dioxide bubbles out when pressure is released (as in "sparkling" or seltzer water). Most well waters contain from 2 to 50 ppm. Carbon dioxide is also formed when bicarbonates are destroyed by acids, coagulants, or heating the water. The formation or absorption of carbon dioxide can reduce the pH of an alkaline cleaning solution, especially if air-sparging is used to agitate the cleaner. Carbon dioxide is corrosive and accelerates the corrosive properties of oxygen.

Oxygen: Oxygen is found in surface and aerated waters. Deep wells contain very little oxygen. The oxygen content of water is inversely proportional to the temperature, meaning that the hotter the water, the less oxygen is present. Note, however, that while water at elevated temperature contains less oxygen, the oxygen that remains is much more aggressive and corrosive. Oxygen is very corrosive to iron, zinc, brass, and other metals. Flash rusting of metals can be a problem when hot parts are rinsed in cold water that contains high amounts of dissolved oxygen.[2]

Water Pretreatment

General

The impurities that cause the most trouble in aqueous cleaning processes are the inorganic salts. Salts are ionic compounds. Ionic compounds are chemicals that have a positively charged species called a cation and a negatively charged species called an anion. It must be pointed out that the term "salt" has commonly meant sodium chloride, i.e., table salt. The chemical definition is "salts are ionic compounds that contain any negative ion except the hydroxide ion (OH^-) and any positive ion except the hydrogen ion." An ionic compound that contains the hydrogen ion (H^+) is called an acid and one that contains the

TABLE 2.1 Dissolved Impurities in Water

Ion Type	Impurity	Property
Cations	Ca^{2+}	Hardness
	Mg^{2+}	Hardness
	Fe^{2+} and Fe^{3+}	Iron stains
	Mn^{2+} and Mn^{4+}	Manganese stains and scales
	Na^+	Too much sodium in rinse water can cause spotting
	K^+	Too much potassium in rinse water can cause spotting
Anions	CO_3^{2-}	Alkalinity—carbonates form hard water deposits with calcium, magnesium, iron, and manganese
	HCO_3^-	Alkalinity—bicarbonates form hard water deposits with calcium, magnesium, iron, and manganese
	PO_4^{3-}	Alkalinity—orthophosphates form hard water deposits with calcium, magnesium, iron, and manganese
	SiO_4^{4-} and SiO_3^{2-}	Silicates can form the most tenacious of deposits, especially in the presence of calcium, magnesium, iron, and manganese
	SO_4^{2-}	Sulfates promote tenacious hard water deposits with calcium and manganese
	Cl^-	Chlorides promote corrosion on aluminum, iron, and steel. Excessive chlorides in rinse water can cause spotting
	NO_3^{2-}	Excessive nitrates in rinse water can cause spotting

hydroxide ion is a base. Specific examples of common ionic impurities that are encountered as impurities in water are given in Table 2.1.

It must be pointed out that silicates are usually discussed/represented as silicon dioxide (SiO_2), when they are actually present in the form of the ions listed on Table 2.1. There are more complex forms of phosphate and silicate ions and numerous other trace impurities that can be present in natural, untreated water, but the above examples represent the bulk of the impurities that the cleaning industry must be concerned with.

Principles of Water Softening

The most problematic impurities are the cations. The most economical method for removing cations is by passing the water through a cation exchange column, better known as a water softener. A standard water softener contains polystyrene beads that have been modified such that the surface of each bead has numerous negatively charged sites. Nature requires that charge must be balanced. The balance of charge is accomplished by pairing each negatively charged site with a sodium or potassium cation. As the impure water passes through the water softener, the hard water ions become attached to the resin beads and displace the sodium cations. The number of displaced sodium cations equals the charge of the hard water ions trapped in the softener. The hard water ions are bound more tightly to the resin because higher positively charged cations bind more strongly to negatively charged surfaces.

Basically, the water softener removes highly charged cations and replaces them with sufficient weaker charged sodium cations to maintain the balance of charge. Thus, sodium ion contamination increases but it does not cause nearly as much trouble as calcium, magnesium, iron, and manganese. There are only a finite number of resin beads in a water softener so there is a point where the softener becomes saturated with hard water cations.

At the saturation point, the softener must be recharged. This is accomplished by passing a saturated salt solution, usually sodium chloride, through the softener. The overwhelming quantity of sodium cations slowly displaces the calcium and magnesium ions; this returns the softener back to working order. Iron and manganese are more difficult to remove from a water softener. Iron and manganese can have very high positive charges, +3 and +4, respectively, which make them bind so tightly to the resin beads that the mass action of the regenerating salt solution cannot knock them off. When a softener becomes saturated with iron, it is referred to as suffering from "iron poisoning." The iron and manganese cations can be washed out of the softener if their charge is lowered first. Reducing agents can be added to the salt recharging solution to lower the charge of the iron and manganese ions. Typical "reducing agents" are sodium sulfite, sodium hydrosulfite, and sodium thiosulfate.

Deionization

Basic water softening removes only the undesirable cations by replacing them with less problematic cations. It is also possible to replace anions by the same technique, only this time the charge on the surface of the resin bead is positive and the charge is balanced by pairing up with an anion. The chloride anion is the most economical choice for balancing charge in an anion exchanger. Unfortunately, excessive chloride content in water can lead to stress cracking of certain stainless steels and promote corrosion on aluminum, magnesium, and mild steels.

When industry has to be concerned with both cation and anion contaminants, it is easier to perform a process called "deionization" than to replace the undesirable cations with sodium and the anions with chloride. Deionization involves passing impure water through a series of cation and anion exchange resins where the cations are replaced by hydrogen ions (H^+, acid) and the anions are replaced by hydroxide ions (OH^-, base). The liberated acid and base then neutralize each other to form water:

$$H^+ + OH^- \rightarrow H_2O$$

When deionization is performed, theoretically an equal amount of acid and base is liberated. While the neutralization reaction should then result in pH neutral water, this is usually not the case. Deionized water typically is slightly acidic with a pH of 4.5–5.5. The mild acidity is caused by the carbonates initially present in the water. As the carbonates pass through the cation exchange column, carbonic acid is formed. The examples below will assume that the carbonates are passing through as sodium salts:

$$Na_2CO_3 + 2H^+ Resin^- \rightarrow H_2CO_3 + 2Na^+ Resin^-$$
$$NaHCO_3 + H^+ Resin^- \rightarrow H_2CO_3 + Na^+ Resin^-$$

Carbonic acid is not stable in water so it self-destructs to form carbon dioxide, CO_2, and water:

$$H_2CO_3 \rightarrow CO_2 + H_2O$$

Carbon dioxide has some solubility in water; so unless the water is boiled to drive off the CO_2 after deionization, the reverse reaction occurs which liberates some acid. This causes the low pH:

$$CO_2 + H_2O \rightarrow H_2CO_3$$
$$H_2CO_3 \rightarrow H^+ + HCO_3^-$$

Both water softening and deionizing systems are very effective at removing impurities but they are not perfect. The exchange of ions is really an equilibrium process, so some impurities can work their way through a column before the resin becomes saturated. Imperfectly sealed control valves and excessively high flow rates can result in some impurities never coming in contact with the resin so that exchange never occurs. In addition, ion exchange columns will not effectively remove nonionic impurities. The incorporation of carbon filters, nanofiltration (NF), or a RO unit will remove the nonionic impurities.

Reverse Osmosis

RO involves separating water from a solution of dissolved solids by forcing the water through a semipermeable membrane. As pressure is applied to the solution, water and other molecules with low molecular weights pass through micropores in the membrane. The membrane retains larger molecules, such as organic dyes, cleaners, oils, metal complexes, and other contaminants. RO membrane systems feature crossflow filtration to allow the concentrate stream to sweep away retained molecules and prevent the membrane surface from clogging or fouling.

In the past, RO applications for industrial operations were mostly limited to the final treatment of combined wastewater streams. Such applications typically involved discharging the permeate (the purified liquid) to a publicly owned treatment works (POTW) and returning the concentrate to the head of the wastewater treatment system. Because of the high flow rates associated with treating combined wastewater streams, large, costly RO units were required. More recent applications in cleaning involve installing RO units in specific process operations (such as a wash tank or for rinse water maintenance), allowing return of the concentrate to the process bath and reuse of the permeate as fresh rinse water.

RO systems have been successfully applied to a variety of industrial operations. By closing the loop, process contaminants are removed and fresh water is recycled. Furthermore, a waste stream is eliminated, one that would otherwise be discharged to the POTW or hauled away. An additional benefit is reducing the cost of waste treatment and disposal.

Reverse Osmosis Components

The essential components of an RO unit include a strainer, a pressure booster pump, a cartridge filter, and the RO membrane modules. The strainer protects the pump by removing large, suspended solids from the feed solution. The booster pump increases the pressure of the feed solution. Typical operating pressures range from 150 to 800 psi. Commercially available cartridge filters are used to remove particulates from the feed solution that would otherwise foul the RO membrane modules. Cartridge filter pore sizes are typically between 1 and 5 µm. The final RO membrane pore size is less than 0.001 µm and it removes impurities down to a molecular weight of about 200 Da.

Cleaning Chemistry[22]

Aqueous cleaners are acidic, neutral, or alkaline. Acid products, which have a pH of less than 6, are used for removal of inorganic scales and to pickle or passivate metallic surfaces. Neutral and alkaline cleaners have a pH range from 6 to above 13. These products are very effective on organic oils and greases.

Additional ingredients are frequently added to provide increased effectiveness on inorganic soils as well. For example, sodium gluconate and glucoheptonate additives turn highly caustic degreasers into alkaline rust removers due to the affinity of the gluconate/glucoheptonate for oxidized forms of iron.

When defined as the removal of soil or unwanted matter from a surface to which it clings, cleaning can be accomplished by one or more of the following methods:

Wetting: The cleaner penetrates and loosens the substrate–soil bond by lowering surface and interfacial tension through the use of surface active agents.

Emulsification: Once wetting occurs, the two mutually immiscible liquids are dispersed via emulsification. Oil droplets are coated with a thin film of surfactant which prevents them from recombining. The emulsion may float to the surface, hover in the solution, or sink to the bottom, depending on the size and density of the emulsion particulate versus the density of the cleaning solution.

Solubilization: Solubilization is the process by which the solubility of a substance is increased in a certain medium. The soil is dissolved in the cleaner bath.

Saponification: The reaction between an ester (can be animal- or vegetable derived as well as synthetic) and free alkali to form soap and an alcohol is *saponification*.

$$\text{Insoluble ester} + \text{alkali} = \text{water-soluble soap} + \text{alcohol}$$

Deflocculation: The process of breaking particulate types of soils into very fine particles and dispersing them in the cleaning media is *deflocculation*. The soil is then maintained as a dispersion and is prevented from agglomerating. This process is similar to emulsification but does not operate on liquid/greasy types of soils.

Displacement: Soil is displaced by mechanical action. The movement of the workpiece or fluid enhances the speed and efficiency of soil removal. Agitation by scrubbing/rubbing is another variation of displacement.

Sequestration: Undesirable ions such as calcium, magnesium, or heavy metals are deactivated via sequestration, thus preventing them from reacting with materials that would form insoluble products (i.e., hard water soap scum).*

Water-based cleaners are generally divided into five major pH groups as follows:

Caustic, pH > 13
Highly alkaline, pH 10–13
Mildly alkaline, pH 8–10
Neutral, pH 6–8
Acidic, pH 1–6

Acid Cleaners

Acid cleaners are generally not used for the removal of organic oily soils. These cleaners are used primarily for the removal of metal oxides and/or scales prior to other surface pretreatment steps or painting. Strongly acidic cleaners typically use dodecylbenzene sulfonic acid (DDBSA), hydrochloric acid, nitric acid, or sulfuric acid. Mildly acidic cleaners typically use citric acid, gluconic acid, glycolic or phosphoric acid. Acid cleaners usually contain some nonionic surfactants for degreasing and wetting purposes.

Systems using acid cleaners generally require constant maintenance because the aggressive chemistry attacks tank walls, pump components, and other system parts as well as the materials to be cleaned. Corrosion inhibitors can be used to reduce this attack. Acid cleaners often suffer from rapid soil loading, particularly metal ion loading. The metal ion loading leads to the need for frequent decanting and dumping of the cleaner solution. Spent acid cleaning solutions almost always are classified as hazardous waste. All of these disadvantages lead to relatively high operating costs compared to alkaline cleaners.

Alkaline and Neutral Cleaners

Ingredients frequently contained in alkaline cleaners include alkaline builders, water conditioners, surface active agents, corrosion inhibitors, fragrances and/or dyes, defoamers or foam stabilizers, and water. Occasionally, hydrocarbon solvents are also added to a formulation.

---

* See Acknowledgment section.

Alkaline Builders

Alkaline builders are selected based on the pH, detergency, corrosion inhibition, and/or cost limitations required for a specific formulation. Environmental or process restrictions must also be considered. These builders may include one or more of the items listed in Table 2.2.

Neutral pH cleaners contain little or no alkalinity builder(s); or the alkalinity reserve is neutralized with an organic or mineral acid.

Water Conditioners

Sequestrants or chelators are frequently used to deactivate undesirable ions such as calcium, magnesium, or heavy metals. The sequestered ions or heavy metals are then no longer free to react with bath substances that would subsequently form undesirable compounds such as hard water soap scum. Some of the more commonly used sequestrants are as follows:

ATMP	Amino trimethylene phosphonic acid
EDTA	Ethylenediamine tetraacetic acid
HEEDTA	*N*-(hydroxyethyl)-ethylenediamine triacetic acid
IDS	Iminodisuccinic acid
NTA	Nitrilotriacetic acid
SHMP	Sodium hexametaphosphate
STPP	Sodium tripolyphosphate
TKPP and TSPP	Tetrapotassium and tetrasodium pyrophosphate, respectively
HEDP	1-Hydroxyethylidene-1,1-diphosphonic acid
Sodium gluconate	
Sodium glucoheptonate	
Low-molecular-weight polyacrylates and polyaspartic acids	

EDTA has maximum effectiveness in tying up calcium, magnesium, and heavy metal ions, thereby softening the water used to dilute the cleaner bath and preventing unwanted electrodeposition onto

TABLE 2.2 Alkaline Builders

Component	Advantages	Disadvantages
Caustic cleaners (pH > 13)		
Hydroxides	Alkalinity, cost-effective	Corrosive
High-alkaline cleaners (pH 10–13)		
Amines	Alkalinity, detergency, corrosion inhibition	More costly, can interfere in wastewater treatment
Carbonates	Detergency, soil anti-redeposition, cost-effective	Consumable
Hydroxides	Alkalinity, cost-effective	Corrosive
Phosphates	Detergency, sequestration, soil anti-redeposition, corrosion inhibition	Environmental restrictions
Silicates	Detergency, corrosion inhibition, soil anti-redeposition	Residues, restricted use
Low-alkaline cleaners (pH 8–10)		
Amines	Alkalinity, detergency, corrosion inhibition	More costly, can interfere in waste water treatment
Borates	Alkalinity, corrosion inhibition, soil anti-redeposition	Limited effect, environmental restrictions
Sulfates	Filler, carrier	Restricted use

Source: Quitmeyer, J., All mixed up: Qualities of aqeous degreasers, *Precision Cleaning*, September 1997.

other substrates. The use of EDTA has been strongly discouraged in the last decade due to its ability to tightly bind toxic heavy metals and because it is not biodegradable. Sodium gluconate and gluco-heptonate have maximum effectiveness in tying up calcium only at high pH but they will tightly bind some heavy metal ions at lower pH. Complex phosphates are cost effective but have come under environmental pressure since the late 1960s due to the problems they cause with eutrophication in lakes and streams. Low-molecular-weight polyacrylates have been growing in use as the result of increasing implementation of phosphate bans. Polyaspartic acids and iminodisuccinic acid have only been around for a few years so their impact is unknown at this time. Many of the water conditioning agents are also alkalinity builders.

Surface Active Agents

Surface active agents, also known as surfactants, are used to reduce the surface or interfacial tension of a water-based solution. The selection of the surfactant package used in cleaner formulations depends on the performance characteristics desired. Today, surface active ingredients frequently used in water-based cleaners are predominantly synthetic surfactants. However, there are still some cleaners that rely on naturally derived fatty acid soaps. Surfactants are classified into four basic types:

1. *Anionic*: negatively charged ions that migrate to the anode
2. *Cationic*: positively charged ions that migrate to the cathode
3. *Nonionic*: electronically neutral
4. *Amphoteric*: ions charged either negatively or positively, depending on the pH

Physical properties affected by surfactants include the cloud point, foaminess, detergency, emulsification, and wetting mechanisms used to facilitate the cleaning process. The nonionic surfactants are the primary work horses for grease and oil removal. They also play a key role in controlling the foam profile of the finished product through the cloud point effect. Anionic, cationic, and amphoteric surfactants are multifunctional; they affect foam profile, assist in grease and oil removal, can provide corrosion inhibition, help prevent redeposition of particulate soils, and can modify the cloud point of the nonionic surfactant. Nearly all cleaners use a combination of surfactants to obtain the specific properties desired.

Corrosion Inhibitors

Corrosion inhibitors are also included in some acid and alkaline cleaners, depending on the application involved. If a wide variety of substrates are involved, a combination of inhibitors may be used. These inhibitors are water soluble and therefore most are removed with a thorough rinse, if desired. Corrosion inhibitors frequently added to aqueous cleaner formulations include, but are not limited to, amines, benzoates, borates, carboxylates, molybdates, nitrites, silicates, thiadiazoles, triazoles, and urea. Phosphates and gluconates could also be added to this list, although their efficacy is more limited. The intended cleaner application dictates the type of inhibitor package selected. As the alloying of metals and composites becomes more complex, there is a greater need for sophisticated inhibitor packages that provide protection over a broad spectrum of substrates. The synergism of chemicals allows the formulator to obtain the inhibiting properties desired. Achieving the desired inhibiting properties may be limited only by the imagination of the formulator and the cost restrictions of the chemicals selected.

One very important rule about using silicate-based cleaners is that they should be rinsed off with heated water, preferably no less than 115°F. Silicates release themselves from metal surfaces at slower rates than most other cleaner ingredients. Rinsing with cool water can result in excessive residue remaining on the parts. This residue could interfere in a later manufacturing step such as alodining, anodizing, or plating. Poor rinse agitation can also lead to excessive silicate residue even if the water is heated.

Additional Ingredients

Aqueous cleaner formulations may also contain a broad spectrum of ingredients designed to affect the appearance, odor, or physical properties of the composition. These ingredients include dyes, fragrances, thickeners, defoamers, foam stabilizers, or fillers for cost reduction. Again, the intended cleaner application will dictate the final composition of a formula.

Solvents

A variety of solvents have been blended with surfactants to make emulsions or semi-aqueous cleaners. Environmental regulations have identified nearly all solvents as volatile organic compounds (VOCs). Formulation chemists must be careful about which solvents they choose during product development in order to conform to environmental regulations while, where possible, avoiding making flammable/combustible products. The few VOC-exempt solvents find nearly zero use in aqueous cleaners due to problems with flammability, odor, and difficulty in emulsifying or cost.

For example, it is easy to make an aqueous cleaner with a flash point that is low enough to be categorized as "flammable" or "combustible" and they do exist in the market. It often takes as little as 1.0%(w/w) D-Limonene to drop the flash point below 140°F. The presence of isopropyl alcohol at around 2%–3% can yield a flash point below 100°F. The VOC-exempt solvents such as acetone, methyl acetate, and the linear methyl siloxanes all have flash point in the 50°F–60°F range. There are several surfactants that are purposely cut with ethanol or methanol to make them more pourable/pumpable for manufacturing purposes. Use too much of these surfactants and you will get a flash point as a result of too much alcohol being present. The cleaners may lose their flash points when diluted for final use applications but these days, environmental health and safety departments usually block purchasing of products with flash points when possible.

Simple aromatic solvent emulsions are great for cutting through greases and carbonized soils but they have very disagreeable odors. Glycol ethers have been added to stabilize formulations and to increase the grease cutting/ink removal efficiency of a composition. In addition, certain glycol ethers, including 2-butoxyethanol or "butyl cellosolve," have been identified as health hazards.

D-Limonene, orange oil, pine oil, and other terpene-based solvents have found widespread use in removing inks and polyurethanes that have not set up or cured. N-methyl-2-pyrrolidone (NMP) is a very polar solvent that can impart the ability to strip some paints in aqueous cleaners. The terpene solvents also have the added benefit of acting as fragrances in some applications. Drawbacks of terpene solvents are that they have low flash points that are very difficult to suppress in the final blend; and their cost is very volatile due to the occasional crop disasters that affect supply.

The recent drive to use as many raw materials derived from renewable resources has resulted in the increased use of solvents such as ethyl lactate, methyl soyate, and other methyl esters. Methyl soyate exhibits solvency that is similar to D-limonene, but it typically requires heating to achieve similar solvency.

Environmental Regulatory Effects on Cleaning Chemistry

Water has been used as a cleaner for centuries. The first water-soluble soaps were blends of lye and animal fat. The chemical reaction of this mixture is a process defined as saponification. The addition of heat made the soap work better at removing the oils and greases of the day which were also made from animal fat.

As industry advanced and metal processing became more sophisticated, various organic and inorganic salts were found to enhance detergency. The salts tie up metal ion impurities that could react with the soap to inhibit cleaning and deposit as soap scum. They also provide alkalinity and improve pH buffering capacity.

Today, traditional neutralized fatty acid soaps make up only a small portion of the surfactants used in industrial and consumer cleaning applications. The development of synthetic detergents as a substitute for soap starting in the 1920s and 1930s has completely changed the cleaning industry. Problems with some synthetic surfactants became obvious in the 1960s when foaming in rivers and wastewater treatment plants indicated that the most popular detergent at that time, branched propylene tetramer-based alkyl benzene sulfonate, was not readily biodegradable. To correct the problem, manufacturers of aqueous cleaners in Europe and the United States voluntarily changed over to biodegradable linear alkyl benzene sulfonates by 1965. Other problems that became apparent over time were eutrophication of lakes and streams from excess phosphate pollution and possible endocrine disruption in amphibians and fish by alkylphenol ethoxylate surfactants (APEs).

By the mid-1970s, government regulations were starting to be felt at the job site. The EPA, California Air Resources Board (CARB), OHSA, and Material Safety Data Sheets (MSDS) became common terms within the industrial arena. Chemicals came under increasing scrutiny for worker safety.

During the following decades, environmental issues played an increasingly important role in chemical evaluation. The terms chlorofluorocarbons (CFCs), the Montreal Protocol, global warming potential (GWP), HAPs, ODS, SARA Reportables, and air quality boards became commonplace.

Bans on some APE surfactants and phosphates have been spreading rapidly. Finding replacements for APEs has been fairly easy but at an increased cost; phosphate replacement has not been as easy, especially in the automatic dish detergent market.[5]

Many new regulatory agencies/programs have appeared since the late 1980s. Organizations and programs such as Canada's Ecologo, the EPA's Design for the Environment (Dfe), Europe's Registration Evaluation Authorization and restriction of Chemicals (REACH), Green Seal, and individual State's own legislation have come together to form a massive body of laws and rules.

It can be very time consuming for the manufacturer of cleaners to determine how best to conform to the assortment of requirements and directives. Some inconsistencies have led to chemical cleaners being restricted from some markets. This results in increased costs due to the need to manufacture more than one type of cleaner to accomplish the same task; there is a decrease in the economy of scale. Because regulations are also changing at increasingly shorter intervals, the lifespan of a chemical cleaner/cleaning process is sometimes shortened.

Increased environmental restrictions have played a role in forcing manufacturers to move their factories from Canada, Europe, and the United States to Mexico, China, and Third World countries in order to avoid the cost increases of complying with the latest environmental regulations.

Ultimately, no cleaning process can completely escape the concerns of health, safety, and the environment. The chemical and manufacturing industries have learned to embrace the changes and have been successfully rising to the challenge of these new regulations.

Guidelines for Cleaning Common Substrates

All metals are sufficiently reactive in the presence of oxygen in the atmosphere that they form a metal oxide layer on the surface. The chemical behavior of the oxide layer plays a significant role in determining the type of aqueous cleaner that can be used in the cleaning process. Metals whose oxide layers readily dissolve in both acidic and alkaline cleaners are classified as amphoteric metals.

Several metals/alloys and their general cleaning chemistry needs are detailed below. Plastics and composites have been added due to the increased use of these as construction materials.

First, a bit of WARNING must be given concerning some of the chemistries required for these substrates. There will be applications where there is no alternative to using a heavy-duty caustic or strongly acidic cleaner. The operator must read the MSDS and product labels to determine the appropriate safety equipment required for each cleaner. Heavy-duty caustic cleaners are very dangerous if they get in your eyes or on your mucous membranes. Acidic cleaners are also dangerous to eyes and mucous membranes. Skin burns are also possible with either type of cleaner. Of particular concern are acids that contain hydrofluoric acid or ammonium bifluoride. Acidic fluorides are highly corrosive but they are also very poisonous. Acidic fluorides can be absorbed through the skin and the symptoms of poisoning do not appear immediately. Thus, extra care must be taken to protect the worker when using these cleaners.

Aluminum

This is the second most common metal used in manufacturing and it is strongly amphoteric. Acidic solutions below pH of ~5 will dissolve the metal oxide surface layer. Once the oxide layer has been removed, the base metal will dissolve in most acids. Phosphoric acid and diluted hydrofluoric acid solutions are the most common of the acid cleaners for general cleaning and brightening of weathered aluminum that do not cause excessive etching. Fresh/unweathered aluminum is not typically exposed to acid cleaners unless the surface requires an etching step or smut formation from a previous process must be removed. Smut removal is typically accomplished with strong nitric acid solutions that contain low levels of hydrofluoric acid. Neutral and mildly alkaline cleaners that are silicate free are safe on aluminum as long as the pH does not exceed ~8.7. Above this pH, the etching of the base metal will occur unless silicate or chromate corrosion inhibitors are present.

The use of chromate corrosion inhibitors has been largely discontinued due to environmental and worker safety concerns. Thus, currently, mild and strongly alkaline cleaners only use silicate corrosion inhibitors to prevent etching. Etching can be stopped up to about a pH of 12 with silicates. Between pH of 12 and 13, etching can still be suppressed if the cleaner contains massive amounts of silicates. At this high a pH, some discoloration of the base metal can be expected. Therefore, these cleaners typically are only used in applications like automatic and manual transmission overhaul where aesthetic appearances are not a big concern.

A fairly common problem with alkaline cleaning of wrought aluminum alloys (e.g., AL2024 and AL6061) is the formation of a fine black particulate smut that forms shortly after the cleaning process. This smut is not readily visible by the naked eye but is easily detected by the white glove test. The problem appears to be related to a combination of magnesium silicate inclusions and submicron particles absorbed at the surface. The smut formation can be stopped by first cleaning the metal with an acid solution (typically phosphoric acid) followed by an alkaline cleaner. The acid cleaner removes the magnesium occlusions and submicron particles from the surface which stops the smut from forming.[7]

Aluminum is sufficiently chemically reactive that temperature is a major consideration, especially with neutral and alkaline cleaners. Neutral and mildly alkaline cleaners that do not contain silicates should be limited to about 150°F due to the tendency of some alloys to discolor. While alkaline silicate-based cleaners have been safely used up to 180°F, it is still preferable to limit the temperature to 160°F. Alodined and anodized aluminum must be limited to a temperature of 145°F–150°F to prevent damage to the conversion coating.

It is possible to clean conversion-coated aluminum parts with silicate-based cleaners as long as the pH is not too high (does not exceed 11.5 at use-dilution) and the temperature limits are still obeyed. The key is that the manufacturer must test the cleaner on their specific parts. There are several types of alodined and anodized coatings and they do not all behave identically. Conversion coatings that incorporate dyes are the most difficult to clean successfully with silicate-based cleaners; blotchy color patterns from leaching of the dye are common.

Cobalt

This metal is not amphoteric. For this discussion, we are dealing only with cobalt-cemented carbide tools such as drills and saw blades. Cobalt containing alloys such as the Haynes series fall under the guidelines given for steel. Many tools use chromium carbide or tungsten carbide to increase hardness and wear resistance. The carbide material cannot be blended into the tool's base metal during casting or forging processes. The carbide material is "glued" to the metal substrate with cobalt metal in an electroplating process. The cobalt glue must not be attacked by cooling fluids and cutting oils during normal use nor must any attack occur when the tools are manufactured or undergo periodic maintenance. Not only will attack weaken the cement bonding leading to shorter tool life, this will allow cobalt metal to build up in the wash tank. Cobalt metal is a known sensitizing agent so keeping cobalt leaching under control is important to the employee's health. Alkaline cleaners are very tricky to formulate for compatibility with cobalt. Typically, the cleaner must not contain any of the ethanol- and isopropanol-based amines commonly used in cleaners, coolants, and lubricants. The cleaner must also not contain citrates, EDTA, gluconates, phosphates, and sulfates.[8–10] One strange contradiction is that on the alkaline side, phosphates are unsafe but acid cleaners based on phosphoric acid cause no problems with cobalt.

Copper (Includes Brass and Bronze)

Copper and bronze are not amphoteric but brass is amphoteric. These metals are sensitive to tarnishing by combinations of elevated pH, moisture, and temperature. The upper pH limit on alkaline use-dilutions is typically about 11.5; cleaning at higher pHs can lead to significant darkening. Discoloration is a problem with neutral and alkaline cleaners when the temperature exceeds 130°F. Discoloration can be suppressed by using cleaners that contain thiadiazole or triazole corrosion inhibitors. These inhibitors can raise the maximum allowable cleaning temperature up to approximately 150°F. Silicate-based cleaners also suppress stock loss on copper alloys, but discoloration is still possible.

Cleaners that contain amines may be problematic with these alloys because the copper oxide and zinc oxide top layers (zinc is a component of brass) readily form water-soluble ammoniacal complexes. The leaching of copper and zinc may change the color of the base metal, typically making the metal lighter in color. Eventually, the dissolved metals content will increase to a level such that the spent tank will be classified as a hazardous waste. Cleaning stressed brass parts with cleaners that contain amines is not recommended because prolonged exposure can lead to stress cracking.

Care must be taken to not overheat the parts if a blown hot air process is used to accelerate drying of parts; excessively high temperatures will darken these alloys. Finding the right neutral or alkaline cleaner may require an exhaustive trial and error process if even the slightest amount of discoloration imparted by the cleaning process is unacceptable (e.g., decorative household fixtures). In some cases, it may be necessary to clean with solvents only.

Copper that does not suffer from significant discoloration can usually be brightened with a citric acid-based cleaner. The brightening of noticeably discolored copper alloys with acid cleaners can only be done if an oxidizer is present; these are more commonly referred to as bright dips. Oxidized copper exists in two states; a reddish-pink cuprous oxide and a black cupric oxide. Alkaline and pH-neutral cleaners that contain amines can only remove the black cupric oxide but are unable to remove reddish-pink cuprous oxide. Oxidizing acid cleaners such as sulfuric acid/hydrogen peroxide and nitric/hydrochloric acid are required to convert the cuprous oxide to a cupric oxide before it can be removed. Nitric acid/hydrochloric acid blends have fallen out of favor due to the generation of chlorine gas which is corrosive and poisonous. Alkaline cyanide cleaners also can remove cuprous oxide; these cleaners are still used in plating facilities.

Magnesium

While this metal is not amphoteric, it is the most chemically reactive metal commonly used in industry. Bare magnesium will slowly corrode in water and eventually dissolve if given enough time.

The trick to cleaning magnesium is to incorporate a chemical that forms a very water-insoluble by-product right on the surface, as soon as the etching starts. The most effective chemicals that do this in alkaline cleaners are sodium and potassium hydroxide. The pH must remain high at use-dilution for these hydroxides to be effective in preventing attack on magnesium. Experience has shown that the pH must stay above 11 to protect magnesium from corrosion.

Magnesium should always be segregated and cleaned with dedicated cleaners/equipment if at all possible. Magnesium is also very prone to discoloration at high pH. If discoloration is objectionable, highly alkaline/caustic cleaners containing high levels of silicate corrosion inhibitors can suppress this problem. Acid cleaners that are safe on magnesium are hydrofluoric acid, ammonium bifluoride blended with other acids, and phosphoric/nitric acid blends. Magnesium forms an insoluble barrier of magnesium fluoride or magnesium phosphate as part of the corrosion inhibition process. Avoid all cleaners that contain chlorides as those will promote corrosion. It is possible to clean magnesium at milder pH. However, to prevent excessive stock loss/discoloration, this is usually only possible at temperatures below 130°F and for limited exposure periods.

Magnesium should be anodized at the end of the manufacturing process to suppress its tendency to corrode in the presence of moisture and dirt.

Plastics and Composites

These materials are not amphoteric. The properties of these substrates are so varied that no one rule is universally applicable, except "Don't clean at a temperature above the substrate's glassy transition point or decomposition point." Exceeding these points will irreversibly damage the substrate.

Materials like Kalrez®, polyethylene, polypropylene, and Teflon® are inert to nearly all chemical conditions that occur in aqueous cleaning. Other materials like the Lexan class of polycarbonates will suffer stress cracking in acidic and alkaline cleaners, especially chlorinated bleach formulas. Polyethylene terephthalate (PETE) will stress crack in the presence of alkaline cleaners. The author once ran into a polyetherimide plastic used as a housing for a medical diagnostic tool that had poor chemical resistance (it softened and cracked) to any ingredient used in water-based cleaning other than the water itself.

The take-home message is that the manufacturer must know what specific plastics and/or composites they are dealing with and provide that information to the cleaner manufacturer. The cleaner manufacturer may already know if their product is safe on the substrate. In some cases, the manufacturer can only find the right cleaner through trial-and-error testing.

Steel (Mild Carbon and Stainless)

These alloys are the most widely used in manufacturing and are not amphoteric. Steel alloys are quite chemically resistant to nearly all pH-neutral, alkaline, and heavy-duty caustic cleaners over a wide temperature range.

The presence of chlorides/halides in alkaline cleaners is undesirable because they promote rusting of mild steels and can cause stress cracking in some stainless steels. Certain high tensile strength steel applications require that the cleaner does not cause hydrogen embrittlement.

Rust and scale deposits can be removed by highly caustic cleaners loaded with sodium gluconate or glucoheptonate. Gluconate-based derusters are preferred over glucoheptonates due to the strong ammonia odor associated with the latter. Acid cleaners are also commonly used to remove rust and scale from mild steels. Mild steels are typically cleaned with inhibited hydrochloric or phosphoric acid cleaners whereas stainless steels use nitric acid spiked with a small amount of hydrofluoric acid.

Mild- and high-chrome-content steels are very prone to flash rusting, especially during transfer from the wash to the rinse tank, during rinsing, and during drying. The transfer time between the wash and rinse tank must be quick enough to minimize drying of the cleaner on the part, otherwise rusting will occur. It may be necessary to maintain a wet spray on the parts to protect them during the transfer. The rinse tank must be heated to reduce the amount of dissolved oxygen present in the water. Dissolved oxygen readily promotes rusting on mild and high chrome steels. A common cause of flash rusting is transferring a part from a heated wash tank to a cool rinse tank; the increased temperature of the metal combined with the higher oxygen content in the cool rinse causes rusting almost immediately. It may be necessary to add a corrosion inhibitor such as an amine or a nitrite to the rinse tank even when the water is heated. The drying step must be quick. Slow drying times promote rust on mild and high chrome-content steels.

Tin

This metal is weakly amphoteric. Tin is less sensitive to high pH than aluminum and zinc; but one must not clean with very high pH cleaners (>12 at use-dilution) such as alkaline derusters because heavy etching will occur. Silicate-based cleaners are excellent at suppressing discoloration and etching. Chromate inhibitors also are effective at stopping etching in alkaline cleaners; but they are rarely used due to the environmental health hazard they pose. While tin is not very reactive to diluted hydrochloric and sulfuric acids, heated and concentrated forms of these acids readily attack the metal. Neutral pH cleaners do not have any compatibility problems with tin.

Titanium

This metal is not amphoteric. Titanium is a fairly robust metal when it comes to aqueous cleaning applications. However, acid cleaners are generally not used except in pickling operations because the metal is readily attacked at low pH. There are specialized acid cleaners used on titanium-based turbine compressor components, but the exposure times are very short.

Alkaline, caustic, and neutral cleaners have very few restrictions due to the chemical inertness of titanium. Exceptions include cleaning critical components of turbine engines used in aircraft and power plants. Formulations used to clean turbine compressor components must have very low halide salt and sulfur content; they must not cause hydrogen embrittlement and their dried residue must not corrode the metal if the turbine is reassembled and operated without rinsing first. Alkaline derusters are commonly used in the overhaul of turbine compressor components.

Some compressor components are made of super alloy steels which can be cleaned at 180°F–200°F. However, titanium becomes too reactive in this range; so cleaning temperatures should be restricted to a maximum of 170°F.

Titanium should have its own separate cleaning tank when undergoing alkaline derusting; the presence of dissolved iron from cleaning cast iron and steel parts in the same tank significantly accelerates attack on titanium.

Medical applications such as titanium hip implants also have cleaner restrictions. The cleaners cannot contain ingredients that are animal derived and/or paraffin based. These medical restrictions are not based on corrosion problems with the base metal. Instead, they must be avoided to prevent the patient from getting infections and/or to avoid their body trying to reject the implant. In general, residue levels must be low, and the residue must not present toxicity issues for the recipient.

Zinc

This metal is amphoteric. Overall, it is very chemically reactive, almost as reactive as magnesium. Cleaning zinc is further complicated by the fact that it loves to form water-soluble complexes with amines. Amines usually do not cause problems with the physical appearance of the metal, but they will cause zinc to build up in the wash tank. This presents a disposal problem. High-pH cleaners (pH > 12 at use dilution) should be avoided because the surface begins to dissolve at an appreciable rate. Zinc is typically brightened in acid cleaners; but because zinc is readily soluble in all acids, the exposure time must be short, and brightening is typically done at room temperature. It is possible to clean at alkaline pH if the cleaner contains silicate corrosion inhibitors. However, the process may still result in discoloration of the metal.

Rinsing

Rinsing is often the most overlooked aspect of cleaning. While a process may employ the best cleaning equipment money can buy and an optimized cleaning solution, without adequate rinsing, the overall result is often unsatisfactory.

Rinsing is no more than a reduction in contamination by dilution. Adding mechanical or thermal energy will enhance rinsing. While it is important to use adequate water in rinsing, the key to economy is to use no more than is necessary to achieve acceptable parts.

Importance of Rinsing

Rinsing is a science by itself. Many factors enter into the design of a rinse system. Let us start with the end result in mind. Below is a partial list of considerations:

- Part cleanliness required
- Production levels required
- Type of contamination to be removed (amount and type of drag-in)
- Incoming water quality
- Treatment capabilities
- Number of rinse tanks, size, layout
- Water usage and disposal

Each of these concerns is discussed in further detail below.

Part Cleanliness Required

This consideration is the controlling factor in the whole process of rinsing. If the manufacturer is only concerned with gross contamination, then perhaps rinsing may not even be necessary. At the other extreme, if the slightest residue leads to failure, then the final rinse typically requires the use of deionized water. Obviously, final cleanliness requirements must be determined before other decisions can be made.

Production Levels and Drag-In

Production levels and drag-in will determine what measures are needed to meet the determined quality requirements. Drag-in depends on many factors, including part configuration, orientation, temperature, drain time, etc. Production levels and the shape of the parts determine the amount of drag-in per time.

Parts that are shaped like cups or bowls must be racked in a manner that allows them to drain completely as they are moved from the wash to the rinse tank. If complex parts are allowed to dragin large volumes of wash solution, the chemical costs of replacing lost cleaner will rise rapidly. The amount of water needed to maintain a quality rinse will also increase massively, leading to further increases in costs. Rinsing equations must deal with these factors to predict final rinse quality.

Incoming Water Quality

Incoming water quality can vary from naturally soft water with few contaminants to water that contains many hundreds of ppm hardness. Water hardness is generally measured either in ppm of equivalent calcium carbonate, $CaCO_3$, or in grains of hardness (*note*: 1 grain = 17.1 ppm). Values greater than 120 ppm are considered to be hard water, with values greater than 180 ppm considered to be very hard.

Hardness can have many deleterious effects in the cleaning process. Hard water can react with the surfactants (mainly anionic) to deactivate them. Consequences include increased corrosion on steel surfaces, and/or deposits on the cleaned surfaces. Impacts of these residues include paint adhesion problems, plating problems, and aesthetic problems, just to name a few. Fortunately, as we have already discussed, there are alternatives to using unsoftened water. Whether softened water, deionized, distilled, or RO water is chosen depends on the application.

Number of Rinse Tanks

One of the most common problems manufacturers face in properly setting up a cleaning line is that they are unable or unwilling to allocate enough funds and/or floorspace to incorporate an adequate rinsing step.

Generally speaking, one large rinse is less effective than multiple small rinses. A single rinse may be adequate for intermediate manufacturing steps. However, as a rinse in the final or critical cleaning process, this is seldom true. Increasing the number of rinses is usually necessary for sensitive manufacturing processes and final manufacturing steps. Unfortunately, multiple rinses lead to higher costs in equipment (i.e., number of tanks needed and the footprint taken up on the plant floor).

The increased equipment cost can be offset by counterflowing the water from the last rinse back into each previous rinse. Counterflowing significantly reduces water consumption; some of the overflow can be used as add-back into the cleaning tank to replace evaporated water. Studies have shown that a counterflowed, triple rinse has the optimum balance of reducing water consumption, obtaining clean parts, and recouping capital costs. Greater than three rinses tends to yield diminishing returns.

Rinse Tank Design and Placement

Rinse tank design and placement can greatly affect rinsing efficiency. Rather than being an afterthought, rinsing needs should be addressed early in the planning of the cleaning line. Considerations should include tank geometry and composition, degree of agitation, rinse flow, rinse temperature, and necessary final quality of the parts.

It should be noted that rinsing will only be effective when the water contacts the whole surface of the parts. Part orientation, loading, and rinse flow dynamics are important and often overlooked considerations.

There must be adequate agitation to sweep away the soil-loaded cleaner, especially with cleaners that contain free caustic and/or silicates. These chemicals are inherently slow at releasing from the parts surface. The rinsing needs are quite different for the manufacturer of circuit board components versus the rebuilder of lawn mower engines. In the first case, the choice might be a heated cascaded triple rinse with a deionized water source. In the second case, perhaps a quick dip in a warm tap water tank is sufficient.

Drag-In and Final Rinse Quality

Knowing and understanding the volume and composition of drag-in from the previous step is crucial to predicting rinsing needs. Quite simply, drag-in over a given time equals the quantity of residue that must be diluted to some lower specified level.

In a dynamic situation at equilibrium, the rate that the rinse water must be overflowed must equal the drag-in times the wash tank chemical concentration divided by rinse tank chemical concentration:

$$\text{Overflow} = \text{Drag-in} \times \left(\frac{\text{wash tank concentration}}{\text{rinse tank concentration}} \right)$$

If this equation is extended to multiple rinse tanks, the dragin is reduced in each successive tank. If the overflow for each rinse tank is cascaded to the previous tank, for all practical purposes, the equation takes the following form:

$$\text{Overflow} = \text{Drag-in} \times \left(\frac{\text{wash tank concentration}}{\text{rinse tank concentration}} \right)^{1/\text{number of rinse tanks}}$$

Cascading obviously reduces water use dramatically over a single rinse and over multiple rinses that are not cascaded while achieving the same final rinse quality.

The final rinse quality should be maintained just as carefully as the processing tanks. After all, the rinse is the last liquid the parts will see. It does not make sense to go to great lengths to clean the parts only to recontaminate them with a low-quality rinse.

Acceptable rinse quality depends on the requirements of the application. As such, the steps to measure rinse quality will vary. Generally speaking, cleaning performance needs to only be sufficient to eliminate subsequent problems.

While rinse quality is relative, some general guidelines may be helpful. Some sources would classify applications into general cleaning, critical, and very critical cleaning. While different standards may be proposed, one suggestion is to use conductivity as a guide and divide these applications as follows:

- General rinsing operations having a residual level of $1000\,\mu S$
- Critical rinsing with a residual of $500\,\mu S$
- Very critical at less than $50\,\mu S$

Most conversion coating and plating operations call for a high-quality final rinse of less than $50\,\mu S$ as the minimum water quality. It can be seen that at these higher quality rinses, higher quality water must be used to achieve the desired cleanliness. These guidelines are suggested as a place to start when determining final rinse quality.

Disposal

Oftentimes, local regulations limit or prohibit the discharge of any process water to sewers. The most common properties that are commonly regulated on spent wash and rinse solutions are

- pH
- Dissolved heavy metals content
- Total solids
- Biological oxygen demand
- Chemical oxygen demand
- Emulsified oil content

Some geographical regions will also impose restrictions on phosphate and/or alkylphenol ethoxylate (APE) surfactant content. Spent wash and rinse solutions that cannot meet allowable discharge purity limits require on-site chemical treatment systems to purify them for discharge. Otherwise, they must be hauled away as hazardous waste. Either option is expensive.

The increasing world population is now beginning to put tremendous strain on nature's ability to supply us with sufficient potable water. For social and economic reasons, bath life and water conservation have become very important issues. The result is that overall planning and coordination includes a cradle-to-grave approach for setting up a cleaning line, including the rinse and chemical disposal factors. Closed-loop systems for water treatment are becoming economical long-term alternatives that can reduce costs and hazardous waste generation. At the very least, water use minimization is an important consideration.

A whole chapter could be easily devoted just to basic rinsing concepts. Certain basic principles can be summarized as follows:

- Multiple rinses are more efficient than single rinses, with the optimum balance being about three rinses.
- Water usage can be minimized by cascading the final rinse overflow into the previous rinse and that rinse into the one before.
- The final rinse quality is the determining factor in final residue on the part.
- Rinsing will not be effective if it does not reach the parts.

Conclusion

Aqueous cleaning is an increasingly important segment of the cleaning industry. This importance will probably increase with time and the development of improved cleaning systems. The emphasis must be on careful design of cleaning systems, because the interactions between parts, soils, equipment, cleaner, and water are much more complex than they appear on the surface.

Environmental, economic, and other business concerns demand that industry obtain acceptable parts with minimal impact on the environment and at the lowest possible costs. The challenge to the cleaner manufacturer and the equipment manufacturer is to develop effective cleaners and equipment that meet those criteria and that are compatible with each other.

The newer generation of aqueous cleaners is designed to reject contaminants rather than emulsify soils. This feature allows the cleaner to be filtered routinely without significant adverse effect on the cleaner chemistry. Many of the newer formulations can be replenished with routine chemical additions of the cleaner concentrate, according to the maintenance procedures recommended by the chemical supplier.

The extension of cleaner bath life obtained with regular bath maintenance results in reduced chemical consumption, reduced waste generation, reduced waste liability, and reduced cleaning costs. Very often, the newer cleaning processes also yield cleaner parts as well.

Clearly, the future progress of aqueous cleaning will require close cooperation between the chemical and equipment industries. It is also apparent that water quality is a make or break issue in critical cleaning, especially as it applies to rinsing. This area is consistently the most neglected aspect of aqueous cleaning.

Aqueous cleaning is changing to meet the economic and environmental needs of the times. Its future is clear.

Acknowledgment

Portions of the text were reprinted from Ref. [6], Copyright 1995, with permission from Elsevier Science.

References

1. Archer, W. Reactions and inhibition of aluminum in chlorinated solvent systems. *Corrosion 1978 Conference Proceedings*.
2. Betz Laboratories, Inc. *Handbook of Industrial Water Conditioning*, 6th edn. Betz Laboratories, Inc., Trevose, PA, 1980.
3. Peterson, D.S. *Practical Guide to Industrial Metal Cleaning*. Hanser Gardner Publishing, Cincinnati, OH, 1997.
4. Spring, S. *Industrial Cleaning*. Prism Press, London, U.K., 1974.
5. Murphy, K. The dirty truth: They're smuggling soap in spokane. *L.A. Times*, April 6, 2009.
6. Quitmeyer, J.A. The evolution of cleaning technology. *Metal Finishing*, September 1995, 93(9), 34–39.
7. Cherepy, N.J., Shen, T.H., Esposito, A.P., and Tillotson, T.M. Characterization of an effective cleaning procedure for aluminum alloys: Surface enhanced Raman spectroscopy and zeta potential analysis. *Journal of Colloid and Interface Science*, 282(1), 2005, 80–86.
8. Mosher, E., Peterson, L., and Sköld, R. The chemical control of cobalt leaching from cemented carbide tooling. *Materials Performance*, 1986.
9. McChesney, J.M. and Landers, P.E. Method of reducing leaching of cobalt from metal working tools containing tungsten carbide particles bonded by cobalt. U.S. Patent 4,315,889, February 16, 1982, Assignee is Ashland Oil, Inc.
10. Johansson, I. and Sköld, R. Method for mechanically working cobalt-containing metal. U.S. Patent 4,976,919, December 11, 1990, Assignee is Berol Kemi AB.

Bibliography

David, R.L. *CRC Handbook of Chemistry and Physics*, 77th edn. CRC Press, Boca Raton, FL, 1996.
Durkee, J.B. *The Parts Cleaning Handbook*. Gardner Publications Inc., Cincinnati, OH, 1994.
Farrell, R. and Horner, E. Metal cleaning. *Metal Finishing*, 97(1), 1999, 122–135.
Gruss, B. Cleaning and surface preparation. *Metal Finishing*, 96(5A), 1998.

Hanson, N. and Zabban, W. *Plating*, 46, 1959.

Hirsch, S. Deionization for electroplating. *Metal Finishing*, 1999, 97(1), 149–155.

Kanegsberg, B.F. Aqueous cleaning for high-value processes. *A²C² Magazine*, 2(8), 1999.

Metals Engineering Quarterly, November 1967, 7(4).

Mohler, J.B. The rinsing ratio applied to practical problems. Part 1, *Metal Finishing*, 1972.

Nelson, W. The key to successful aqueous cleaning is…water. *Precision Cleaning*, 1996.

Permutit. *Water and Waste Treatment Data Book*, 18th printing, U.S. Filter, Permutit, Bridgewater, NJ, 1993.

Quitmeyer, J. All mixed up: Qualities of aqueous degreasers. *Precision Cleaning*, September 1997.

Schrantz, J. Rinsing: A key part of pretreatment. *Industrial Finishing*, 1990.

Wolf, K. and Morris, M. Ozone depleting solvent alternatives: Have you converted yet? *Finishers' Management*, 1996.

Zavadjancik, J. Aerospace manufacturer's program focuses on replacing vapor degreasers. *Plating and Surface Finishing*, 1992.

3

Cleaning Agent Chemistry

JoAnn Quitmeyer
Kyzen Corporation

Introduction

Legislative changes have forced industry to reevaluate cleaning needs. Today "green technology" and waste minimization are driving forces in process optimization. Once process requirements have been defined, chemical choices must be made.

Cleaning agent chemical options include hydrocarbon solvents, semiaqueous or emulsion cleaners, or aqueous cleaning technologies. These cleaning agents may be single-component materials or they may be blends containing over 20 different chemical substances classified as solvents, surfactants, acids, alkalinity builders, inhibitors, stabilizers, viscosity modifiers, colorants, or fragrances.

Depending on the soils involved and the mechanical method of fluid application, the formulator may customize a composition to meet exact specifications required to meet production needs. An informed manufacturer can then optimize his or her cleaning process for maximum output and quality.

Cleaning Agent Base Solvents

A solvent is a liquid or gas that dissolves a solid, liquid, or gaseous solute, resulting in a solution. The most universal solvent is water, a natural resource covering over 70% of the Earth's surface. Other solvents, called organic solvents, are carbon-containing fluids. These aqueous or hydrocarbon solvents form the basis for cleaning agents.

Water is a compound containing two parts hydrogen (H) and one part oxygen (O), with the chemical formula H_2O. The term refers to the liquid state although the compound also exists in solid and gaseous states as ice and steam. Water is a tasteless, odorless, transparent polar solution. In the liquid state, it is miscible with many other polar substances forming a single homogenous solution. It is immiscible with aliphatic, aromatic, and halogenated hydrocarbons and some oxygenated solvents and it forms an azeotrope (a mixture that boils with a constant composition) with others. When combined with acids, alkalinity builders, or surfactants, aqueous-based cleaning agents are effective on a wide range of polar and nonpolar soils.

As a major solvating material, vast amounts of this natural resource are used as cleaning agents, especially in industrialized countries. Environmental awareness is making aqueous cleaning processes the preferred means of soil removal.

$$H—C\equiv C—H$$

FIGURE 3.1 Aliphatic hydrocarbon structure.

Organic solvents are compounds composed of the elements carbon (C) and hydrogen (H). These hydrocarbon solvents are further classified as aliphatic, aromatic, halogenated, or oxygenated materials.

Aliphatic hydrocarbon solvents are carbon atoms that are joined together in straight or branched chains, as shown in Figure 3.1, or in alicyclic rings by single, double, or triple bonds.

Alicyclic rings are closed chains. Alicyclic compounds may have double bonds. In contrast to a benzene ring, for example, they are not aromatic; so alicyclic compounds behave similarly to aliphatic compounds. They are classified as flammable or combustible depending on the flash point. The vapor pressure is an indication of how rapidly the solvent will evaporate. Water has a vapor pressure of approximately 17.5 mmHg at 20°C. As the vapor pressure increases, the faster the solvent will evaporate. Materials with a vapor pressure less than 17.5 will take longer to evaporate than water. Examples of aliphatic solvents are listed in Table 3.1.

The kauri-butanol (Kb) value[1] is a standardized measure of solvent power governed by ASTM D1133 giving a scaleless index; more precise, complex descriptions of solvency such as the Hansen solubility parameters are discussed in other chapters. As Kb value increases, the more solvent power the material has. The Kb of a solvent shows the maximum amount of hydrocarbon that can be added to a kauri resin solution without causing cloudiness. Strong solvents can be added in greater amounts so they have a higher value than weaker solvents. Unlike hydrocarbon solvents, aqueous cleaners are measured by their pH and the concentration of hydrogen ions in the solution. Since aliphatic hydrocarbons do not use water, they cannot have a pH; they instead are classified by Kb value.

This group of nonpolar solvents is most effective on nonpolar soils including mineral and vegetable oils and grease. All are hydrophobic; water solubility is negligible.

Aromatic hydrocarbon solvents are molecular structures where one or more planar sets of six carbon atoms (a benzene ring) are connected by delocalized electrons numbering the same as if they consisted of alternating single and double covalent bonds. Figure 3.2a and b shows the chemical structures of this type of hydrocarbon and Table 3.2 lists examples of these solvents often used in immersion and manual industrial cleaning operations.

Aromatic solvents are derived from petroleum and coal tar.[2] They have a low to moderate flash point and dry rapidly without leaving residues. These nonpolar hydrophobic solvents are effective at dissolving nonpolar soils including oils, grease, and some resins and rubbers.

TABLE 3.1 Aliphatic Hydrocarbon Solvents

Common Name	Chemical Formula	Flash Point, °C	VOC, g/L	Vapor Pressure, mmHg at 20°C	Kauri-Butanol (Kb) Value
Mineral spirits	Mixture	21–54	770	2.0	33–37
Mineral seal oil	Mixture	52	847	<0.01	27.5
Kerosene	Mixture	65	810	0.4	34
Lacquer thinner	Mixture	7	758	38.0	46
Heptane	C_7H_{16}	<−6	696	45.0	29
VM&P naptha	Mixture	11.1	739	52.0	34
Hexane	C_6H_{14}	<−6	676	137.0	30
Turpentine	$C_{10}H_{16}$	34	862	13.4	64
Pine oil	$C_{10}H_{16}$	46	853	1.3	62
D-limonene	$C_{10}H_{16}$	43	841	<2.0	67

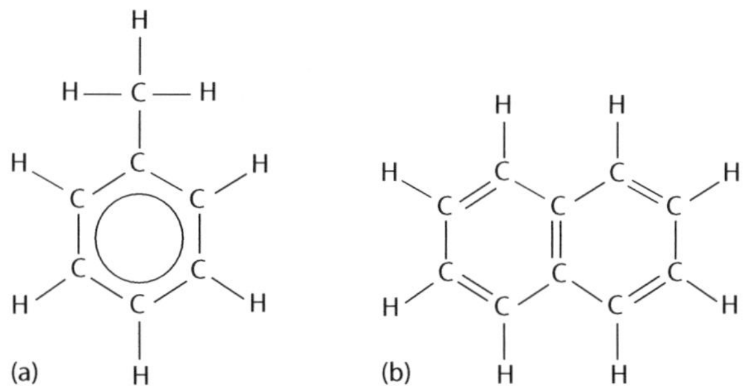

FIGURE 3.2 Aromatic HCs: (a) toluene and (b) naphthalene.

TABLE 3.2 Aromatic Hydrocarbon Solvents

Common Name	Chemical Formula	Flash Point, °C	VOC, g/L	Vapor Pressure, mmHg at 20°C	Kauri-Butanol (Kb) Value
Toluene	C_7H_8	7	872	22.0	105
Xylene	C_8H_{10}	27	868	5.1	98
Heavy aromatic naptha	Mixture	82	930	<10.0	112

Aliphatic and aromatic solvents are safe to use on virtually all metals, glass, and ceramics. With adequate ventilation, they can be used in immersion and manual cleaning applications and spent solutions can be reclaimed by distillation.

Halogenated solvents have one or more of the hydrogen atoms replaced with chlorine (Cl), fluorine (F), or bromine (Br). Many, but not all, of these solvents are self-extinguishing, have low surface tension for good penetration in tight spaces, and evaporate rapidly without leaving a residue. They are nonpolar and solubilize or swell nonpolar soils including oils, grease, flux, adhesive, and paint. They have worker exposure limits as defined by the Occupational Safety and Health Administration (OSHA) and some have ozone-depleting potential (ODP).

Examples of halogenated solvents are listed in Table 3.3. These solvents are typically used in vapor, immersion, or manual cleaning applications for the removal of nonpolar soils including oils, grease, resins, waxes, pitch, and tar; they are ineffective on inorganic and polar soils. A typical chemical structure is shown in Figure 3.3.

Halogenated cleaning agents are compatible with virtually all metals, glass, and ceramic substrates. They clean by solubilization or swelling and may be reclaimed by distillation. For many halogenated

TABLE 3.3 Halogenated Solvents

Common Name	Chemical Formula	Flash Point, °C	VOC, g/L	Vapor Pressure, mmHg at 20°C	Kauri-Butanol (Kb) Value
Trichloroethane	CH_3CCl_3	None	1340	100	124
Chloroform	$CHCl_3$	None	1484	159	115
Trichloroethylene	$ClCH=CCl_2$	None	1470	57.8	129
Perchloroethylene	C_2Cl_4	None	1620	18.2	92
Methylene chloride	CH_2Cl_2	None	1327	350.0	136
n-Propyl bromide	C_3H_7Br	None	1350	111.0	129
Decafluoropentane	$C_5H_2F_{10}$	None	Exempt	226.0	NA
Methyl nonafluoroether blend	Mixture	None	Exempt	202.0	NA

Note: NA, not available.

solvents, inhibitors are added at the point of manufacture to forestall the formation of acids (HCl, HBr); care must be taken to insure that redistillation leaves an adequate level of inhibitor to make sure corrosion and stability problems do not arise during reuse.

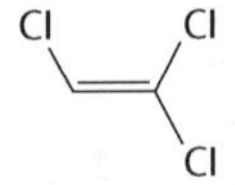

Oxygenated solvents are a series of petroleum-derived oxidates composed of organic acids and esters. They contain oxygen (O) in addition to carbon and hydrogen as shown in Figure 3.4a and b. Examples listed in Tables 3.4 through 3.7 are polar solvents miscible with water; they have a broad range of flash points and all dry residue-free.

FIGURE 3.3 Halogenated HC.

- Alcohols are a broad class of these hydroxyl (OH) containing organic solvents that occur naturally in plants or are made synthetically from petroleum derivatives. Being polar solvents, these fluids are effective in removing polar soils including finger prints, synthetic coolants, flux, and some inks.
- Depending on the chemical blend, alcohols may be used in immersion or manual processes when applied straight. When mixed with water, they may also be sprayed. Table 3.4 lists examples and comparative chemical data of these chemicals. These alcohols are flammable or combustible and must be used with adequate ventilation for optimum worker safety. All evaporate completely and dry residue-free. They are miscible with water and can be diluted further with water or water-rinsed depending on the process involved.
- Glycol ethers are a group of solvents based on alkyl ethers of ethylene glycol or derived from diethylene glycol. They have excellent solvency, chemical stability, and are compatible with water and many organic solvents. Examples are given in Table 3.5. Shown structurally in Figure 3.4c, these solvents have a characteristically sweet odor and taste. They are hygroscopic, compatible with virtually all metals, glass, and ceramics, and are effective on polar and many nonpolar soils. Used straight, they can be applied by immersion and manual methods. When diluted with water they can also be applied by spray-in-air processes.
- Glycols are classified as the "E-Series" of ethylene glycol based solvents or the "P-Series" based on propylene glycol.

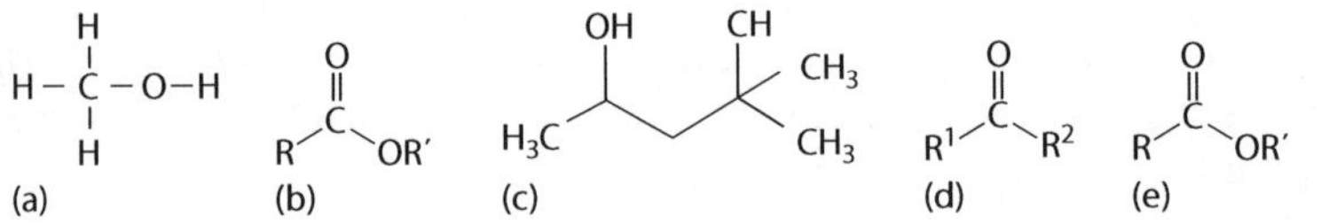

FIGURE 3.4 (a) Oxygenated solvent, alcohol; (b) oxygenated solvent, ester; (c) glycol ether; (d) ketones; and (e) esters.

TABLE 3.4 Alcohol Solvents

Common Name	Chemical Formula	Flash Point, °C	VOC, g/L	Vapor Pressure, mmHg at 20°C	Kauri-Butanol (Kb) Value	HAPs
Isopropanol	C_3H_8O	12	Exempt	44.00	50	No
Methanol	CH_4O	12	791	125.00	NA	Yes
Benzyl alcohol	C_7H_8O	93	1050	0.15	NA	No
Cyclohexanol	$C_6H_{12}O$	63	960	0.98	NA	No
Ethanol	C_2H_6O	13	789	60.00	NA	No
Butanol	$C_4H_{10}O$	37	810	5.00	NA	No
Phenol	C_6H_6O	79	1070	0.35	NA	Yes

Note: HAPs, hazardous air pollutants; NA, not available.

TABLE 3.5 Glycol Solvents

Series	Common Name	Chemical Formula	Flash Point, °C	VOC, g/L	Vapor Pressure, mmHg at 20°C	HAPs
E	Glycol ether EB	$C_6H_{14}O_2$	72	902	0.60	No
E	Glycol ether DB	$C_8H_{18}O_3$	106	955	0.06	No
E	Glycol ether DB acetate	$C_{10}H_{20}O_4$	116	980	0.01	No
P	Propylene glycol	$C_3H_8O_2$	99	1036	0.13	No
P	Glycol ether TPM	$C_{10}H_{22}O_4$	121	962	0.03	No
P	Glycol ether DPnB	$C_{10}H_{22}O_3$	100	912	0.02	No
P	Glycol ether DPnP	$C_9H_{20}O_3$	88	922	0.05	No

TABLE 3.6 Ketones

Common Name	Chemical Formula	Flash Point, °C	VOC, g/L	Vapor Pressure, mmHg at 20°C	HAPs
Acetone	C_3H_6O	−20	Exempt	400	No
Methyl ethyl ketone (MEK)	C_4H_8O	−9	810	78	Yes
Methyl isobutyl ketone (MIBK)	$C_6H_{12}O$	14	800	16	No

TABLE 3.7 Esters

Common Name	Chemical Formula	Flash Point, °C	VOC, g/L	Vapor Pressure, mmHg at 20°C	HAPs
Methyl acetate	$C_3H_6O_2$	−10	Exempt	170	No
Amyl acetate	$C_7H_{14}O_2$	23	876	5	No
n-Butyl acetate	$C_6H_{12}O_2$	26	882	15	No
n-Propyl acetate	$C_5H_{10}O_2$	14	900	23	No
Dibasic ester	$C_7H_{12}O_4$	100	1092	0.20	No

- Ketones are compounds that contain a carbonyl group (C=O) bonded to two other carbon atoms. They are derived by oxidation of secondary alcohols, have low flash points, and are fast evaporating with no residues. Typically used in immersion or manual cleaning applications, ketones are safe to use on virtually all metals, glass, and ceramics and are effective on a broad spectrum of soils including glues and adhesives, inks, resins, and waxes. These solvents tend to be hygroscopic and effectively solubilize polar and some nonpolar soils. A typical chemical structure is shown in Figure 3.4d and examples are listed in Table 3.6.
- Esters are organic compounds that are derived from acids by the exchange of the replaceable hydrogen atom with an organic radical, usually a reaction of an acid with an alcohol. Where an inorganic or organic acid where one hydroxyl group (⁻H) is replaced with an alkyl group (–O–), the preparation process is called a condensation reaction or esterification.

Many naturally occurring fats and oils are the fatty acid esters of glycerol. Typical esterified acids include carboxylic acids, phosphoric, sulfuric, nitric, and boric acids. The ester structure is shown in Figure 3.4e and typical examples are listed in Table 3.7.

Esters are multimetal safe and effective at removing reactive soils including inks, glues, and adhesives. They are typically not water soluble and have boiling points higher than similar molecular weight hydrocarbons. Most are VOC reportable and many have a characteristic fruity odor; the low-viscosity essential oils are frequently used as fragrances in cleaning agents.

Local municipalities govern the manner of disposal of spent cleaning solutions containing oxygenated solvents. Depending on the contaminants introduced into the wash bath, most wash bath solutions are disposed of as nonhazardous wastes by licensed waste haulers. Frequently rinse waters are permitted to be disposed of by local publically owned treatment works (POTWs).

Cleaning Agent Chemical Components

Cleaning agents contain typical building blocks; components are selected depending on the contaminants involved, the degree of cleanliness needed, and the market segment being served. In addition to the base solvent, chemical component categories include surfactants, pH modifiers, stabilizers, conditioners, inhibitors, viscosity modifiers, odorants, dyes, and surface sanitizers.

Surfactants

Surface-active agents, also called surfactants, are a family of organic compounds, both natural and synthetic, that are added to cleaning agents to reduce surface and interfacial tension to enhance cleaning efficiency, lubrication, wetting, solvency, and rinsing properties of a fluid.[3] As shown in Figure 3.5, surfactants reduce surface tension and enhance the wetting properties of a cleaning agent, allowing the agent to penetrate tight areas and undercut soils.

A surfactant or amphiphile is a linear molecule with a hydrophilic (water loving) head and a hydrophobic (water hating) tail, as shown in Figure 3.6. This characteristic enables them to be soluble in both polar and nonpolar solutions.

During the washing process, once an amphiphile stops functioning as a monomer, it starts functioning as a micelle. This point is called the critical micelle concentration (cmc). The hydrophobic tails form the core that encapsulates nonpolar substances while the hydrophilic heads form an outer shell that maintains contact with water or other polar substances. The average number of amphiphile molecules in a single micelle is described as the aggregate number (m). The (cmc) and (m) together characterize the micelle the amphiphile will form under a given set of conditions, spherical or planar, as shown in Figure 3.7.

Some generalizations made about micelles are the following:

1. As chain length increases, water solubility decreases.
2. Single micelle chain amphiphiles have a higher cmc than two carbon chain micelles.
3. Ionic solutions have greater solubility and higher cmc than nonionic solutions.
4. Ionic solutions have greater rejection forces between polar groups than nonionic solutions.

Water molecules dissociate into H^+ and OH^-, allowing the positive ions to equilibrate among protonatable groups (those groups of molecules capable of adding protons) and making them convenient ions for the creation of electrical potential differences. Hydrogen bonds are fairly strong so in air–water mixtures, molecules orient themselves with higher bonding potential in the center and lower bonding potential on the edges. To increase the entropy of the system, water minimizes its surface area, resulting in high surface tension.

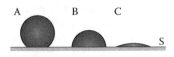

FIGURE 3.5   Wetting agent effect.

Nonpolar molecules do not form hydrogen bonds. The bonding potential of neighboring water molecules is negligible. They do not attract each other; rather they are pushed together as they are mutually rejected from the water. This is called the hydrophobic effect.

Nonionic solutions form micelles with smaller surface areas per amphiphile. Increase in ionic strength decreases rejection forces between polar groups in an ionic solution while m increases.

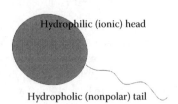

Hydrophilic (ionic) head

Hydropholic (nonpolar) tail

FIGURE 3.6   Amphiphile.

(a)   (b)

FIGURE 3.7   Spherical and planar micelle structures.

In addition to surface tension modification, surfactants may also be used as antistats, antifoaming agents, bacteriastats, corrosion inhibitors, detergents, dispersants, emulsifiers, foaming agents, or soaps.

There are four general classifications of synthetic surfactants: anionic, cationic, nonionic, and amphoteric.

Nonionic surfactants are materials that carry no electrical charge. Their water solubility is driven by the presence of polar functionalities capable of hydrogen bonding with water. Accounting for 45% of industrial production, nonionic surfactants offer formulation flexibility and include the following list of materials:

Glycerides	Glycol esters	Fatty amine ethoxylates
Block copolymers	Glycerol esters	Ethoxylated propoxylated copolymers
Alkanolamides	Polyglycosides	Alkylphenol ethoxylates
Amine oxides	Alcohol ethoxylates	Polyoxyethylene-based materials
Polyglycerides	Fatty acid ethoxylates	Polyglycerol polyol derivatives
Glucosides	Sucrose esters	Sorbitan esters (ethoxylates)

Materials can be selected based on foaming characteristics and cloud points to customize cleaning agent compositions as desired. Often a combination of materials is used.

Anionic surfactants are a group of materials that carry a negative charge on the active portion of the molecule. These surfactants make up approximately 50% of the world production and include the following list of materials:

Sulfate esters	Aliphatic sulfonates	Sulfated fatty acid condensation products
Fatty alcohol sulfates	Alkylaryl sulfonates	α Sulfocarboxylic acids/derivatives
Sulfated ethers	Carboxylate soaps	Alkyl glyceryl ether sulfonates
Sulfated fats and oils	Lignosulfonates	Miscellaneous sulfo esters and amides
Sulfonic acid salts	Phosphate esters	

As a general rule, anionic surfactants exhibit superior wetting and emulsifying properties while tending to be higher foaming materials.

Cationic surfactants are a group of materials carrying a positive charge on the active portion of the molecule. They include the following:

Imidazoline derivatives	Pyridines	Quaternary ammonium compounds
Betaines	Morpholines	

These materials have excellent antibacterial properties, provide corrosion protection, and are used as demulsifiers.

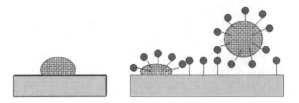

FIGURE 3.8   The cleaning process.

Amphoteric surfactants are a group of materials that can be either cationic or anionic depending on the solution pH. This group includes Zwitteronic types that possess permanent charges of each. Included are the following materials:

Imidazoline derivatives	Amine condensates	Quaternary ammonium compounds
Betaines	Sulfobetaines	Phosphatides

At low or acid pH, these materials function as cationic surfactants and at high alkaline pH, they are anionic. When formulated into cleaning agents, these surfactants function as bacteriastats and tend to be high foaming.

The formulator has many hundreds of surfactants to choose from to customize the wetting, emulsifying, and foaming characteristics most required to meet any cleaning need. Once the cleaning agent reaches the soil, it encapsulates and lifts the soil from the surface and dissolves or disperses it in the cleaning solution as shown in Figure 3.8. Most aqueous cleaning agents contain one or more surfactants to aid the cleaning process and to facilitate water rinsing.

Key considerations when selecting a surfactant package include the pH of the solution, the soils to be removed, the water source identified, and the foaming properties that are compatible with the cleaning application selected.

Stabilizers and Inhibitors

Stabilizers and corrosion inhibitors[4] are often included in a cleaning agent formulation. Solvent stabilizers serve to prevent solvent breakdown and to chemically inhibit reactions that may degrade solvent properties. They are used as pH acceptors, metal inhibitors, and antioxidants. Stabilizers commonly found in hydrocarbon solvent blends are listed in Table 3.8.

Many of these stabilizers are present at volumetrically inconsequential levels to be considered upon disposal, however when the solvents are distilled for reuse, sump life may be shortened due to lack of a sufficient stabilizer. Major manufacturers of hydrocarbon solvents take great care to ensure that even trace components meet specification, especially with recycled material.

Inhibitors are defined as agents that slow or interfere with a chemical reaction. In a cleaning agent, they are materials that when added to a fluid, decrease the corrosion rate of a metal. This is accomplished

TABLE 3.8   Hydrocarbon Stabilizers

1,4-Dioxane	2-Methyxy phenol	Pentene oxide	Butoxy methyl oxirane
1,3-Dioxolane	Morpholine	Furanidine	Diisopropylamine
Nitromethane	Thiazoles	*Sec*-Butanol	Butadiene oxide
1,2-Butylene oxide	Glycidyl acetate	Amyl alcohol	Methyl pyrrole
Oxolane	Stearates	Trioxane	Epichlorohydrin
Triethylamine	Ethyl acetate	Aniline	Cyclohexene oxide
1,4-Epoxybutane	Propanol	Isocyanates	

TABLE 3.9 Inhibitors

Hexamine	Silicates	Phenylenediamine
Hydrazine	Nitrites	Carboxylates
Ascorbic acid	Chromates	Dimethylethanolamine
Zinc oxide	Morpholine	Condensation products of aldehydes
Sulfonates	Phosphates	Monoethanolamine
Complex amines	Triazoles	Triethanolamine

by the formation of a passivation layer or thin film, by inhibiting oxidation or reduction of the redox system, or by scavenging the dissolved oxygen. Inhibitors may be grouped as water soluble, water displacing, or emulsion chemistries.

The most practical inhibitor chemistries provide indoor protection up to 6 months. These compounds are compatible with aqueous cleaning agents and often the inhibitor package is formulated directly into the cleaner composition. The amount of inhibitor present in the bath dictates the length of protection enjoyed. Sometimes inhibitors are added sump-side to enhance protection of an existing composition or to provide inhibition to a rinse solution. Some examples are listed in Table 3.9.

Barrier films like paints, lacquers, and oils are also used for short- or long-term corrosion during storage.

Conditioners

Aqueous cleaning agents contain water or are designed to be diluted further with water prior to use. Use solutions may contain up to 99% water. For the most part, aqueous concentrates are made with distilled or deionized water to minimize the amount of dissolved solids added to the mixture.

Water that contains dissolved minerals will leave a residue upon evaporation. Minerals such as calcium (Ca) or magnesium (Mg) can react with carbonates and soaps to form insoluble scum and scale. To prevent this reaction from occurring, cleaning agents are formulated with sequestering agents or chelators, materials used to tie up the offending elements.

Water hardness varies considerably across the United States and the world. Figure 3.9 shows a map[5] of these variations. Water solids will increase dramatically over time, doubling every 4 weeks. These accumulated solids can cause spotting on polished surfaces and can cause scale build up in the washing equipment and on plumbing surfaces. Chlorides, sulfates, and dissolved copper act as catalysts and can enhance the corrosion potential of the water.

Depending on the water source used, hard water salts or other solids may need to be managed for optimum performance. Common chelating materials used to condition aqueous cleaning fluids are listed in Table 3.10. Conditioners may be selected based on their performance at managing hard water salts or they may be chosen for their ability to manage metal oxides. Often a combination of components is used depending on the cleaning job involved.

Water conditioners and metal chelators control oxidation and scale formation. They may, however, interfere with subsequent effluent treatment if heavy metals are involved. The local POTW may have regulations concerning the use of chelation materials; it is best to check with local officials concerning waste disposal issues.

pH Modifiers

pH[6] is a measure of acidity or basicity of a solution that contains water. It is the co-logarithm of the activity of dissolved hydrogen (H^+) ions. The pH scale is not an absolute scale; it is relative to a set of standard solutions whose pH is established by international agreement. It is generally thought that "p" is a constant that stands for "negative logarithm." "H" stands for hydrogen. It can be measured potentiometrically if

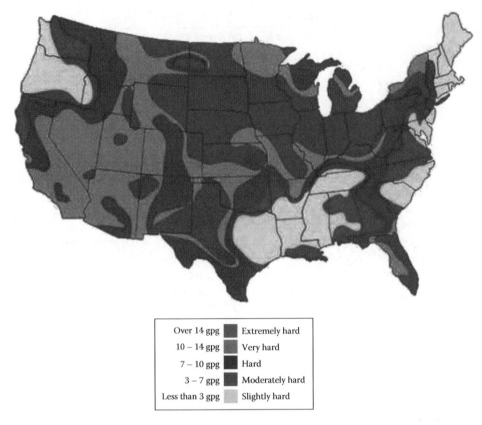

Over 14 gpg		Extremely hard
10 – 14 gpg		Very hard
7 – 10 gpg		Hard
3 – 7 gpg		Moderately hard
Less than 3 gpg		Slightly hard

FIGURE 3.9 Water hardness, map of the United States.

TABLE 3.10 Common Conditioners and Sequestration Rates

Conditioner	Calcium Ion Sequestration Rate, mg/g	Calcium Carbonate Sequestration Rate, mg/g	Iron Sequestration Rate, mg/g
EDTA, 30%	41	103	
NTA	63	158	
HEEDTA, 41%	48	120	14.5
STPP	52	250	
ATMP, 50%	183	325–590	
Sodium gluconate	13	33	334
Sodium glucoheptanate	22	55	393

Notes: All percentages are expressed in w/v. EDTA, ethylenediamine tetraacetic acid; NTA, nitrilotriacetic acid; HEEDTA, hydroxyethyl ethylenediamine triacetic acid; STPP, sodium tripolyphosphate; ATMP, amino tri(methylene phosphonic acid).

an electrode is calibrated with a solution of known hydrogen ion concentration. pH paper and indicator solutions can be used for rough estimates of pH. The simplest (although not complete) differential between an acid and a base is that an acid is a proton donor and a base is a proton acceptor.

Pure water is said to have a neutral pH of 7. Solutions with a pH less than 7 are acidic and liquids having a pH greater than 7 are basic. Materials that do not contain water cannot be measured by pH; Kb value is used instead.

Aqueous cleaning agents contain acids or alkalinity builders depending on the pH and application of the cleaner. Neutral cleaners will have a pH in the 5–9 range. Acid cleaners will have a pH less than 5. Alkaline cleaners will have a pH greater than 9. Organic or mineral acids are used to lower the pH while alkalinity builders and hydroxides are added to raise the pH.

Acid cleaners are used to remove inorganic soils like rust, scale, and metal oxides, usually by solubilizing or chemically digesting the soils. Some acids have more than one pK_a because they are polyprotic; they have more than one proton per molecule. These cleaners may contain mild organic acids, caustic mineral acids, or a combination of acids selected to target a specific set of soil conditions. Some of these acids are listed in Tables 3.11 and 3.12. As a general rule, strong acids have a low pH and a low pK_a while weak acids have a higher pH and a higher pK_a value. Multiple pK_a values offer a wider range of strength and buffering potential, thus making them effective on a broader range of soils while providing greater soil holding potential, resulting in longer bath life.

Alkaline cleaners contain bases, alkalis, or hydroxides. A base is any chemical compound that, when dissolved in water, gives a solution having a pH greater than 7. It is the opposite of an acid; the hydronium ion (H_3O^+) concentration is reduced. An alkali[7] is a basic ionic salt of an alkali metal or alkaline earth metal that is water soluble and forms hydroxide ions (OH^-). A strong base hydrolyzes completely raising the pH toward 14. In chemistry hydroxides refer to any inorganic compounds that contain the hydroxyl group. While acids are "corrosive," strong bases are "caustic," capable of burning, corroding, dissolving, or eating away by chemical action.

Some general properties of alkaline cleaners include the following:

1. They are moderately concentrated with a pH 10 or greater.
2. Some concentrated solutions may cause caustic burns.

TABLE 3.11 Mineral Acids

Acid Type	Acid	Formula	pH, 1.0 M Approx.[a]	pK_a[b]
Mineral	Sulfuric	H_2SO_4	<1	−3, 1.99
	Phosphoric	H_3PO_4	1.5	2.15, 7.20, 12.35
	Hydrochloric	HCl	<1	−4
	Hydrofluoric	HF	<1	3.2
	Nitric	HNO_3	<1	−1
	Chromic	H_2CrO_4	<1	0.74, 6.49

[a] All pH values are based on 1 M solutions at ambient temperature.
[b] pK_a is the acidity constant or acid dissociation constant, a quantitative measure of the strength of an acid; a logarithmic scale $-\log 10\ K_a$.

TABLE 3.12 Organic Acids

Acid Type	Acid	Formula	pH, 1.0 M Approx.[a]	pK_a[b]
Organic	Citric	$C_6H_8O_7$	2.2	3.13, 4.76, 6.40
	Acetic	$C_2H_4O_2$	2.4	4.75
	Sulfamic	H_3NSO_3	1.2	1.04
	Boric	H_3BO_3	5.1	9.27, 12.7, 13.8
	Ascorbic	$C_6H_8O_6$	3	4.10
	Formic	H_2CO_2	1.9	3.75
	Carbonic	H_2CO_3	2.1	6.37, 10.25

[a] All pH values are based on 1 M solutions at ambient temperature.
[b] pK_a is the acidity constant or acid dissociation constant, a quantitative measure of the strength of an acid; a logarithmic scale $-\log 10\ K_a$.

3. They are slippery to the touch due to saponification of fatty acids on the skin.
4. They are water soluble.
5. They are measured on the pH scale.

Ammonia and amine materials are bases but not alkalis.

Alkalinity builders are a group of basic materials used to raise the pH of a cleaning agent. They are commonly found in aqueous and some oxygenated hydrocarbon compositions to provide detergency. Some of the commonly used builders include those listed in Table 3.13. The formulator can select the alkalinity builder package best suited for the performance objectives of the cleaning agent. Often a combination of compounds is built into a composition to provide optimum detergency on the substrates involved. Alkalinity builders are selected based on the substrates involved, the soils being removed and the regulatory concerns governing the cleaning process.

Amines are a class of organic compounds containing nitrogen that are derived from ammonia (NH_3) when one or more of the hydrogen atoms is/are replaced with an alkyl group. They may be primary, secondary, tertiary, or complex depending on the number of hydrogen atoms replaced.

Amine-based cleaning agents are alkaline having a pH ranging from 7.5 to 13. Amine-based cleaners can fall in the neutral pH range, generally classified as having a pH up to 9. And neutral pH cleaners can contain a limited amount of alkalinity builders as well. In fact, most do. Each will chemically react with organic acids to form amine soaps. When used in aqueous and semiaqueous compositions, these materials are used to saponify or neutralize acidic soils such as flux, paste, animal and vegetable oils, grease, etc. They also contribute to pH control, buffering, and corrosion inhibition. Unlike mineral or metallic soaps formed during saponification with hydroxides, organic amine soaps have a higher hydrophilic-lipophilic balance (HLB) and therefore are more water-soluble and easier to rinse with water.

Alkaline salts commonly used in the formulation of cleaning agents include phosphates, carbonates, and silicates. Sodium chloride and sodium sulfate are occasionally used as filler materials in powdered

TABLE 3.13   Alkalinity Builders

Alkalinity Builder	Chemical Formula	pH, 1 M Approx.	pK_b[a]
Calcium hydroxide	$Ca(OH)_2$	12.2	2.43, 1.4
Lithium hydroxide	$LiOH$	14	−0.36
Potassium hydroxide	KOH	>13	0.5
Sodium hydroxide	$NaOH$	>13	0.2
Tetra ammonium hydroxide	Mixture	>13	0.0
Ammonia	NH_3	11.6	4.75
Triethanolamine	$C_6H_9NO_3$	10.5	6.2
Diethanolamine	$C_4H_{11}NO_2$	11.0	4.75
Monoethanolamine	C_2H_7NO	11.7	4.5
2-Amino-2-methyl-1-propanol	$C_4H_{11}NO$	11.5	9.8
Sodium carbonate	Na_2CO_3	11.6	6.4, 10.3
Potassium carbonate	K_2CO_3	11.6	6.37, 10.25
Sodium silicate	Na_2SiO_3	11–12.5	Not available
Potassium silicate	K_2SiO_3	11.7	1.92
Sodium metasilicate	Na_2SiO_3	13	1.0
Sodium tripolyphosphate	$Na_5P_3O_{10}$	9.8	Not available
Tetrapotassium pyrophosphate	$K_4P_2O_7$	10–11	Not available
Trisodium phosphate	Na_3PO_4	12–14	2.23

Note: The values given in Tables 3.11 through 3.13 were taken from various books and Internet sources. pH values are based on 1 M solutions.

[a] pK_b is the alkalinity constant or alkalinity dissociation constant; a quantitative measure of the strength of the base solution ($-\log 10\ K_b$).

compositions and some aqueous blends. These cleaning agents may range from neutral to highly alkaline, all are nonvolatile and do not contribute to reportable volatile organic compounds (VOC). All of these cleaners require water rinsing for complete removal of all product residues. Typical properties are listed in Tables 3.14 through 3.16.

Inorganic hydroxides include the sodium and potassium salts. They are highly basic and corrosive. Some nonferrous alloys including aluminum and zinc are attacked by free hydroxides. Hydroxides are hygroscopic and nonvolatile with no VOC. When mixed with fatty acids, they form mineral or metallic soaps and they require water rinsing for complete removal of all product residues. Some typical properties are listed in Table 3.17.

TABLE 3.14 Phosphates

MW	Common Name	Formula	pH, 1%	pH, 0.1%	P_2O_5/P, %	H_2O, %
380	TSP-12	$Na_3PO_4 \cdot 12H_2O$	11.77	11.23	18.7/8.16	52.1
182	TSP. H_2O	$Na_3PO_4 \cdot H_2O$	11.62	11.41	39.0/17.0	0
164	ANHYD. TSP	Na_3PO_4	11.75	11.43	43.3/18.9	0
120	ANHYD. MSP	NaH_2PO_4	4.66	5.15	59.2/25.8	0
142	ANHYD. DSP	Na_2HPO_4	9.05	8.16	50.0/21.8	0
368	TRIPOLY	$Na_5P_3O_{10}$	9.74	9.70	58.0/25.3	0
266	PYRO	$Na_4P_2O_7$	10.00	9.92	53.3/23.3	0
330	TKPP	$K_4P_2O_7$	10.08	9.82	43.0/18.8	0

Note: MW, molecular weight.

TABLE 3.15 Carbonates

MW	Common Name	Formula	pH, 1%	pH, 0.1%	CO_2, %	Na_2O, %
106	Light ash	Na_2CO_3	10.84	10.73	41.5	58.5
106	Dense ash	Na_2CO_3	11.05	10.82	41.5	58.5
84	Bicarbonate	$NaHCO_3$	8.13	8.19	52.5	37.0
226	Sesquecarbonate	$NaHCO_3 \cdot 2H_2O$	9.85	10.05	38.9	41.2
138	Potassium carbonate	K_2CO_3	11.15	10.73	31.9	68.1 K_2O

Note: MW, molecular weight.

TABLE 3.16 Silicates

MW	Common Name	Formula	pH, 1%	pH, 0.1%	SiO_2, %	Na_2O, %
122	Metasilicate	Na_2SiO_3	12.17	11.66	49.1	50.8
212	Metasilicate pentahydrate	$Na_2SiO_3 \cdot 5H_2O$	12.15	11.5	28.3	29.2
184	Orthosilicate	Na_4SiO_4	12.29	11.83	32.6	67.4
252	Sesquisilicate $\cdot 5H_2O$	$Na_3HSiO_4 \cdot 5H_2O$	12.12	11.61	23.8	36.9

Note: MW, molecular weight.

TABLE 3.17 Hydroxides

MW	Common Name	Formula	pH, 1%	pH, 0.1%	Na_2O, %
40	Caustic soda	NaOH	12.40	12.06	77.4
56	Caustic potash	KOH	12.45	12.02	83.9 K_2O

Note: MW, molecular weight.

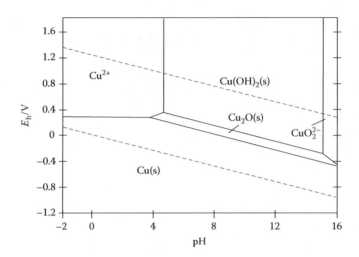

FIGURE 3.10 Pourbaix diagram.

Pourbaix diagrams, also called potential/pH diagrams, map out possible equilibrium phases of an electrochemical system. Named after Marcel Pourbaix (1904–1998), a Russian chemist who invented them, these diagrams are read like standard phase diagrams with a different set of axes. The vertical axis is labeled E_h for the voltage potential and hydrogen. The horizontal axis is labeled pH. Metal-binding agents, temperature, and concentration can shift the equilibrium lines somewhat. An example is given in Figure 3.10.

These Pourbaix diagrams are important in achieving the appropriate surface quality because they indicate the regions of immunity, corrosion, and passivity of a particular element in a given environment. In the immunity range, the element is unaffected, in the corrosion range the metal is attacked, and in the passivation range the element forms a stable coating on the surface providing a layer of protection.

Aluminum is passive in solutions having a pH less than 8.5, magnesium is passive in solutions above pH 10.5, and copper is passive in liquids in the 4–14 pH range. Cleaning agents having a pH outside the passive range could cause compatibility issues with that substrate. The alkalinity builder package can be custom-blended for cleaning compounds safe on ferrous, nonferrous, and multimetals.

Miscellaneous Additives

In addition to the abovelisted compositional choices, cleaning agents may be further customized with additions of viscosity modifiers, colorants, odorants, and surface sanitizers as desired.

Viscosity modifiers or thickening agents change the viscosity of a fluid without changing other properties. These materials provide body, increase stability, and improve suspension properties of a cleaning agent. Examples include polymers, vegetable gums, and cellulose materials.

Colorants include polar and nonpolar dyes, pigments, or other coloring agents added to a cleaning compound to provide a unified distinct appearance.

Odorants include polar and nonpolar materials having distinct aromas like essential and citrus oils, to add a distinct odor to the cleaning solution or mask the chemical odor of one or more of the constituents.

Biostats formulated into industrial cleaning agents are compounds that control microbial growth on hard surfaces, also called surface sanitizers. Examples include quaternary ammonium compounds, alcohols, oxidizers, and biocides.

Cleaning agents are a specialized classification of solutions used to remove contaminants from surfaces. They are available in concentrates as aerosols, gels, liquids, powders, and tablets; as applied, the powders and tablets are liquefied. These agents include cleansers, degreasers, detergents, abrasives, and strippers. They also include process additives like rinse aids, inhibitors, and surface modifiers to enhance bonding. Disinfectants and sanitizers control microbial activity and passivators deactivate chemically active layers. Etchants attack surfaces and descalers remove inorganic residues. All of the above chemistries are used singly or in combination to prepare a workpiece for further processing, assembly, or packaging.

Chemical Blends

The industrial manufacturer has a broad range of chemical options available when selecting the optimum industrial cleaning agent to meet his or her cleaning needs. Factors to consider are the type of equipment that will be used, the substrates to be cleaned, and waste treatment of spent fluids. Governing performance specifications and environmental regulations must also be considered. Because such a large number of component options are available, chemical blends are most frequently used in industrial cleaning processes.

Formularies listed below include the various raw materials contained in the blend and whether that material is present as a major component (>10%), minor component (1%–10%), or trace component (<1%). Published safety codes for each material are also given according to the guidelines specified by the Hazardous Materials Identification System developed by the National Paint and Coatings Association (HMIS/NPCA) and the National Fire Protection Association (NFPA). This information can be used in comparing the hazards that may or may not be associated with the use of that chemical blend. All other things being equal, lower numbers are more favorable.

Hazards Coding

Two separate systems are used to codify the hazards associated with varying chemistries. The NFPA is concerned with short-term or acute exposure concerns related to handling issues during a fire. Four colored sections are used to codify the health (blue), flammability (red), reactivity (yellow), and special concerns (white) on a scale of 0–4, where 0 has minimal hazard and 4 is very hazardous. Guidelines in assigning codes include the following:

Health (Blue)

4: Very short exposure could cause death or serious residual injury even though prompt medical attention is given.

3: Short exposure could cause death or serious residual injury even though prompt medical attention is given.

2: Intense or continued exposure could cause temporary incapacitation or possible residual injury unless prompt medical attention is given.

1: Exposure could cause irritation but only minor residual injury even if no treatment is given.

0: Exposure under fire conditions would offer no hazard beyond that of ordinary combustible materials.

Flammability (Red)

4: Will rapidly or completely vaporize at normal pressure and temperature, or is readily dispersed in air and will burn readily.

3: Liquids and solids that can be ignited under almost all ambient conditions.

2: Must be moderately heated or exposed to relatively high temperatures before ignition occurs.

1: Must be preheated before ignition will occur.

0: Material will not burn.

Reactivity (Yellow)

4: Readily capable of detonation or of explosive decomposition or reaction at normal temperatures and pressures.

3: Capable of detonation or explosive reaction, but requires a strong initiating source or must be heated under confinement before initiation, or reacts explosively with water.

2: Normally unstable and readily undergoes violent decomposition but does not detonate. Also, may react violently with water or may form potentially explosive mixture with water.

1: Normally stable but can become unstable at elevated temperatures and pressures or may react with water with some release of energy, but not violently.

0: Normally stable even under fire exposure conditions and does not react with water.

Special Hazards (White)

This section is used to denote special hazards. One most common is unusual reactivity with water. The letter W with a horizontal line through it indicates a potential hazard using water to fight a fire involving this material. Other symbols, abbreviations, or words may appear here to indicate unusual hazards.

HMIS, Health (Blue)

The health system conveys the health hazards associated with the use of the material. In the latest version of the HMIS˚ code the blue health bar has two spaces, one for an asterisk and one for a numerical rating. If present, the asterisk signifies a chronic health hazard, meaning that long-term exposure could cause a health problem. NFPA lacks this important information because the NFPA system is meant only for emergency or acute, short-term exposure. On a qualitative level, the numbering systems are more or less identical with a 0–4 scale where 0 indicates minimal hazard and 4 indicates an extreme hazard.

Flammability (Red)

According to NCPA, the criteria used to assign numeric values are identical to those used by NFPA.

Physical Hazard (Orange)

Reactivity hazards are assessed using OSHA criterion of physical hazards. Included are water reactives, organic peroxides, explosives, compressed gases, pyrophoric materials, oxidizers, and unstable reactives. The old yellow section titled "reactivity" is now obsolete. The level of hazard is also ranked on a 0–4 scale.

Personal Protection (White)

This is the largest area of difference between the NFPA and HMIS systems. In the HMIS system, this section is used to convey what personal protective equipment (PPE) should be used when handling the material.

The NFPA and HMIS codes are included for the chemical blends listed below for comparative purposes. Individual components may have one set of hazards while the compounded material may be totally different. No one formula can be used to calculate a blended NFPA/HMIS rating. This is best done by the formulator who understands all the chemical reactions that may take place in the mixture. Experience, past history, with similar compositions and actual laboratory testing may influence the ratings assigned.

Hydrocarbon Blends

Hydrocarbon-based cleaners are effective on organic soils and are multimetal safe. They may attack elastomers and seals and may be costly if concentrates are used. Most contribute to VOC and chemical oxygen demand (COD) values and have worker exposure limits. They become viable when parts or subsequent processing preclude the use of water. The cleaning process usually involves a single wash stage, and since parts dry residue-free, air drying is usually adequate.

Halogenated hydrocarbon blends include chlorinated, fluorinated, and brominated compounds. Due to the ozone-depleting properties of trichloroethane (TCA), chlorofluorocarbons (CFCs), and hydrofluorochlorofllurocarbons (HCFCs), these solvents, often collectively referred to as ozone depleting compounds (ODCs), are being replaced by solvent mixtures containing lower levels of these solvents or containing other halogenated compounds. Typically used in sealed vapor or immersion systems or manual applications, these blends are supplied as liquids or aerosols.

- Formula HC-1 is a chlorinated solvent blend that was broadly used in vapor degreasing applications until the mid-1990s when production was banned in the United States and most first world countries. Due to its ODP, it has subsequently been phased out and replaced with alternative materials including trichloroethylene (TCE), *n*-propyl bromide (*n*PB), and fluorinated solvents. When in use, it was extremely effective on organic soils including oils, greases, and waxes encountered in industrial manufacturing processes. This blend dissolves nonpolar soils, is multimetal safe, has no flash point, and dries residue free.

  Although rarely used today, TCA is included here for comparative purposes.

Component	HMIS Code	NFPA Code	Formula HC-1 Concentration
Trichloroethane	3-1-0	2-1-0	Major
Stabilizers and inhibitors	2-1-0	1-0-0	Trace

- Formula HC-2 is a chlorinated blend that has been used to replace TCA in some vapor-degreasing applications. This solvent blend is effective at removing reactive soils including paints and adhesives while solubilizing nonpolar oils and grease. This mixture is multimetal safe, has no flash point, and dries residue free. It is, however, classified as a potential carcinogen and must be used in tightly sealed equipment to minimize health risk.

Component	HMIS Code	NFPA Code	Formula HC-2
Methylene chloride	2-1-0	2-1-0	Major
Stabilizers and inhibitors	1-0-0	1-0-0	Trace

- Formula HC-3 is another chlorinated solvent material widely used as an alternative for TCA and CFC in vapor degreasing equipment. This mixture dissolves nonpolar soils and is especially effective on mineral oils and greases; it is nonflammable and dries residue-free.

Component	HMIS Code	NFPA Code	Formula HC-3
Trichloroethylene	2-1-1	2-1-0	Major
Stabilizers and inhibitors	1-0-0	1-0-0	Trace

- Formula HC-4 is the preferred alternative solvent used in dry cleaning applications. It also is chlorinated and when used in industrial applications, it dissolves nonpolar soils, evaporates completely, and will not support a flame.

Component	HMIS Code	NFPA Code	Formula HC-4
Perchloroethylene (Perc)	2-0-0	2-0-0	Major
Stabilizers and inhibitors	1-0-0	1-0-0	Trace

- Formula HC-5 contains *n*-propyl bromide. It is a cost-effective alternative for TCA and Freon 113 and the similar Kb value makes it effective on a similar range of soils. Safe to use on ferrous, aluminum, copper, nickel, titanium, glass, and ceramics, this formulation is self-extinguishing and dries completely.

  Brominated solvents are very effective at removing oils, grease, flux, shop dirt, and wax using existing vapor degreasing systems however low worker exposure limits make these materials less desirable for immersion and manual applications.

Component	HMIS Code	NFPA Code	Formula HC-5
n-Propyl bromide	2-1-0	2-1-0	Major
Alcohol	2-3-2	1-3-0	Minor
Epoxybutane	3-3-2	3-3-2	Trace

- Formulas HC-6 and HC-7 contain fluorinated hydrocarbon solvents. These materials are the most costly of all the halogenated solvent cleaning agents. With an extremely low surface tension, these blends readily penetrate tight spaces and clean by wetting, displacement, and dissolution. They are multimetal-safe, nonflammable, and fast evaporating, making them ideal for vapor degreasing or manual applications requiring clean and dry surfaces. These mixtures have a lower health rating than other halogenated solvent making them safer to use. Since the vapors can displace oxygen, adequate ventilation is essential when working with these products.

Component	HMIS Code	NFPA Code	Formula HC-6
2,3-Dihydroperfluoropentane	1-0-1	1-0-1	Major
Inhibitors and stabilizers	1-0-0	1-0-0	Minor

Component	HMIS Code	NFPA Codes	Formula HC-7
Trans, 1,2-Dichloroethylene	2-3-2	2-3-2	Major
1,1,1,2,2,3,4,5,5,5-Decafluoropentane	1-0-1	1-0-1	Major

Table 3.18 displays the chemical and physical properties of the above halogenated hydrocarbon blends. A lower boiling point material requires less heat to reach the vapor phase in vapor and vacuum systems. Those blends having higher vapor pressure evaporate most rapidly.

All of the blends in Table 3.18 are self-extinguishing and do not support a flame, making them ideally suited for use in vapor and vacuum systems where the solvents are used at or above the boiling point.

When not in use, all containers of these products should be tightly sealed to minimize evaporation. Since vapors will displace oxygen, adequate ventilation should be used when storing, transporting, transferring, and using these materials.

Nonhalogenated hydrocarbon cleaning agents are solvent mixtures that are free of all halogenated materials including chlorine, fluorine, or bromine. They are selected for use where water is ineffective or cannot be used and where complete evaporation is essential for optimum part efficiency and performance. These cleaning agents are effective at removing a broad spectrum of polar and nonpolar soils

TABLE 3.18 Physical and Chemical Properties of Halogenated Hydrocarbon Blends

Typical Properties	Formula HC-1	Formula HC-2	Formula HC-3	Formula HC-4	Formula HC-5	Formula HC-6	Formula HC-7
Boiling point, °C	74.1	39.8	87	122	68	55	39
Specific gravity at 20°C	1.340	1.318	1.470	1.619	1.250	1.580	1.410
Evaporation rate, BuAc = 1	1.9	27.5	4.46	2.1	>1	Not available	Not available
Vapor pressure, mmHg at 20°C	100	350	57.8	13	139	226	464
Volatile organic compound (VOC), g/L	1340	1318	1470	1619	1250	1580	528
Kauri-butanol value	124	136	129	92	>15	Not available	Not available
Surface tension, dynes/cm	25.4	28.1	32	32	24.2	14.1	15.2
Flash point, °C	None[a]	None[a]	None[a]	None[a]	None[a]	None[a]	None[a]
Water solubility	No	No	No	No	No	No	No
HMIS code	3-1-1	2-1-0	2-1-1	2-0-0	2-1-0	1-0-1	1-1-1
NFPA code	2-1-0	2-1-0	2-1-0	2-0-0	2-1-0	1-0-1	2-1-2

[a] Self-extinguishing.

from virtually all types of metal, glass, and ceramics. They are used in vapor cleaning, vacuum degreasing, and immersion cleaning operations. Parts emerge clean and dry; no separate rinsing and drying stages are typically needed.

- Formula NH_1 is a blend of aromatic and aliphatic solvents specially designed for removal of oils, grease, ink, and adhesives using immersion or manual methods. Soils are typically dissolved in the cleaning bath. A spent solution can be reclaimed by distillation or it may be used as an alternate fuel. This product is combustible and mildly irritating to eyes, skin, and mucous membranes.

Component	HMIS Code	NFPA Code	Formula NH_1
Mineral spirits	1-2-0	1-2-0	Major
Stoddard solvent	1-2-0	1-2-0	Major
Aromatic naptha	1-2-0	1-2-0	Minor
Benzene, 1,2,4-trimethyl	3-4-0	3-4-0	Minor

- Formula NH_2 is an aromatic, oxygenated, and aliphatic blend of solvents. This mixture is designed for all-purpose cleaning of reactive and nonreactive soils including inks, adhesives, oil, and grease from multimetals, glass, ceramics, and composites. Soils are dissolved or dispersed and spent fluids can be reclaimed by distillation or used as an alternate fuel. This product is combustible and mildly irritating. It should be used only in well ventilated areas and eye protection should be worn.

Component	HMIS Code	NFPA Code	Formula NH_2
Aromatic naptha	1-2-0	1-2-0	Major
Stoddard solvent	1-2-0	1-2-0	Major
2-Butoxyethanol	2-2-1	2-2-0	Minor
Methyl acetate	2-3-1	1-3-0	Minor

- Formula NH_3 contains aromatic, aliphatic, and oxygenated solvents for the removal of a broad spectrum of reactive soils including ink, glue, adhesive, elastomers, and rubbers. Soils will be

TABLE 3.19   Physical and Chemical Properties of Nonhalogenated Hydrocarbon Blends

Typical Properties	Formula NH1	Formula NH2	Formula NH3
Boiling point, °C	157	>145	100
Specific gravity at 20°C	0.800	0.850	0.910
Evaporation rate, BuAc = 1	Not available	Not available	Not available
Vapor pressure, mmHg at 20°C	2	<1	Not available
Volatile organic compound (VOC), g/L	800	825	910
Kauri-butanol value	33	Not available	Not available
Surface tension, dynes/cm	32	29	23
Flash Point, °C	41	46	>42
Water solubility	Negligible	Negligible	Negligible
HMIS code	1-2-0	2-2-0	2-2-0
NFPA code	1-2-0	1-2-0	2-2-0

dissolved, dispersed, swelled, or displaced. The mixture has a low surface tension for good penetration in press-fit areas. It is combustible, mildly irritating, and reclaimable by distillation.

Component	HMIS Code	NFPA Code	Formula NH$_3$
Aromatic naptha	1-2-0	1-2-0	Major
Benzene, 1,2,4-trimethyl	3-4-0	3-4-0	Major
2-Butoxyethanol	2-2-1	2-2-0	Minor
Xylene	2-3-0	2-3-0	Minor

Table 3.19 depicts the chemical and physical properties of the nonhalogenated hydrocarbon blends listed above. When comparing the data, it is evident that these properties do not differ significantly even though individual components may vary considerably. As a whole, the total chemical mixture must be considered and the supplier's suggested use recommendations should be followed exactly.

None of the hydrocarbon solutions in Table 3.19 contain water, nor will they mix with water. All are combustible, evaporate completely, and are classified as reportable VOCs. They are effective at removing nonpolar and some polar soils including oils, greases, fingerprints, glues, and adhesives. All of these compositions are multimetal safe and can be reclaimed by distillation. Spent mixtures may be used as alternative fuels.

Aqueous hydrocarbon blends contain water, are diluted further with water, or are designed to be rinsed with water. Some products completely evaporate while others rely on a water rinse or hand wipe for complete removal of all soils and cleaner residues.

- Formula AH 1 is used as a concentrate or diluted with water as desired. It dissolves polar and some nonpolar soils using immersion, spray-under-immersion, or manual methods of application. The flash point is high and rinsing is required to remove all traces of soils and cleaner residues. This mixture is safe to use on virtually all metals, glass, and ceramic substrates.

Component	HMIS Code	NFPA Code	Formula AH 1
Glycol ether DPM	1-2-0	0-2-0	Major
n-Methyl pyrrolidone	2-2-0	2-2-0	Major
Glycol ether TPM	1-1-0	1-1-0	Major
Nonionic surfactant	1-0-0	1-0-0	Trace
Anionic surfactant	1-0-0	1-0-0	Trace

- Formula AH 2 is especially effective at dissolving asphalt, tar, pitch, adhesive, wax, and ink. It evaporates completely and when mixed with water effectively removes polar soils as well. Safe to use on ferrous and nonferrous metals, glass, and ceramics, it is designed for immersion and manual applications. A water rinse is optional.

This cleaning agent has a health rating of 2 and may be irritating to eyes, skin, and mucous membranes. It is combustible and care should be taken to insure that adequate ventilation is available when handling or using this product.

Component	HMIS Code	NFPA Code	Formula AH 2
Aromatic naptha	1-2-0	1-2-0	Major
2-Butoxyethanol	2-2-1	2-2-0	Major
Toluene	2-3-1	2-3-0	Major
Nonionic surfactant	1-0-0	1-0-0	Minor

- Formula AH 3 cleans polar and nonpolar soils by dissolution, saponification, and wetting. A water rinse is required to remove all residual materials left behind. Designed for use in immersion, spray-under-immersion, or manual methods, this mixture is very cost effective when used as a water dilution.

  This product is safe to use on ferrous, nickel, copper, titanium, aluminum, gold, silver, platinum, glass, and ceramic surfaces for the removal of flux, paste, oils, fingerprints, and shop dirt encountered in the manufacturing process. With a health rating of 1, this product may be mildly irritating; eye protection should be worn when handling industrial grade chemicals.

Component	HMIS Code	NFPA Code	Formula AH 3
Water	0-0-0	0-0-0	Major
D-Limonene	1-2-0	1-2-0	Major
Diethanolamine	1-1-0	1-1-0	Minor
Anionic surfactant	1-0-0	1-0-0	Minor
Isopropanol	2-3-0	1-3-0	Minor
Tetrasodium EDTA	2-0-0	2-0-0	Minor

- Formula AH 4 is classified as a semiaqueous or emulsion cleaning agent. Using immersion or manual methods, this mixture is applied as a concentrate and rinsed with water for effective removal of oils and flux residues from multimetals, glass, and ceramic substrates. An inhibitor may be added to the rinse to control oxidation of the cleaned metal surfaces as needed. This mixture may be mildly irritating.

Component	HMIS Code	NFPA Code	Formula AH 4
Stoddard solvent	1-2-0	1-2-0	Major
Dibasic ester blend	1-1-0	1-1-0	Major
Nonionic surfactant	1-0-0	1-0-0	Minor

- Formula AH 5 is a noncombustible solvent blend effective at dissolving and displacing polar soils using spray, immersion, or manual methods. A water rinse is required. With a health rating of 1, this mixture is mildly irritating and standard eye protection should be worn during handling and use.

Component	HMIS Code	NFPA Code	Formula AH 5
Methyl esters	1-1-0	1-1-0	Major
Glycol ether DPnP	1-2-0	0-2-0	Major
Nonionic surfactant	1-0-0	1-0-0	Minor
Anionic surfactant	1-0-0	1-0-0	Minor

TABLE 3.20   Chemical and Physical Properties of Aqueous Hydrocarbon Blends

Typical Properties	Formula AH 1	Formula AH 2	Formula AH 3	Formula AH 4	Formula AH 5
Boiling point, °C	120	100	107	221	265
Specific gravity at 20°C	1.030	0.910	0.990	0.850	0.910
Evaporation rate, BuAc = 1	Not available	Not available	Not available	Not available	<1
Vapor pressure, mmHg at 20°C	3	Not available	17.5	<0.1	<1
Volatile organic compound (VOC), g/L	1000	844	500	807	864
Kauri-butanol value	Not available	Not available	Not available	64	Not available
Surface tension, dynes/cm	28	28	32	33	30
Flash point, °C	>100	41	71	96	121
Water solubility	Complete	Moderate	Complete	Slight	Emulsifies
HMIS code	1-1-0	2-2-0	1-1-0	1-1-0	1-0-0
NFPA code	1-1-0	2-2-0	1-1-0	1-1-0	1-0-0

The chemical and physical properties of these aqueous hydrocarbon blends are compared in Table 3.20. Those mixtures having a flash point below 100°C are classified as combustible and care must be taken to insure that adequate ventilation is available when using them.

Each of the above aqueous hydrocarbon formulations incorporate water in the cleaning process, either as a component of the cleaning agent, as a diluent or as a rinsing agent. Local POTWs will dictate whether the spent solutions can be sewered; large volumes of heavily contaminated fluids may require disposal by a licensed waste hauler. Evaporators may be used to reduce effluent volume. Redistillation is not a viable means of treating these spent mixtures.

Aqueous Blends

Aqueous cleaning agents contain a variety of building blocks including alkalinity builders, surface active agents, and corrosion inhibitors. The formulator selects one or more of each building block to customize a product to meet specific cleaning requirements. All components must be compatible with all exposed substrates and the chemistry selected must have the appropriate chemical and physical properties so it will work within the limitations of the equipment selected. All regulatory requirements must be met while staying within the constraints of waste management.

Formulations listed here are generic compositions used to illustrate the impact differing components have on the chemical and physical properties of a cleaning agent. They may be used as starting points for customized formulations designed to meet specific cleaning applications. Raw material choices are vast; each experienced formulator will have a preferred list of materials under each building block category.

Acid aqueous blends typically contain organic and/or mineral acids, chelators, and surfactants in a water base. They have a pH ranging from 0 to 5 depending on the strength of the acids used; organics are less aggressive than mineral acids and have a more neutral pH.

- Acid 1 contains organic and mineral acids and is used to descale and brighten metal substrates using spray, immersion, or manual methods. Light oils are removed by wetting or emulsification while metal oxides and scale are dissolved or chemically digested. A water rinse is needed to remove all cleaner residues and an inhibitor may be added to the rinse to prevent reoxidation.

   This mixture has a health rating of 2; it is irritating to eyes, skin, and mucous membranes. Protective eye and skin protection should be worn when handling this product.

Component	HMIS Code	NFPA Code	Acid 1
Water	0-0-0	0-0-0	Major
Hydroxyacetic acid	3-1-0	3-1-0	Major
Phosphoric acid	3-0-0	3-0-0	Major
Anionic surfactant	1-0-0	1-0-0	Trace

- Acid 2 is a strong, heavy duty mineral acid blend specially designed to chemically digest carbonaceous soils and scale. This acid blend is highly corrosive and must be limited to use in 316 stainless or polypropylene containers. Proper handling practices should be in place and recommended protective apparel should be worn at all times when handling or using this material. An inhibited water rinse will remove all cleaner residues while neutralizing the substrate surface and controlling reoxidation.

Component	HMIS Code	NFPA Code	Acid 2
Water	0-0-0	0-0-0	Major
Sulfuric acid	3-0-2	3-0-2	Major
Sulfamic acid	3-0-0	3-0-0	Minor
Ammonium bifluoride	4-0-2	3-0-0	Minor
Nonionic surfactant	1-0-0	1-0-0	Minor

- Acid 3 is designed to chelate inorganic soils and carbonaceous materials from ferrous and nonferrous metals, glass, and ceramics. The surfactant present in this composition will promote wetting while emulsifying light oils and fingerprints. The chelating properties of this mixture make it especially effective on mineral scale and carbonate build-up. Since it is low foaming, it can be applied by spray, immersion, or manual methods. A water rinse is required to remove all product residues.

  This strong acid mixture is corrosive and will burn human tissue; eyes and skin should be protected during the handling and use of this chemical mixture.

Component	HMIS Code	NFPA Code	Acid 3
Water	0-0-0	0-0-0	Major
Phosphoric acid	3-0-0	3-0-0	Major
Gluconic acid	2-0-0	2-0-0	Minor
Sulfamic acid	3-0-0	3-0-0	Minor
Nonionic surfactant	1-0-0	1-0-0	Minor

- Acid 4 is a chromic acid blend that may be incompatible with some waste treatment processes. Effective on a broad spectrum of soils including scale and carbonaceous residues, it provides excellent corrosion protection but may be restricted due to environmental concerns. This mixture is moderate foaming and works effectively in immersion and manual processes.

  Depending on the concentration used, this mixture can be used on ferrous, copper, titanium, nickel, and some aluminum alloys. Water rinsing is needed to completely remove all soils and cleaner residues.

  This is a strong corrosive mixture that can cause severe burns upon contact with eyes, skin, or mucous membranes. Protective apparel should be worn whenever handling or using this product.

Component	HMIS Code	NFPA Code	Acid 4
Water	0-0-0	0-0-0	Major
Sulfuric acid	3-0-2	3-0-2	Minor
Nonionic surfactant	1-0-0	1-0-0	Trace
Sodium dichromate	3-0-2	3-1-1	Trace

- Acid 5 is a mild product designed to impart a conversion coating on ferrous and aluminum substrates. It is effective on light oils and may eliminate the need for a preclean step prior to coating. This product may be applied using spray, immersion, and manual methods for improved paint adhesion. A water rinse may be optional. With a health rating of 2, this product may be irritating to eyes, skin, and mucous membranes. Eye protection and gloves should be worn when using this product.

Component	HMIS Code	NFPA Code	Acid 5
Water	0-0-0	0-0-0	Major
Phosphoric acid	3-0-0	3-0-0	Major
Monoethanolamine	3-1-0	3-1-0	Minor
Nonionic surfactant	1-0-0	1-0-0	Minor
Anionic surfactant	1-0-0	1-0-0	Trace
Hydrofluoric acid	4-0-2	4-0-1	Trace

The chemical and physical properties of these acid mixtures are listed in Table 3.21. All are nonflammable and as a result of water dilution, the recommended health rating is the same.

The above cleaning agents are effective on a wide range of inorganic soils including metal oxides, scale, and carbon. When formulated with surfactants, acids may also be effective on light oil films. They clean by wetting, dissolution, or chemical digestion. All of these acid solutions will leave nonvolatile residues behind if not followed by pure water or inhibited water rinse steps as part of the cleaning process.

All of the acid mixtures listed here are classified as corrosive. They may cause burns to eyes and skin and may cause irreparable damage if ingested or inhaled. Use these products with adequate ventilation and wear protective apparel when handling these materials.

Neutral aqueous solutions are liquids that have a pH ranging from 5 to 9 and typically contain surfactants, conditioners, and chelating agents in a water base. Because these compositions tend to be mild, they are frequently chosen for use in manual applications where there may be human contact.

Depending on the surfactants used in the formula, neutral pH products may be used in spray, immersion, and manual processes. They are multimetal safe and can be used on most glass, ceramic, and plastic surfaces. A water rinse is optional.

- Formula NAq 1 is a very mild alkaline liquid containing surfactants and low levels of glycol and alkalinity builder for the removal of light oils, grease, and fingerprints. It is a moderate foaming blend, making it ideal for immersion and ultrasonic applications. It is too foamy for mechanical spray processes. A water rinse is needed to remove all cleaner residues.

TABLE 3.21 Physical and Chemical Properties of Aqueous Acid Blends

Typical Properties	Acid 1	Acid 2	Acid 3	Acid 4	Acid 5
Boiling point, °C	110	>100	>100	110	>100
Specific gravity at 20°C	1.060	1.145	1.210	1.840	1.215
pH at 10%	2.2	1.9	<1	<1	3.2
Vapor pressure, mmHg at 20°C	17.5	17.5	17.5	17	17
Volatile organic compound (VOC), g/L	0.0	0.0	0.0	0.0	0.0
Surface tension, dynes/cm	32	32	30	36	29
Flash point, °C	>110	>100	>100	>110	>100
Water solubility	Complete	Complete	Complete	Complete	Complete
HMIS code	2-0-0	3-0-0	3-0-0	3-0-0	2-0-0
NFPA code	2-0-0	3-0-0	3-0-0	3-0-0	2-0-0

Component	HMIS Code	NFPA Code	NAq 1
Water	0-0-0	0-0-0	Major
Glycol ether DPM	1-2-0	0-2-0	Minor
Anionic surfactant	1-0-0	1-0-0	Minor
Nonionic surfactant	1-0-0	1-0-0	Minor
Trisodium phosphate (TSP)	2-0-0	2-0-0	Minor

- Formula NAq 2 is a nonphosphate mixture. This composition cleans by solubilization and wetting; it cannot saponify fatty acid soils or chelate inorganic oxides or salts. This mixture is moderate to high foaming and is ideal for immersion cleaning processes. Care should be taken to minimize aeration. Parts and equipment should be water rinsed on a routine and regular basis to minimize scale build-up. When needed, a mild acid descaler should be used to keep equipment in good operating order.

Component	HMIS Code	NFPA Code	NAq 2
Water	0-0-0	0-0-0	Major
Anionic surfactant	1-0-0	1-0-0	0%–5%
Nonionic surfactant	1-0-0	1-0-0	0%–5%
Sodium carbonate	2-0-1	2-0-1	0%–5%

- Formula NAq 3 is a light duty composition containing surfactant and alcohol for removal of light oils, shop dirt, and fingerprints from precision parts and delicate surfaces. It can be applied as a concentrate or diluted with water; rinsing is optional. Since it is low foaming, it can be used in spray, immersion, or manual applications.

Component	HMIS Code	NFPA Code	NAq 3
Water	0-0-0	0-0-0	Major
2-Butoxyethanol	2-2-1	2-2-0	Major
Nonionic surfactant	1-0-0	1-0-0	Minor
Tetrasodium EDTA	2-0-0	2-0-0	Minor

- Formula NAq 4 is a low-foaming blend of surfactant and chelators. It is specially designed for spray applications. A water rinse is suggested for removal of all product solids.

Component	HMIS Code	NFPA Code	NAq 4
Water	0-0-0	0-0-0	Major
Citric acid	2-1-0	2-1-0	Minor
Nonionic surfactant	1-0-0	1-0-0	Minor
Sodium gluconate	1-0-0	1-0-0	Minor
Tetrasodium EDTA	2-0-0	2-0-0	Minor
Trisodium phosphate	2-0-0	2-0-0	Minor

The performance and physical properties of the above aqueous neutral blends are listed in Table 3.22. Water is a major component of each of these mixtures and all are nonflammable. With a health rating of 1, all may be mildly irritating to eyes, skin, and mucous membranes. Although eye protection is suggested whenever using industrial chemicals, no special apparel is needed when working with the above products.

All of these neutral aqueous blends contain nonvolatile materials that may require rinsing. Typically diluted further with water these mixtures are economical, multimetal safe, and effective on a wide

TABLE 3.22 Physical and Chemical Properties of Aqueous Neutral Blends

Typical Properties	NAq 1	NAq 2	NAq 3	NAq 4
Boiling point, °C	>100	>100	>100	100
Specific gravity at 20°C	1.060	1.040	0.995	1.000
pH at 10%	9.6	9.4	8.8	8.2
Vapor pressure, mmHg at 20°C	17.5	17.5	17	18
Volatile organic compound (VOC), g/L	75	0.0	100	0.0
Surface tension, dynes/cm	30	29	30	29
Flash point, °C	>100	>100	>100	>100
Water solubility	Complete	Complete	Complete	Complete
HMIS code	1-0-0	1-0-0	1-0-0	1-0-0
NFPA code	1-0-0	1-0-0	1-0-0	1-0-0

variety of soils encountered during workpiece manufacture. The mild pH makes them preferred choices for manual cleaning applications where direct worker exposure is a concern.

Alkaline aqueous blends contain water in the concentrate or as a diluent. They have a pH greater than 9, have low or no VOC, and since there is a wide variety to choose from, a formula can be selected to meet unique cleaning needs.

High alkaline cleaning agents may contain alkalinity builders, chelating agents, inhibitors, and surfactants; some also contain low levels of free caustic alkalis. Depending on the composition and pH, some blends may have alloy compatibility issues. Ferrous alloys can tolerate caustic environments. However, since iron is highly reactive, flash rusting may occur if all wash and rinse solutions are not inhibited. Copper-containing alloys can tolerate caustic solutions but may tarnish if exposed to amines and high temperatures. Aluminum alloys, especially 2000 and 7000 series, are attacked by caustic solutions. The chemical supplier should specify what substrates are compatible with the cleaning agent identified.

All of the following mixtures may be mildly irritating to eyes, skin, and mucous membranes. No special protective apparel is required, however it is good practice to always wear eye protection whenever handling industrial chemicals.

- Formula AAq 1 is a multimetal-safe all-purpose cleaner especially effective on oils and greases encountered during the production process. Soils are removed by dissolution, displacement, dispersion, saponification, or wetting. The moderate foaming surfactant package makes it ideal for immersion and ultrasonic applications. An abundant rinse is needed to prevent a silicate bloom from forming on part surfaces. This is a very mild product and no special handling requirements are needed.

Component	HMIS Code	NFPA Code	AAq 1
Water	0-0-0	0-0-0	Major
Anionic surfactant	1-0-0	1-0-0	Major
Nonionic surfactant	1-0-0	1-0-0	Minor
Gluconic acid	2-0-0	2-0-0	Minor
Sodium metasilicate	3-0-0	3-0-0	Minor
Sodium hydroxide	3-0-0	3-0-0	Trace

- Formula AAq 2 is a heavy-duty composition effective on most oils, greases, and light metal oxides. It is low foaming and can be applied in immersion and spray systems. This formulation will tolerate hard water, but requires a thorough water rinse for complete removal of all soils and cleaner residues.

Component	HMIS Code	NFPA Code	AAq 2
Water	0-0-0	0-0-0	Major
2-Butoxyethanol	2-2-1	2-2-0	Minor
Anionic surfactant	1-0-0	1-0-0	Minor
Nonionic surfactant	1-0-0	1-0-0	Minor
Sodium gluconate	1-0-0	1-0-0	Minor
Sodium hydroxide	3-0-0	3-0-0	Minor
Tetrasodium EDTA	2-0-0	2-0-0	Minor
Tolytriazole	2-0-0	2-0-0	Minor

- Formula AAq 3 is multimetal safe and heavy duty with good soil-holding properties for long bath life. This mixture is low foaming and may be used in spray and immersion systems. It is hard water–tolerant and effective on a broad spectrum of soils including oils, grease, wax, fingerprints, shop dirt, light carbonaceous soils, and metal oxides. Depending on the application involved, a water rinse may be needed for complete removal of product residues.

Component	HMIS Code	NFPA Code	AAq 3
Water	0-0-0	0-0-0	Major
Nonionic surfactant	1-0-0	1-0-0	Minor
Sodium gluconate	1-0-0	1-0-0	Minor
Citric acid	2-1-0	2-1-0	Minor
Tetrasodium EDTA	2-0-0	2-0-0	Minor
TKPP	2-0-2	2-0-2	Minor
Triethanolamine	2-1-0	2-1-0	Minor

- Formula AAq 4 is a mild chelated cleaner for ferrous and copper alloys. This mixture is low foaming and may be used in spray and immersion applications. A water rinse is optional. The low level of caustic may negatively impact some aluminum alloys. It is mildly irritating and care should be taken to insure all recommended personal protection practices are being followed. Eye protection should be worn whenever handling industrial chemicals.

Component	HMIS Code	NFPA Code	AAq 4
Water	0-0-0	0-0-0	Major
Anionic surfactant	1-0-0	1-0-0	Major
Nonionic surfactant	1-0-0	1-0-0	Minor
Sodium gluconate	1-0-0	1-0-0	Minor
Potassium hydroxide	3-0-0	3-0-0	Minor
Tolytriazole	2-0-0	2-0-0	Minor
Diethanolamine	1-1-0	1-1-0	Minor

- Formula AAq 5 is a highly silicated caustic all-purpose cleaner for ferrous and copper alloys. It can be used in spray and immersion applications for the removal of shop dirt, metal fines, oils, grease, light metal oxides, and wax. An abundant water rinse is needed to prevent white residues.

  This mixture has a 2 health rating; protective eye and skin protection should be worn for optimum operator safety. Use only in well-ventilated areas and follow standard industrial practices when handling this or any other industrial chemical.

Component	HMIS Code	NFPA Code	AAq 5
Water	0-0-0	0-0-0	Major
Sodium metasilicate	3-0-0	3-0-0	Major
Nonionic surfactant	1-0-0	1-0-0	Minor
Potassium hydroxide	3-0-0	3-0-0	Minor
Sodium hydroxide	3-0-0	3-0-0	Minor
Tetrasodium EDTA	2-0-0	2-0-0	Minor
Tolytriazole	2-0-0	2-0-0	Minor

- Formula AAq 6 is a highly chelated all-purpose cleaner specially designed for removal of carbonaceous soils and oxides using spray or immersion applications. This product is recommended for use on ferrous and copper alloys. The low levels of caustic may adversely affect some aluminum substrates.

  This mixture may be irritating to eyes, skin, and mucous membranes; protective apparel should be worn to prevent direct contact.

Component	HMIS Code	NFPA Code	AAq 6
Water	0-0-0	0-0-0	Major
Tetrasodium EDTA	2-0-0	2-0-0	Major
Nonionic surfactant	1-0-0	1-0-0	Minor
Potassium hydroxide	3-0-0	3-0-0	Minor
Sodium metasilicate	3-0-0	3-0-0	Minor
2-Butoxyethanol	2-2-1	2-2-0	Minor
Tolytriazole	2-0-0	2-0-0	Minor

Table 3.23 lists the physical and chemical data on the above aqueous alkaline blends. Again water is a major component of all of these mixtures. They are all nonflammable and have low or no VOC. With a health rating of 2 or less, these products may be irritating or mildly irritating. Protective eyewear is suggested and standard handling practices should be followed when using these industrial cleaning agents.

Caustic solutions are very effective on oils, greases, carbonized soils, and light rust. They are generally safe for use on ferrous, copper, titanium, nickel, and magnesium substrates. With a health rating of 3, these solutions are corrosive and will burn skin so they should be restricted for use in industrial cleaning equipment; they are not recommended for use in manual applications where there may be operator contact. The more mild alkaline solutions should be used in manual applications where the agent may come in contact with skin.

TABLE 3.23 Physical and Chemical Properties of Aqueous Alkaline Blends

Typical Properties	AAq 1	AAq 2	AAq 3	AAq 4	AAq 5	AAq 6
Boiling point, °C	>100	>100	100	100	100	100
Specific gravity at 20°C	1.040	1.125	1.000	1.000	1.140	1.008
pH at 10%	12.3	12.5	10.8	13.0	12.8	11.5
Vapor pressure, mmHg at 20°C	>17	>17	>17	24	17.5	17.5
Volatile organic compound (VOC), g/L	0.0	75	<25	<25	<25	42
Surface tension, dynes/cm	32	29	30	30	29	30
Flash point, °C	>100	>100	>100	>100	>100	>100
Water solubility	Complete	Complete	Complete	Complete	Complete	Complete
HMIS code	1-0-0	1-0-0	1-0-0	2-0-0	2-0-0	2-0-0
NFPA code	1-0-0	1-0-0	1-0-0	2-0-0	2-0-0	2-0-0

- Formula CCA 1 is a moderate strength low-foaming caustic cleaner specially designed for use on ferrous, titanium, nickel, and copper alloys. It also can be used to lightly etch aluminum. This blend contains a conditioner for hard water control and a low level of surfactant to aid in surface wetting. An inhibited rinse should be used with ferrous alloys to avoid flash rusting.

  This mixture is very cost effective at removing general production soils from industrial parts and assemblies. This product is corrosive; wear eye and skin protection when handling or using this cleaning agent.

Component	HMIS Code	NFPA Code	CCA 1
Water	0-0-0	0-0-0	Major
Sodium hydroxide	3-0-0	3-0-0	Major
Nonionic surfactant	1-0-0	1-0-0	Minor
Tetrasodium EDTA	2-0-0	2-0-0	Minor

- Formula CCA 2 is a low-foaming chelated caustic heavy duty cleaning agent for spray applications. It is effective at removing carbonized oils, grease, metal oxides and scale from cast iron, steel, copper, brass, nickel, and titanium. An inhibited rinse will remove all soil and cleaner residues while preventing flash rusting of ferrous substrates. This product will etch aluminum and will cause burns upon direct contact with skin, eyes, and mucous membranes. Protective apparel should be worn.

Component	HMIS Code	NFPA Code	CCA 2
Water	0-0-0	0-0-0	Major
Sodium hydroxide	3-0-0	3-0-0	Major
Sodium gluconate	1-0-0	1-0-0	Minor
Anionic surfactant	1-0-0	1-0-0	Minor
Nonionic surfactant	1-0-0	1-0-0	Minor

- Formula CCA 3 is a low-foaming caustic all-purpose cleaner containing phosphates. An inhibited water rinse will readily remove cleaner residues. It also will attack aluminum and cause burns.

Component	HMIS Code	NFPA Code	CCA 3
Water	0-0-0	0-0-0	Major
Sodium hydroxide	3-0-0	3-0-0	Major
Trisodium phosphate	2-0-0	2-0-0	Major
Nonionic surfactant	1-0-0	1-0-0	Minor

The physical and chemical properties of the above caustic alkali mixtures are listed in Table 3.24. Water is a major component of these blends and all have low or no VOC. High caustic levels cause each of these products to have a health rating of 3. They are classified as corrosive and extreme care should be taken when using these materials. Face protection, gloves, and aprons should be worn whenever handling caustic mixtures.

The above are just a small sampling of the commercial cleaning agents available to the industrial manufacturer. Once a chemistry partner has been identified, chemistry selection is based on the equipment, substrates, and soils involved. Spray in air systems require low-foaming chemistries and aluminum substrates cannot be exposed to caustic alkalis without metal attack. Light oils require less aggressive chemistry than do heavy carbonized greases. Fresh soils may be easier to remove than aged residues encountered in remanufacturing processes. Chemical suppliers should work with equipment manufacturers to assist in selecting the optimum process to meet specific cleaning needs.

TABLE 3.24   Physical and Chemical Properties of Aqueous Caustic Blends

Typical Properties	CCA 1	CCA 2	CCA 3
Boiling point, °C	>100	>100	>100
Specific gravity at 20°C	1.247	1.150	1.380
pH at 10%	13.1	14	14
Vapor pressure, mmHg at 20°C	>17	>17	>17
Volatile organic compound (VOC), g/L	<25	<25	<25
Surface tension, dynes/cm	33	33	32
Flash point, °C	>100	>100	>100
Water solubility	Complete	Complete	Complete
HMIS code	3-0-0	3-0-0	3-0-0
NFPA code	3-0-0	3-0-0	3-0-0

Corrosion Inhibitors

Corrosion inhibitors may need to be considered as part of the cleaning process. Inhibitors are a class of chemicals used to inhibit oxidation of metal surfaces. These inhibitors may be designed to provide storage protection indoors or outdoors. Some are used to provide in-process short-term protection during the manufacturing process not to exceed 15–30 days. Others are designed to provide long-term indoor protection up to 6 months. Still others are designed to provide outdoor protection against the elements for 6 months or more.

Inhibitors may be grouped as water-soluble, water-displacing, or emulsion chemistries. The most practical inhibitor chemistry provides indoor inhibition up to 6 months. These chemistries are compatible with aqueous cleaners and often the inhibitor package is formulated directly into the cleaner composition. The amount of inhibitor present in the bath dictates the length of protection expected. Examples are listed in Table 3.25.

All of the inhibitors in Table 3.25 can be applied by immersion or manual methods. The first three also can be applied by spray processes as a final rinse solution or added directly to the wash stage for

TABLE 3.25   Corrosion Inhibitor Compositions

Composition	Inhibitor Type	Protection	Alloys Protected
Water 2-aminoethanol Amine carboxylate	Water soluble	Short term	Ferrous
Water Amine carboxylate Tolytriazole	Water soluble	Short term	Ferrous, copper, brass
Water Amine carboxylate Tolytriazole Sodium silicate Sodium hydroxide	Water soluble	6 months indoors	Ferrous, copper, brass, aluminum
Water Benzotriazole Anionic surfactant Calcium sulfonate	Emulsifiable	6 months indoors	Ferrous, copper, aluminum
Mineral spirits Calcium sulfonate	Water displacing	6 months indoors	Ferrous, copper, aluminum
Cosmolene®	Water displacing	6 months outdoors	Ferrous, copper, aluminum

in-process protection. Some may be formulated directly into a cleaning compound while others are stand-alone products designed to be added sump-side to wash or rinse stages.

Inhibition potential may be measured in any number of laboratory methods including but not limited to

- Cast iron chip tests (CICT), ASTM D-4627
- Closed cast iron chip tests (CCICT), ASTM D-4627
- Sandwich corrosion, ASTM F-1110
- Immersion corrosion, ASTM F-483
- Hydrogen embrittlement, ASTM F-519
- Humidity tests, ASTM D-1748
- Salt spray testing, ASTM B-117
- Weatherometer testing, ASTM G-153
- Environmental test chamber testing, ASTM D-1748

Each of these methods is used to predict the inhibition properties of an agent when applied to a specific substrate. When compared to a control solution, one can determine whether inhibition will be adequate for the time frame during which the part will be exposed to a specific type of environment.

Cleaner Chemistry Selection

Chemistry selection is based on the equipment, substrates, and soils involved. Spray-in-air and other high turbulence systems require low-foaming chemistries. Ultrasonic applications require low surface tension technologies. Some precision and press-fit assemblies cannot tolerate water so hydrocarbon solvents must be used. Vapor degreasers clean in the vapor phase; halogenated hydrocarbons are used because they can be heated up to and above the boiling point without flammability concerns.

With all cleaning agents, potential problems must be considered. Caustic alkalis will chemically attack aluminum alloys and they are corrosive to human tissue. Amines will tarnish copper alloys and are odorous. Neutral and acid solutions will attack magnesium. High silicate fluids will cause white bloom if not thoroughly rinsed. Carbonates can cause soap scum and scale buildup. Surfactants and solvents may have objectionable odors. Solvents may have worker exposure limits. All of these issues must be considered when selecting a cleaning chemistry.

The soils requiring removal will also dictate the aggressiveness of the chemistry needed. Hydrocarbon solvents work well on organic soils but are ineffective on inorganic oxides and salts. Light manufacturing soils may be removed by mild chemistries while aged carbonized soils will require more aggressive technologies.

Corrosion inhibitors and rinse aids may also be needed to complete the cleaning process.

Your chemical partner can assist in identifying the best cleaning agents, inhibitors, and processing aids required to meet your manufacturing needs.

Most users of ozone-depleting compounds (ODCs) have now converted their cleaning processes away from regulated materials. Some have replaced ODCs with alternate "drop-in" replacement hydrocarbons while others have completely changed the cleaning process away from a single vapor degreaser to multiple stage systems that contain some water. Alternate hydrocarbons include cost-effective brominated solvents that have low worker exposure limits or petroleum distillates that are flammable. Fluorinated solvents are expensive and some alternate chlorinated solvents are classified as potential carcinogens. All of these solvents are used as ODC alternatives for industrial cleaning applications where water cannot be tolerated.

Semiaqueous or emulsion cleaning systems still use hydrocarbons as part of the cleaning chemistry. They may be diluted with water or they are rinsed with water. These agents typically will not completely evaporate like straight hydrocarbon solutions; a rinse or wipe is required for complete removal of soils and cleaner residues.

The majority of former ODC users have converted to aqueous alternatives. Processes are optimized to minimize the footprint needed; mechanical action is identified to enhance cleaner performance; filtration and skimming are add-ons to maximize bath life; and rinses are incorporated to provide inhibition when needed and to remove all residual materials from the part surface. Water dilution reduces VOC and COD values of the chemistry and also helps reduce overall cleaning costs.

Aqueous and semiaqueous systems increase the demand for water. Local POTWs may assess surcharges if effluents exceed regulated limits. To minimize water consumption and reduce waste, the least amount of chemistry should be used to obtain the cleaning results needed. Recycling practices like skimming and filtration should be incorporated to maximize bath life and reduce the amount of chemistry consumed and/or disposed of. Cascading rinses also help minimize the overall water consumed and the amount of contaminated waste generated.

The overall goal is the efficient production of quality parts. An informed manufacturer can better understand the options available when selecting a cleaning process.

References

1. Passaponti metal cleaning technology, Florence, Italy, November 22, 2007.
2. Quitmeyer, J.A. Going green. *Process Cleaning Magazine*, 3(4) July/August 2008.
3. Quitmeyer, J.A. Surfactants in aqueous cleaners. *Metal Finishing Magazine*. Elsevier, Inc., 2005.
4. Mohr, T.K.G. Solvent stabilizers white paper. Santa Clara Valley Water District, San Jose, CA, June 14, 2001.
5. www.qualitywatertreatment.com, Quality Water Treatment © 2006.
6. www.wikipedia.org/pH
7. www.wikipedia.org/alkali

4

Solvents and Solubility

John Burke
*Oakland Museum
of California*

Overview

The mixing behavior of dissimilar liquids is a complex thermodynamic phenomenon that can be described, and even predicted, by a number of theoretical models. In the area of nonionic solutions, models based on the Hildebrand solubility parameter have been most successful, in large part due to their relative simplicity. Such models, which include Hansen's three-component system and the related triangular Teas graph, have been widely used in areas as diverse as the development of paint formulations, toxicity testing of barrier films, and art conservation treatments. Of equal importance is their suitability as didactic tools to facilitate a conceptual grasp of otherwise elusive physical behavior.

In the early twentieth century, Joel H. Hildebrand defined a measurement of solubility behavior, quantified by what he termed cohesive energy density, that enabled solubility parameters to be derived by measurements of evaporation. This resulted in a relative scale that could be used to predict the solubility behavior of solvents and mixtures through numeric calculation. Hansen refined this approach by separating the single Hildebrand parameter into three values for dispersion, polarity, and hydrogen bonding energies, which Teas then abstracted into fractional contributions, thus allowing these three coordinates to be plotted on a simple triangular graph (a development that sacrificed some precision for the ability to clearly describe, calculate, and predict complex solubility behavior).

Historical Background

With the advent of the industrial age (specifically, one might argue, World War I) the issue of solubility assumed increasing economic importance. Manufacturers needed precise control over the costs and end properties of paints, plastics, and other industrial materials and wanted to avoid excessive trial and error experimentation. This commercial need to predict solubility behavior led some industries, especially paint and varnish makers, to develop unique methodologies specific to their areas of concern.

Early solubility schemes were largely empirical, relying on observed changes in mixtures upon dilution. One example is a clear solution of kauri resin in butyl alcohol that becomes cloudy upon the addition of too much hydrocarbon solvent. Accordingly, the kauri-butanol (KB) value of a hydrocarbon is the maximum amount that can be added to a standard KB solution without causing cloudiness. "Stronger" solvents such as toluene can be added in a greater amount (and thus have higher KB values) than "weaker" solvents such as mineral spirits. Similar solvent measurements include the aniline cloud point (aniline is soluble in aromatic solvents but less so in paraffins), the wax number (the amount of solvent a benzene/beeswax solution will tolerate), and the heptane number (the amount of heptane that could be added to a solvent/resin mixture before becoming cloudy, a method for ranking the relative solubility of the resin rather than the solvent), and so on. In this way, various solvents and other materials could be organized and ranked according to solubility "strength." But since all of these approaches were based on measuring an observed behavior of specific solvents and resins, they were each applicable only to a narrow range of materials. For example, the KB number is not appropriate for alcohols or ketones where spuriously high KB numbers would be generated. The question of underlying principles among these methods, along with the feasibility of a universal descriptor of solubility behavior, remained elusive.

In the first half of the twentieth century, Dr. Joel Henry Hildebrand, in researching the color of iodine solutions, noticed some solutions that appeared to deviate from ideal behavior. These intriguing anomalies led to research that he described in monographs entitled *The Solubility of Nonelectrolytes*. He suggested that the deviations he observed were due to differences in what he described as the "cohesive energy density" of the components, a property he derived from a material's energy of vaporization. Dr. Hildebrand's critical breakthrough was to quantify a hitherto unmeasurable quantity (cohesive energy density) by direct measurements of vaporization behavior. The values he derived we now call Hildebrand solubility parameters, and they form the bedrock of most commonly used solubility models. To understand how such models are used requires an exploration of solubility at the molecular level, and an introduction to the intriguing relationship between vaporization and solubility behavior.

Solutions

When two liquids are combined, simply stated, they either stay together or they don't. They may stay together because of a chemical reaction that essentially fuses the original components into a new compound (involving an exchange of electrons), which makes separation of the original components difficult. Or, they may stay together because the two components are mutually soluble, such as gin and vermouth. This kind of simple solubility implies that the individual components remain essentially unchanged even when mixed, and could conceivably be separated again by distillation.

For this discussion, we need to ignore ionic reactions of the first type, since solubility parameter theory specifically concerns solutions of nonelectrolytes, thus excluding even mixtures containing water. But simple solutions of the second type are easier to characterize and are exactly the kind of nonelectrolytes that Hildebrand described.

But to call this kind of interaction simple may be misleading. Not just any two liquids can be successfully combined. When asked why oil and water do not mix, most people will reply that the oil is "lighter" than the water, yet this immiscibility is not at all due to gravity. And even apparently miscible solutions can become cloudy or even separate out again after some time or following some change in concentration or temperature. The complexity of this behavior can be accounted for by considering the process of evaporation—but to understand this first requires a closer look at what is happening along the surface of molecules.

Unlike gases, the molecules of liquids (and solids) tend to stick together and resist the tendency to evaporate into space. This implies that the molecules that make up liquids (and solids) possess some kind of intermolecular cohesive force that prevents them from easily separating. These cohesive forces were first described by Johannes van der Waals in 1873, and thus bear his name. Originally thought to be small gravitational attractions, van der Waals forces are actually due to electromagnetic forces. In fact, molecules tend to behave as if they were small magnets, or dipoles, with positive and negative charges that attract and repel each other.

The outer shell of each atom is composed entirely of a cloud of negatively charged electrons, completely enclosing the positively charged nucleus within. Molecules, therefore, are also surrounded by a cloud of electrons. In both cases, we can imagine a positively charged center surrounded by a negatively charged outer shell. In nonionic molecules, these positive and negative charges essentially balance out, resulting in a geometry that is neutral or magnetically balanced overall.

But, the electron cloud is never evenly distributed. From instant to instant, random fluctuations in electron distribution give rise to small localized charges that shift about a molecule's surface. Areas of the molecule with greater electron density will temporarily be more negatively charged, while electron-deficient areas will be more positively charged. When two molecules are in proximity, the random polarities in one molecule tend to induce corresponding polarities in the other. This allows the electrons of one molecule to be temporarily attracted to the nucleus of the other, and vice versa, resulting in a play of attractions between the molecules. These induced attractions between otherwise nonpolar molecules are called dispersion forces. The amount of dispersion forces in a molecule is related to surface area: the larger the molecule, the greater the number of temporary dipoles and the greater the potential for intermolecular attractions. Molecules with straight chains, for example, have more surface area and thus greater dispersion forces than more compact branched-chain molecules of the same molecular weight.

But, in addition to these weak fluctuating dispersion forces, stronger and more stable polarities can also exist when atoms share electrons asymmetrically within a molecule. This is especially true, for example, when a hydrogen atom is connected to a strongly electron-hungry atom such as oxygen. This causes the hydrogen's sole electron to be drawn toward the electronegative atom, leaving the positively charged hydrogen nucleus exposed. Such an arrangement can result in molecules with tremendous mutual attraction, a force called hydrogen bonding. Such stable polar forces, especially hydrogen bonding, also play an important role in solubility theory.

It is the interplay of these polar molecular structures that accounts for the cohesion holding liquids and solids together. And it is the different strengths of these van der Waals forces, the relative polarities of molecules overall, that result in the anomalies that Hildebrand observed. But how are these van der Waals forces related to evaporation?

Cohesive Energy

To cause a liquid to boil, we usually add energy in the form of heat. At some point, as the temperature increases, the liquid begins to boil away into a gas. But once the liquid reaches its boiling point, the temperature of the liquid will not continue to increase. Any subsequent heat that is added will be used up in the process of evaporating the molecules of the liquid. Only when all the liquid has been completely vaporized will the temperature of the system again begin to rise (Figure 4.1).

If we measure the amount of energy that is added between the moment when boiling starts until the point when all the liquid has boiled away, we would have a direct indication of the energy required to separate the liquid into a gas. Interestingly, this is also a measure of the van der Waals forces that held the molecules of that liquid together, since those are the forces that needed to be overcome to separate the molecules of the liquid into a gas. It is important to note here that the temperature at which the liquid begins to boil is not important, but rather the amount of heat that has to be added to separate the molecules. A liquid with a low boiling point may require considerable energy to vaporize, while a liquid with a higher boiling point may vaporize quite readily, or vice versa. Regardless of the temperature at which boiling begins, a liquid that vaporizes readily has less intermolecular cohesion than a liquid that requires considerable heat to vaporize. The energy required to vaporize the liquid is called, not surprisingly, the heat of vaporization, and reflects the cohesive forces that exist between its molecules.

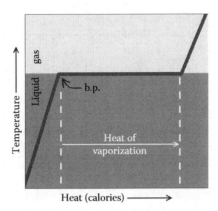

FIGURE 4.1   The heat of vaporization is the amount of heat energy that it takes to transform a liquid into a gas.

Here is the connection we have been looking for: vaporization and solubility are similar because the same intermolecular forces have to be overcome to vaporize a liquid as to dissolve it. In order to combine into a mixture, the closely attracted molecules of each component have to be physically separated in the process of mixing. This is exactly the same as the molecular separations that need to occur during vaporization. The same intermolecular van der Waals forces must be overcome in both cases.

So it stands to reason that for two materials to be soluble in each other, their internal cohesive energies must be similar. The molecules of a strongly polar liquid (such as water) with a high degree of internal cohesive forces (mostly due to hydrogen bonding), will simply not allow the molecules of a nonpolar liquid (such as oil) with weak dispersion forces to interpenetrate. Their cohesive energies (as shown by their heats of vaporization) are too dissimilar.

If we wanted to give a number to these attractions independent of temperature (basically to put everything on an even playing field), we could derive the cohesive energy density by the heat of vaporization (in calories) needed per molar volume. In 1936, Joel H. Hildebrand, in his landmark book on the solubility of nonelectrolytes, proposed the square root of the cohesive energy density as a numerical value indicating solubility:

$$\delta = \sqrt{c} = \left[\frac{\Delta H - RT}{V_m} \right]^{1/2} \tag{4.1}$$

where
    c is the cohesive energy density
    H is the heat of vaporization
    R is the gas constant
    T is the temperature
    V_m is the molar volume

It was not until the third edition in 1950 that the term solubility parameter was proposed for this value, represented by the symbol δ, the units of which are termed Hildebrands. Table 4.1 lists solvents arranged according to their Hildebrand solubility parameter.

Such a table of Hildebrand parameters can be viewed as a kind of solvent spectrum, with solvents arranged in order of relative "strength." In this way, if a certain material were soluble in acetone, it may also be soluble in neighboring solvents with similar internal energies, such as diacetone alcohol or methyl ethyl ketone. Conversely, it may not be soluble in solvents with very different internal energies further away on the list, such as ethyl alcohol or cyclohexane. Theoretically, there would be a contiguous

TABLE 4.1 Hildebrand
Solubility Parameters

Solvent	Parameters
n-Pentane	(7.0)
n-Hexane	7.24
Freon® TF	7.25
n-Heptane	(7.4)
Diethyl ether	7.62
Cyclohexane	8.18
Amyl acetate	(8.5)
1,1,1-Trichloroethane	8.57
Carbon tetrachloride	8.65
Xylene	8.85
Toluene	8.91
Ethyl acetate	9.10
Benzene	9.15
Chloroform	9.21
Trichloroethylene	9.28
Tetrahydrofuran	9.52
Cellosolve acetate	9.60
Acetone	9.77
Ethylene dichloride	9.76
Methylene chloride	9.93
Diacetone alcohol	10.18
Butyl Cellosolve	10.24
Morpholine	10.52
Pyridine	10.61
Cellosolve	11.88
n-Butyl alcohol	11.30
Ethyl alcohol	12.92
Dimethyl sulfoxide	12.93
n-Propyl alcohol	11.97
Dimethylformamide	12.14
Methyl alcohol	14.28
Propylene glycol	14.80
Ethylene glycol	16.30
Glycerol	21.10
Water	23.5

Source: Hildebrand values from Hansen, C.M., *J. Paint Technol.,* 39(505), 104, 1967.

Note: Values in parenthesis from Crowley et al. (1966).

group of solvents that will dissolve a particular material, while the rest of the solvents in the spectrum would not. Some materials may dissolve in a larger range of solvents, while others may be soluble in only a few, but the cohesive energies of solvent and solute would theoretically need to be more or less the same for solubility to occur.

An interesting aspect of this model is that it attempts to predict solubility numerically. In other words, once the solubility of a material was determined in a few solvents, the theory predicts that any other solvent with a similar solubility parameter should also be effective.

In the same way, the behavior of mixtures can be predicted. The parameter of a mixture is the average of its component parameters. For example, using the Hildebrand list, a mixture of two parts toluene and one part acetone will have a Hildebrand value of 9.2 ($\frac{2}{3}$(8.91) + $\frac{1}{3}$(9.77)), or about the same as chloroform. Theoretically, such a 2:1 toluene/acetone mixture should have similar solubility characteristics to chloroform. Thus, for example, if a material was soluble in chloroform, it should also be soluble in the 2:1 toluene/acetone mixture, even though neither acetone nor toluene is near chloroform on the list. What's attractive about this approach is that it attempts to predict the properties of a mixture using only the parameters of its components. No empirical information about the mixture is required. Essentially, the mixture performs like a single solvent whose solubility value is the net contribution of the mixture's individual values.

Unfortunately, inconsistencies do arise. For example, Figure 4.2 plots the swelling behavior of a dried linseed oil film in various solvents arranged according to Hildebrand number. Of the solvents listed, chloroform swells the film to the greatest degree, about 6 times as much as ethylene dichloride, and over 10 times as much as toluene. Solvents with greater differences in Hildebrand value have less swelling effect, and the range of peak swelling occupies less than 2 Hildebrand units. Theoretically, one would expect any solvent with a Hildebrand value between 19 and 20 to behave similarly.

But careful examination of the graph reveals an anomaly. Two solvents with Hildebrand values right in the middle of the severe swelling range, methyl ethyl ketone (19.3) and acetone (19.7), cause very little swelling behavior. Theoretically, liquids with similar cohesive energy densities should have similar solubility characteristics, but the observed behavior in this instance clearly does not bear this out. In the previous discussion about van der Waals forces, it was pointed out that some molecules exhibit only weak dispersion forces while others, especially those capable of hydrogen bonding, can be strongly polar. It is precisely these different types of polarity that must be taken into account in order to improve solubility predictions.

The inconsistencies in Figure 4.2 stem from differences in hydrogen bonding between chlorinated solvents and ketones. The material that was being affected, linseed oil, has cohesion forces due primarily to dispersion forces, with little hydrogen bonding involved. This kind of polarity is perfectly matched by the chloroform molecules, thus encouraging interpenetration and swelling of the linseed oil polymer. The smaller acetone and methyl ethyl ketone molecules on the other hand, have less surface area and therefore less dispersion forces, but instead have stronger hydrogen bonding potential. Even

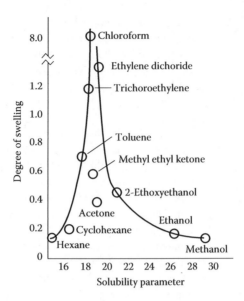

FIGURE 4.2 The swelling behavior of linseed oil films in solvents arranged according to solubility parameter. (Adapted from Feller, R.L. et al., *On Picture Varnishes and Their Solvents*, National Gallery of Art, Cleveland, OH, 1971.)

though the overall Hildebrand solubility parameter values are similar, the differences in polarity, primarily hydrogen bonding, lead to differences in solubility behavior. Basically, the ketone molecules are more attracted to each other than to the linseed oil film.

A scheme to overcome the inconsistencies caused by polarity was proposed by Harry Burrell in 1955. His solution was to segregate the solvent spectrum into three separate lists depending on hydrogen bonding strength. This is briefly summarized as follows:

1. Weak hydrogen bonding: hydrocarbons, chlorinated, and nitrocompounds
2. Moderate hydrogen bonding: ketones, esters, ethers, and glycol monoethers
3. Strong hydrogen bonding: alcohols, amines, acids, amides, and aldehydes

Accordingly, for solvents to exhibit similar solubility behavior, they need to have both similar Hildebrand numbers and similar polarity. This system of classification does in fact improve the prediction of solvent behavior, and is still widely used in some practical applications.

But it was only a matter of time before researchers, looking for greater precision in quantifying solvent behavior, began examining other flavors of van der Waals forces. In 1966, Crowley, Teague, and Lowe of Eastman Chemical developed a three-component approach using the Hildebrand parameter, a hydrogen bonding number, and the overall molecular dipole moment. While this approach further improves solubility predictions, one drawback was the need to work in three dimensions to describe solubility behavior. In fact, this was literally what the researchers did, by using a large box, with a scale representing each of the three values assigned to the three vertices. Solvent locations, represented by a small ball supported on a rod, were then located inside the box at the intersection of their appropriate numeric values (Figure 4.3).

Once a wide range of solvents had each been positioned within the cube, solubility tests could then be conducted on a material such

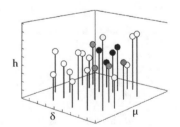

FIGURE 4.3 A three-dimensional box used to plot solubility information. δ, Hildebrand value; μ, dipole moment; h, hydrogen bonding value. (After Crowley, J.D. et al., *J. Paint Technol.*, 38(496), 269, 1966.)

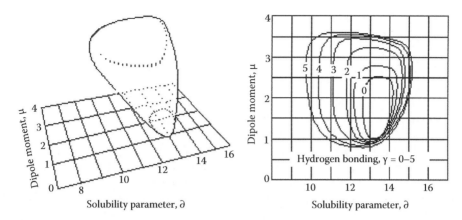

FIGURE 4.4 Approximate representations of solid model and solubility map for cellulose acetate. (From Crowley, J.D. et al., *J. Paint Technol.*, 38(496), 269, 1966. With permission.)

as a polymer. If a solvent successfully dissolved the polymer being tested, that property was indicated using by placing a black ball at the solvent location; if a solvent did not exhibit solubility, its location was indicated by a white ball. Marginal solubilities, such as swelling, were indicated by gray. After enough solvents were tested, a contiguous three-dimensional (3D) zone of solubility would be revealed for the polymer in the form of a solid region of black balls, perhaps surrounded by a few gray, with everything outside that zone being white. Theoretically, even though not every available solvent might be tested, any other solvent or solvent mixture that fell within the solubility zone should also be a good solvent for that polymer.

Representing such data in a publication was problematic. Often such a 3D polymer solubility zone might be represented as a graph using two of the parameters, restricted to a single slice through the zone at one value of the third scale. Alternatively, a topographic map might be used to indicate several values of the third parameter at the same time (Figure 4.4). Because these solubility zones for a polymer might have an unusual shape, several graphs might be required if the total solubility area was to be shown.

Maps such as these could be used in conjunction with a table of three-component parameters for individual solvents, and in this way provide useable information about solvent–polymer interactions that would facilitate the formulation of polymer or solvent blends to suit specific applications with less empirical front-end testing. For this reason, such solubility maps are often included in technical reports and manufacturer's product data sheets. How these graphs are actually used to accomplish the development of formulations purposes will be described later in terms of the triangular Teas graph, where such procedures are greatly simplified.

Hansen Parameters

In 1966, Charles M. Hansen developed what is now the most widely used three-component system. While Crowley, Teague, and Lowe used three independent scales for each of the three parameters, Hansen took the step of relating them all to the single Hildebrand value, such that when the three Hansen parameters are added together, their sum is the Hildebrand value for that solvent.

His calculations were made in several steps. First, the dispersion force (*d*) of a molecule was determined by matching the molecule to an alkane of similar structure. This is called a molecule's "homomorph" (for example, the homomorph of ethanol is ethane). Polarity (*p*) was then determined by measuring molecular dipole moments, and by comparison with similar functional groups in smaller molecules. Finally, the degree that hydrogen bonding (*h*) contributed to the overall polarity was determined. The relative contribution of these independent values was then calculated as fractions of the single Hildebrand value such that

$$\delta_d^2 + \delta_p^2 + \delta_h^2 = \delta^2 \tag{4.2}$$

Perhaps most importantly, these calculated parameters were subsequently fine-tuned and adjusted by exhaustive empirical testing on a range of polymers and other solvents to determine relative distributions within the Hansen space. This final step resulted in some solvent locations being moved considerably from their purely calculated locations. Hansen parameters are an approximation derived from the Hildebrand parameter and adjusted to better correspond to empirical behavior.

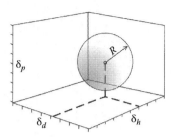

FIGURE 4.5 HSP area for typical solvent, designated by polarity, dispersion, and hydrogen bonding parameters, and a radius of interaction *R*.

Looking at three-dimensional solubility areas for polymers, Hansen also found that by doubling the dispersion axis, a spherical zone of solubility would result (Figure 4.5). This made it possible to describe solubility behavior using four numbers: the three Hansen solubility parameters and a radius of interaction, *R*. The three solubility parameters indicated the center of the solubility sphere, and *R* indicated the sphere's size (the center of the sphere was located after the solubility area had been determined).

The Teas Graph

As a three-dimensional system, Hansen space still requires more than one view to adequately represent all of the data for a specific polymer, since in two dimensions a solvent might appear within the sphere while its actual position could be in front or behind. Furthermore, complex calculations of mixtures remain difficult without computer assistance. In 1968, given the practical difficulties of working with three-component data, Jean P. Teas devised a triangular graph on which Hansen's solubility data could be plotted in two dimensions. Because of its clarity and ease of use, the Teas graph has found wide application in problem solving, documentation, and analysis, and is an excellent vehicle for understanding complex solubility behavior. However, in order to plot the three Hansen parameters on a single planar graph, a further departure had to be made from established theory: Teas constructed his graph on the imaginary hypothesis that all materials have the same Hildebrand value.

Once this leap has been made, solubility values can be assigned, not by relative contributions to a specific material's Hildebrand value, but by the percent that the three component forces (dispersion, polar, and hydrogen bonding) contribute overall. But because Hildebrand values are not the same for all liquids, it should be remembered that the Teas graph is a further departure from Hansen's values, and therefore even less accurate overall. Hansen derived his parameters from the Hildebrand value: when all three Hansen parameters for a solvent are added together, their sum will be the Hildebrand value for that solvent. Teas parameters (δ_d, δ_p, δ_h in Table 4.2), also called fractional parameters, are mathematically derived from Hansen values by calculating the relative amount that each Hansen parameter contributes to the sum of all three Hansen values:

$$f_d = \frac{\delta_d}{\delta_d + \delta_p + \delta_h} \quad f_p = \frac{\delta_p}{\delta_d + \delta_p + \delta_h} \quad f_h = \frac{\delta_h}{\delta_d + \delta_p + \delta_h} \tag{4.3}$$

In other words, when all three of Teas' fractional parameters are added together, their sum will always be 100. The intersection of three fractional values can thus be located on a triangular grid (Figure 4.6).

This further abstraction also resulted in additional loss of accuracy so that Teas, just as Hansen before him, had to adjust the location of many solvents on the graph to correspond with observed behavior. In

TABLE 4.2 Hansen Fractional Parameters

δ	Solvent Name	δ_d	δ_p	δ_h	δH_d	δH_p	δH_h	PEL	TLV	OEL
1	*n*-Pentane	100	0	0	—	—	—	1000	600	600
1	*n*-Hexane	100	0	0	—	—	—	500	50	50
1	*n*-Heptane	100	0	0	—	—	—	500	400	400
1	*n*-Octane	100	0	0	—	—	—	500	300	300
2	Cyclopentane	—	—	—	90	0	10	—	600	600
3	Cyclohexane	94	2	4	99	0	1	300	100	100
4	Methyl cyclohexane	94	0	6	94	0	6	500	400	400
5	Mineral spirits (Stoddard)	90	4	6	—	—	—	—	100	100
6	VMP naphtha	94	3	3	—	—	—	—	300	300
7	Naphtha (high flash)	—	—	—	84	3	8	—	—	100
8	Benzene	78	8	14	90	0	10	1	0.5	10[a]
9	Toluene	80	7	13	84	7	9	200	20	20[a]
10	Xylene (all isomers)	83	5	12	81	5	14	100	100	100
11	Ethylbenzene	87	3	10	90	3	7	100	100	100
12	*p*-Diethylbenzene	97	0	3	94	1	5	—	—	—
13	Naphthalene	70	8	22	71	7	22	10	10	10
14	Styrene	78	4	18	78	4	17	100	20	20[b]
15	Methylene dichloride	59	21	20	65	13	22	25	50	50[a]
16	Ethylene dichloride	67	19	14	62	24	13	50	10	10[a]
17	1,1,1-Trichloroethane	70	19	11	73	19	9	350	350	350
18	Chlorobenzene	65	17	18	75	17	8	75	10	10
19	Chloroform	67	12	21	67	12	21	50	10	10
20	Trichloroethylene	68	12	20	68	12	20	100	10	10[a]
21	Carbon tetrachloride	85	2	13	66	34	0	10	5	5[a]
22	Diethyl ether	64	13	23	64	13	23	400	400	400
23	Tetrahydrofuran	55	19	26	55	19	26	200	50	50
24	2-Methoxyethanol (methyl cellosolve, EGME)	39	22	39	39	22	39	25	0.1	0
25	2-Ethoxyethanol (cellosolve, EGEE)	42	20	38	41	23	36	200	5	5
26	2-Butoxyethanol (butyl cellosolve, EGBE)	—	—	—	48	15	37	50	20	20
27	1-Methoxy-2-propanol (PGME)	—	—	—	47	19	35	150	100	100
28	2-(2-Ethoxyethoxy)ethanol (carbitol, DEGEE)	48	23	29	43	25	33	—	—	—
29	2-(2-Butoxyethoxy)ethanol (butyl carbitol, DEGBE)	48	21	32	48	21	32	—	—	—
30	1-Methoxy-2-propanol acetate (PGMEA)	—	—	—	47	18	35	—	—	—
31	Dioxane	67	7	26	67	6	26	100	20	20
32	Acetone	47	32	21	47	32	21	1000	500	500
33	Methylethyl ketone (2-butanone)	53	30	17	53	30	17	200	200	200
34	Diethyl ketone	56	27	17	56	27	17		200	200
34	Methylpropyl ketone (2-pentanone)	—	—	—	57	27	17	200	200	200
36	Methylisobutyl ketone (hexone)	58	22	20	60	24	16	200	50	50
37	Methylisoamylketone	62	20	18	62	22	16	100	50	50
38	Diisobutyl ketone	—	—	—	67	16	17	50	25	25
39	Isophorone	51	25	24	52	25	23	25	5	5
40	Mesityl oxide	55	24	21	57	21	21	25	15	15
41	Cyclohexanone	55	28	17	61	22	17	50	20	20

(*continued*)

TABLE 4.2 (continued)　　Hansen Fractional Parameters

δ	Solvent Name	δ_d	δ_p	δ_h	δH_d	δH_p	δH_h	PEL	TLV	OEL
42	Ethyl lactate	44	21	35	44	21	35	—	—	—
43	Butyl lactate	49	20	31	49	20	31	—	5	5
44	Methyl acetate	45	36	19	51	24	25	200	200	200
45	Ethyl acetate	51	18	31	56	19	25	400	5	5
46	*n*-Propyl acetate	57	15	28	56	16	28	200	200	200
47	*n*-Butyl acetate	60	13	27	61	14	24	150	150	150
48	Isobutyl acetate	60	15	25	60	15	25	150	150	150
49	*sec*-Butyl acetate	—	—	—	57	14	29	200	200	200
50	*tert*-Butyl acetate	—	—	—	55	22	22	200	200	200
51	*n*-Amyl acetate	—	—	—	63	13	24	100	50	50
52	Isoamyl acetate	60	12	28	60	12	28	100	50	50
53	2-Ethoxyethyl acetate (cellosolve acetate, EGEEA)	51	15	34	51	15	34	100	5	5
54	Dimethyl carbonate	53	13	33	53	13	33	—	—	—
55	Diethyl carbonate	64	12	24	61	25	14	—	—	—
56	Propylene carbonate	48	38	14	48	43	10	—	—	—
57	Trimethyl phosphate	39	37	24	39	37	24	—	—	—
58	Isobutyl isobutyrate	63	12	25	63	12	25	—	—	—
59	Methyl formate	46	22	32	45	25	30	100	100	100
60	Diethyl sulfate	42	39	19	42	39	19	—	—	—
61	Acetonitrile	39	45	16	39	46	15	40	20	20
62	Butyronitrile	44	41	15	47	38	16	—	—	—
63	Nitromethane	40	47	13	40	47	13	100	20	20
64	Nitroethane	44	43	13	44	43	13	—	—	—
65	2-Nitropropane	50	37	13	50	37	13	25	10	10
66	Nitrobenzene	52	36	12	61	26	13	1	1	1
67	Pyridine	56	26	18	56	26	18	5	1	1
68	Morpholine	57	15	28	57	15	28	20	20	20
69	Aniline	50	19	31	56	15	29	5	2	2
70	N-methyl-2-pyrrolidone	48	32	20	48	33	19	—	—	—
71	Diethylenetriamine	38	30	32	38	30	32		1	1
72	Cyclohexylamine	—	—	—	64	12	24	—	10	10
73	Formamide	28	42	30	28	42	30	—	10	10
74	Dimethylformamide	41	32	27	41	32	27	10	10	10
75	Carbon disulfide	88	8	4	76	22	2	20	1	1[a]
76	Dimethylsulfoxide	41	36	23	41	36	23	—	—	—
77	Methanol	30	22	48	30	25	45	200	200	200
78	Ethanol	36	18	46	36	20	44	1000	1000	1000
79	1-Propanol (*n*-propanol)	40	16	44	40	17	43	200	100	100
80	2-Propanol (isopropanol)	—	—	—	41	16	43	400	200	200
81	1-Butanol (*n*-butyl alcohol)	43	15	42	43	15	42	100	20	20
82	2-Butanol (*sec*-butanol)	—	—	—	44	16	40	150	100	100
81	*tert*-Butanol	—	—	—	43	15	42	100	100	100
84	Benzyl alcohol	48	16	36	48	16	36	—	—	—
85	Cyclohexanol	50	12	38	50	12	39	200	50	50
86	*n*-Amy alcohol (1-pentanol)	46	13	41	45	17	39	—	—	—
87	Diacetone alcohol	45	24	31	45	24	31	—	50	50
88	2-Ethylhexanol	50	9	41	46	17	37	—	5	5

TABLE 4.2 (continued)   Hansen Fractional Parameters

δ	Solvent Name	δ_d	δ_p	δ_h	δH_d	δH_p	δH_h	PEL	TLV	OEL
89	2-Ethylbutanol	48	10	42	—	—	—	—	—	—
90	Glycerol	25	23	52	30	21	50	—	—	—
91	Ethylene glycol	30	18	52	31	20	48	—	100	100
92	Diethylene glycol	31	29	40	34	24	42	—	—	—
93	Propylene glycol	34	16	50	34	19	47	—	—	—
94	Phenol	46	15	39	46	15	38	5	5	5
95	Methylal (dimethoxymethane)	59	7	34	59	7	34	1000	1000	1000
96	Benzaldehyde	61	23	16	60	23	17	—	—	—
97	Dipentene (limonene)	75	20	5	74	8	18	—	—	—
98	Turpentine	77	18	5	—	—	—	100	20	20
99	Trichlorotrifluoroethane (Freon 113)	—	—	—	90	10	0	—	1000	1000
100	Dibasic esters (DBE)	54	15	31	54	15	31	—	—	—
Binary azeotropic mixtures										
a1	Acetone/hexane (59:41)	69	19	12	—	—	—			
a2	Acetone/methanol (88:12)	45	31	24	45	31	24			
a3	Isopropanol/toluene (58:42)	57	12	30	59	12	29			
a4	Isopropanol/hexane (33:77)	—	—	—	86	4	10			
a5	Isopropanol/heptane (54:46)	—	—	—	68	9	23			
a6	Isopropanol/MEK (32:68)	—	—	—	49	26	25			
a7	Isopropanol/cyclohexane (22:78)	—	—	—	80	5	13			
a8	Ethanol/cyclohexane (30:70)	77	7	17	80	6	14			
a9	Ethanol/toluene (68:32)	50	14	35	51	16	33			
a10	Ethanol/hexane (21:79)	87	4	10	87	4	9			
a11	Ethanol/heptane (49:51)	69	9	23	69	10	22			
a12	Ethanol/MEK (40:60)	46	25	29	46	25	28			
a13	Acetone/cyclohexane (67:33)	63	22	15	64	21	14			

Source: Hansen, C.D., *Hansen Solubility Parameters: A User's Handbook*, 2nd edn., CRC Press, Inc., Boca Raton FL, 2007, Table A-1, pp. 347–483, using Equation 3. OSHA PEL, ACGIH TLV, and recommended OEL provided by James Unmack.

Note: CIH. numbers in column 1 refer to solvent locations on Teas Solubility Chart, Table 4.3.

[a] Carcinogen, OSHA Table Z-2; From 29CFR1910.1000.

[b] Carcinogen, OSHA Table Z-2; From 29CFR1910.1052.

fact, when fractional parameters are calculated directly from Hansen values (δH_d, δH_p, δH_h in Table 4.2), it can be seen that Teas shifted over a third of the solvent locations from their theoretical locations (the larger adjustments are indicated in gray on the Teas chart, Table 4.3). Given the errors inherent in any simplification, this is a reminder not to expect too much precision in any graphical representation of complex solubility behavior. On the other hand, this does not prevent the two-dimensional triangular Teas graph from being a useful tool, and perhaps the most convenient method by which solubility information can be illustrated.

Visualizing Solvents

Examining a Teas graph (Table 4.3) containing a wide variety of solvents is very educational. The alkanes, whose only intermolecular bonding is due to dispersion forces, are located in the far lower right corner corresponding to 100% dispersion forces. Moving toward the lower left corner are solvents with increasing hydrogen bonding contribution. Moving from the bottom of the graph upward are solvents with polarity due less to hydrogen bonding than to an increasing dipole moment of the molecule

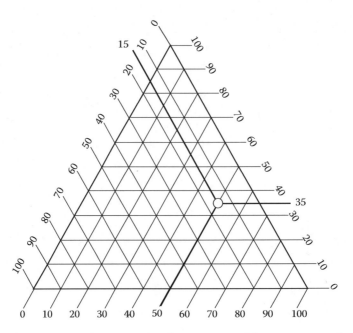

FIGURE 4.6   Any point on a triangular Teas graph is the intersection of three percentage values.

as a whole. It can also be seen that most solvents are grouped closer to the lower right apex than the others. This is because dispersion forces are present in all molecules, polar or not, and determining the dispersion component is the first calculation in assigning Hansen parameters, from which fractional parameters are derived. Unfortunately, this tends to overemphasize the dispersion force relative to the more significant polar forces, especially hydrogen bonding interactions, and this is one reason why Teas needed to adjust positions.

It is also apparent that increasing molecular weight within each solvent class shifts the relative position of solvents closer to the bottom right apex (Figure 4.7). This is because, as molecular weight increases, the polar part of the molecule that causes the specific character identifying it with its class, called the functional group, is increasingly "diluted" by progressively larger, nonpolar "aliphatic" molecular segments. This gives the molecule as a whole relatively more dispersion force and less of the polar character specific to its class.

This trend toward less polarity with increasing molecular weight within a class also accounts for the observation that lower molecular weight solvents are often "stronger" than higher molecular weight solvents of the same class (although determinations of solvent strength must really be made relative to the solubility of a material). Another reason low molecular weight solvents may seem more active is that smaller molecules tend to disperse throughout solid material more rapidly than larger relatives.

The only class in which increasing molecular weight places the solvent further away from the lower right corner is the alkanes. As previously stated, the intermolecular attractions between alkanes are due entirely to dispersion forces, and accordingly, Hansen parameter values for alkanes show zero polar contribution and zero hydrogen bonding contribution. Since fractional parameters are derived from Hansen parameters, one would expect all the alkanes to be placed together at the extreme right apex. Observed behavior, however, indicates that different alkanes do exhibit unique solubility characteristics, perhaps because of the tendency of larger dispersion forces to mimic slightly polar interactions. For this reason, Teas adjusted the locations of the alkanes to correspond to empirical evidence.

The position of water on the chart is uncertain, due to the ionic character of the water molecule, and the placement in this paper is according to Teas (1971). As a strongly ionic molecule, the presence of water in a solvent blend can alter dramatically the accuracy of solubility predictions.

TABLE 4.3 Teas Solubility Chart

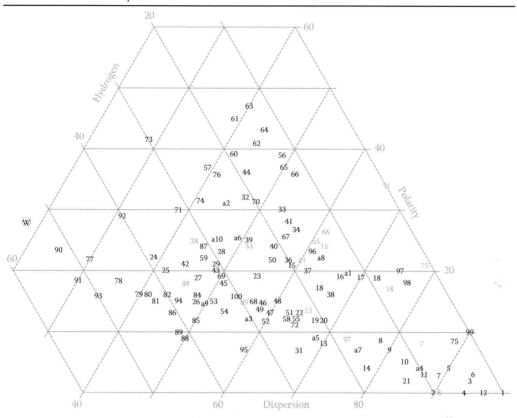

#	Solvents				
1	n_Pentane	21 Carbon tetrachloride	45 Ethyl acetate	69 Aniline	93 Propylene glycol
1	n_Hexane	22 Diethyl ether	46 n-Propyl acetate	70 Methyl_2_pyrrolidone	94 Phenol
1	n_Heptane	23 Tetrahydrofuran	47 n_Butyl acetate	71 Diethylenetriamine	95 Benzaldehyde
1	n-Octane	24 2-Methoxyethanol	48 Isobutyl acetate	72 Cyclohexylamine	96 Benzaldehyde
2	Cyclopentane	25 2-Ethoxyethanol	49 sec-Butyl acetate	73 Formamide	97 Dipentene (limonene)
3	Cyclohexane	26 2-Butoxyethanol	50 tert-Butyl acetate	74 Dimethylformamide	98 Turpentine
4	Methylcyclohexane	27 1-Methoxy 2-propanol	51 n-Amyl acetate	75 Carbon disulfide	99 Trichlorotrifluoroethane
5	Mineral spirits	27 DPGME and PGMEA	52 Isoamyl acetate	76 Dimethylsulfoxide	100 Dibasic esters (DBE)
6	V M P naptha	29 Butyl carbitol	53 2-Ethoxyethyl acetate	77 Methanol	
7	Naptha (high flash)	31 Dioxane	54 Dimethyl carbonate	78 Ethanol	# Azeotropes
8	Benzene	32 Acetone	55 Diethyl carbonate	79 1_Propanol	a1 Acetone/hexane (59:41)
9	Toluene	33 Methyl ethyl ketone	56 Propylene carbonate	80 2_Propanol	a2 Acetone/methanol (88:12)
10	Xylene	34 Diethyl ketone	57 Trimethyl phosphate	81 1_Butanol	a3 2-Propanol/toluene (58:42)
11	Ethylbenzene	34 Methyl propyl ketone	58 Isobutyl isobutyrate	82 2_Butanol	a4 2-Propanol/hexane (23:77)
12	p-Diethylbenzene	36 Methyl isobutyl ketone	59 Methyl formate	81 tert-Butyl alcohol	a5 2-Propanol/heptane (54:46)
13	Naphthalene	37 Methyl isoamyl ketone	60 Diethyl sulfate	84 Benzyl alcohol	a6 2-Propanol/MEK (32:68)
14	Styrene	38 Diisobutyl ketone	61 Acetonitrile	85 Cyclohexanol	a7 2-Propanol/cyclohexane (22:78)
15	Methylene dichloride	39 Isophorone	62 Butyronitrile	86 n-amyl alcohol	a7 Ethanol/cyclohexane (30:70)
16	Ethylene dichloride	40 Mesityl oxide	63 Nitromethane	87 Diacetone alcohol	a8 Acetone/cyclohexane (67:33)
17	1,1,1-Trichloroethane	41 Cyclohexanone	64 Nitroethane	88 2_Ethyl_1_hexanol	a9 Ethanol/toluene (68:32)
18	Chlorobenzene	42 Ethyl lactate	65 2_Nitropropane	89 2_Ethyl_1_butanol	a4 Ethanol/hexane (21:79)
19	Chloroform	43 Butyl lactate	66 Nitrobenzene	90 Glycerol	a5 Ethanol/heptane (49:51)
20	Trichloroethylene	44 Methyl acetate	67 Pyridine	91 Ethylene glycol	a10 Ethanol/MEK (40:60)
				68 Morpholine	92 Diethylene glycol	W = Water (theoretical)

Note: Gray numbers = locations calculated from Hansen values when >5% discrepancy with Teas locations.

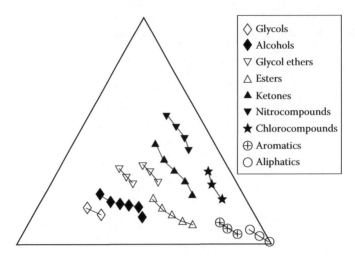

FIGURE 4.7 Within each class, the positions of solvents with higher molecular weights tend to be closer to the lower right axis.

Visualizing Solubility

With solvents located on the Teas graph, the model can be used to graphically illustrate complex solubility behavior. For example, a material such as a polymer can be tested for solubility in several solvents and the results indicated on the chart, similar to the work of Crowley et al. Effective solvents might be marked with stars, marginal solvents with + signs, and non-solvents with circles. Once this is done, a contiguous region of stars would be visible, possibly edged by + signs, and surrounded by solvents in circles. Not every solvent would need to be tested, only enough to get a general picture of the solubility window for that particular polymer (Figure 4.8). Once a few boundary solvents were located, it is unlikely that solvents further away from the solubility area would be effective (unless the material in question was a combination of materials each with a different solubility zone).

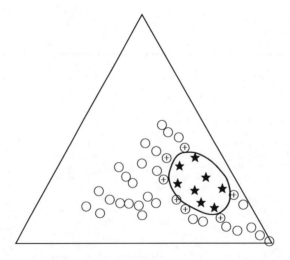

FIGURE 4.8 An example of a solubility window for a hypothetical material, with solvents marked according to solubility behavior: stars indicate successful solubility, crosses poor solubility, and empty locations non-solubility.

The boundaries of the polymer solubility window could be more precisely defined by testing with mixtures of two liquids, one clearly a good solvent and the other a poor one. By mixing various proportions of these two liquids, the concentration that seems to define the boundary can be found and indicated. A few tests of this kind would be sufficient to reveal the general curve and extent of the solubility window.

Once the solubility window of material is indicated on a Teas graph, a number of further extrapolations become available, such as the ability to formulate, entirely within the model, solvent mixtures with specific properties, such as evaporation rate, cost, or environmental impact, as will be discussed in the following section. But it should be pointed out that because the Teas graph is an empirically adjusted abstraction of Hansen's parameters, which themselves are an abstraction of the Hildebrand parameters, too much precision or accuracy should not be expected.

A number of other factors can also affect the reliability of calculations based on solubility parameter theory. Solvents with smaller molecular size will have more accurate chart locations and wider solubility ranges than larger molecules. And, since most published solubility data on materials such as polymers are usually derived from 10% concentrations at room temperature, entropy effects may result in more dilute solutions having decreasing solubility, especially with solvents near the solubility boundary. Increases in temperature can increase solubility area, thus allowing a "boundary" solvent to be more active at higher temperatures. The hydrogen bonding parameter is most sensitive to this effect, so that performance of solvents on the left side of the chart may increase as temperatures increase.

The solubility window for a material can reveal other characteristics of polymer solutions. The viscosity of a solution can vary depending on where its solvent is located in the solubility window. One might expect viscosity to be at a minimum with a solvent near the center of a window, but this is not the case. Solvents near the center of a polymer's solubility window may dissolve the polymer so effectively that individual molecules are allowed to uncoil and extend, thus increasing dispersion forces by increasing molecular surface area.

When dissolved in solvents slightly off center in the solubility window, polymer molecules may remain grouped together in clumps with smaller overall surface attractions, resulting in solutions of slightly lower viscosity. As solvents nearer the edge of the window are used, these clumps become progressively larger and more connected, resulting in viscosity increases until immiscibility is reached at the window boundary (Figure 4.9).

The relative location of a solvent in the solubility window of a polymer also has a marked effect on the dried film characteristics of a polymer. Because of the uncoiling of polymer molecules, films cast from solvent solutions near the center of the solubility window may exhibit greater adhesion to compatible substrates, due to an increase in polymer surface area in contact with the substrate. For the same reason, gas permeability characteristics may also decrease when solvents more central to a polymer's solubility window are employed.

Solvent Mixtures

As mentioned above, the Teas graph is a particularly useful tool for formulating solvent mixtures with less environmental impact and/or lower toxicity, as well as for controlling other desirable performance characteristics such as evaporation rate, adhesive properties, and material costs. Solvent mixtures can also be tailored to exhibit selective solubility if the solubility windows of the various substrates do not completely overlap. In this way, the use of the Teas graph can reduce trial and error experimentation to a minimum by allowing the solubility behavior of mixtures to be predicted in advance.

These theoretical manipulations are possible by virtue of the fact that when two or more solvents are combined, the mixture essentially behaves like a single solvent whose solubility values lie at the intersection of the mixture values. This can be demonstrated by combining two solvents, each of which are ineffective solvents for a material by themselves, in such proportion that the mixture location falls within

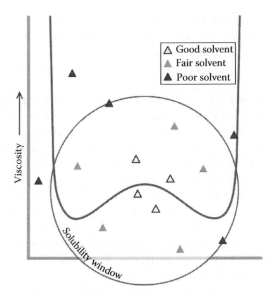

FIGURE 4.9   The lowest viscosity for a polymer would be found in solvents that are closer to the edge of the solubility window, rather than the center.

the solubility window. In this case, the mixture should behave as an effective solvent, even though its components may not be (Figure 4.10).

There are two ways to determine the solubility parameters of a mixture, the numeric method where parameters are calculated from the individual components, and the graphic method where a plot like the Teas graph is used to connect the components and measure distances. The latter method is easiest when dealing with mixtures of no more than two components.

To calculate a mixture's solubility parameters numerically, the solubility parameters for each component liquid are multiplied by the fraction that that liquid occupies in the mixture, and the individual results added together.

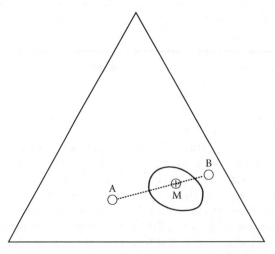

FIGURE 4.10   A mixture, M, of two solvents, A and B, can often act as a true solvent for a material, even if the individual components are non-solvents, if the parameters of the mixture falls within the material's solubility window.

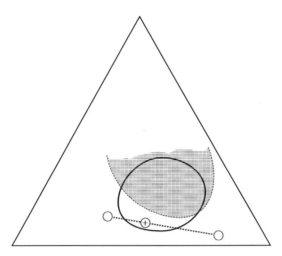

FIGURE 4.11 If the solubility windows of two materials do not overlap completely, a solvent mixture can sometimes be formulated to attack one material while leaving the other relatively unaffected.

The graphic method, which is the easiest way to find the parameters for two-part mixtures, involves simply drawing a line between the two component solvents on a Teas graph, and measuring the location on the line that represents the ratio of their concentration. In other words, the location of a 50:50 mixture of two solvents would lie exactly in the center of the line connecting them on the chart. A greater concentration of one solvent would move the point proportionately closer to that solvent on the line, etc.

Calculating solvent blends in this way is an effective first step at developing formulations for targeted application. We already mentioned the ability to blend non-solvents for a specific polymer into a mixture that acts as a true solvent by ensuring that the parameters of the mixture are located inside the solubility window of the polymer. This is most effective when the distance between the component solvents is not too great. This same approach can be used to replace an undesirable solvent, in terms of toxicity, cost, or other characteristics, with a mixture that has essentially the same solubility performance but improved features.

This ability to blend solvents that target a specific solubility parameter is also valuable when selective solvent action is required, such as dissolving one material while leaving other materials unaffected. If the solubility windows of the materials do not completely overlap, a solvent mixture can be designed to attack one material while leaving the other relatively unaffected (see Figure 4.11).

In formulating mixtures like this, it is important to remember that along with the other variables discussed above, other differences between the components may cause the parameters of the mixture to drift over time. This is best illustrated in terms of evaporation, where differences between components may shift the solubility parameter of the blend if the solvents evaporate at different rates. This problem can be avoided by using solvents of similar vapor pressure, mixtures where the least volatile component is inside the solubility window, or even by the addition of a small amount of slow-evaporating but very effective solvent than can assure consistent solubility throughout the process. Regardless, even though the formulation of mixtures is greatly facilitated by solubility parameter models, the effect of variables that could shift the performance of the mixture over the life of the process must always be considered.

Solvents and Health

Governmental regulations on environmental pollution and greenhouse gas formation have led to an increasing use of "green" replacement solvents. These green solvents may be replacements for traditional solvents that have long atmospheric lifetimes or that are volatile organic compounds (VOCs). Other motivations for producing green replacement solvents include improved safety performance and more favorable recycling characteristics. Sometimes, though not always, the green replacement solvents may have improved human toxicity characteristics. But, because the development of green replacement solvents is driven largely by environmental regulation such as VOC levels and other issues such as flammability, it is not safe to assume that green solvents are necessarily less toxic than their traditional counterparts.

An increasingly important use of cohesion parameter theory is the search for solvents or blends with decreased environmental impact and lower toxicity. Using the Teas chart, solvents and mixtures are easily selected on the basis of likely toxicity characteristics, for example by substituting larger branched molecules which have lower evaporation rates and are slower at passing the skin barrier for smaller straight chain molecules with similar solubility parameter values. Table 4.2 includes OSHA permissible exposure limits (PEL), ACGIH threshold limit values (TLV), and recommended occupational exposure limits (OEL) for many of the solvents listed.

When unfamiliar solvents are encountered, a careful examination of the material's material safety data sheet (MSDS) is advised, including the recommended exposure limits and safety recommendations. In addition, locating their position on the Teas chart can also help clarify their solubility behavior compared to traditional solvents, and facilitate their potential utility as substitutions in various formulations.

Furthermore, it is important to remember that individuals can have different sensitivities to any solvent, and the long-term effects of replacement solvents may be less well understood. It may be prudent to check the worker exposure levels recommended by several sources including manufacturers, professional organizations, and government agencies. For example, there is an ongoing controversy regarding long-term exposure to orange oil solvents (dipentene), and the biological effects of fluorocarbons and siloxanes may not yet be fully appreciated.

Finally, formulating mixtures of less toxic solvents to mimic the behavior of more toxic chemicals is not a guarantee of safety. Such mixtures may present unknown hazards that could differ from the toxicity of the individual components. On the other hand, using the Teas graph to develop formulas that substitute solvents of lower vapor pressure and/or greater threshold limit values may well serve to decrease the risks of cumulative exposure.

Conclusion

Cohesion parameter theory is a developing, though still not exact, science and there are caveats to consider. With the evolution from the single Hildebrand parameter to the three-component Hansen solubility parameter model, practical accuracy has dramatically increased, but largely through empirical adjustments. The triangular Teas fractional model may be convenient but is a further abstraction. This means that anomalies should not be surprising, especially with increasing mixture complexity. Furthermore, the solubility behavior of polymers can be heavily influenced by factors not fully incorporated in the theoretical model (such as crystalline/amorphic regions, degree of cross-linking, temperature, concentration, etc), especially with aged samples. Finally, the current solubility toolkit specifically excludes ionic systems, and the presence of water in a formulation can sometimes dramatically alter solubility behavior.

On the positive side, cohesion parameter systems are still capable of accurately predicting a wide range of solubility behavior, and the Teas graph, even with its theoretical limitations, can be a powerful tool for visualizing and manipulating solvents in many real-world applications. Solubility parameters, and the models derived from it, are often used to assist in formulating mixtures where complex solubility conditions exist. With increasing concern over environmental impacts and worker safety, these models can be an invaluable guide in categorizing the behavior of new replacement solvents alongside familiar ones, or in the selection and development of lower toxicity formulations. This can be of great practical use in saving time, increasing precision, lowering material costs, and improving worker health and safety when involving solvent-based operations.

Perhaps even more importantly, because of its simplicity, solubility parameter theory can serve as a unique didactic tool, enabling a clear grasp of otherwise complex chemical interactions. Once the underlying theory, and its limitations, are appreciated, the broad concepts represented in a tool such as the Teas graph can essentially be internalized to serve as a model for visualizing general solubility behavior. Thus solubility decisions can be informed and guided by a sound foundation that reduces guesswork and greatly improves the possibility of success.

References

Barton, A.F.M. *Handbook of Solubility Parameters and Other Cohesion Parameters*. Boca Raton, FL: CRC Press, Inc., 1983.

Burke, J. Solubility parameters: Theory and application. In: *AIC Book and Paper Group Annual*, vol. 3, C. Jensen, Ed., 1984, pp. 13–58.

Burrell, H. The challenge of the solubility parameter concept. *Journal of Paint Technology*, 40(520), 197–208, 1968.

Crowley, J.D., G.S. Teague Jr., and J.W. Lowe Jr. A three dimensional approach to solubility. *Journal of Paint Technology*, 38(496), 269–280, 1966.

Feller, R.L. The relative solvent power needed to remove various aged solvent-type coatings. In: *Conservation and Restoration of Pictorial Art*, Bromelle and Smith, Eds., London, U.K.: Butterworths, 1976.

Feller, R.L., N. Stolow, and E.H. Jones. *On Picture Varnishes and Their Solvents*. Cleveland, OH: The Press of Case Western Reserve University, 1971.

Hansen, C.M. The three-dimensional solubility parameter—Key to paint component affinities: I. Solvents plasticizers, polymers, and resins. *Journal of Paint Technology*, 39(505), 104–117, 1967.

Hansen, C.M. The three dimensional solubility parameter—Key to paint component affinities: II. Dyes, emulsifiers, mutual solubility and compatibility, and pigments. *Journal of Paint Technology*, 39(511), 505–510, 1967.

Hansen, C.M. and Skaarup, K. The three dimensional solubility parameter—Key to paint component affinities: III. Independent calculations of the parameter components. *Journal of Paint Technology*, 39(511), 511–514, 1967.

Hansen, C.M. The universality of the solubility parameter concept. *Industrial & Engineering Chemistry Product Research and Development*, 8(1), 2–11, 1969.

Hansen, C.M., Ed. *Hansen Solubility Parameters: A User's Handbook*, 2nd edn. Boca Raton, FL: CRC Press/ Taylor & Francis, 2007.

Hedley, G. Solubility parameters and varnish removal: A survey. *The Conservator*, (4), 12–18, 1980.

Hildebrand, J.H. *The Solubility of Non-Electrolytes*. New York: Reinhold, 1936.

Teas, J.P. Graphic analysis of resin solubilities. *Journal of Paint Technology*, 40(516), 19–25, 1968.

Teas, J.P. *Predicting Resin Solubilities*. Columbus, OH: Ashland Chemical Technical Bulletin, Number 1206, 1971.

Torraca, G. *Solubility and Solvents for Conservation Problems*, Rome, Italy: ICCROM, 1978.

Hydrofluoroethers

John G. Owens
3M Company

Introduction

Cleaning applications are critical to many manufacturing processes, which means that the selection of the appropriate cleaning solvent can be important to the success of the entire production process. Each year, however, it seems that an increasing number of environmental and safety requirements necessitate changes in the materials used in these cleaning processes. Concerns over flammability, toxicity, and environmental issues including ozone depletion and climate change have displaced many of the solvents that were once used in cleaning.

Hydrofluoroethers (HFEs) are one solvent technology that addresses all of these concerns. HFEs are a class of fluorinated liquids possessing a unique combination of properties that have proven to be useful in a variety of critical cleaning applications. The HFE compounds consist of carbon chains interrupted by an ether oxygen with one or more hydrogen atoms replaced by fluorine. The commercially available HFE compounds contain enough fluorine atoms on the carbon chain to render the materials nonflammable. HFEs were originally developed as alternatives to ozone-depleting compounds. While HFEs have been used in many applications to replace chlorofluorocarbons (CFCs) and hydrochlorofluorocarbons (HCFCs), they have now been used to replace many other solvents with one or more undesirable characteristic such as perfluorocarbons (PFCs), hydrofluorocarbons (HFCs), and several bromine and chlorine containing solvents.

The first-generation replacements for ozone-depleting solvents were partially or fully fluorinated alkanes (HFCs or PFCs). It was determined, however, that the insertion of an ether oxygen atom into the backbone of the molecule often modified the physical properties of a compound in important ways, particularly with regard to environmental properties. Research on a number of HFE compounds demonstrated that they could have significantly shorter atmospheric lifetimes and as a result lower global warming potentials (GWPs) when compared to HFCs and PFCs.[1] This is particularly true for segregated HFEs, those in which all of the hydrogen atoms reside on carbons with no fluorine substitution and are separated from the fluorinated carbons by the ether oxygen, that is, $C_xF_yOC_mH_n$. This research also showed that segregated HFEs combined these environmental attributes with the performance and safety properties required in a sustainable alternative solvent.

All of the HFEs have physical properties similar to the solvents they replace. Although their inherent solvent power is relatively low, HFEs are readily mixed with other components to produce nonflammable, high-strength solvents. As a result, HFEs have been extensively used as replacements for ozone-depleting substances (ODSs), high-GWP solvents, as well as chlorinated and brominated solvents in a variety of industrial cleaning processes.

Properties of Segregated HFEs and Their Impact on Cleaning Processes

General Characteristics

It is important for chemists and engineers to have an understanding of the physical and chemical properties of cleaning solvents in order to select the optimum material for their application. The properties displayed by a solvent are directly related to their chemical structure. The unique structure of the HFEs results in a combination of properties that are very useful in cleaning applications. HFEs are clear, colorless liquids that have very little odor. This class of compounds offers a wide range of boiling points from 34°C to well over 100°C. The materials have very low freezing points, often well below −100°C. The commercially available HFEs are nonflammable, noncorrosive, thermally stable, and electrically nonconductive. Like the solvents they replace, these fluids have high densities but low viscosity and surface tension.

Four HFEs have been commercialized for use in solvent cleaning applications. The structures and several identifying names and numbers are listed in Table 5.1. Two of the products contain inseparable isomers with essentially identical properties. The commercial HFEs are produced at high purity with very low nonvolatile residue, making them suitable for precision cleaning applications.

Physical Properties

Effective solvent cleaning processes make use of many of the unique properties of fluorinated solvents. The key physical properties that determine the performance of solvents in typical end users are listed in Table 5.2.[2–4] The HFE solvents have boiling points higher than many of the materials they replace, meaning that they have lower vapor pressures and are easier to contain in cleaning systems. These fluids have very low freezing points, which allow the use of low temperature cooling coils to further enhance fluid containment. The wide liquid range of the HFEs also enables their use in other applications such as heat transfer fluids.

The solubility of water in the HFEs is very low, which limits their ability to absorb humidity from the air. Their solubility in water is also extremely low, minimizing the potential to contaminate contacting water streams. The high density of the HFEs combined with their very low viscosity and surface tension are important properties in cleaning applications, particularly for penetrating and cleaning components having complex geometry. A useful parameter for assessing the potential performance of a precision cleaning solvent is the "wetting index," which is defined as the ratio of the solvent's density to its viscosity and surface tension.[5] A higher wetting index indicates an increased ability for the solvent to wet component surfaces and penetrate into tight spaces, especially for the removal of particulate contamination. The HFEs possess a combination of properties that lead to high wetting indices, indicating that they are well suited for precision cleaning applications. Another advantage of superior wetting capability is that it provides good drainage of the solvent from components at the end of the cleaning cycle. This property, combined with the low heat of vaporization, allows the HFEs to rapidly dry from part surfaces and minimizes fluid drag out from the cleaning process.

TABLE 5.1 Structures and Names of Commercially Available HFEs Used in Cleaning Applications

Chemical structures	$C_4F_9OCH_3$	$C_4F_9OC_2H_5$	$(CF_3)_2CFCF(OCH_3)CF_2CF_3$	$CHF_2CF_2OCH_2CF_3$
CAS numbers	163702-07-6 163702-08-7	163702-05-4 163702-06-5	132182-92-4	406-78-0
CAS names	2-(difluoromethoxymethyl)-1,1,1,2,3,3,3-heptafluoropropane 1,1,1,2,2,3,3,4,4-nonfluoro-4-methoxybutane	1-ethoxy-1,1,2,2,3,3,4,4-nonfluorobutane 2-(ethoxydifluoromethyl)-1,1,1,2,3,3,3-heptafluoropropane	1,1,1,2,2,3,4,5,5,5-decafluoro-3-methoxy-4-(trifluoromethyl)pentane	1,1,2,2,-tetrafluoroethyl-2,2,2-trifluoroethylether
Halocarbon numbers	HFE-449sccc1 HFE-449scym1	HFE-569sfccc2 HFE-569sfcym2	HFE-64-13m(m)yy(s)c3	HFE-347pcf2
Simplified halocarbon numbers	HFE-449s1	HFE-569sf2	HFE-64-13	HFE-347pcf2
Commercial names	3M™ Novec™ 7100 engineered fluid	3M™ Novec™ 7200 engineered fluid	3M™ Novec™ 7300 engineered fluid	ASAHIKLIN™ AE-3000

Notes: 3M and Novec are registered trademarks of 3M Company, St. Paul, Minnesota. ASAHIKLIN is a registered trademark of Asahi Glass Company, Ltd., Tokyo, Japan.

TABLE 5.2 Physical Properties of HFEs

Structure	HFE-449s1 $C_4F_9OCH_3$	HFE-569sf2 $C_4F_9OC_2H_5$	HFE-64-13 $(CF_3)_2CFCF(OCH_3)CF_2CF_3$	HFE-347pcf2 $CF_2HCF_2OCH_2CF_3$
Boiling point (°C)	61	76	98	56
Freezing point (°C)	−135	−138	−38	−94
Flash point (°C) Open or closed cup	None	None	None	None
Solubility for water (ppmw)	95	92	67	900
Solubility in water (ppmw)	12	3	0.6	100
Vapor pressure (atm)	0.266	0.143	0.059	0.305
Heat of vaporization (cal/g)	30	30	24	39
Density, ρ (g/mL)	1.52	1.43	1.66	1.47
Viscosity, μ (cp)	0.6	0.6	1.18	0.65
Surface tension, γ (dynes/cm)	14	14	15	16.4
Wetting index $(1000 \cdot \rho/\mu \cdot \gamma)$	181	170	94	138

Solvency and Mixtures Including Azeotropes

HFEs display an interesting combination of solvency characteristics. Since they are highly fluorinated, HFEs have very high solubility for fluorocarbons and other halogenated compounds. Fluorinated oils and greases are typically completely miscible in an HFE solvent. However, the pure HFEs typically have only limited solubility for other organic substances such as hydrocarbon or silicone oils. As indicated by the solvent parameters listed in Table 5.3, the pure HFEs are considered to be relatively mild solvents. They do not have the hydrocarbon solvency to match the ODS solvents they have typically replaced. However, since the HFEs can be readily mixed with other solvents to form nonflammable, high solvency mixtures, or multi-solvent systems, they can be just as effective in cleaning.

TABLE 5.3 Solvency Properties of HFEs

	HFE-449s1	HFE-569sf2	HFE-64-13	HFE-347pcf2	Azeotrope 1[a] 71DE	Azeotrope 2[b] 71DA	Azeotrope 3[c] 72DE	Azeotrope 4[d] 72DA
Hildebrand solubility parameter	6.5	6.3	6.1	7.3	7.7	7.8	7.8	8.4
Kauri-butanol (KB) value	10	10	20	13	27	33	52	58
Solubility for mineral oil (weight %)	<1	<1	<1	<1	20	20	M	45
Solubility for fluorinated oil (weight %)	M	M	M	<5	M	M	<1	<1

Note: Azeotropes are 3M products, e.g., Novec 72DA. M, Miscible in all proportions.

[a] Mixture of 50% by weight of *trans*-1,2-dichloroethylene in $C_4F_9OCH_3$.

[b] Mixture of 44.6% by weight of *trans*-1,2-dichloroethylene and 2.7% by weight of ethanol in $C_4F_9OCH_3$.

[c] A mixture of azeotropes containing 70% by weight of *trans*-1,2-dichloroethylene, 20% by weight of $C_4F_9OC_2H_5$, and 10% by weight of $C_4F_9OCH_3$.

[d] A mixture of azeotropes containing 68% by weight of *trans*-1,2-dichloroethylene, 20% by weight of $C_4F_9OC_2H_5$, 10% by weight of $C_4F_9OCH_3$, and 2% by weight of isopropanol.

Many of the solvent mixtures used in cleaning applications are azeotropes. The derivation of the term azeotrope is from the Greek word "zeotrope," which means to be separable by boiling. Since the prefix "a" indicates the negative, an azeotrope means *not* separable by boiling. Following this nomenclature, all mixtures that are not azeotropes are called zeotropes. Since formation of an azeotrope requires specific conditions to exist, it is often thought that azeotropes are a rare occurrence. In reality, azeotropic mixtures are fairly common, with thousands of them reported in the literature. The HFEs are miscible with a wide range of organic solvents and form azeotropic mixtures with numerous compounds.[6–9]

Azeotropes are classified according to their characteristics, including

1. The number of components in the mixture: two is considered a binary azeotrope, three is a ternary azeotrope, four a quaternary, etc.—both binary and ternary azeotropes containing HFEs have been identified.
2. Whether it is a minimum boiling or maximum boiling azeotrope: the boiling point of the azeotrope is below or above the boiling points of the individual components—all of the azeotropes identified containing HFEs are minimum boiling azeotropes.
3. Whether the mixture forms one or more liquid phases: a single phase is a homoazeotrope and those mixtures forming two or more liquid phases are considered heteroazeotropes—all of the commercially available HFE azeotropic mixtures are homogeneous azeotropes.

The most advantageous feature of an azeotrope is the fact that it is inseparable by boiling. When an azeotrope is boiled, the vapor generated will have the same composition as the liquid. As a result, an azeotrope is often described to behave as if it were a "single compound." This is in sharp contrast to what happens to ordinary mixtures (zeotropes), which change in composition when boiled. A common misconception is that an azeotropic mixture will maintain constant in composition under any condition. Unfortunately, this is not true. The ability to remain constant in composition occurs only at the boiling point of the azeotrope. If an azeotrope is held at conditions other than its boiling point, its composition could change.

For example, if an azeotrope is placed in an open container at room temperature and allowed to evaporate, it is likely that the liquid composition will eventually change. In fact, this has been shown to be true experimentally for a number of different azeotropes. An example of this phenomenon is shown in Figure 5.1. This change in composition is often so slow that it has no practical significance on the use of an azeotrope in applications below its boiling point. Nevertheless, there is no guarantee that an azeotrope will maintain its composition if it is allowed to evaporate without boiling. It is necessary for a solvent manufacturer to evaluate azeotropic mixtures prior to commercialization to ensure that any change in composition will not result in the formation of a flammable mixture.

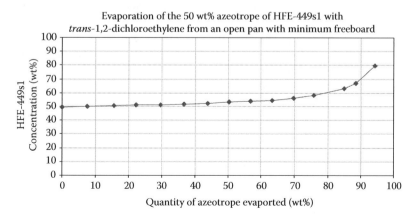

FIGURE 5.1 Change in mixture composition due to evaporation of an azeotrope below its boiling point.

Another fact often overlooked is that azeotropes are not ideal mixtures (thermodynamically speaking). The reason minimum boiling azeotropes form is because the components are chemically very different, resulting in the formation of a nonideal mixture. A consequence of this nonideality is that azeotropic mixtures will often phase split at reduced temperatures. Most binary, homoazeotropes will phase split into two liquids when cooled to a low enough temperature. This is another situation in which the mixture composition will change since two liquids of different compositions will have been formed. The temperature at which this begins to occur is typically called the critical solution temperature.

In some cases, addition of other materials to an azeotrope can also cause the mixture to phase split, particularly if the new material added to the azeotrope has greater solubility in one azeotrope component over the other. This effect is observed with HFE azeotropes containing an alcohol when water is added to them. Care has to be taken to limit contact with water to prevent a phase split and change in composition.

Some materials, if present in the azeotrope in a large enough concentration, can cause a shift in the vapor composition when the mixture is boiled. Instead of producing the true azeotropic composition, the solution can shift to somewhat higher concentrations of one component in the vapor. This phenomenon is known as extractive distillation. It is likely to be an issue in a cleaning application only if the system is allowed to reach very high soil loading levels.

Azeotropes comprising HFEs have found significant use in cleaning processes. Several nonflammable, azeotropic mixtures containing HFEs are commercially available, a number of which are identified in Table 5.3. All of these are single phase, homoazeotropes including both binary and ternary mixtures. These mixtures have kauri-butanol (KB) values similar to the ODS solvents and consequently are used in many of the same cleaning applications.

Safety Considerations

Safe use of a cleaning solvent requires consideration of a number of safety factors. A thorough safety assessment should consider the hazards associated with the solvent during a potential spill or upset condition as well as during normal operation.

Flammability

As indicated in Table 5.2, the commercially available HFE compounds are nonflammable, having no open or closed cup flash point. These products as well as the commercially available azeotropes have been assigned NFPA (National Fire Protection Association) flammability indices[10] of zero to one similar to the ODS solvents. These products do not become flammable under any condition of normal use.

Toxicity

In addition to flammability, the toxicity of the solvent is a critical parameter in determining if it can be safely used in a specific cleaning application. Workers may be exposed to a small amount of a solvent during normal use. The risk of far greater exposure also exists in the event of a solvent spill, leak or equipment failure. A solvent manufacturer needs to conduct extensive toxicological tests to determine if a solvent is safe in its intended applications. A number of the key findings for the HFEs[11–14] are listed in Table 5.4.

The segregated HFE solvents have very high acute lethal concentrations, meaning that even at high concentration they produced very little toxicological effect.[11–13] The Hodge and Sterner scale classifies materials into toxicity ratings ranging from 1 (extremely toxic) to 6 (relatively harmless) based on acute (short-term) oral, inhalation, and dermal exposure.[15] The segregated HFEs are classified as practically nontoxic on this scale. All of the segregated HFEs were found to be nonmutagenic and are dermally and occularly nonirritating.[11–13] The high exposure guidelines and lack of exposure ceiling indicate that the segregated HFEs have a wide margin of safety.

TABLE 5.4 Toxicological Properties of HFEs

	HFE-449s1 $C_4F_9OCH_3$	HFE-569sf2 $C_4F_9OC_2H_5$	HFE-64-13 $(CF_3)_2CFCF(OCH_3)CF_2CF_3$	HFE-347pcf2 $CF_2HCF_2OCH_2CF_3$
Acute lethal conc. 4 h LC_{50} (ppmv)	>100,000	>92,000	>30,000	3010
Mutagenicity	Negative	Negative	Negative	Negative
Ocular irritant	No	No	No	Slight
Dermal irritant	No	No	No	No
Exposure guideline 8 h TWA (ppmv)	750	200	100	50
Exposure ceiling (ppmv)	None	None	None	1000
Toxicity classification (Hodge and Sterner scale)	Practically Nontoxic	Practically Nontoxic	Practically Nontoxic	Slightly Toxic
Vapor hazard ratio, Calculated at 25°C	350	720	590	6100

Vapor Hazard Ratio

A useful parameter in assessing the margin of safety associated with various cleaning solvents is the vapor hazard ratio (VHR). The VHR was introduced by Popendorf to compare solvent vapor hazards.[16] The vapor pressure of a liquid provides a measure of the saturated concentration of the material in the air immediately above the surface of the liquid. The VHR is the ratio between this saturation concentration at the vapor source and the airborne concentration of concern for the material. For an occupational hazard assessment, this is calculated as the ratio of the vapor pressure at the temperature of handling or use to the exposure guideline for the compound under consideration. Quantitatively, the VHR provides the minimum factor by which the vapors at the source must be diluted in order to maintain a safe working environment. Comparison of VHR values for different solvents can illustrate the level of industrial hygiene control that will need to be employed when using these materials. The VHR values for HFE solvents at 25°C (calculated as vapor pressure in atm × 10^6/exposure guideline in ppmv) are shown in Table 5.4. A higher VHR value indicates that increased ventilation and containment practices will be required to maintain a safe work environment. The low VHR values for the segregated HFEs indicate that their exposure guidelines will be relatively easy to maintain in comparison to other solvents.

Thermal and Chemical Stability

Another important safety consideration is the thermal stability of the solvent. The HFEs are capable of being continuously refluxed without degradation. Even in the presence of air and metals there has been no evidence of peroxide formation which is common to many hydrocarbon ethers.

All solvents can degrade if severely overheated. This degradation can produce by-products, which are more hazardous than the original solvent. For example, it was known that CFC-113 could produce decomposition products such as HCl and HF if overheated.[17] Similarly, the HFEs can generate hazardous decomposition products such as HF if severely overheated (e.g., exposure to temperatures 150°C or higher). Conventional cleaning equipment has safety interlocks incorporated into their design to prevent overheating the solvents. In addition, the high temperatures required to decompose an HFE solvent (90°C or more above their boiling point) provide a wide margin for safe use. Solvent manufacturer's recommendations should be followed when recycling and recovering solvent for reuse.

The HFEs also have a high degree of chemical stability. The materials are hydrolytically and oxidatively stable under normal use conditions. The pure HFEs are stable when refluxed in the presence of water or strong aqueous base. Contact with many other relatively strong acids and bases produces little, if any, reaction. Exceptions are reactions with amines such as piperidine. As with all halogenated solvents, the HFEs should not be contacted with finely divided active metals, alkali, or alkaline earth metals (i.e., Groups IA and IIA of the periodic table).

Environmental and Regulatory Considerations

An increasing number of environmental properties need to be considered when selecting cleaning solvents. The properties most frequently considered when assessing the environmental sustainability of a solvent are listed in Table 5.5. These properties affect both local environmental issues such as smog formation as well as global issues such as ozone depletion and climate change. Since the HFEs contain no chlorine or bromine, they have no ozone depletion potential.[18] The HFEs are not considered hazardous air pollutants since they are low in toxicity. The atmospheric lifetimes of the HFE solvents are in a very desirable range. They are long enough lived to prevent contribution to formation of photochemical smog. Consequently, many of the HFE compounds are exempt from volatile organic compound (VOC) regulations.[19] It should be noted, however, that the azeotropic blends contain components that are typically regulated as VOCs. The atmospheric lifetime of the HFEs are still short enough to preclude concerns with accumulation in the atmosphere. These shorter atmospheric lifetimes lead to lower GWPs.[20]

Many countries are making efforts to reduce greenhouse gas emissions. While some regions have already implemented regulations, many others are considering legislation that would restrict the use of high-GWP materials. HFEs are among the lowest-GWP, nonflammable compounds that are commercially available. The lower GWPs of the HFEs can lead to significant reduction in greenhouse gas emissions from some critical cleaning applications. For example, U.S. EPA projects emission reduction of 82%–96% on a CO_2 equivalent basis when using HFEs to replace higher-GWP solvents.[21]

In many cleaning applications, the use of HFEs will also lead to a reduction in the quantity of solvent and energy consumed. The HFEs typically exhibit lower loss rates from cleaning equipment due to the combination of lower fluid drag out (a result of the higher wetting indices) with higher boiling points and vapor densities.[22] The reduced loss rates along with the lower heats of vaporization decrease the energy consumption of most HFE cleaning systems, particularly when compared to aqueous processes. HFEs also demonstrate a high degree of stability that allows their recovery and recycle for repeated use. Used HFE solvents are not classified as hazardous waste; however, the soils present in them may change that classification (e.g., metals from a defluxing operation would typically be considered hazardous waste). For this reason, the leading manufacturer of HFEs has established a return program to assist users in disposal of used solvents.

Materials Compatibility

The pure HFE solvents are compatible with essentially all common metals, most plastics, and a number of elastomers. Table 5.6 lists a number of the specific materials that have been tested with the HFEs.[4,23] Test coupons of the materials listed in Table 5.6 were exposed to the HFE solvents for 3 or 7 days at the

TABLE 5.5 Environmental Properties of HFEs

Structure	HFE-449s1 $C_4F_9OCH_3$	HFE-569sf2 $C_4F_9OC_2H_5$	HFE-64-13 $(CF_3)_2CFCF(OCH_3)$ CF_2CF_3	HFE-347pcf2 $CF_2HCF_2OCH_2CF_3$
Atmospheric lifetime (years)	3.8	0.77	3.8	7.1
Ozone depletion potential, [CFC-11 = 1]	0	0	0	0
Global warming potential,[a] [CO_2 = 1]	297	59	210	580
Photochemical smog precursor	No	No	No	No

[a] GWPs calculated using the IPCC 2007 method.[20]

TABLE 5.6 Materials Compatibility with HFE Solvents

	HFE-449s1 $C_4F_9OCH_3$	HFE-569sf2 $C_4F_9OC_2H_5$	HFE-347pcf2 $CF_2HCF_2OCH_2CF_3$	Azeotrope 1 HFE-449s1 t-CClH=CClH
Aluminum	1	1	1	1
Copper	1	1	1	1
Carbon steel	1	1	1	1
302 Stainless steel	1	1	1	1
Brass	1	1	1	1
Zinc	1	1		1
Molybdenum	1	1		1
Tantalum	1	1		1
Titanium	1	1		1
Tungsten	1	1		1
Acrylic	1	1	2	3
Polyethylene	1	1	1	2
Polypropylene	1	1	1	2
Polycarbonate	1	1	1	3
Polyester	1	1	1	2
Nylon	1	1	1	1
Epoxy	1	1	1	2
PVC	1	1	1	3
PET	1	1		3
ABS	1	1	1	3
PTFE	1	1	1	1
Butyl rubber	1	1		2
Natural rubber	1	1	1	3
Nitrile rubber	2	2	2	3
EPDM	2	2	1	2
Fluoroelastomer	2	2	2	2
Polychloroprene	2	2	1	2

Note: 1, compatible over extended exposures to the solvent with less than 5% changes in weight or volume; 2, compatible with limited exposure to the solvent (i.e., short time exposures such as a cleaning cycle); 3, typically incompatible.

boiling point of the solvent and subsequently examined for weight, volume, and appearance changes. The HFE azeotropes containing *trans*-1,2-dichloroethylene exhibit compatibility similar to the pure HFEs with all metals but due to their higher solvency have limited compatibility with most polymeric materials. Compatibility of the components to be cleaned as well as materials of construction for the cleaning equipment should be evaluated with the various cleaning fluids during process selection.

Cleaning Processes and Equipment

Several different cleaning processes exist that employ HFE solvents. These processes are designed to cover a range of potential substrates and soils that can be encountered in cleaning applications. The processes in commercial use include:

1. Neat cleaning systems using pure HFEs
2. Azeotrope cleaning systems using azeotropic mixtures
3. Multi-solvent cleaning systems using zeotropic mixtures

TABLE 5.7   HFE Cleaning Process Selection Guidelines

	Neat Cleaning Process	Azeotrope Cleaning Process	Co-Solvent Cleaning Process
Kauri-butanol value of cleaning solvent	10–20	27–58	20 to >150
Substrates		Soils	
	Light hydrocarbon and silicone oils, halogenated oils, particulate	Medium oils, lubricants, release agents, some waxes and fluxes	Heavy oils, greases, buffing compounds, heavy flux
Plastic parts[a]	OK	Not appropriate	OK
Metal parts	OK	OK	OK
Circuit boards	Not appropriate	OK	OK
Coils	OK	OK	OK
Ball or roller bearings	OK	OK	OK
Catheters[a]	OK	May be applicable	Not appropriate
Elastomer parts[a]	OK	May be applicable	OK
Glass parts	OK	OK	OK

[a] Careful evaluation of the compatibility of the parts with the various cleaning fluids should be evaluated during process selection.

The appropriate cleaning system is selected based upon the soil to be removed, the materials of construction of the part to be cleaned and a number of other factors as indicated in Table 5.7. All of the HFE cleaning processes can be conducted in conventional cleaning equipment such as a vapor degreaser (Figure 5.2) as well as in-line cleaning systems.

Neat Cleaning Systems

The neat cleaning process, employing pure HFE solvents, is used when a single-component, mild solvency cleaning fluid is required. This process is conducted in a conventional vapor degreaser (Figure 5.2) and can effectively clean light hydrocarbon and silicone oils, particulate contamination

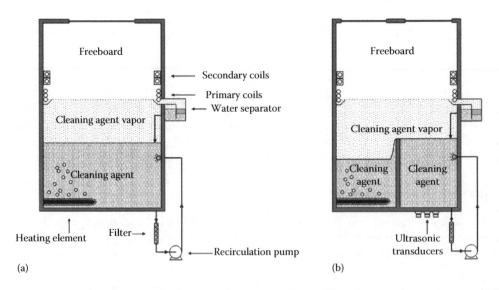

FIGURE 5.2   HFE cleaning equipment for neat and azeotropic solvents: (a) single-sump cleaning system and (b) multi-sump cleaning system.

and halogenated lubricants, oils and greases from parts. The HFE solvent can be used in a single or multiple sump system. Parts' cleaning with this process is conducted in a manner similar to that used with conventional vapor degreasing solvents. Both vapor phase and liquid phase cleaning are possible depending upon the parts and soils to be removed. The mechanism of cleaning with neat HFE systems can be by dissolving or displacement of the soil.

Azeotrope Cleaning Systems

The azeotrope cleaning process uses an HFE azeotropic mixture, such as those listed in Table 5.3, as the cleaning fluid. This process is used in applications requiring a stronger solvent mixture. Since the solvents used in this process are azeotropes or azeotrope-like mixtures, the compositions of the cleaning and rinse sumps remain essentially constant throughout use. The mechanism of cleaning with an azeotrope system is almost exclusively via dissolving the soil. The equipment required is similar to the conventional degreasers used with ODSs and their azeotropes. The HFE azeotropic mixtures can be used in single or multiple sump equipment in the same manner as the neat cleaning system described above. This process effectively cleans many oils, waxes, greases, and fluxes depending upon the azeotropic solvent used.

Multi-Solvent Cleaning Systems

Co-Solvent Cleaning Process

A co-solvent process combines two different fluids to conduct the cleaning process. The first fluid is a low volatility, organic solvent that is referred to as a solvating agent. This high-solvency fluid is used to dissolve the soil from a part's surface. The second fluid, for example the HFE, functions as a rinsing agent since it is used to remove the solvating agent from the part. This process typically uses solvating agents that are miscible with the HFE. These mixtures are not azeotropes (i.e., they are zeotropes) since the components separate when boiled, resulting in very different compositions in the cleaning sump and rinse sump. The process operates analogously to a two-sump vapor degreaser with a mixture of the solvating agent and rinsing agent in the boil sump and pure HFE in the rinse sump (see Figure 5.3). Because of the large difference in boiling points between the two components, the mixture acts as a zeotrope and separates during boiling with very little solvating agent distilled into the rinse sump. The rinse sump contains essentially pure HFE throughout the process. The mechanism of cleaning in this process is most often by dissolving the soil into the mixture of solvating and rinsing agents in the first sump.

A wide variety of high boiling, combustible solvents can be used as solvating agents in this process, provided that their flash points are well above the operating temperature of the system. The materials used as solvating agents typically have flash points in excess of 90°C. The HFE rinse agent renders the solvating agent nonflammable during use. Higher solvency mixtures are created by increasing the ratio of solvating agent to rinsing agent in the boil sump. The process is controlled by monitoring the boiling temperature of this mixture since the operating temperature is determined by composition in the boil sump as shown in Figure 5.4. Temperature and composition control can be accomplished manually through periodic measurements of the boiling temperature followed by solvent additions when necessary or via an automated system.

Since this process requires immersion into the solvent mixture in the cleaning sump, vapor cleaning is not possible. The HFE co-solvent process is capable of effectively cleaning a very wide variety of soils including heavy oils, greases, fingerprints, waxes, and flux. Co-solvent processes offer a great deal of flexibility by selecting a solvating agent and rinsing agent combination, which best meets the needs of a particular cleaning application. The process can provide sufficient solvency for a given soil while maintaining compatibility with the part's materials of construction. The higher solvency mixtures of the solvating and rinsing agents can accommodate higher soil loading levels than the neat or azeotrope systems.

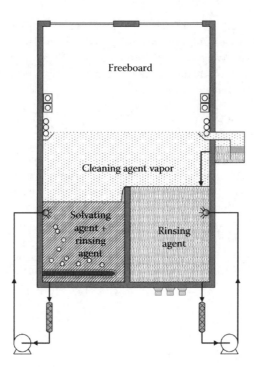

FIGURE 5.3   HFE co-solvent cleaning equipment.

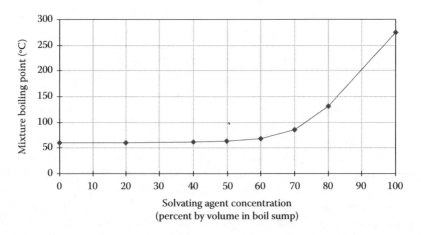

FIGURE 5.4   Operating temperature of HFE co-solvent cleaning process. Example using $C_4F_9OCH_3$ with isopropyl myristate as a solvating agent.

Bi-Solvent Cleaning Process

A bi-solvent cleaning system is similar to the co-solvent cleaning process except that the high-solvency organic solvent is not mixed with the HFE. In this process, the solvating agent is contained in a separate container. As a result, the bi-solvent cleaning process utilizes the full solvent power of the solvating agent without dilution by the rinsing agent. The separation of fluids also reduces the carryover of solvating agent into the rinsing agent, thereby maintaining a higher purity rinse. An example of equipment for a bi-solvent cleaning process is shown in Figure 5.5. This process design

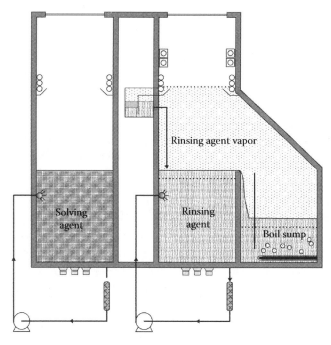

FIGURE 5.5 HFE bi-solvent cleaning equipment.

increases the variety of solvating agents that can be used in conjunction with HFEs, further extending the range of soils that can be effectively removed. Many of the solvating agents used in the co-solvent or bi-solvent cleaning process have vapor pressures low enough to be defined as exempt from VOC regulations. Consequently, when these solvating agents are used with the VOC-exempt HFEs, the cleaning process becomes compliant with the regulations found in even the most restrictive areas in the United States. However, since regulations vary, users are advised to consult local regulations prior to using any material.

Drying/Water Removal Processes

HFE solvents can be used as a drying fluid following processes such as aqueous cleaning or metal plating. The drying processes can be conducted in equipment similar to vapor degreasers used for cleaning applications. Water can be absorbed from the surface of a component using HFE/alcohol solutions such as the azeotrope that forms between $C_4F_9OCH_3$ and isopropanol. These mixtures are typically useful for applications with relatively low water removal needs since the solutions can become saturated with water, reducing their efficacy. The solutions can be distilled to remove the water but care must be taken to avoid extracting the alcohol into the water and removing it from the drying solution.

Applications demanding larger water removal capability require the use of an HFE/surfactant mixture to displace the water from the component's surface. Parts to be dried are immersed into a dilute solution of a surfactant in the HFE solvent. The surfactant allows the solvent to preferentially wet the surface. This wetting action causes the water to bead up and displace from the surface, floating to the top of the solution. The part is then immersed in one or more sumps of pure HFE solvent to rinse the surfactant from its surface. This process is capable of efficiently removing water from complex components in short cycle times.

Aerosol Cleaning

HFEs are used as solvents in a number of aerosol cleaners. The pure HFEs as well as several of the azeotropic mixtures are used in pressurized aerosol cans for spot cleaning, rework, and other manual cleaning operations. The HFEs are also used in a number of custom aerosol formulations. The unique combination of properties such as rapid evaporation rate, low surface tension, and lack of electrically conductivity make the HFEs useful for cleaning electrical components and other precision devices. Often, this allows device manufacturers to use the same solvent formulation for manual cleaning as they use in their production-scale cleaning process. Dispensing through an aerosol can allows the user to take advantage of the solvency of the mixture as well as enhance the cleaning via displacement by effective use of the pressurized spray.

Summary: The Role of HFEs in Critical Cleaning

HFEs have been shown to have a combination of properties that make them a sustainable technology for use in critical cleaning applications. As a result, HFEs currently provide a useful range of solutions for solving many critical cleaning challenges. The key qualities provided by the HFEs include

- Performance—HFEs are used in a variety of cleaning processes that take advantage of their solvency, wetting capabilities, and rapid drying characteristics. The formulations that use HFEs in azeotropes and multi-solvent cleaning systems further enhance their cleaning and rinsing capabilities. As a result, cleaning processes using HFEs are effective at removing a wide range of soils from nearly any type of substrate.
- Safety—HFEs and their commercial azeotropes are nonflammable. In addition, the HFEs are used in many multi-solvent systems that are nonflammable. HFEs are also low in toxicity with most materials classified as practically nontoxic. These materials provide a wide margin of safety when used in cleaning applications.
- Environment—HFEs are non-ozone depleting and produce low climate impact due to their low GWPs. The pure HFEs as well as many of the materials used in the multi-solvent cleaning processes are VOC exempt.

HFEs combine effective cleaning performance with a wide margin of safety and a favorable environmental profile. This combination of features has allowed a variety of industries to meet the stringent requirements of critical cleaning applications with an environmentally sustainable technology. While HFEs are not capable of solving every cleaning challenge, ongoing solvent research continues to expand this class of compounds and the mixtures and azeotropes that can be formed with them. With these developments, it is expected that HFEs will provide an even broader range of cleaning solutions in the future.

References

1. Owens, J.G. Segregated hydrofluoroethers: Low GWP alternatives to HFCs and PFCs, presented at Joint IPCC/TEAP Expert Meeting on Options for the Limitation of Emissions of HFCs and PFCs, Petten, NL, May 26–28, 1999.
2. Novec 7200 Engineered fluid—Product information sheet. 3M Company, St. Paul, MN, 2005.
3. Novec 7300 Engineered fluid—Product information sheet. 3M Company, St. Paul, MN, 2005.
4. *AE-3000 Technical Bulletin*. Asahi Glass Company, Ltd., Tokyo, Japan, 2004.
5. Kenyon, W.G. New ways to select and use defluxing solvents, presented at Nepcon West, Anaheim, CA, 1979.
6. Flynn, R.M., Milbrath, D.S., Owens, J.G., Vitcak, D.R., and Yanome, H. U.S. Patent 5827812 to 3M Company, 1998.

7. Flynn, R.M., Milbrath, D.S., Owens, J.G., Vitcak, D.R., and Yanome, H. U.S. Patent 5814595 to 3M Company, 1998.

8. Rajtar, P.E. and Owens, J.G. U.S. Patent 7071154 to 3M Company, 2006.

9. Mochizuki, Y., Takayuki, I., Tetayuki, S., and Sekiya, A. Japanese Patent JP10324897 to the Agency of Industrial Science and Technology, Japan, 1998.

10. NFPA (National Fire Protection Association) 49, Hazardous Chemicals Data, 1994 edition, NFPA, Quincy, MA.

11. Workplace Environmental Exposure Level, HFE-7100. American Industrial Hygiene Association, Fairfax, VA, 1999.

12. Toxicology assessment: 3M Novec 7200 engineered fluid. 3M Company, St. Paul, MN, 2008.

13. Toxicology assessment: 3M Novec 7300 engineered fluid. 3M Company, St. Paul, MN, 2009.

14. Material Safety Data Sheet: AE-3000. Asahi Glass Company, Ltd., Tokyo, Japan, 2007.

15. Hodge, H.C. and Sterner, J.H. *American Industrial Hygiene Association Quarterly*, 10, 93–96, 1949.

16. Popendorf, W. *American Industrial Hygiene Association Journal*, 45(10), 719–726, 1984.

17. Ellis, B.N. *Cleaning and Contamination of Electronics Components and Assemblies*. Electrochemical Publications Limited, Ayr, Scotland, p. 175, 1986.

18. Wallington, T.J., Schneider, W.F., Sehested, J., Bilde, M., Platz, J., Nielsen, O.J., Christensen, L.K., Molina, M.J., Molina, L.T., and Wooldridge, P.W. *The Journal of Physical Chemistry A*, 101, 8264–8274, 1997.

19. U.S. Federal Register, 72, January 18, 2007, 2193. Environmental Protection Agency, 40 CFR Part 51, Air quality: Revision to definition of volatile organic compounds.

20. IPCC 2007. *Climate Change 2007: The Physical Science Basis. Contribution of Working Group I to the Fourth Assessment Report of the Intergovernmental Panel on Climate Change*, Solomon, S., Qin, D., Manning, M., Chen, Z., Marquis, M., Averyt, K.B., Tignor, M., and Miller, H.L. (eds.), Cambridge University Press, Cambridge, U.K. and New York, 996 pp., 2007.

21. U.S. Environmental Protection Agency. Analysis of HFC production and consumption controls, October, 2009.

22. Warren, K.J. Use of hydrofluoroethers in electronics cleaning applications, presented at the International Conference on Ozone Protection Technologies, Baltimore, MD, November, 1997.

23. 3M Novec Engineered Fluids Technical Bulletin, *The Science of Precision and Electronics Cleaning*, 3M Company, St. Paul, MN, 2003.

6

Hydrofluorocarbons

Joan E. Bartelt
*DuPont Chemicals
and Fluoroproducts*

Abid Merchant

Introduction

Hydrofluorocarbons (HFCs) are a family of compounds containing carbon, fluorine, and hydrogen. The absence of chlorine in the hydrofluorocarbon molecules makes them non-ozone-depleting substances (ODSs); they are, therefore, suitable replacements for chlorofluorocarbons (CFCs) and hydrochlorofluo-rocarbons (HCFCs). Additionally, the presence of hydrogen atom(s) reduces atmospheric life and there-fore these compounds have significantly lower global warming potential (GWP) than fully fluorinated molecules (perfluorinated or PFCs) and CFC compounds. The presence of a large number of fluorine atoms tends to make these compounds nonflammable, low in toxicity, stable to heat, low in reactivity, and compatible with most materials of construction.

At present, the Montreal Protocol–dictated phaseouts of CFCs and HCFCs are well underway. The phaseout of the once high-volume chemicals, CFC-113, methyl chloroform, and HCFC-141b, is essen-tially complete for noncritical applications. In terms of cleaning applications, one ODS, HCFC-225ca/cb, is scheduled for a production phaseout in 2015.

Examples of HFCs, which have been commercialized for critical cleaning and related solvent applica-tions, are HFC-43-10mee (manufactured by DuPont), HFC-365mfc (manufactured by Solvay), HFC-245fa (manufactured by Honeywell), and HFC-c447 (heptafluorocylcopentane, manufactured by Zeon Corp.). The chemical solvency of HFC class of materials lies between that of CFC-113 and the PFCs. Solvency can be enhanced significantly by the use of appropriate azeotropes and blends with alcohols, hydrocarbons, esters, and hydrochlorocarbons. Many of these blends are commercially successful. This chapter focuses on aspects of HFCs important to critical cleaning applications and primarily on HFC-43-10mee.

Research in the fluorinated solvent market is very active today. While HFCs have been extremely important for quick replacement of CFCs and HCFCs, the goal of current research is to reduce the GWP of these materials. Global warming is predominately a result of CO_2 emissions from energy-producing activities worldwide, but all global warming gases are facing possible increased regulation. It is expected that new fluorinated materials will be available in the future with lower GWPs than any of the similar chemicals on the market today.

Structure, Toxicity, and Properties of HFC-43-10mee

Chemical Structure and Nomenclature

HFC-43-10mee is a straight-chain HFC. Its molecular structure is provided in Figure 6.1. In the Chemical Abstracts nomenclature, its name is 1,1,1,2,2,3,4,5,5,5-decafluoropentane. It is marketed under the trade name of DuPont™ Vertrel® XF. The commercial material consists of a pair of diasteriomers with similar properties. Also in Figure 6.1 is the structure of HFC-365mfc, which is named 1,1,1,3,3-pentafluorobutane.

Toxicity of HFC-43-10mee

HFC-43-10mee has a low order of acute oral, dermal, and inhalation toxicity. The rat 4h LC50 is 11,100 ppm. The rabbit dermal and rat oral LD50 values are >5000 mg/kg. Sustained inhalation of 2000 ppm and higher HFC-43-10mee were associated with transient central nervous system effects, including tremors, seizures, and convulsions in rats. Acute central nervous system effects precluded inhalation exposures of 10,000 ppm and above in cardiac sensitization studies. No cardiac sensitization was observed in epinephrine-challenged dogs exposed to 5000 ppm. HFC-43-10mee was slightly irritating to skin and eyes during laboratory tests.

HFC-43-10mee was not genotoxic when evaluated in laboratory tests. Genetic toxicity tests include an in vitro Ames assay, a chromosomal aberration study with human lymphocytes, and an in vivo mouse micronucleus. No mutagenic changes were observed in any of these studies.

HFC-43-10mee was evaluated in repeated inhalation studies. In a 90 day study, adverse effects were observed at concentrations of 2000 ppm and higher. The adverse effects were similar to those observed in acute inhalation studies, and consisted of transient central nervous system effects, tremors, and seizures. HFC-43-10mee is not a developmental or reproductive toxin in laboratory animals. Adverse effects observed during developmental and reproductive inhalation tests were similar to those observed in acute inhalation studies.

The safe workplace exposure concentration to which nearly all workers could be exposed throughout a working lifetime without experiencing adverse effects was set by DuPont at 200 ppm 8h time-weighted

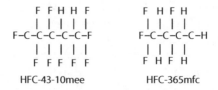

FIGURE 6.1   Structures of HFC-43-10mee and HFC-365mfc.

average, with a 400 ppm ceiling. This worker safety value is reflective of the relatively low order of toxicity associated with HFC-43-10mee.

Properties of HFC-43-10mee

HFC-43-10mee is a clear, colorless, dense liquid with a faint ethereal solvent odor. Table 6.1 gives a list of physical, transport, and thermodynamic properties of HFC-43-10mee along with that of HFC-365mfc. HFC-43-10mee has a boiling point of 55°C and a freezing point of −80°C and therefore can be used as a liquid over a broad temperature range. The high liquid density of the HFCs helps to displace soils and particulate from the surfaces of parts being cleaned and to float these soils to the surfaces of the solvent. One of the clear advantages of HFC-43-10mee is its low surface tension compared to other solvents such as chlorocarbons, hydrocarbons, alcohols, and water. A low surface tension makes it easier to wet surfaces and thus assist in the removal of soils from the small crevices and openings found in surface mount printed wiring boards (PWBs) and the close tolerances of precision inertial guidance components. The energy consumption of a recirculating solvent system in a degreaser is a direct function of the heat of vaporization of the solvent. The heats of vaporization of the HFCs are much lower than that of alcohols, hydrocarbons, chlorocarbons, and water, making it more energy efficient in use. HFC-43-10mee has no flash point by both open and closed cup methods and no flammable limits in air. HFC-365mfc is flammable but can be inerted by the use of blends containing HFC-43-10mee.

TABLE 6.1   Physical Properties of HFC-43-10mee and HFC-365mfc

Property[a]	HFC-43-10mee	HFC-365mfc[b]
Molecular weight	252	148
Boiling point, °C (°F)	55 (130)	40 (104)
Vapor pressure, mm Hg (psia)	226 (4.4)	353 (6.8) (at 20°C)
Freezing point, °C (°F)	−80 (−112)	Not available
Liquid density, g/cc (lb/gal)	1.58 (13.2)	1.27 (10.6) (at 20°C)
Surface tension, dyn/cm	14.1	Not available
Viscosity, cPs	0.67	0.45 (at 20°C)
Solubility in water, ppm	140	840 (at 23°C)
Solubility of water, ppm	490	Not available
Heat of vaporization (at boiling point), cal/g (Btu/lb)	31.0 (55.7)	Not available
Specific heat at 20°C (68°F), cal/g·°C (Btu/lb·°F)	0.27 (0.27)	Not available
Diffusivity, cm²/s	0.066	Not available
Thermal conductivity, Btu/h ft °F (mW/m K)		
Vapor	0.0057 (9.9)	0.0061 (10.6)
Liquid	0.036 (62)	Not available
Refractive index	1.24	Not available
Flash point, closed cup[c]	None	≤−27°C[d]
Flammable range in air	None	3.6%–13.3%
Auto ignition point in air	None[e]	Not available

[a] At 25°C (77°F) except where indicated.
[b] Obtained from Solvay Fluor literature.
[c] Tag closed cup tester (ASTM D56).
[d] DIN 51755, Tiel 2.
[e] None detected up to 540°C.

TABLE 6.2 Density and Vapor Pressure
Change with Temperature of HFC-43-10mee

Temperature, °C (°F)	Density, g/cc (lb/g)	Vapor Pressure, mm Hg (psia)
−20 (−4)	1.70 (14.2)	16 (0.3)
−10 (14)	1.68 (14.0)	36 (0.7)
0 (32)	1.66 (13.8)	62 (1.2)
10 (50)	1.62 (13.5)	109 (2.1)
20 (68)	1.60 (13.3)	176 (3.4)
30 (86)	1.57 (13.1)	284 (5.5)
40 (104)	1.55 (12.9)	434 (8.4)
50 (122)	1.51 (12.6)	641 (12.4)
60 (140)	1.49 (12.4)	921 (17.8)
70 (158)	1.46 (12.2)	1288 (24.9)
80 (176)	1.43 (11.9)	1753 (33.9)
90 (194)	1.40 (11.7)	2343 (45.3)
100 (212)	1.38 (11.5)	3072 (59.4)
110 (230)	1.34 (11.2)	3961 (76.6)
120 (248)	1.32 (11.0)	5032 (97.3)
130 (266)	1.30 (10.8)	6309 (122.0)

Table 6.2 gives both vapor pressure and specific gravity data as a function of temperature. The relationship between vapor pressure and temperature for a monomolecular compound is described by the Antoines equation. The Antoines equation for vapor pressure of HFC-43-10mee, including the compound-specific constants (*a*, *b*, and *c*) is

$$\log_{10} P = a - b/(c + T)$$

where
  $a = 7.03668$
  $b = 1093.094$
  $c = 208.3936$
  P is the millimeter of mercury
  T is in °C

Electrical Properties

Electrical properties are given in Table 6.3. The dielectric constant is slightly higher than that of CFC-113 and the breakdown voltage is lower than that of CFC-113. Thus, the dielectric strength of the HFC-43-10mee is lower but still considered acceptable. The volume resistivity is in the most desirable range to minimize electrostatic discharge (ESD) discharge. ESD has become very important in the computer and electronic industry. The high-density disk drives use new technology consisting of giant magnetoresistive (GMR) heads which are extremely sensitive to small changes in current flow through the devices. ESD can result in a momentary current surge which could, as a minimum, cause severe damage to the magnetic properties and could lead to a complete failure by the fusion of the GMR head. Therefore, it is desirable for all materials used in the manufacturing process to have a low propensity to generate ESD.

TABLE 6.3 Electrical Properties
of HFC-43-10mee

Resistivity	2.9×10^9 Ω-cm
Dielectric constant	7
Dissipation factor	0.14
Breakdown voltage	27 kV

The ESD properties of a solvent can be measured by measuring its volume resistivity. The resistivity can be classified per EIA-541 (Electronics Industries Alliance, Packaging Material Standards for ESD Sensitive Items) and other specifications as

Conductive	Less than 1×10^5 Ω-cm
Dissipative	Between 1×10^5 and 1×10^{11} Ω-cm
Insulative	Above 1×10^{12} Ω-cm

A lower value of the resistivity means a lower propensity to generate ESD. As can be seen, the HFC has a resistivity number in the range called the "dissipative," which is considered most desirable to minimize ESD generation and at the same time have adequate insulative properties.

Compatibility of Materials of Construction

Two types of compatibility testing are conducted, short term and long term. Short-term compatibility tests, generally a 15 min exposure under a given set of conditions, are useful for evaluating compatibility with materials of construction of articles being cleaned. Longer-term compatibility tests, generally a 1–2 week period, at or above the boiling point, are useful for evaluating compatibility of the materials of construction of cleaning equipment. HFC-43-10mee is a milder solvent than CFC-113 and its compatibility with most plastics, elastomers, and metals is as good as or better than CFC-113.

Plastics and Elastomers

The effect of solvents on plastics or on elastomers (e.g., swelling, shrinkage, extractables, and hardness) depends on the nature of polymer, the compounding method, the curing or vulcanizing conditions, the presence of plasticizers or extenders, and other factors. For these reasons, it is difficult to make generalizations on the effects of any solvents on these materials. Testing in the proposed application is particularly important.

Table 6.4 summarizes the effects of HFC-43-10mee on a large variety of plastics in the unstressed state exposed at 50°C in a sealed tube test for a period of 2 weeks. The plastic specimens were examined before and after the exposure tests for physical change as well as weight gain, and were assigned an empirical rating from 0 to 2; i.e., 0 means no effect, 1 means borderline, 2 means incompatible. These tests show that the HFC has minimal effects on most commonly used plastics, except acrylic. Fluoropolymer plastics show high weight gain; however, weight returns to normal after air-drying. The 2 week tests simulate exposures for the vapor degreaser construction material and are significantly more stringent than the exposure during a normal cleaning cycle. The tests also suggest that the HFC may dissolve and extract plasticizer from flexible, highly plasticized PVC tubing and contaminate the solvent. Therefore, flexible PVC tubing should not be used to transfer the solvent.

Table 6.5 gives the effects of the HFC on a large variety of elastomers exposed at 50°C in a sealed tube test for a period of 1–2 weeks. The elastomer specimens were examined before and after the tests for linear swell, hardness change, and physical appearance. The samples were assigned 0–2 empirical ratings similar to the plastics, as discussed above. The fluoroelastomers as a group, and two others ("Vamac" and "Alcryn"), were considered incompatible due to excessive linear swells and hardness change. Elastomer swelling and shrinking will, in most cases, revert to within a few percent of the original size after air-drying.

Metals Compatibility

Table 6.6 summarizes the effect of HFC-43-10mee on most commonly used metals exposed at 100°C in a sealed tube test for a period of 2 weeks. The tests were conducted under two conditions: dry and

TABLE 6.4 Plastic Compatibility with HFC-43-10mee, 2 Weeks at 50°C (122°F) in a Sealed Tube Test

Plastic	Common Brand Name	Rating	Weight Gain, %
HDPE	"Alathon"	0	0.3
PP	"Tenite"	0	0.5
PS	"Styron"	0	0.3
PVC		0	0.1
CPVC		0	0.1
PTFE	Teflon®	1[a]	3.5
ETFE	Tefzel®	1	1.4
PVDF	"Kynar"	0	0.4
Ionomer	Surlyn®	0	0.5
Acrylic	Lucite®	2	—[b]
ABS	"Kralastic"	0	0.0
Phenolic		0	0.0
Cellulosic	"Ethocel"	1[c]	4.7
Epoxy		0	0.0
Acetal	Delrin®	0	0.2
PPO	"Noryl"	0	0.2
PEK	"Ultrapek"	0	−0.1
PEEK	"Victrex"	0	−0.1
PET	Rynite®	0	0.2
PBT	"Valox"	0	0.0
Polyarylate	Arylon®	0	0.0
LCP		0	0.1
Polyimide			
A	Vespel®	0	0.0
PB	"Ultem"	0	0.1
PAI	"Torlon"	0	0.0
PPS	"Rython"	1	2.7
Polysulfone	"Udel"	0	−0.1
Polyaryl sulfone	"Rydel"	0	−0.1

Note: Rating: 0, compatible; 1, borderline; 2, incompatible.
Physical change.
[a] More flexible.
[b] Sample dissolved.
[c] Some extraction.

fully saturated with water. The metal coupons were examined for change in appearance and weight, and solvent was tested for any reaction products. All metals were judged to be compatible, although a slight discoloration of zinc, brass, and copper were seen with the solvent saturated with water.

Chemical and Thermal Stability

HFC-43-10mee is extremely stable to most soils and chemicals commonly encountered in solvent uses. It is also quite stable to water and air. Therefore, no stabilizer is required.

Like all hydrogen containing halocarbons, it is not stable to strong bases and amines, particularly in the presence of alcohol. Therefore, contact with such materials should be avoided.

TABLE 6.5 Elastomer Compatibility with HFC-43-10mee, 2 Weeks at 50°C (122°F) in a Sealed Tube Test

Elastomer	Rating	Linear Swell, %	Units Hardness Change
Natural rubber	0	−0.6	−1
Butyl rubber	0	1.0	−1
Nordel® EDPM	0	−1.0	−2
Neoprene CR	0	0.2	1
Buna-S	0	0.7	0
Nitrile rubber			
Buna-N	0	−0.6	2
NHBR	0	3.9	−8
Vamac® EA	2[a]	13.9	−12
Hypalon® CSM	0	1.3	0
Fluoroelastomer			
Viton® A	2	17.3	−14
Viton® B	2	22.8	−34
Zalak®	2[a]	13.7	−13
Kalrez®	2	21.6	−20
Fluorinated silicone	2	14.1	−11
Silicone	0	0.5	−4
Epichlorohydrin			
Homopolymer	0	−0.5	1
Copolymer	0	0.0	2
"Adiprene" U	1[a]	2.7	−2
FA polysulfide	0	1.5	0
Thermoplastic			
Alcryn®	2[a]	−1.2	13
"Santoprene"	0	0.1	0
"Geoplast"	1[a]	−0.5	−3
Hytrel® polyester	0	0.3	0

Note: Rating: 0, compatible; 1, borderline; 2, incompatible.
[a] Noticeable extraction affecting rating.

It is extremely stable to heat. In laboratory tests using an accelerating rate calorimeter, no decomposition could be detected at temperature reaching 300°C.

Selective Solvent Power

HFC-43-10mee is a relatively mild solvent, with selective solvency for soils, oils, and greases. Some common barometers to measure the solvency are Hansen and Hiderbrandt solvency parameters and kauri-butanol number (Kb). Several laboratory tests[1] were run to determine the Hansen and Hilderbrand solubility parameters as well as those for PFC-C_6F_{14} and CFC-113. The results are given in Table 6.7. Based on these data, it is apparent that HFC-43-10mee has overall solvency between C_6F_{14} and CFC-113. Unlike C_6F_{14}, the HFC is completely miscible with most esters, ketones, ethers, ether-alcohols, and the lower

TABLE 6.6 Metal Compatibility with HFC-43-10mee, 2 Weeks at 100°C in a Sealed Tube Test

Metal	Rating	
	Dry	Wet
Zinc (1)	0	0[a]
Stainless steel	0	0
Brass	0	0[a]
Aluminum	0	0
Copper	0	0[a]

[a] Slight discoloration.

TABLE 6.7   Solubility Parameters[a]

Compound	Dispersive	Polar	Hydrogen Bonding	Hildebrand
C_6F_{14}	5.6	0	0	5.6
CFC-113	7.2	0.8	0	7.2
HFC-43-10mee	6.3	2.2	2.6	7.8

[a] Units are $(cal/cm^3)^{1/2}$.

alcohols such as methanol, ethanol, and isopropanol. The lower hydrocarbons, such as pentane, hexane, and heptane also have good solubility and are fully miscible with most chlorocarbons and fluorocarbons including high-molecular-weight fluorocarbon lubricants such as "Krytox®" and "Fomblin®." Therefore, HFC-43-10mee can be used as an application carrier fluid or to remove these kind of compounds.

Unlike CFC-113, the neat HFC-43-10mee has limited solvency for many higher-molecular-weight materials such as hydrocarbon oils, silicon oils, fluxes, waxes, and hydrocarbon greases. However, many HFC formulations with chlorocarbons, hydrocarbons, esters, and alcohols have enhanced solubility and cleaning efficiency.

Applications of HFC-43-10mee, Neat

Carrier Fluid

HFC-43-10mee has excellent solvency for high-molecular-weight fluorocarbon lubricants and is currently the preferred solvent by far to apply fluorolubricant to computer hard disks. The lube application process consists of dipping the disks in a solvent bath containing a small amount (~2%) of the high-molecular-weight, high-boiling fluorocarbon lubricant. The disks are either removed from the bath or the bath is pumped out of the chamber. The solvent from the disk surface evaporates, leaving a fine film of the lubricant on the surface. A similar lubrication process is also used in applying a thin film of fluorolubricant to bearings as well as in applying lubricant to a specific joint, shutter, or moving parts in precision equipment such as cameras, videos, tape decks, etc.

Particulate Removal

Low surface tension, low viscosity, and high density make HFC-43-10mee very efficient for particulate removal. One common measure of particulate removal efficiency is wetting index, and it is defined as a ratio of density over viscosity and surface tension. Table 6.8 gives wetting indexes for the most common solvents. A high wetting index means the solvent readily wets the soil and easily rinses away contaminant such as particulate. The wetting index for HFC-43-10mee, as can be seen, is higher than many common solvents.

TABLE 6.8   Wetting Indexes of Solvents

Cleaning Agents	Density	Surface Tension	Viscosity	Wetting Index
Freon® TF (CFC-113)	1.48	17.4	0.7	122
1,1,1-Trichloroethane (TCA)	1.32	25.9	0.79	65
Isopropyl alcohol (IPA)	0.79	21.7	24	15
DI water	0.997	72.8	1.00	14
Saponifier solution (6% water)	0.998	29.7	1.08	31
HFC-43-10mee	1.58	14.1	0.67	167

HFC-43-10mee is tested and commercially used in several particulate removal applications. One such application is the removal of particulates from camera-original motion picture negatives prior to making a large number of prints for use in theaters. An aerospace company successfully demonstrated particulate removal (size range from 0.5 to 60 μm) from miniature ion pumps used in the Shuttle Hardware Program.[2] Wipes are used on cleaned parts of the shuttle hardware to remove particulates before final assembly, and by electricians to clean cable surfaces prior to splicing the wires.

Semiconductor Applications: CF Removal

The superior ability of HFC-43-10mee to dissolve certain fluorinated species has resulted in a new application of CF removal in the semiconductor industry. Low-molecular-weight CF species form during certain cleaning or etching steps of some microelectromechanical systems' manufacturing processes. Solvency for CF, the high density and low viscosity of HFC-43-10mee, and fast drying make this material preferred over other solvents.

Rinsing Agent

As discussed earlier, HFC-43-10mee has limited solvency for high-molecular-weight hydrocarbon oils and greases. However, it is fully miscible with many more aggressive organic solvents such as hydrocarbons, esters, ester–ethers, and ether–alcohols. Unfortunately, many of these solvents are flammable, are combustible, or have high boiling points. By contrast, a mixture of HFC-43-10mee with one of these solvents can provide increased solvency, suppression of flammability, and rapid drying. The co-solvent process uses this concept and can be carried out in a conventional two-sump degreaser with minimum modification. Typically, the first sump of the degreaser is charged with a mixture of HFC-43-10mee with solvating agent, and the second sump is filled with pure HFC as a rinsing agent. The first sump primarily removes the soil and the second sump provides a clean rinse of fluorocarbon solvent to remove traces of high boiling solvent left on the components to be cleaned. Table 6.9 gives a list of potential solvating agents.

Displacement Drying Application

CFC-113-based fluid was commonly used for spot-free drying after aqueous cleaning. HFC-43-10mee, with physical properties similar to those of CFC-113, makes an attractive replacement in this application. HFC-43-10mee drying fluid with a proprietary surfactant is now commercially available and effectively used in many drying applications.

The solvent drying process is carried out by dipping the wet parts in a boiling solvent containing a small amount of surfactant. The surfactant and the dense solvent lifts the water from the surface of the parts. Water floats to the top of the solvent and is pushed over into a compartment by the circulating solvent where it is decanted. The parts are subsequently rinsed in one or two baths of the pure solvent to remove residual surfactant. The whole process typically takes 3–10 min. This solvent drying process is more energy efficient than conventional oven or hot air drying processes.

Two example case studies[3] indicate the effectiveness of HFC-based displacement drying. One involved drying of optical glass products, changing from a conventional oven drier to the solvent displacement dryer. The new process reduced the reject rate (depending on the part size) from 23% to 65%, and

TABLE 6.9 Solvating Agents for HFC-43-10mee

Dibasic esters (DBE)	Aliphatic hydrocarbons
Diisobutyl DBE	Aliphatic alcohols
Methyl decanoate	Dipropylene glycol butyl ether
Isopropyl myristate	Propylene glycol *n*-propyl ether
N-Methyl-2-pyrrolidone (NMP)	Dipropylene glycol monomethylether
Tetrahydrofurfuryl alcohol (THFA)	

reduced the drying cycle time from 45 to 5 min. The second case study involved drying of microprocessor lids where the HFC-43-10-based formulation replaced PFC-C_6F_{14} (a high global warming fluid).

HFC-43-10mee Formulations

Since the chemistry of soils varies greatly, the chemistry of a cleaning agent must be appropriately adjusted for effective soil removal. HFC-43-10mee can be effectively blended with a variety of other solvents to form specialized cleaning agents for widely divergent cleaning applications. The resulting formulations are used in vapor degreasing, aerosol, or cold cleaning processes.

A careful and thorough evaluation was conducted in selecting formulation materials for commercialized HFC-43-10mee-based formulations. The materials qualified for blending must have properties that would enhance the performance of the neat HFC-43-10mee for removal of certain soils and must also have acceptable environmental and toxicity profiles. The candidates were further screened to pick those that would either form azeotropes or behave like azeotropes. Based on these criterion, the candidates that met these requirements were alcohols (methanol, ethanol, and isopropanol), hydrocarbons (cyclopentane and heptane), and *trans*-1,2-dichloroethylene (*t*-DCE), surfactants, and hexamethyldisiloxane were chosen for some specialty blends.

There are three basic types of HFC-43-10mee formulations: azeotropes, azeotrope-like blends, and specialty mixtures. Table 6.10 gives their trade names, components, and key applications. Table 6.11 gives their physical, thermodynamic, and environmental properties.

TABLE 6.10 Formulations of HFC-43-10mee

Blend	Components	Applications
Azeotropes		
MCA	43-10/*t*-DCE	Oxygen system cleaning
		Precision cleaning
		Cleanliness verification
SMT	43-10/*t*-DCE/methanol	Defluxing
		Precision cleaning
XM	43-10/methanol	Ionic and particulate removal
		Pneumatic system cleaning
XE	43-10/ethanol	Ionic and particulate removal
XP	43-10/IPA	Ionic and particulate removal
Azeotrope-like Formulations		
SFR	HFCs/*t*-DCE/methanol	Defluxing
		Precision cleaning
SDG	HFCs/*t*-DCE	Degreasing
		Silicone deposition and removal
MCA Plus	43-10/*t*-DCE-cyclopentane	Precision cleaning
XMS Plus	43-10/*t*-DCE/	Defluxing
	MeOH/cyclopentane	Precision cleaning
Non-Azeotropes		
XSi	43-10/OS-10	Silicone deposition and removal
		Swelling of silicone tubing
X-P10	43-10/IPA	Ionic and particulate removal
		Absorption drying
X-DA	43-10/surfactant/antistatic	Drying

TABLE 6.11 Formulations, Physical, Thermodynamic, and Environmental/Safety Properties of HFC-43-10mee

Blend → Property	MCA	SMT	SFR	SDG	MCA Plus	XMS Plus	XM	XE	XP	X-P10	XSi	X-DA
Physical												
Boiling point (°C)	39	37	41	43	38	38	48	52	52	54	56.6	55
Liquid density (g/mL)	1.41	1.37	1.28	1.29	1.33	1.34	1.49	1.52	1.53	1.42	1.0473	1.58
Liquid density (lb/gal)	11.76	11.43	10.7	10.8	11.12	11.15	12.43	12.68	12.76	11.84	8.73	13.18
Kb	20	38	101	95	29	32	9.5	9.4	9.4	10.5	17	
Vapor pressure (psia at 25°C)	9.0	9.1	8.4	7.5	8.90	9.10	5.8	4.8		4.6	2.6	4.4
Surface tension (dyn/cm)	15.2	15.5	19.9	21.2	16.1	14.9	14.1	14.1	15.1	14.1	14	14.1
Freezing point (°C)	<−50	<−50	<−50	<−50	<−50	<−50	<−80	<−80	<−80	<−80	<−50	−80
Heat of vaporization at bp (cal/gm)	43	53	68	67	51	54	43	35			38	31
Heat capacity at 25°C (cal/g °C)	0.27	0.28	0.28	0.27	0.28	0.29	0.27	0.27		0.27		0.27
Viscosity (cps)	0.49	0.47	0.58	0.59	0.49	0.46	0.63	0.73	0.7	0.75	0.6	0.67
Molecular wt.	157	128	109	107	136	125	178	212	228	191	203	252
Environmental/Safety												
AEL (ppm)	200	192	187	193	214	197	200	235	213	238	200	186
ODP	0	0	0	0	0	0	0	0	0	0	0	0
GWP (100 year ITH)	806	688	264	148	650	662	1222	1248	1258	1170	741	1292
VOC (g/L)	536	645	1063	1150	665	658	89	61	50	142	Exempt	8
Flashpoint (closed cup)	None	None	None	None	None	None	None	None	None	None	None	None
LEL (vol %)	None	7	7	7	6	4	9	None	None	5	5	None
UE (vol %)	None	15	15	14	11	14	11	None	None	11	a	None

Note: AEL, acceptable exposure limit, set by Du Pont; a lower governmentally set number would take precedence; tbd, to be determined.
a Could not determine due to condensation.

Binary Azeotropes

Three binary azeotropes of HFC-43-10mee with alcohol have been commercialized. XM is the methanol azeotrope, XE is the ethanol azeotrope, and XP is the isopropanol (IPA) azeotrope.

The presence of alcohol makes these azeotropes excellent in removing particulates and ionic contamination. The methanol-based azeotrope is used for cleaning laser disks and flushing pneumatic lines. The ethanol azeotrope is used for precision cleaning as well as removal of trace moisture from thin tubes. The isopropanol azeotrope is effectively used for particulate removal from the large satellite mirrors.

Although *t*-DCE contains chlorine, the presence of hydrogen and the double bond makes the molecule reactive in the lower stratosphere. Therefore, it has near-zero ozone depletion potential. However, it is a volatile organic compound (VOC). *t*-DCE has excellent solvency for higher molecular hydrocarbon oils and greases and for many fluxes. The permissible worker exposure limit of *t*-DCE is 200 ppm, the same as that of HFC-43-10mee. *t*-DCE is flammable, but the presence of the HFC in the azeotrope makes the azeotropic mixture nonflammable. The azeotrope is MCA.

This azeotrope is effective for precision cleaning of mechanical components, bearings, and compressor parts. It is also used as a verification fluid in the NASA shuttle hardware program and other aerospace applications.[4] It has found extensive use in the oxygen service (equipment) industry, where even trace residues of organic contamination can burn in the presence of oxygen.

Because *t*-DCE is somewhat aggressive to plastics and elastomers, it is important that the compatibility of the components materials to be cleaned should be tested.

Multicomponent Azeotrope Blends

Additional azeotrope blend formulations of HFC-43-10mee are available for precision cleaning, defluxing, and degreasing. HFC-43-10mee, *t*-DCE, and methanol form a ternary azeotrope, SMT. A small amount of nitromethane is added to prevent free radical reaction of alcohol with active metals. This azeotrope is effective in removing ionic impurities in defluxing printed circuit boards and precision cleaning electronic components. Processes using this azeotrope have met or exceeded the cleaning requirement for ionic impurities for the military and the surface insulation resistance.[5]

Another solvent formulation is a blend of two HFC/*t*-DCE azeotropes, called SDG. This formulation has high solvency for hydrocarbon-based soils, greases, etc. and exhibits a very high Kb value. It has been developed as an alternative to HCFC-225, *n*-propyl bromide, trichloroethylene, methylene chloride, and other high Kb solvents. SFR is similar to SDG but also contains a small amount of alcohol for improved cleaning of ionic contaminates and defluxing. This formulation has higher solvency for no-clean and Pb-free fluxes than SMT, and has one of the highest Kb values in the solvent market today.

In some cases, it is desirable to have good solvency for both fluorinated oils or greases and hydrocarbon-based oils or greases. To meet this requirement, MCA Plus was developed, and is a ternary blend consisting of HFC-43-10mee, *t*-DCE, and cyclopentane. This formulation has higher solvency for hydrocarbon oils and greases than MCA.

XMS Plus is a four component azeotrope-like mixture consisting of HFC-43-10mee, *t*-DCE, cyclopentane, and methanol. A small amount of nitromethane is added to protect against free radical reaction of alcohol with active metals. This formulation has a slightly better solvency for the fluxes and improved compatibility with PWB components as compared with MCA Plus. It is used in defluxing and precision cleaning applications.

Non-Azeotropic Blends

There are three non-azeotropic blends of HFC-43-10mee commercially available for specialty applications.

1. A proprietary blend of HFC-43-10mee, XSi, contains hexamethyldisiloxane. It has excellent solvency for the high-molecular-weight silicones and at the same time good compatibility with polycarbonates and polyurethane. The solvent is used as a carrier fluid to apply or remove the silicone lubricant from the surfaces of medical devices used for insertion into the human bodies. It is also used as a swelling media for silicone rubber tubing.

   Both components of this formulation are VOC exempt and therefore the blend is a non-VOC (U.S. EPA). This blend is flammable and in-use can separate into flammable compositions, and hence should be used in flammable rated equipment or with hand-wiping.

2. X-P10 is a proprietary blend of HFC-43-10mee with 10% isopropanol. The higher amount of alcohol helps in the removal of water and ionic impurities from nonporous surfaces and for absorption drying. Although the blend is nonflammable as formulated, because it is non-azeotropic, its composition may shift in use and could become flammable. It is used primarily in the drying sump of absorption dryers, with XP in the rinse sumps.

3. X-DA is a proprietary blend of HFC-43-10mee with a small amount of fluorosurfactant additive (<1% by wt.), an antistatic additive (0.5%). This formulation works extremely well as a displacement drying fluid. As discussed earlier, this blend offers one-step, low-energy, spot-free drying that is efficient, safe to use, and environmentally responsible.

Equipment Design Considerations for Low Emission

With increased environmental awareness and increased cost of solvent, it is imperative that the equipment designed to operate with the HFC solvent have state-of-the-art emission reduction technology. The primary enhancement features[6] recommended are as follows:

- An extended deep freeboard: freeboard to width ratio of 1.2–2.0
- A secondary condenser for vapor diffusion control operating at −20°F to −30°F
- Piping systems containing welded or soldered joints to minimize joint leaks
- Hoods and/or sliding doors on the top entry machines

Some additional enhancements, although costly, may further reduce emission:

- Automated work transport facilities
- Facilities for superheated vapor drying

The emission measurement tests[7] run with an unmodified degreaser and with the primary enhancement features showed emission rate reduction from 0.075 to 0.0155 lb/h-ft^2, or about 80% reduction in emission loss. A rough calculation indicates that the cost of retrofitting an existing degreaser can be recovered in reduced solvent consumption in as little as 6 months of operation.

Emission Measurement Data

Several vapor-in-air concentration measurements have been made for HFC-43-10mee solvents in the working space immediately adjacent to the degreasers with and without enhancement features.[7] In all cases, the emission has been well below the allowable exposure limit of 200 ppm. Table 6.12 gives the actual emission measurement data under various operating conditions around an enhanced Ultronix-modified Barons-Blakeslee 120 model degreaser (higher freeboard and a secondary low temperature coil). As one can see, the average emission is less than 10 ppm.

TABLE 6.12   Vapor Concentration in Air of HFC-43-10mee Measured Adjacent to a
Modified BBI-MSR-120 Vapor Degreaser

Sample Location	Operating Conditions	Vapor-in-Air Conc., ppm, HFC-43-10mee	
		Range of Values	Average Values
Front and behind	Degreaser idling with lid open	2.3–18	7
Front	Degreaser idling with lid open	2.3–12	6
24″ behind degreaser	Degreaser idling with lid open	4.9–18	9.6
24″ behind degreaser	Cleaned 6 basket loads of parts/lid open at all times	1.6–31	8.4
24″ behind degreaser	Cleaned 4 basket loads of parts/lid closed during idling period	0.6–38	8.5

Replacement of Ozone-Depleting Substances and High Global Warming Gases: Opportunities, Alternatives, and Current Research

Ozone-Depleting Substances: CFCs and HCFCs

HFC-43-10mee successfully replaced ODSs CFC-113 and methyl chloroform in applications where not-in-kind alternatives and other alternative solvents were not acceptable due to safety, process, or material incompatibility problems. With the implementation of the Montreal Protocol, the manufacture of both of these ODSs for solvent applications ceased in January 1996 (except in a few Article-5 countries), and are now essentially out of use. Similarly, HCFC-141b ceased broad production in 2004 while HCFC-225 will cease production or import in the United States by 2015. It was estimated that the 2003 pre-phaseout production volumes of HCFC-141b and HCFC-225 were 12.6 and 0.6 million pounds, respectively.[8] Users are actively seeking alternatives to avoid the higher cost of these materials as stockpiles are reduced.

High Global Warming Gases: PFCs

Perfluorocarbons (PFCs) such as C_5F_{12}, C_6F_{14}, C_7F_{16}, and C_8F_{18} were introduced as substitutes for CFC-113 from the late 1980s to the early 1990s. Prior to this, PFCs were used in small quantities in niche applications. The U.S. EPA, via the Significant New Alternatives Program (SNAP), has restricted the use of PFCs in cleaning applications to where reasonable efforts have been made to ascertain that other alternatives are not technically feasible[9] and thus the current volume used in the cleaning market is small.

HFCs as Alternatives to CFCs, HCFCs, and PFCs

The replacement of CFCs and HCFCs in electronic applications, especially defluxing, has been primarily by three not-in-kind technologies: aqueous, semi-aqueous, and "no-clean." However, there are instances where these not-in-kind alternatives have not performed satisfactorily or have been rejected due to compatibility, safety, flexibility, or reliability reasons. Other solvent alternatives are considered. In these cases, neat HFC-43-10mee or HFC-43-10mee blends and azeotropes are preferred. Blends of HFCs, such as HFC-43-10mee and HFC-365mfc, with *t*-DCE, have found broad use to replace HFC-141b, particularly in aerosol (spray) formulations for defluxing circuit boards following rework operations. The most common azeotrope or azeotrope-like blends are of HFCs with hydrochlorocarbons, alcohols, and hydrocarbons. These will be discussed in greater detail in other sections of this chapter.

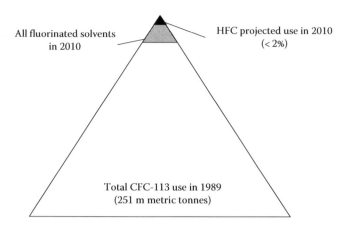

All fluorinated solvents in 2010

HFC projected use in 2010 (< 2%)

Total CFC-113 use in 1989 (251 m metric tonnes)

FIGURE 6.2 Fluorinated solvent market size from 1989 to 2010. (From DuPont estimate.)

The HFCs are ideal substitutes for highly global warming PFC (C_6F_{14}, C_7F_{18}) compounds and are successfully replacing them in many applications. HFC-43-10mee is a preferred substitute where excellent solubility for fluorinated materials is required and where good thermal stability and non-flammability are also important.

The volume of HFCs as a solvent substitute is small. Figure 6.2 gives an estimate of the 2010 fluorinated solvents and HFC solvent markets relative to peak CFC-113 production in 1989.

Environmental, Regulatory Considerations

The environmental properties of HFC-43-10mee, HFC-365mfc, HCFC-225ca/cb, and PFC-C_6F_{14} are given in Table 6.13 for discussion purposes.

Because the HFCs have no chlorine, they have zero ozone depletion potential. The U.S. EPA, under the SNAP, accepted HFC-43-10mee, HFC-365mfc, HFC-c447, and HFC-245fa as substitutes for ODSs like CFC-113 for various cleaning categories. HCFC-225 is SNAP approved but subject to the Montreal Protocol phaseout.

The presence of hydrogen in the molecule reduces the atmospheric lifetime significantly relative to 3200 years for the PFC-C_6F_{14}. In addition, the reactivity with hydroxyl ions in the lower atmosphere is very low, and therefore, emissions to the atmosphere do not contribute to the formation of smog or air pollution. For this reason, the U.S. EPA granted VOC-exemption to several HFCs.

HFC-43-10mee is not regulated under the U.S. EPA hazardous waste regulations (RCRA) and is not considered a hazardous air pollutant (HAP) and therefore not subject to NESHAP regulations or SARA-III Title reporting requirements. Table 6.14 summarizes regulatory status for HFC-43-10mee.

TABLE 6.13 Environmental Properties of Example HFCs, HCFC, and PFC

Property	HFC-43-10mee	HFC-365mfc	HCFC-225ca/cb	PFC
Formula	$C_5H_2F_{10}$	$C_4H_5F_5$	$C_3HF_5Cl_2$	C_6F_{14}
Ozone depletion potential (ODP)	0	0	0.03	0
Atmospheric lifetime, years[a]	15.9	8.6	1.9/5.8	3200
GWP (100 year TH)[a]	1640	794	122/595	9300
VOC exempt—EPA	Yes	Yes	Yes	No
VOC exempt—CARB[b]	No	No	No	No

[a] IPCC fourth assessment report, 2007.

[b] California Air Resource Board.

In effect in 2005, the Kyoto Protocol addresses global warming by reducing greenhouse gases such as CO_2, CH_4, PFCs, SF_6, N_2O, and HFCs. As of 2010, the Kyoto Protocol has not been ratified by the United States. However, legislation on greenhouse gas reduction is under consideration by the U.S. Congress, with a phase-down schedule and ultimate cap for HFCs.

In the proposed legislation, neither production nor use of HFCs would be phased out and no applications would be banned. It would provide for predictable use of HFCs going forward and also reflect the value of these societally important molecules by providing for their ongoing use. It would allow transitioning over time to next generation products in large volume applications such as air conditioning and refrigeration.

The proposed legislation also gives the U.S. EPA the authority to regulate additional greenhouse gases under the HFC cap. Therefore, gases such as hydrofluoroethers (HFEs) could be covered in the future.

The European F-gas regulation, with regard to the solvent market, also does not ban HFCs. The regulation formalizes good-use practices across all European Union members, requires recovery of solvent after solvent change-outs, and specifies training of personnel performing these tasks. When the F-gas regulation is updated (currently anticipated in 2011), it is expected that a number of additional greenhouse gases will fall under regulation as well.

TABLE 6.14 Regulatory Status for HFC-43-10mee

Regulations	Status
SARA III	No
HAP	No
RCRA	No
Halogenated NESHAP	No
SNAP	Approved
VOC	Exempt (EPA)
ODP	Zero

Current Research in Fluorinated Solvents and Gases

Research in the fluorinated solvent and gases market is very active today. An example of a new class of chemicals in development is the hydrofluoroolefins, or HFOs. These structures, with partial unsaturation, have inherently lower atmospheric lifetimes due to a higher reactivity to hydroxyl radicals than the corresponding HFC (saturated) structures. It is expected that new fluorinated materials will be available in the future that have lower GWPs than other fluorinated chemicals on the market today. Already, some HFC gases are being replaced by HFO structures.

Summary

HFCs and their blends, introduced within the last 15 years, have been successful in many applications due to their combination of good safety and health attributes, zero ozone depletion, low-to-moderate GWP, and superior properties such as solvency, low viscosity, and fast, spot-free drying.

While recent environmental concerns have shifted from ozone depletion to global warming, no HFCs have been banned in cleaning applications. It is the expectation of the authors that future environmental legislation on global warming will adopt a phasedown approach (for all fluorinated gases, not only HFCs) rather than outright phaseouts as was the approach in the Montreal Protocol for ODSs. This approach should result in market-driven, long-term viability for some or all the HFCs, including HFC-43-10mee.

Current research has shown new classes of compounds, such as hydrofluoroolefins or HFOs, have very low GWPs while maintaining the good solvency, no flammability, and low toxicity that make the HFCs an important class of fluorinated materials. The extent to which these new materials can replace the successful HFCs is the subject of current work.

References

1. H.L. Fritz. DuPont internal report, February 1994.
2. D.W. Jones. Evaluations of ODS-free particulate removal from miniature ion pumps. *Int. CFC Halon Altern. Conf.*, Washington, DC, October 1995.

3. G.A. Westbrook et al. Solvent-based drying system offer fast spot-free results. *Precision Cleaning*, June 1999.

4. C.L. Wittman et al. The search for a replacement for CFC-113 in the precision cleaning and verification of shuttle hardware. *Int. CFC Halon Altern. Conf.*, Washington, DC, October 1995.

5. M.R.B. Hanson et al. Performance testing of HFC azeotropes for precision cleaning. *Int. CFC Halon Altern. Conf.*, Washington, DC, October 1995.

6. R.B Ramsey et al. Considerations for the selection of equipment for employment with HFC-43-10mee. *Int. CFC Halon Altern. Conf.*, Washington, DC, October 1995.

7. R.B. Ramsey et al. Environmental Vapor Concentrations Adjacent to Ultronix, Inc.—Modified Baron-Blakeslee MSR-120 Vapor Degreaser Charged with HFC-43-10mee, DuPont internal report, KSS-9155, February 1995.

8. The U.S. Solvent Cleaning Industry and the Transition to Non Ozone Depleting Substances. Prepared for U.S. Environmental Protection Agency Significant New Alternatives Policy (SNAP) Program. Prepared by ICF Consulting, Washington, DC, September 2004.

9. U.S. Environmental Protection Agency. Significant new alternatives policy program list of alternatives for solvent use (available at www.epa.gov.)

10. U.S. EPA Office of Atmospheric Programs. Global mitigation of non-CO_2 greenhouse gases, EPA 430-R-06-005, June 2006.

7

n-Propyl Bromide

John Dingess

Richard Morford
Enviro Tech
International, Inc.

Ronald L. Shubkin

Update

Since this book was last published in January of 2001, much experience has been gained using *normal*-propyl bromide (*n*-propyl bromide or *n*PB) in cleaning applications. The bromine-based solvent has been found to have unique cleaning properties, while closely matching the physical properties of one of the most successful solvents that is no longer produced, 1,1,1-trichloroethane. By and large, many users in various fields such as aerospace, electronics, automotive, medical oxygen equipment, optics, paint stripping, and other application where precision cleaning is required have used stabilized *n*PB safely and with effective removal of soil. The latest field to use stabilized *n*PB safely and successfully is the dry cleaning market as an alternative to perchloroethylene (PCE). Furthermore, the brominated solvent has been EPA approved as an alternative to ozone-depleting solvents in vapor degreasing and related precision cleaning applications.

Other Solvent Alternatives

The cleaning of complex parts to exacting specifications has become a major consideration in the manufacture of a wide variety of machines, appliances, and instruments. The introduction of chlorinated solvents provided manufacturers and fabricators a convenient and economical way to remove a host of difficult soils that contaminate strategic parts. Efficient cleaning, rapid drying, low flammability, residue-free parts, and relatively low solvent costs all contributed to the popularity of fluids containing fluorine and chlorine. However, many chlorine-containing solvents have now been restricted because of environmental and/or health considerations. With the mandated elimination of the most popular cleaning solvents, many manufacturers switched to aqueous or semiaqueous cleaning systems. While these proved to be viable solutions in many applications, they were not suitable for all situations. In the search to find more appropriate alternatives, a wide variety of new solvents were developed.[1] Many of the new solvent cleaners do excellent jobs, but still suffer from one or more deficiencies relative to the overall cost/performance of the chlorinated materials they replaced. Some of the newer solvents, such as some hydrochlorofluorocarbons (HCFCs), have been shown to have environmental problems and are scheduled for phaseout. In other cases, the newly introduced solvents meet the environmental and toxicological requirements, but do not meet the performance standards. The need of specialized applications for a high-performance cleaning agent, which could be used in a safe and efficient manner, led to the development of cleaning systems based on the solvent *n*PB.

Historical Development

In 1991, the total U.S. market for 1,1,1-trichloroethane (TCA) as a cleaning solvent was 700 million pounds. Another 200 million pounds were sold as emissive solvents. TCA was and still could be a very effective and popular cleaning solvent, but it suffers from two important environmental drawbacks. It has a relatively high ozone depletion potential and a relatively high global warming potential. The Montreal Protocol banned the manufacture of TCA effective January 1, 1996.

A host of new cleaning systems have been introduced in recent years to meet the challenges presented by today's industrial cleaning requirements. All of the new systems offer advantages in specified niche applications, but none have met all of the requirements of the marketplace. Some of the important cleaning options are the following:

- *Alternative chlorocarbon solvents*: Most chlorinated solvents are effective cleaning agents. However, most either have been banned from manufacture, restricted to specified applications, scheduled for phaseout, or have regulations restricting emissions and requiring extensive reporting.
- *Hydrocarbons and oxygenated hydrocarbons*: These solvents have the obvious advantage of a historical low cost. However, most are readily flammable and present serious hazards when used in cleaning operations. Many also have toxicological problems.
- *Hydrochlorofluorocarbons*: As with many chlorocarbon solvents, the most popular of these (HCFC-141b) has been restricted to noncleaning applications. HCFC-225 is still being used in cleaning, but it is scheduled for phaseout in the United States by 2015. Other HCFC molecules are very volatile, have only moderate cleaning ability, or are not considered cost-effective in specific application areas.
- *Fluorocarbons*: Fluorocarbons are nontoxic, nonflammable, and very safe to use. However, they tend to be expensive and have poor solvency for most soils. The notable exception is their ability to solubilize highly fluorinated oils and greases.
- *Hydrofluorocarbons (HFCs)*: HFCs have low or moderate solvency for most soils of interest and historically have been expensive. They are excellent for niche applications, and they are often blended with more aggressive cleaning solvents. The environmental acceptability of the blend tends to be dependent on the second component.

- *Hydrofluoroethers (HFEs)*: HFEs are similar to HFCs in solvency, cost, and overall performance. Again, blends are sometimes used to boost cleaning performance.
- *Volatile methyl siloxanes (VMSs)*: Linear VMSs, such as hexamethyldisiloxane, are low in toxicity and contain no halogen atoms. They are chemically very stable. On the other hand, they have flash points and only moderate solvency.
- *Semiaqueous systems*: The problem of proper disposal of semiaqueous systems is often overlooked. Because of the high organic content, it is not appropriate (or legal in most cases) to dispose of these systems down the drain. Separation and recycle of the organic phase is usually difficult and is not cost-efficient. Slow drying and potential corrosion problems may also come into play.
- *Aqueous systems*: Aqueous systems are very inexpensive in terms of detergent cost, but they are not suitable for all applications. High capital investment, multistep processing, a large equipment footprint, and high-energy costs are often reported. Residues on the clean parts and difficult drying are also problems. Corrosion of metal parts may become a factor. Finally, electrical and electronic applications often cannot tolerate the presence of remaining traces of water.
- *No clean systems*: Some manufacturers have eliminated the need to clean at various stages of manufacture. Sometimes this requires a change in the manufacturing process or the order of assembly.

n-Propyl Bromide

In the mid-1990s, a new solvent/cleaner based on *n*PB was developed to meet the needs of those who require the cleaning efficiency of the chlorinated solvents but who must meet the strict environmental standards for a replacement solvent.[2] The new solvent/cleaner does not suffer from many of the shortcomings of other alternative solvents that have been offered to the market. Stabilized *n*PB is an effective cleaning agent. It is safe to use under the proper conditions, has a low ozone depletion potential (ODP), and a low global warming potential (GWP), and it is not regulated under most environmental and worker safety rules. It is compatible with metals, has a low tendency to cause corrosion and may be used in most current vapor degreasing equipment. It is easily recycled and is moderately priced.

Physical Properties

Table 7.1 compares the physical properties of *n*PB to trichloroethylene, and HCFC-225, which is a mixture of the two isomers, $CF_3-CF_2-CHCl_2$ and $CF_2Cl-CF_2-CHClF$.

Cleaning Power

Indicators that relate to the cleaning ability of a solvent are the solubility parameters. These are the Hildebrand parameter, the kauri-butanol number, and the Hansen parameters. As indicated by the data in Table 7.2, the values for *n*PB compare quite well to the common chlorocarbons. Solubility information for TCA is included to provide perspective. Given production phaseouts and regulatory constraints, TCA is largely unavailable for most applications.

Although solubility parameters provide a guide to cleaning power, they do not take the place of experimental results. In order to compare the cold cleaning ability of neat *n*PB to neat chlorinated solvents, a simple test procedure was devised.[3] Solutions of typical soil contaminants (30 wt. %) were prepared in solvent. Steel wool wedges were weighed and then soaked in the contaminated solvent, drained, and dried at 100°C for 30 min. The wedges were reweighed and the weight of retained soil recorded. The impregnated wedges were then placed in short glass tubes and washed with 3 mL of the test solvent. The wedges were drained, dried, and weighed as before. The grams of soil lost per milliliter test solvent gives a measure of cleaning power.

TABLE 7.1   Physical Property Comparisons

	n-Propyl Bromide	Trichloroethylene	HCFC-225
Boiling point, °C	71	87	54
Specific gravity, 25°C	1.35	1.46	1.55
Viscocity, 25°C, cP	0.49	0.54	0.59
Vapor pressure, 20°C, torr	110.8	57.8	285
Specific heat, 25°C	0.27	0.22	0.25
Latent heat of vap., cal/g	58.8	57.2	33
Sol. in water, g/100 g water	0.24	0.11	0.033
Sol. of water, g/100 g solv.	0.05	0.03	0.03
Surface tension, 25°C, dynes/cm	25.9	26.4	16.2
Flash point, TCC, °C	None	None	None
Flammability limits,[a] volume %	4–7.8	8–10.5	None

[a] The *n*PB molecule has three carbon atoms, seven hydrogen atoms, and one bromine atom. Like the chlorinated and fluorinated solvents (tetrachloroethylene and carbon tetrachloride are exceptions), the hydrogen atoms enable *n*PB to have flammability limits, but *n*PB does not have a flash point (i.e., it does not sustain a flame), and bromine is a natural flame retardant. *n*PB suppresses the flash point of various flammable solvents (acetone with its high vapor pressure may be an exception) and stabilizers with which it can be formulated.

TABLE 7.2   Solubility Parameters

	n-Propyl Bromide	1,1,1-TCA	Trichloroethylene	Methylene Chloride	Perchloroethylene
Hildebrand parameter[a]	18.2	17.4	18.8	19.8	19.0
Kauri-butanol no.	125	14	129	136	90
Hansen parameters:					
Non-polar[a]	16.0	17.0	18.0	18.2	19.0
Polar[a]	6.5	4.3	3.1	6.3	6.5
Hydrogen bonding[a]	4.7	2.1	5.3	6.1	2.9

Source: Barton, A.F.M. (Ed.), *CRC Handbook of Solubility Parameters and Other Cohesion Parameters*, 2nd edn., CRC Press, Boca Raton, FL, 1991.
[a] Data presented as $\delta/(MPa)^{1/2}$.

The cleaning power of chlorinated solvents relative to that of neat *n*PB (Figure 7.1) shows that *n*PB-based cleaners have fractionally lower cleaning ability than TCA when used to clean mineral oil, equivalent performance on grease and silicone oil, and superior performance in removing polyol esters. These experimental results are consistent with the Hansen parameters, which show that *n*PB has a slightly lower value for nonpolar materials and higher values for polar and hydrogen bonding compounds.

The cleaning power of *n*PB-based cleaners is clearly equivalent to the popular chlorinated solvents that have been phased out of production or restricted. Comparisons to the new alternative solvents that have been introduced to replace the chlorinated materials, however, are more relevant to today's needs. Again, it is instructive to first compare the relative solvency power of *n*PB to some of the solvents that are that are currently offered in the cleaning market.

Table 7.3 compares the kauri-butanol number of *n*PB to the several alternative solvents that have been introduced in recent years. By this one measure, superior performance is expected from the

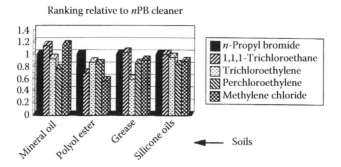

FIGURE 7.1   Relative cleaning ability, at ambient temperature, of *n*PB and chlorinated solvents.

TABLE 7.3   Comparisons of Solvency Characteristics

Solvent	Kauri-Butanol Number
n-Propyl bromide	125
HFC 43–10 mee	9
HFE (methyl ether)	10
HCFC-225	31
VMS (hexamethyldisiloxane)	17

cleaning formulations based on *n*PB. The abbreviations used here and throughout the rest of this chapter are as follows:

- *n*PB: *n*-Propyl bromide
- *HCFC-225*: Dichloropentafluoropropane (two isomers)
- *HFC-43-10mee*: Decafluoropentane
- *HFE*: Perfluorobutyl alkyl ethers (especially the methyl ether)
- *VMS*: Volatile methyl siloxanes (especially the dimer, hexamethyl disiloxane)

Of course, the kauri-butanol number is only an indication of cleaning performance. Figure 7.2 shows the results of experiments that were conducted in the same manner as those carried out to compare *n*PB

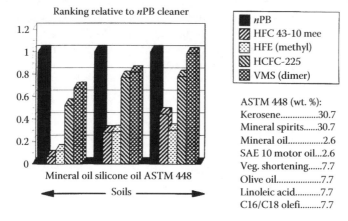

FIGURE 7.2   Relative cleaning ability, at ambient temperature.

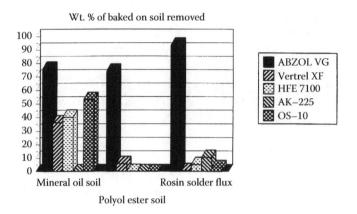

FIGURE 7.3   Cleaning of tough, baked-on soils. Coupon immersed in boiling solvent for 5 min.

with the chlorinated solvents. For these experiments, performance was compared for the removal of mineral oil, silicone oil, and a complex mixture of common soils that simulates the ASTM 448 Standard Soil (some minor substitutions were made). These experimental results indicate clearly superior cleaning performance for nPB-based cleaners. It is for these reasons that the other solvents are often sold as blends with more aggressive solvents that for one reason or another cannot be sold by themselves as cleaning solvents.

Cold cleaning is suitable for a variety of cleaning needs, but the very difficult soils usually require more severe conditions. For this reason, difficult soils are often removed using vapor-degreasing equipment. In this equipment, the parts may be cleaned in the hot vapors of the solvent. For the most difficult soils, especially those that have been "baked on," the part to be cleaned is often immersed in the boiling solvent. Because cold cleaning performance does not always reflect performance at higher temperatures, a new cleaning test was designed.

Steel coupons (C1010) were weighed and coated with the selected soil (ca. 0.1 g). The coupons were placed in an oven at 250°C for 1 h, cooled and reweighed. The coupons were then immersed in boiling solvent for 5 min, dried, and reweighed. Each trial was conducted in triplicate. The weight percent of soil removed from each coupon was recorded and averaged for the three trials.

Because formulation plays a part in total cleaning efficiency, commercial cleaning solvents were used for this comparison. Five formulated solvents were chosen. The solvents were a stabilized nPB (Albemarle's VG Cleaner), HFC 43-10 mee (DuPont's Vertrel™ XF), HFE (3M's HFE-7100), HCFC-225 (Asahi Glass's AK-225), and VMS (Dow Corning's OS-10). The soils chosen were mineral oil, a polyol ester, and a rosin-based solder flux. The results that are shown in Figure 7.3 indicate that the VG Cleaner removes the baked-on mineral oil much more efficiently than the other cleaning solvents. The baked-on polyol ester and the rosin-based solder flux remain nearly untouched by the other solvents, but are mostly removed after 5 min in the boiling VG Cleaner.

Drying

From the standpoint of raw material costs, the most economical class of substitutes for chlorinated solvents would appear to be that of aqueous based systems. However, aqueous cleaning procedures suffer from a number of difficulties, one of which is the slow drying. Drying rates can be improved by the installation of specialized drying equipment, but this is costly and usually contributes significantly to the process time. Cleaning systems based on nPB have drying rates that are comparable to the chlorinated solvents. To compare evaporation rates, loss of weight from 2 mL of solvent at room temperature

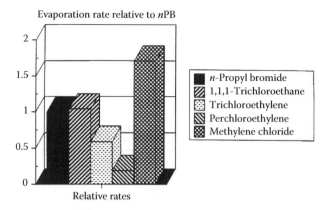

Evaporation rate relative to *n*PB

- ■ *n*-Propyl bromide
- ▨ 1,1,1-Trichloroethane
- ▦ Trichloroethylene
- ◩ Perchloroethylene
- ▦ Methylene chloride

Relative rates

FIGURE 7.4 Relative evaporation rates.

(20°C) was measured after 5 min. Figure 7.4 compares the relative rate of evaporation of *n*PB to four common chlorinated solvents.

An alternative, but popular, benchmark is the evaporation rate relative to butyl acetate at room temperature. For 1,1,1-TCA, the rate is 5.8 times that of butyl acetate. For *n*PB, the relative evaporation rate is 6.2—marginally faster than 1,1,1-TCA.

Compatibility

Because *n*PB is a very aggressive cleaner base, care must be taken to ensure that it is compatible in the short term and at the appropriate temperature with all of the components of the parts being cleaned. Likewise, care must be taken to ensure that the stabilized *n*PB is compatible for the long term with the materials of construction for the cleaning equipment and the storage and handling facilities.

Plastics and Elastomers

Plastics and elastomers that pass short-term compatibility tests with *n*PB (immersion in boiling solvent for 15 min) include the following:

Plastics	Elastomers
Acculam™_epoxy glass	Adiprene™_polyurethane
Alathon™_HDPE	Aflas™_PTFE
Delrin™_acetal[a]	Buna-N™_rubber
Kynar™_polyvinyl fluoride[a]	Kalrez™_fluorelastomer[a]
Nylon™_(6 and 6.6)	Neoprene™_polychloroprene
Phenolics[a]	Viton-A™_fluoroelastomer[b]
Polyester (filled and unfilled)	Viton-B™_fluoroelastomer[b]
Polypropylene	
Teflon™_PTFE[a]	
Tefzel™_ethylene/PTFE[a]	
XLPE™_crosslinked PE	

[a] These materials are also compatible for long-term (2 months) immersion at elevated temperature (65°C).

[b] The Viton™_fluorelastomers are marginal for long-term (2 months) immersion at elevated temperature (65°C).

Plastics and elastomers that were found to be unsuitable (U) or marginal (M) for contact with *n*PB at elevated temperature for short periods (15 min) include the following:

Plastics	Elastomers	
Low-density polyethylene (M)	Butyl rubber (M)	EPDM-60 (U)
Ultem™_polyether imide (M)	NBR nitrile rubber (M)	Silicone (U)
Polycarbonate (U)		
ABS (U)		
Acrylic (U)		

Metals

*n*PB must be formulated with passivating agents in order to be used with many metals. The type and concentration of the passivators determines the performance. Stabilized *n*PB (e.g., ABZOL VG or EnSolv from Enviro Tech) was tested for compatibility with metals according to Mil-T-81533A 4.4.9. This is a metal corrosion test that was originally designed to test the suitability of TCA for military applications. The metal coupon is held half-submerged in the refluxing cleaning fluid for 24 h. It is then examined for signs of corrosion. All of the following metals passed this test:

Nickel	Inconel	Titanium
Brass	Copper	Zinc
Monel	Aluminum	Tantalum
Stainless steel 316L	Carbon steel 1010	Magnesium
Stainless steel 304L		

Fresh aluminum surfaces react immediately with 1,1,1-TCA at room temperature. *n*PB is much less reactive toward aluminum. If an aluminum coupon is scratched beneath the surface of TCA, which contains no metal passivators, there is an immediate formation of a dark, brownish red color. If the test is repeated using neat *n*PB, no color formation is observed. At reflux, some small dark spots are observed on the edges of the aluminum coupons after 3–4 h. Similar tests were done with magnesium. Stabilized *n*PB is safe for use with aluminum, magnesium, and other active metals.

From the periodic table of the elements, fluorine is more electronegative (i.e., it has a stronger affinity for electrons) than chlorine. Chlorine is more electronegative than bromine. Consequently, *n*PB is less stable than the chlorinated and fluorinated solvents, because the carbon–bromine bond is not as strong as the carbon–chlorine or the carbon–fluorine bond. The lower electronegativity of bromine also helps to explain why bromine is more oil soluble (less polarizing) than chlorine, which is more oil soluble than fluorine. Conversely, however, HBr is not as corrosive toward metals as HCl or HF: the dissociation of hydrogen and bromine is not as strong as it is for the other halogens.

Drums and Drum Linings

In addition to the corrosion test, two-month immersion studies were carried out at 130°F with carbon steel 1010, stainless steel 316, and high-baked phenolic linings. All of these materials were shown to be suitable for long-term storage of cleaning fluids containing *n*PB. Most perfluorinated plastics are also suitable for storage.

Thermal Stability

The knowledge of the thermal stability and thermal degradation products of a new solvent is important for safety considerations during use. Buildup of contaminants on heating elements, for instance, can cause localized hot spots that may degrade the solvent. It is important to know at what temperature this will occur and to insure that the products of the degradation are not dangerous or highly toxic. Two different approaches were taken to determine thermal stability.

Columbia Scientific conducted thermal degradation studies using the accelerating rate calorimetry (ARC) method. This method identifies the temperature for the onset of degradation by detecting the accompanying exotherm. For stabilized nPB (VG), a significant exotherm occurred at 226.5°C. The test was terminated at 347°C. For nPB containing no stabilizers, no exotherm was reported up to 395°C. However, the final pressure of the bomb after it was cooled was considerably higher than for the VG cleaner. Scrutiny of the temperature and pressure data shows that the temperature curve for the neat nPB flattened briefly at 226.5°C and that the rate of increase in pressure increased simultaneously. A possible interpretation of the data is that neat nPB thermally degrades at 226.5°C, but the event is either endothermic or it is slightly exothermic and is not detected by the ARC instrument. In the case of the VG cleaner, the products of the degradation react exothermically with one or more of the stabilizers present.

The degradation products formed in the experiments described above were trapped in a stainless steel bomb and analyzed by gas chromatography/mass spectrometry (GC/MS). The products are essentially the same from both the stabilized and the unstabilized nPB, although the ratios of the products were somewhat different. No free bromine or HBr was detected. Although trace amounts of methyl bromide and benzene were found, no products of a highly toxic nature formed in significant quantities. Unlike chlorinated solvents, it is chemically impossible to produce an extremely toxic compound such as phosgene, which contains two chlorine atoms.

The second approach to determining thermal stability was to simulate a real-world failure of a heating element in a vapor degreaser.[4] A coiled nichrome wire was immersed in VG cleaner in the bottom of a 250 mL flask. The flask was connected to a dry ice/acetone trap, which was vented to a laboratory hood. Electric current was passed through the nichrome wire until the exposed part glowed red-hot. The cleaning solution boiled vigorously. The vaporized products were collected in the trap and analyzed by GC/MS. Unlike the ARC experiment that was conducted in the absence of air, some of the products formed in this experiment contain oxygen. The decomposition products detected after the two experiments were as follows:

ARC Method	Submerged Nichrome Wire
Propane	Propene
Isobutane	Methyl bromide
Butane	Ethyl bromide
Methyl bromide	Benzene
2-Methyl butane	Toluene
Pentane	Dipropyl ether
Ethyl bromide	1,3,5-Trioxacycloheptane
Branched C_6H_{14} isomers	4-Bromo-2-butanol
Isopropyl bromide	4-Bromo-1-butanol

Excessively high-temperature "hot spots" on the interior walls of a vapor degreaser can occur if the heating elements short circuit. The thermal degradation studies indicated that the use of stabilized nPB creates no risks that are not normally encountered in the event of catastrophic equipment failure.

Hydrolytic Stability

Laboratory tests show that nPB is subject to a small degree of hydrolysis when contacted with water for extended times, particularly at elevated temperatures. In laboratory tests, stabilized nPB (VG) was compared to a 1,1,1-TCA cleaning solvent. After refluxing with water for 164 h, the layers were separated and analyzed. The *n*PB formulation showed two to three times as much hydrolysis as the TCA. Other tests were then performed to determine the relative corrosivity of HBr and HCl. In these tests, dilute HBr was six to seven times less corrosive at 50°C than was an equimolar concentration of HCl.[2] Hydrolysis of *n*PB is less likely to cause corrosion than hydrolysis of TCA.

Comparison of *n*PB Properties to Other New Cleaning Solvents

Table 7.4 is a brief summary of the physical performance characteristics of the relatively new entries in the cleaning solvent application area.[5] Data are for the neat solvents, not the commercial blends.

Special Formulations: Dry Cleaning

The dry cleaning market has primarily used PCE to clean fabrics of all kinds. But for years, PCE has been under scrutiny for its health and environmental shortcomings. PCE is a possible carcinogen, and it is an EPA listed waste. It does have the advantage of having neither flash point nor flammability limits. PCE is relatively stable and is not as aggressive on some plastic buttons as stabilized *n*PB.

In 2005, a stabilized *n*PB was introduced to the dry cleaning market (DrySolv, Enviro Tech). The new dry cleaning solvent has been found to have some unique advantages over PCE. The *n*PB-based formulation has: a lower boiling point than PCE, requires less energy to heat in the dry cleaning still, has greater oil solubility than PCE, gives better cleaning results and more natural fabric-softening ability than perchloroethylene, and allows for shorter cycle times.

The *n*PB-based formulation should not be used in old transfer machines, but is used successfully in existing third, fourth, and fifth generation dry cleaning machines after a small number of modifications are made to the equipment.

Normal modifications involve replacing aluminum parts with stainless steel, replacing gaskets, and an overall "tightening" of the machine. The replacement of aluminum parts with stainless steel is important because moisture combined with heat tends to deplete the acid acceptor in all stabilized *n*PB products, resulting in the formation of hydrobromic acid. Hydrobromic acid is aggressive toward cast

TABLE 7.4 Comparison of Properties for New Cleaning Solvents

	*n*PB	HCFC-225	HFC-43-10	HFE	VMS
Boiling pt, °C	71	54	55	60	100
Flash point, °C	None	None	None	None	−2.77
Flammability limits, %vol	3–8	None	None	None	1.2–18.6
Density, g/mL	1.35	1.55	1.58	1.43	0.76
Cleaning					
Wide range of soils	G[a]	G[b]/F	G[b]/P	G[b]/P	G/F
High soil loading in liquid	G	G[b]/F	G[b]/P	G[b]/P	G/F
Fluorinated oils and greases	G[b]/P	G	G	G	P
Silicone oils and greases	G	G[b]/P	G[b]/P	G[b]/P	G
Fast drying	G	G	G	G	F
Compatibility					
Metals	G	G[c]	G	G	G
Plastics and elastomers	G/F/P[d]	G/F/P	G/F[e]	G/F[e]	G/F/P
Liq. oxygen—direct contact	P	P	G	G	P

[a] G, good; F, fair; P, poor.
[b] Applies to blend or azeotrope, not neat solvent.
[c] Must be properly formulated for metals compatibility.
[d] G/F/P, compatibility varies significantly with different plastics and elastomers.
[e] Fluorinated solvents may have poor compatibility with fluoroplastics and elastomers.

aluminum components in a dry cleaning machine, especially in the still. Therefore, it is important to monitor the stabilizer level by using an Acid Acceptance Kit. The nPB-based system includes the use of ancillary products, such as detergents, which helps maintain the correct level of acid acceptors. Dry cleaning machines built specifically for the nPB-based formulation are entering the market.

Although spent stabilized nPB by itself is not classified as a hazardous waste, it may be necessary to treat it as a hazardous waste depending on what products are deposited in it during the cleaning process. It has been shown that PCE at or slightly above RCRA levels are common in the waste stream, due to residual PCE in clothes from previous cleaning and from other dry cleaning materials, such as spotting compounds. Since proper ongoing testing of each waste load is likely to be economically prohibitive, it is recommended that all nPB dry cleaning be manifested and disposed of by licensed hazardous waste haulers.

Special Formulations: Electronics

The electronics industry faces some cleaning challenges not found in other types of cleaning applications. In particular, it is of utmost importance that ionic residues remaining on integrated circuit boards after soldering operations be removed to very low levels. In addition, conventional formulations for stabilized nPB solvents have a tendency to tarnish silver and silver-plated leads. To address these specific challenges, formulations have been introduced that are designed specifically for the electronics industry. New formulations[6] are efficient at the removal of ionic residues.

Compatibility with Metals

The stabilized nPB (EG) was tested for compatibility with metals using the Mil-T-81553A protocol. Metal coupons were polished with a fine emery cloth and then exposed to boiling solvent (liquid and vapor) for 24 h. The coupons were examined for signs of corrosion. The metals that passed this test were the following:

Zinc	Monel	Titanium-2
Copper	Magnesium	Tin plate
Brass	Inconel 6000	Nickel 2000
C1010 steel	Galv. steel	316L Steel
Al 2024	304 Steel	

The only metal that failed this test was C12L14 leaded carbon steel.

Compatibility with Plastics and Elastomers

Plastics and elastomers that pass short-term compatibility tests (immersion in boiling solvent for 15 min) with stabilized nPB (EG) are shown below. Those that were (M)arginal or (U)nsatisfactory after 2 weeks at 65°C are marked accordingly. The marked materials may be cleaned but should not be incorporated in equipment where they may contact the hot cleaning solvent for extended periods.

Plastics		Elastomers
Acculam™_epoxy glass (M)	Teflon™_PTFE	Adiprene™_polyurethane (U)
Alathon™_HDPE	Teflon™_PTFE	Aflas™_PTFE
Delrin™_acetal	Tefzel™_ethylene/PTFE	Lucite™_(U)
Kynar™_polyvinyl fluoride	Ultem™_polyether imide	Natural rubber (U)
Low Density PE	XLPE™_cross-linked PE	NBR polyether imide
Phenolics		Neoprene™_polychloroprene (U)
Polypropylene		Viton-A™_fluoroelastomer (M)

Removal of Ionic Residues from Integrated Circuit Boards

The stabilized *n*PB (EG) testing was carried out in conjunction with Contamination Studies Laboratories Inc. (CSL). CSL prepared IPC B-24 boards fluxed with Higrade RMA liquid flux and processed through a standard wave solder system. The boards were then shipped to the Albemarle Technical Center where they were cleaned using a Branson 5 gal laboratory vapor degreaser. The cleaning cycle was as follows:

1. Vapor zone—2 min
2. Boil sump—3 min
3. Rinse sump—1 min
4. Vapor zone—2 min

The cleaned boards were returned to CSL for evaluation. Three techniques were used:

1. Resistivity of solvent extract (ROSE) using an Omega Meter.
2. Ion chromatography (IC) after extraction.
3. Surface insulation resistance.

The results of the ROSE (Omega Meter) tests and the ion chromatography tests are given in Table 7.5. Note that the IC values are for chloride ions only. Bromide ions were also found, but control tests showed that these were most likely extracted from the brominated flame retardants incorporated in the board material. The IC test employs a rigorous extraction technique that involves heating the boards at 80°C for 1 h in a 3/1 IPA/water solution.

The CSL report concludes that, "The ionic residue testing of ROSE (Resistivity of Solvent Extract—Omega Meta) and Ion Chromatography show that the samples have low residues." They also report, "The SIR (Surface Insulation Resistance) data all pass the SIR testing protocol for good product performance."

At a presentation for the IPC/SMTA Electronics Assembly Expo in October of 1998, Howard Feldmesser of The Johns Hopkins University Applied Physics Laboratory reported that both the VG and EG formulations were superior to AK-225 AES in the removal of ionic residues from circuit boards. In addition, the EG formulation was superior to the three other stabilized *n*PB formulations that he included in his evaluation.[7]

Prevention of Silver Tarnish

Reports from the field indicated that stabilized *n*PB formulations were tarnishing silver and silver-plated contacts. A study of the additive technology resulted in a solution, which was then incorporated into the stabilized *n*PB (EG) formulation.[8,9] Silver-plated steel coupons were placed vertically into beakers of several commercial stabilized *n*PB formulations such that the solvent came approximately one-third of the way up the coupon. The beaker was then heated on a hot plate to boiling and held at the boiling temperature for 10 min. The heat-up time was approximately 5 min. The coupons were removed from the boiling solvent, dried and examined. All of the commercial grades badly tarnished the coupons, but the EG cleaner showed no tarnish.

TABLE 7.5 Removal of Ionic Residues with Electronic Grade *n*PB Cleaner

Board Number	Omega Meter, g/in.^{2}	Ion Chromatography Chloride ion, g/in.^{2}
1	4.3	2.37
2	5.2	2.49
3	4.1	2.08
Average	4.5	2.31

A second experiment involved the cleaning of lead frames that are made of copper and are tipped with silver. One frame was cleaned with a typical stabilized *n*PB formulation and the other frame with the EG cleaner. The cleaning cycle included 3 min, in an ultrasonic sump, 3 min, in the boil sump, and 4 min, in the vapor zone. The difference was dramatic and obvious. A solvent cleaner based on *n*PB and formulated in the usual way for vapor degreasing severely tarnishes silver and silver plate. The EG formulation, on the other hand, does not tarnish silver when used in a normal cleaning cycle (including ultrasonics and/or submersion in the boiling solvent).

It has been found that nitromethane, used as a stabilizer, tarnishes silver.

Health, Safety, Environmental, and Regulatory Issues[10]

Toxicology: Acute

*n*PB is somewhat toxic, but can be handled safely if reasonable precautions are taken. Below is a compilation of the current acute toxicology data for *n*PB. Current Material Safety Data Sheets should be obtained from the manufacturer for additional and/or the latest information.

Mammalian Genetic Toxicity

*n*PB is negative for dominant lethal activity in rats at 400 mg/kg/day given for 5 days to male rats prior to breeding once weekly for 8 successive weeks. No difference in mating performance was noted in treated males. Their frequency of fertile matings, mean numbers of corpora lutea, number of implants per female, number of live embryos per female, and the dominate lethal index was comparable to the negative control group at weeks 1, 2, 3, 4, 5, 6, 7, and 8 after treatment. The frequency of dead implants was higher at week 8 of treatment compared to the control group, but no increase was observed in the dominant lethal index at that or any other time. The frequency of dead implants in the treated group was comparable to the control group at weeks 1, 2, 3, 4, 5, 6, and 7.

Mammalian Metabolism

The half-life of *n*PB in the rat is very short (approximately 2 h). The majority of the administered dose is eliminated rapidly in expired air as the unchanged parent compound. The remainder is metabolized and excreted in the urine (predominant route) or in the expired air as CO_2 (minor route).

Following a single intraperitoneal dose (200 mg/kg), the initial rate of excretion of the unchanged ^{14}C-labeled parent compound in the expired air of the rat was rapid. Two hours after administration, 56% of the administered dose was exhaled as the parent compound. After 4 h, 60% had been exhaled: only trace amounts were detected in expired air after this time. An earlier study also reported the elimination of the unchanged parent compound in expired air. Oxidation to CO_2 occurred only to a minor extent. Only 1.4% of the total dose (or 3.5% of the metabolized dose) was exhaled as CO_2 over 48 h. Approximately 40% of the total IP-administered dose was available for metabolism in the rat and excretion in the urine.

Aquatic Acute Toxicology

The solubility of *n*PB in is approximately 0.25 g/100 mL water at 20°C. The 96 h LC50 in flathead minnows is 67,300 µg/L.

Environmental and Health Regulatory Status

Worker safety, public safety, and environmental protection are paramount in the development of any new product. The status of *n*PB and solvent/cleaner systems based upon it is given in Table 7.6. The table includes information on 1,1,1-TCA and TCE for comparison. Explanations of the regulations follow the table.

TABLE 7.6 Environmental and Health Regulatory Status

Regulation	n-Propyl Bromide	1,1,1-Trichloroethane	Trichloroethylene
SARA[a]	No	Yes	Yes
HAP[b]	No	Yes	Yes
NESHAP[c]	No	Yes	Yes
RCRA[d]	No	Yes	Yes
HGWP[e]	0.0001	0.023	Almost zero?
ODP[f]	0.016–0.019	0.1	Almost zero?
Atmospheric lifetime[g]	19 days[10]	5.4 years	14 days
PEL[h]	No OSHA PEL 25 ppm recommendation of manufacturers 10 ppm ACGIH TLV 5 ppm California PEL (2010)	350-ppm	50-ppm (25-ppm in CA) 10 ppm ACGIH TLV
VOC[i]	Yes	No	Yes
SNAP[j]	Acceptable	Unacceptable	Acceptable

Note: Bromine is more than twice as heavy as chlorine and more than four times as heavy as fluorine. The bromine-carbon bond is also not as strong as that of chlorine and fluorine as a consequence of bromine's lower electronegativity, which helps to explain why nPB is not as stable in the atmosphere as chlorinated and fluorinated solvents.

[a] *SARA:* Superfund Amendments and Re-authorization Act. This act requires reporting of inventories and emissions of listed chemicals and groups.

[b] *HAP:* Hazardous Air Pollutant: A listing of chemicals that the EPA has declared hazardous.

[c] *NESHAP:* National Emission Standard for HAP. Sets standards for use of HAPs.

[d] *RCRA:* Resource Conservation Recovery Act. Defines hazardous wastes and how to manage them.

[e] *GWP* and *HGWP:* Global warming potential, atmospheric lifetime and ozone depletion potential calculations were carried out by Atmospheric and Environmental Research, Inc. (Independent study prepared for Albemarle Corporation, 1995.) GWP is calculated relative to CO_2, while HGWP (Halocarbon GWP) is calculated relative to CFC-11. GWP calculations were done using different integration time horizons. By the HGWP method, CFC-11 is 10,000 times more detrimental as a global warming agent than is nPB. By the GWP method, CFC-11 is 14,000 times higher than, and CO_2 is 10 times higher than nPB.

Compound	HGWP	GWP (20 Years)	GWP (100 Years)	GWP (500 Years)
CFC-11	1.0	4500	3400	1400
nPB	0.0001	1.01	0.31	0.1

[f] *Atmospheric lifetime:* Once emitted into the atmosphere, chemical compounds break down into degradation products and finally dissipate altogether, at different rates ranging from hours to years depending on the compound. Atmospheric lifetime is the calculation of how long the chemical takes to break down. A chemical compound must be able to reach the stratosphere before it can effect ozone residing there. When emitted at surface level in northern latitudes (e.g., United State and Europe), it normally takes about 6 months for a chemical compound to reach the stratosphere, whereas emission at the equator requires much less time to reach the stratosphere. It is generally accepted that compounds with short atmospheric lifetimes (less than 6 months) will not reach the stratosphere whereas compounds with long lifetimes (over 6 months) will reach the stratosphere. (Wuebbles, D. Effects of short atmospheric lifetimes on stratospheric ozone. EPA-HQ-OAR-2002-0064-0114.) Atmospheric lifetime then is an important data point in determining the Ozone Depletion Potential of a chemical compound. The atmospheric lifetime of nPB is 19 days. In comparison, TCA has an atmospheric lifetime of 5.4 years.

[g] *ODP:* Ozone depletion potential. Ozone residing in the stratosphere protects the earth from harmful UV rays. The depletion of ozone in the stratosphere has been shown to have adverse affects on human and animal health. Emission of certain chemical compounds at surface level have been shown to reduce the amount of ozone in the stratosphere. Computer models have been developed to predict the potential for any chemical compound emitted into the atmosphere to destroy stratospheric ozone, which is called the ozone depletion potential. Chemical compounds which have been shown by the computer models to deplete the stratospheric ozone layer have been banned for use and production by the international treaty called the Montreal Protocols. For comparison, TCA has an ODP of 0. Production of TCA was banned and its use is also banned in most instances. NPB has an ODP of 0.016. (Wuebbles, D. et al., *J. Geophys. Res.*, 106, 14551–, 2001.)

[h] *PEL:* Permissible exposure limit. Determined under formal Federal rulemaking by OSHA, a PEL is a legally enforceable maximum limit to which a worker can be exposed to a given chemical compound. The most common PEL is generally stated in parts per million (ppm), as an inhalation limit. As of the time of publication, OSHA has not set a PEL for nPB.

In June of 2003, USEPA recommended an acceptable exposure limit (AEL) of 25 ppm as part of a proposal under the USEPA's Significant New Alternatives Program (SNAP). *Federal Register*/68(106) Tuesday, June 3, 2003. The term AEL is used by USEPA to differentiate USEPA opinions and recommendations from OSHA's legally binding PEL. In the final SNAP rule published in 2007, USEPA made no formal recommendation of an AEL for nPB. *Federal Register*/72(103)/Wednesday, May 30, 2007.

In 2005, the ACGIH adopted a threshold limit value (TLV) of 10 ppm for nPB. TLV is the registered trademark of the ACGIH and its practical purpose is essentially the same that of a PEL and AEL; a maximum limit, generally stated in ppm, of the amount of chemical a worker may inhale. We note that the USEPA stated its formal agency position as AACGIH's TLV for nPB of 10 ppm has significant limitations as a reliable basis for an acceptable exposure limit. Details can be found in *Federal Register*/72(103)/Wednesday, May 30, 2007, p. 30149.

Chemical exposure limits can also be set by individual States. California has set a California State workplace exposure level of 5 ppm to be effective in late 2010. In the absence of a PEL, a number of manufacturers have set their own Workplace Exposure Limit. Users should check the MSDS of the product they are using.

[i] *VOC:* Volatile organic compound. Under the Clean Air Act, all carbon compounds are classified by default as volatile organic compounds (VOCs) so by default, nPB is currently classified as a VOC. Emission of VOCs are regulated in some parts of the United States (generally urban areas) because it is expected that those emissions will cause urban smog. If scientific data shows that a compound is negligibly reactive, it may be granted an exemption from the definition of a VOC and is therefore usable without complying with VOC regulations. Studies have been conducted to determine the reactivity of nPB and the data has been supplied to the EPA in a Petition to exempt nPB from the definition of a VOC. At the time of this writing, the classification is still under review.

[j] *SNAP:* Significant New Alternatives Policy. Administered by the USEPA, SNAP is the program by which the United States implements the Montreal Protocol, an international treaty. SNAP reviews chemical compounds which have been proposed as ozone depleting substances. SNAP approves the use of those alternatives which do not present a substantially greater risk to public health and the environment than the compound they replace or than other available substitutes. USEPA has determined that n-propyl bromide (nPB) is an acceptable substitute for methyl chloroform (TCA) and chlorofluorocarbon (CFC) 113 in the solvent cleaning sector. *Federal Register*/72(103)/Wednesday, May 30, 2007.

TABLE 7.7   Comparison of Environmental Parameters for New Cleaning Solvents

	*n*PB	HCFC-225	HFC-43-10	VMS
Atmospheric lifetime	19 days	2.7–7.9 years	20.8 years	<30 days
Volatile organic cmpd	Yes	No	No	No
Ozone depletion potential	0.016–0.019	0.03	None	None
Global warming potential[a]	0.31	170/690[b]	1000	Low
Halocarbon GWP[c]	0.0001	0.04/0.06[b]	0.25	Low
Exposure guideline, TWA	25 ppm	50 ppm	200 ppm	200 ppm

     [a] GWP (CO_2) = 1, 100 year ITH.
     [b] Different values for the two isomers.
     [c] HGWP (CFC 11 = 1.0).

Environmental and Health-Related Parameters Compared to Other New Solvents

It is useful to compare the environmental impact and safe usage factors of the various new solvents. As one can see by studying the data on Table 7.7, there remain difficult conflicts in deciding which solvent has the least impact on environmental, health, and safety issues. The ODP and the global warming potential (GWP and HGWP) are often in conflict. One must also consider whether the compound is a volatile organic compound (VOC).

Case Studies

Electronic Equipment

A company that produces electronic components for clinical equipment used in biomedical applications had used CFC-113 blends in both liquid and vapor phase defluxing.[12] Exacting performance standards, rate of throughput, space limitations, limited capital equipment budget, and lack of an adequate industrial water system were major constraints on a changeover to a new cleaning system.

The company attempted to switch to a system based on D-limonene. Residue from the cleaning agent, buildup of rosin flux, and reactivity to produce assorted oxidation products made the electronic components unsuitable for use. An unacceptable green residue remained on the assemblies and there was an increase in product failures.

After switching to VG cleaner, the company found that the new cleaner performed better at removing flux than the CFC-113 blend. There were no residue problems as there were with D-limonene. Some of the plastic components did show some discoloration, but shortening the exposure time, an added benefit that increased production throughput, solved this problem.

Electric Motor Stators and Refrigeration Coils

An aerospace company needed to remove burnt oils from used electric motor stators. They sent two such stators to the *n*PB supplier for test cleaning. One was clean and the other was covered in oil. Both were cleaned using VG cleaner. They were first lowered into the vapor zone of a vapor degreaser and held there until condensation of the vapor on the parts ceased. This required about 5 min. They were then lowered into the ultrasonic bath for 10 min, and back into the vapor zone for 5 min. The parts were returned for examination.

The aerospace company reported that the no residual oils or other contaminants were found, and there was no damage to the electrical wiring or casings. This finding led the manufacturer to examine other applications for the VG cleaner. They have since devised a flushing station for cleaning the long coils of copper tubing used in refrigeration units. An outline of the extensive selection process has been presented.[13,14]

Implantable Body Parts

A company manufactured artificial body parts, such as hip joints, for implantation. The parts consist of a titanium bone replacement and an ultrahigh molecular weight polyethylene (UHMWPE) cartilage replacement. Standards for cleanliness are, of course, very high. In addition, the company expressed concern about retention of solvent in the UHMWPE parts. A final criterion was that the cleaning solvent must kill at least 50% of the bacterial spores on an artificially inoculated UHMWPE part. The company wanted to replace two cleaning solvents in their process. The first was based on HCFC-141b. The second was based on trichloroethylene.

An initial set of vapor degreasing tests were performed by Baron-Blakeslee, Inc. UHMWPE parts were exposed to VG vapors for 90 s. Both cleaned and uncleaned parts were returned to the company, who found the cleaning to be satisfactory. Acetabular components and hip stems were then contaminated with buffing compound. After a cleaning cycle of 30 s in the vapor, 5 min immersion in the ultrasonic sump at 130°F, 2 min, drying and 2 min freeboard dwell, the equipment supplier observed the parts to be completely dry with no solvent drag out. The device manufacturer determined that the cleaning was satisfactory. In a third experiment, a femoral with fingerprints was exposed to vapor for 90 s. Evidence of fingerprints remained after this test.

The second company concern was retention of the solvent by the UHMWPE parts. Stabilized *n*PB (VG) was compared directly with commercial cleaning grades of HCFC-141b and trichloroethylene. The test procedure was to place five parts in the boiling solvent for 3 min. The parts were removed and placed in the same solvent at ambient temperature for 2 h. The parts were then removed and placed in an open dish. The dish was left in a fume hood with the fan going until it was time to take the measurements. To obtain the quantity of headspace vapors, the parts were placed in a sealed glass chamber and allowed to equilibrate for 1 h. The vapors were then analyzed using gas chromatography/mass spectrometry and quantified against a standard. Headspace concentrations of *n*PB were determined at 24 and 96 h. For the other two solvents, the headspace concentrations were determined at 24 and 106 h. The vapor in the headspace is reported in ppm by volume.

	Concentration of Solvent in Headspace (ppm)	
	<24 h	96 or 106 h
*n*PB (VG)	30	3.7
HCFC-141b	169	4.4
Trichloroethylene	469	27

These experiments indicate that the *n*PB is retained in the UHMWPE parts to a lesser extent than either the HCFC-141b or the trichloroethylene.

The third criterion is that the solvent/cleaner reduce bacterial spore counts on the UHMWPE parts by at least 50%. The company supplied two sets of parts (six parts each) contaminated with Spordex™ *Bacillus subtilis* (*globigii*) spores. Each set was divided into two sets of three parts each—one set to-be-cleaned and one set as a control. The cleaning procedure was as follows:

1. Tests parts were placed in the basket of a laboratory vapor degreaser and covered with a metal screen to prevent the parts from floating to the surface. The degreaser contained stabilized *n*PB (VG).
2. The basket was lowered into the vapor zone and remained there until condensation stopped.
3. The basket was then lowered into the boil-up sump (7°C) for 3 min, (Set A) or 1.5 min, (Set B).
4. The parts were placed in the rinse sump for 1 min., followed by the vapor zone for 1 min. The basket was allowed to hang in the free board zone for an additional minute to assure that the parts were dry.

The two sets of cleaned parts along with the accompanying control sets were returned to Smith & Nephew for analysis. The bioburden validation was done at Axios, Inc. (Kennesaw, Georgia). The results were as follows:

Sample	Spore Count	% Reduction
Sample A		
Control	5.5×10^6	
Cleaned	1.5×10^6	73
Sample B		
Control	8.8×10^6	
Cleaned	2.8×10^6	68

Aluminum Parts for Optical Applications

A manufacturer of optical equipment that uses anodized aluminum components requested these tests. There were two areas of concern. First, aluminum is a very active metal. It is especially active toward halogenated materials. The manufacturer was concerned about possible interactions of the aluminum with *n*PB-based cleaners. The second concern involved the lettering and other markings that the parts have on them. The customer requested an evaluation to make sure that the markings would not be damaged in the normal cleaning process. They sent four sets of components, each set containing six different parts. Two sets were cleaned in stabilized *n*PB (VG) by immersing in the boil-up sump for 10 min, followed by 1 min, in the rinse sump. The other two sets were immersed for only 3 min, in the boil sump and then 1 min, in the rinse sump. All 24 parts were returned to the customer for examination. No damage to the parts was observed.

High-Performance Inertial Navigation Systems

The Guidance & Control Systems Division of Litton Industries builds high-performance navigation systems. Systems include gyroscope instruments and associated electronics assemblies. Producing inertial navigation systems involves exacting and multistep cleaning of complex subassemblies. Each subcomponent of a system may require various cleaning steps. Fluxes, oils, and flotation fluids must be removed from an array of materials of construction including a wide assortment of metals, plastics, and epoxies. For this application, soil residues, cleaning agent residues, and water are not acceptable. When faced with the prospect of having to replace 1,1,1-TCA and CFC-113, Litton undertook an ambitious program to evaluate co-solvent systems, fluorinated solvents, and stabilized *n*PB cleaners.[15]

Litton listed a number of requirements for new cleaning systems. These included the following:

- Replacement of ozone depleting chemicals
- Process improvement
- Maintenance of superior performance
- Minimization of contamination
- Cost-effective operation
- Efficient processing
- Compliance with national and local regulatory requirements
- Assurance of worker safety
- Production to meet exacting customer standards

A variety of systems were evaluated. Based on feedback from the assemblers in the plant, the cleaning sequences were refined and implemented. The approach initially adopted was a co-solvent system consisting of initial cleaning with a hydrocarbon blend containing various alcohols followed by two to three rinses with isopropyl alcohol (IPA). This new process allowed elimination of TCA cleaning. Some perfluorinated material continued to be used as a final rinse to assure thorough removal of fluorolube. Overall, the number of process steps was reduced. In some cases, 18 steps were reduced to 4 to 6 steps. There were still problems, however. Because IPA was found to react with beryllium periodically, intermittent residues were found. Eventually, the IPA was replaced with VMS.

The co-solvent system developed to replace TCA was still far from optimal. The subassemblies are very complex, with close tolerances and blind holes. While the cleaning agent can be removed with careful process control, extreme and constant care is required to assure that no cleaning agent residue is left. In addition, the hydrocarbon blends were costly; some of the operators found the odor to be disagreeable, and the blends were flammable.

Cleaners based on *n*PB had been introduced to the cleaning market by this time, and Litton undertook an extensive evaluation. One consideration was that the *n*PB is a very aggressive solvent that could be expected to have cleaning performance similar to TCA. In the end, the VG cleaner was chosen as the most suitable for the Litton requirements. In addition to the cleaning capability, the reliability and the environmental and safety acceptability of the VG, Litton found improved processing time and lower cleaning agent usage. Litton has reported that the processing time has been reduced by over 40%. The cleaning agent usage has been reduced to one-third of the previous amount.

Conclusions

Solvent/cleaner systems based on *n*PB are replacements for chlorinated solvents in cleaning applications. As with all aggressive solvents, *n*PB must be used with appropriate worker safety and environmental controls. Stabilized *n*PB is an aggressive, fast-drying solvent that is suitable for a variety of difficult cleaning and degreasing applications. The use of *n*PB-based solvents is not regulated under SARA, HAP, NESHAP, or RCRA, and they are approved under SNAP for many applications. *n*PB has low potentials for ozone depletion and for global warming, but it is currently classified as a VOC. In comparative performance testing, *n*PB-based formulations have been demonstrated to be as effective as chlorinated solvents and more effective than HCFCs, hydrofluorocarbons, hydrofluorocarbon ethers, and VMSs.

Recent developments in *n*PB-based cleaning solvents include formulations to meet specific requirements in specialty niche applications such as electronics, aerospace, medical oxygen equipment, optics, dry cleaning, paint stripping, etc. In addition to the expected advantages for stabilized *n*PB, appropriate formulations provide enhanced efficiency in the removal of ionic residues while not tarnishing silver.

The unique properties of the bromine-based solvent far outweigh the stability issues. The advantages are that bromine is more oil soluble than chlorine. *n*PB is softer on fabric and requires less energy to heat. Although bromine breaks down ozone more readily than chlorine, bromine is not as stable in the atmosphere (breaks down ozone in our breathing zone, but does not reach the stratosphere), because bromine is a heavier atom than chlorine and fluorine, and bromine is not as strongly bonded to carbon. Even though *n*PB is less stable than some chlorinated solvents, hydrobromic acid is less corrosive to soft metals than hydrochloric acid. Furthermore, *n*PB will never breakdown to form phosgene gas.

References

1. Kanegsberg, B., Precision cleaning without ozone depleting chemicals. *Chemistry and Industry*, no. 20, 787, October 21, 1996.
2. Shubkin, R.L., A new and effective solvent/cleaner with low ozone depletion potential. *1996 International Conference on Ozone Protection Technologies, Proceedings and Presentation*, Washington, DC, October 21–23, 1996.
3. Unless otherwise noted, experimental procedures and results referred to in this chapter were performed at the Albemarle Technical Center, Baton Rouge, LA, or by contract laboratories under the direction of Albemarle technical personnel.

4. Shubkin, R.L. and Liimatta, E.W., A new cleaning solvent based on *n*-propyl bromide. *NEPCON West '97 Conference*, Anaheim, CA, February 23–27, 1997.
5. Shubkin, R.L., Solvent cleaning into the next century and beyond. Presented at *CleanTech99*, Rosemont, IL, May 19, 1999.
6. Shubkin, R.L. and Liimatta, E.W., *N*-propyl bromide based cleaning solvent and ionic residue removal process. U.S. Patent 5,792,277 (August 11, 1998, to Albemarle Corp.).
7. Feldmesser, H.S., Loyd, K.M., Clausen, M., and Karvar, P., Examining the compatibility of electronic assembly materials with cleaning solvents. Presented at *IPC/SMTA Electronics Assembly Expo*, Providence, RI, October 25–29, 1998, and at the *Ninth Annual Solvent Substitution Workshop*, Scottsdale, AZ, December 1–4, 1998.
8. Shubkin, R.L., Method for inhibiting tarnish formation when cleaning silver with ether stabilized, *n*-propyl bromide based solvent systems. U.S. Patent 5990071 (1999) applied for (to Albemarle Corp.).
9. Shubkin, R.L., Method for inhibiting tarnish formation during the cleaning of silver surfaces with ether stabilized, *n*-propyl bromide based solvent systems. U.S. Patent 6165284 (2000) applied for (to Albemarle Corp.).
10. Shubkin, R.L. and Smith, R.L., *normal*-Propyl bromide: Formulation technology and product stewardship for the electronics industry. *NEPCON West '99 Conference*, Anaheim, CA, February 23–25, 1999.
11. Nelson Jr., D.D, Wormhoudt, J.C., Zahniser, M.S., Kolb, C.E., Ko, M.K.W., and Weisenstein, D.K., OH reaction kinetics and atmospheric impact of 1-bromopropane. *Journal of Physical Chemistry A*, 101(27), 4987–4990, 1997.
12. Kanegsberg, B., Cleaning high value components for biomedical and other applications. *1996 International Conference on Ozone Protection Technologies, Proceedings and Presentation*, Washington, DC, October 21–23, 1996.
13. Petrulio, R. and Kanegsberg, B.F., Practical solutions to cleaning and flushing problems, *CleanTech98*, Rosemount, IL, May 19–21, 1998.
14. Petrulio, R., Kanegsberg, B.F., and Chang, S.-C., A practical search solves aerospace cleaning quandary. *Precision Cleaning Magazine*, VI (8), August, 1998.
15. Carter, M., Anderson, E., Chang, S.-C., Sanders, P.J., and Kanegsberg, B.F., Cleaning high precision inertial navigation systems, a case study and panel discussion. Presented at *CleanTech99*, Rosemont, IL, May 18–20, 1999.
16. Barton, A.F.M. (Ed.), *CRC Handbook of Solubility Parameters and Other Cohesion Parameters*, 2nd edn., CRC Press, Boca Raton, FL, 1991.

8

Vapor Degreasing with Traditional Chlorinated Solvents

Stephen P. Risotto
*American Chemistry
Council*

Overview

Today's manufacturing engineers and plant managers can face a difficult challenge when choosing among the many available surface cleaning options. Aqueous, semi-aqueous, flammable solvents, and new fluorinated and brominated solvents are just a few of the possibilities. Among these many choices, however, one process stands out for its ability to produce a clean, dry part at a reasonable price—vapor degreasing with the chlorinated solvents.

Trichloroethylene (TCE, TRI), perchloroethylene (PCE, PERC), and methylene chloride (MC, METH) have been the standard for cleaning performance in precision parts cleaning for more than 50 years. Today, the development of new equipment and processes that minimize emissions and maximize solvent recovery makes TCE, PCE, and MC more effective than ever.

Despite their superior performance attributes, however, some companies have replaced TCE, PCE, or MC with other solvents or processes. Their decision often was based on misperceptions about the chlorinated solvents' regulatory status, continued availability, and safety in use. The facts are

- TCE, PCE, and MC have not been banned. Among the commonly used chlorinated solvents, only 1,1,1-trichloroethane (methyl chloroform) was phased out of production, due to its ozone depletion potential. Meanwhile, the U.S. Environmental Protection Agency (EPA) has issued a 1994 decision under its Significant New Alternatives Policy (SNAP) program that the other three chlorinated solvents are viewed as acceptable substitutes for ozone-depleting solvents.
- Chlorinated solvents will continue to be available. TCE and PCE demand have remained steady in recent years as a result of their use as raw materials in the production of refrigerant alternatives to

CFCs. MC continues to be used in a wide variety of applications. The producers of these solvents remain committed to serving their markets for many years to come.

- TCE, PCE, and MC can be used safely. From the point of view of health and the environment, the chlorinated solvents are among the most thoroughly studied industrial chemicals. Animal tests and epidemiological studies indicate that when the solvents are handled, used, and disposed of in accordance with recommended and mandated practices, they do not cause adverse health or environmental effects.
- The potential impacts of the solvents can be minimized. Environmental, health, and safety regulations governing the chlorinated solvents are strict but manageable. In complying with these regulations companies can get help from several sources—EPA, Occupational Safety and Health Administration (OSHA), and state and local agencies, producers and distributors of solvents and degreasing equipment, and organizations like the Halogenated Solvents Industry Alliance (HSIA).

Physical and Chemical Information

TCE, PCE, and MC are clear, heavy liquids with excellent solvency. All are virtually nonflammable, since they have no flash point as determined by standard test methods. Each has its own advantages for specific applications, based on its physical profile (see Table 8.1). These solvents work well on the oils, greases, waxes, tars, lubricants, and coolants generally found in the metal processing industries. They are widely used in the vapor degreasing process.

TABLE 8.1 Typical Properties of the Chlorinated Solvents

	Methylene Chloride	Perchloroethylene	Trichloroethylene
Chemical formula	CH_2Cl_2	C_2Cl_4	C_2HCl_3
Molecular weight	84.9	165.8	131.4
Boiling point °F (°C) at 760 mmHg	104 (40)	250 (121)	189 (87)
Freezing point °F (°C)	−139 (−95)	−9 (−23)	−124 (−87)
Specific gravity at 68°F (g/cm³)	1.33	1.62	1.46
Pounds per gallon at 77°F	10.99	13.47	12.11
Vapor density (air = 1.00)	2.93	5.76	4.53
Vapor pressure at 77°F (mmHg)	436	18.2	74.3
Evaporation rate at 77°F			
Ether = 100	71	12	30
n-Butyl acetate = 1	14.5	2.1	4.5
Specific heat at 68°F (BTU/lb/°F or cal/g/°C)	0.28	0.205	0.225
Heat of vaporization (cal/g) at boiling point	78.9	50.1	56.4
Viscosity (cps) at 77°F	0.41	0.75	0.54
Solubility (g/100 g)			
Water in solvent	0.17	1.01	0.04
Solvent in water	1.70	0.015	0.10
Surface tension at 68°F	28.2	32.3	29.5
Kauri-butanol (KB) value	136	90	129
Flash point			
Tag open cup	None	None	None
Tag closed cup	None	None	None
Flammable limits (% solvent in air)			
Lower limit	13	None	8
Upper limit	23	None	11

TCE has been long recognized for its cleaning power. TCE is a heavy substance (12.11 lb/gal) with a high vapor density (4.53 times that of air) that allows for relatively easy recovery from vapor degreasing systems. The solvent's ability to provide constant pH and to protect against sludge formation has helped make it the standard by which other degreasing solvents are compared. Its high solvency dissolves soils faster, providing high output.

TCE is used extensively for degreasing zinc, brass, bronze, and steel parts during fabrication and assembly. It is especially suited for degreasing aluminum without staining or pitting the work, because its stabilizer system protects the solvent against decomposition. For cleaning sheet and strip steel prior to galvanizing, TCE degreases more thoroughly and several times faster than alkaline cleaning, and it requires smaller equipment that consumes less energy.

PCE has the highest boiling point, weight (13.47 lb/gal), and vapor density (5.76 times that of air) of the chlorinated solvents. PCE's high boiling point gives it a clear advantage in removing waxes and resins that must be melted in order to be solubilized. The higher temperature also means that a larger temperature difference could exist between solvent and part, allowing more vapors to be condensed on the work than with other solvents, thus washing the work with a larger volume of solvent.

PCE is effective in cleaning lightweight and light-gauge parts that would reach the operating temperature of lower-boiling solvents before cleaning is complete. When cleaning parts with fine orifices or spot-welded seams—especially if there is entrapped moisture—PCE's high boiling point is essential for obtaining good penetration.

Inherently more stable than other chlorinated (and brominated) solvents, PCE also incorporates a multicomponent stabilizer system that provides the greatest resistance to solvent decomposition available in the industry. While it can be used to degrease all common metals, PCE is especially applicable to cleaning those which stain or corrode easily, including aluminum, magnesium, zinc, brass, and their alloys.

MC has the lowest boiling point of the chlorinated solvents, as well as the lightest vapor density (2.93 times that of air) and weight (10.98 lb/gal). MC is uniquely suited for use as a vapor degreasing solvent in applications where low vapor temperatures and superior solvency are desirable. The low boiling point of MC makes it a popular choice for cleaning temperature-sensitive parts such as thermal switches or thermometers.

Vapor degreasing with MC allows more rapid processing and handling, particularly when cleaning large, heavy parts. The more aggressive nature of MC is especially useful when degreasing parts soiled with resins, paints, or other contaminants that are difficult to remove.

Environment and Worker Health Considerations

The potential health effects of TCE, PCE, and MC have been very well studied. Each can cause acute health effects at elevated exposure levels, but these effects have been found to be reversible. The primary concern with these solvents has been their potential to cause cancer, based on the results of laboratory animal tests showing an increase in certain tumors following lifetime exposure to the solvents. Scientific questions have been raised as to the relevance of these animal tumors to human health, however, and epidemiology studies of workers exposed to the chlorinated solvents over extended periods of time have failed to produce a consistent pattern of increased cancer incidence. The animal data, nevertheless, has traditionally been viewed by regulatory agencies as indicating a potential for risk in humans.

Environment

Chlorinated solvents have been found as contaminants in soil and groundwater as a result of historic handling and disposal practices. Releases of the chlorinated solvents to land and water are minor in volume but longer lasting, in comparison to atmospheric emissions. The residence times of TCE, PCE, and MC in the atmosphere are very short and, despite many years of use, concentrations in the ambient air are very low. TCE and PCE are oxidized to carbon dioxide, water, and hydrogen chloride in the lower atmosphere by reaction with either oxygen (ozone) or hydroxyl (OH) radical. MC is oxidized by OH only, forming the same naturally occurring organic breakdown products. Because of their relatively short lifetimes in the atmosphere, TCE, PCE, and MC are not considered to contribute to depletion of the stratospheric ozone layer. Similarly, the chlorinated solvents have negligible global warming potential.

TCE is photochemical reactive and is believed to contribute to the formation of ozone (smog) in the lower atmosphere under certain conditions. The decomposition of PCE and MC contributes only negligibly to the formation of ozone.

In reviewing the acceptability of TCE, PCE, and MC in its SNAP review, EPA noted that these compounds are regulated under several other environmental laws and regulations, including the occupational limits and national emission standards described elsewhere in this chapter. The agency concluded that compliance with these regulations will significantly reduce the potential for environmental releases and worker exposure from degreasing operations. As a result, the SNAP program did not impose further use restrictions on the three solvents in degreasing.

Worker Health

As with all industrial chemicals, occupational exposure to the chlorinated solvents should be kept as low as practical. The OSHA has set permissible exposure limits (PELs) for chlorinated solvents. The PEL for PCE and TCE is 100 parts per million (ppm) for an 8 h time-weighted average (TWA). The limits for MC are 25 ppm for an 8 h TWA and 125 ppm for a 15 min short-term exposure limit (STEL). In addition to the TWA and STEL, the OSHA standard for MC imposes several additional requirements. The American Conference of Governmental Industrial Hygienists (ACGIH) also recommends exposure limits for the chlorinated solvents. The solvent producers recommend maintaining workplace exposure levels within the OSHA limits or the ACGIH* levels, whichever is lower (see Table 8.2).

OSHA's Hazard Communication (HAZCOM) standard specifies a minimum element of training for people working with hazardous materials, including the chlorinated solvents (Table 8.3). This includes how to detect the presence or release of a solvent, the hazards of the solvent, and what protective measures should be used when handling it. OSHA's HAZCOM standard also regulates the labeling of all hazardous chemicals. Labels must contain a hazard warning, the identity of the chemical, and the name and address of the responsible party.

TABLE 8.2 Workplace Limits for the Chlorinated Solvents (in Parts per Million, or ppm)

	Methylene Chloride	Perchloroethylene	Trichloroethylene
OSHA permissible exposure limits[a]			
8 h TWA	25	100	100
15 min STEL	125	—	—
Ceiling	—	200	200
Peak	—	300	300
ACGIH® threshold limit values[b]			
8 h TWA	50	25	10
15 min STEL	—	100	25

[a] An 8 h TWA is an employee's permissible average exposure in any 8 h work shift of a 40 h week. The STEL is a 15 min TWA exposure that should not be exceeded at any time during the day. The acceptable ceiling concentration is the maximum concentration to which a worker may be exposed during a shift, except that brief excursions to the acceptable maximum peak are permissible.

[b] Threshold limit values (TLVs®) are established by the ACGIH.

TABLE 8.3 Operating Parameters for the Chlorinated Solvents

	Methylene Chloride	Perchloroethylene	Trichloroethylene
Vapor thermostat setting °F (°C)	95 (35)	180 (82)	160 (71)
Boil sump thermostat setting[a] °F (°C)	110 (43)	260 (127)	195 (91)
Steam pressure (psi)	1–3	40–60	5–15
Solvent condensate temperature[b] °F (°C)	100 (38)	190 (88)	155 (68)
Cooling coil outlet temperature range (°F)	75–85	100–120	100–120

[a] Maximum boiling temperature, based on 25% contamination with oil.
[b] To facilitate effective separation of the solvent from the water.

Regulatory Overview

U.S. federal regulations affecting the use, handling, transportation, and disposal of chlorinated solvents can be found under the Clean Air Act, the Clean Water Act, the Resource Conservation and Recovery Act (RCRA), and the Comprehensive Environmental Response Compensation and Liability Act (CERCLA, or Superfund). State and local regulations also exist for the purpose of controlling emissions. Though numerous, these regulations are manageable and companies can obtain compliance assistance from numerous sources. The major federal regulations pertaining to the chlorinated solvents are summarized below.

General

Volatile organic compound (VOC) regulations under the Clean Air Act apply to TCE and limit its emissions in order to reduce smog formation, particularly in ozone non-attainment areas. Exact requirements vary by state, but generally include obtaining a permit allowing a specific amount of VOC emissions from all sources within a facility. PCE and MC, however, are exempt from VOC regulations in most states.

The Clean Air Act also calls for the three chlorinated solvents to be regulated as hazardous air pollutants (HAPs). EPA has issued National Emission Standards for Hazardous Air Pollutants (NESHAP) for solvent cleaning with halogenated solvents, which are discussed below. Other NESHAPs govern dry cleaning with PCE and the use of MC in aerospace manufacture and rework, wood furniture manufacture, and polyurethane foam manufacture.

The Clean Water Act defines chlorinated solvents as toxic pollutants and regulates their discharge into waterways. Under RCRA, wastes containing chlorinated solvents from solvent cleaning operations are considered hazardous. Generators, transporters, and disposers of such hazardous waste must obtain an EPA ID number.

The Superfund law requires that if a reportable quantity of a chlorinated solvent or other hazardous chemical is released into the environment in any 24 h period, the federal, state, and local authorities must be notified immediately. Reportable quantities are 1000 lb for MC and 100 lb for PCE and TCE.

NESHAP Requirements

EPA's NESHAPs for new and existing halogenated solvent cleaning operations governs emission standards for chlorinated solvent degreasing operations. These standards cover both vapor degreasing and cold cleaning with TCE, PCE, and MC.

Originally promulgated in 1994, the degreasing NESHAP was revised in May 2007 to establish annual facility-wide emissions limits for facilities operating cleaning equipment using one of the three chlorinated solvents. The EPA rule revises the requirements of the original 1994 national emission standard for degreasers to further limit emissions of these solvents from new and existing batch and in-line machines by establishing facility-wide limits for each of the solvents.

The 2007 revision requires that facilities ensure that annual emissions from all degreasing activities subject to the 1994 standard are less than or equal to the following limits—60,000 kg (132,000 lb) for MC, 14,100 kg (31,000 lb) for TCE, and 4,800 kg (10,500 lb) for PCE. The chemical-specific limits are based on EPA's current assessment of the relative toxicity of the three chlorinated solvents.

Higher facility-wide limits exist for federal facilities involved in maintenance of military vehicles—100,000 kg for MC, 23,500 kg for TCE, and 8,000 kg for PCE. Facilities in three industry sectors—aerospace manufacture, narrow tubing manufacture, and facilities that use continuous web cleaning machines—were exempted from the facility-wide limits when the revisions were promulgated. EPA has announced its intent to reconsider these exemptions, however, and the agency requested comment in October 2008 on options for further regulating facilities in these three sectors. The options included facility-wide limits and percent reduction requirements.

The 1994 NESHAP requirements remain applicable for all cleaning machines. Companies subject to the newly established facility-wide limits are required to determine the 12 month rolling total emissions of the three chlorinated solvents for all cleaning machines at the facility. If a facility's solvent emissions from its degreasing operations exceed the applicable limit, the facility is required to implement means to comply with the limit. There are no additional equipment monitoring or work practice requirements associated with the facility-wide emissions limits.

In developing the original 1994 standards, EPA focused on equipment and work practice requirements that permit a level of control between 50% and 70%. Companies operating batch or in-line degreasers are given three options for compliance: (1) installing one of several combinations of emission control equipment and implementing automated parts handling and specified work practices; (2) meeting an idling-mode emission limit, in conjunction with parts handling and work practice requirements; or (3) meeting a limit on total emissions.

When a company chooses the first option, it may choose from a series of combinations of two or three procedures, which include freeboard ratio of 1.0, freeboard refrigeration device, reduced room draft, working-mode cover, dwell, and superheated vapor. In addition to these options, solvent cleaning processes must include an automated hoist or conveyor that carries parts at a controlled speed of 11 ft/min or less through the complete cleaning cycle.

Compliance with one of the control options for batch or in-line vapor equipment is demonstrated by periodic monitoring of each of the control systems chosen. Work practices are also required as part of the new EPA standards. Rather than require direct monitoring of work practice compliance, however, EPA has developed a qualification test, included as an appendix to the standard. The test is to be completed by the operator during inspection, if requested.

A company choosing to comply with the second option, the idling emission limit (0.045 lb/ft^2-h for batch vapor equipment, 0.021 lb/ft^2-h for in-line equipment), is required to demonstrate initial compliance by using EPA's idling reference test method 307. Data from the equipment manufacturer may be used, provided the unit tested is the same as the one for which the report has been submitted. Compliance with the idling emission limit also requires installation of automated parts handling system and compliance with work practice requirements. In addition, the company must show that the frequency and types of parameters monitored on the solvent cleaning machine are sufficient to demonstrate continued compliance with the idling standard.

Complying with the third option, the limit on total emissions requires the company to maintain monthly records of solvent addition and removal. Using mass balance calculations, the company calculates the total emissions from the cleaning machine, based on a 3 month rolling average, to ensure they are equal to or less than the established limit for the cleaner (30.7 lb/ft^2-month for small batch vapor machines, 31.4 lb/ft^2-month for large batch vapor machines, 20.3 7 lb/ft^2-month for in-line machines). For new machine designs without a solvent/air interface, EPA has established an emission limit based on cleaning capacity ($=330 \times (vol)^{0.6}$).

Companies meeting the total emission limit requirements do not need to conduct monitoring of equipment parameters, but must maintain records of their solvent usage and removal of waste solvent. According to EPA, this compliance option provides an incentive for innovative emission control strategies to limit solvent use. For some cleaning machines, EPA calculates that the alternative total emission limit could be more stringent than the equipment specifications. In particular, EPA expects that this alternative standard will be more difficult to meet for larger machines, for machines operating more than one shift, and for machines' cleaning parts with difficult configurations.

Life Cycle Assessment

Substitution of chlorinated solvents, for cleaning metal parts, is frequently proposed by regulators and others. Alternatives, such as aqueous cleaning with detergents, are often perceived as having less environmental impacts than cleaning with chlorinated solvent use in a vapor cleaning process.

Objectives and Methodology

A life cycle assessment (LCA) was conducted by the European Chlorinated Solvents Association (ECSA) to provide robust data relating to the environment impact of metal parts cleaning in TCE vapor degreasing and aqueous processes. Five major environmental indicators were considered: nonrenewable resource depletion, greenhouse effect, air acidification, eutrophication (water impacts), and solid waste.

For a viable comparison, the cleaning processes were studied assuming the same performance (the "functional unit") and a clear definition of what is and is not included in the system (the "system boundary"). As requested by standard LCA methodologies, the study also incorporated the environmental impacts caused by the manufacturing of the cleaning agents, upstream of their use in cleaning installations.

The functional unit adopted in this study was the complete removal of 46.8 g (1.6 oz) of grease from 1 m² (10.7 ft²) of metal parts. Thus, a single level of contamination is assumed in all the cases studied, as well as the removal of the whole quantity of grease by the cleaning step.

In order to allow for viable comparisons, the same boundaries were used for all the cleaning processes assessed, as required by the standard methodology for LCA. In particular, both for solvent cleaning and for aqueous cleaning, the following were included in the LCA:

- The environment impacts incurred by the manufacturing of the cleaning agents
- The environmental impacts of the cleaning steps themselves
- The environmental burdens associated with the treatment of the cleaning residues or effluents

Seven cleaning scenarios were chosen, three for solvent cleaning and four for aqueous cleaning. Data on the environmental impacts of metal parts cleaning were collected at five cleaning sites in Europe. The sites were selected to be representative of various technologies across different countries.

Interpretation of the Results

Each cleaning technology was found to have potentially significant environmental impact (see Figures 8.1 through 8.6). The primary disadvantage of TCE, air pollution (i.e., air acidification), can be minimized with emission controls. The water pollution disadvantages of aqueous cleaning, however, remain significant even after significant physiochemical + biological treatment of the cleaning residues. With aqueous cleaning, impact on water was between 200 and 2000 times higher than with TCE degreasing, depending on the site under consideration.

For cleaning and drying metal parts, solvent technology has a lower overall environmental impact than aqueous technology. This is true even without the use of carbon recovery of solvents in the vapor phase, provided that equipment meets the NESHAP requirements and is operated to best practice.

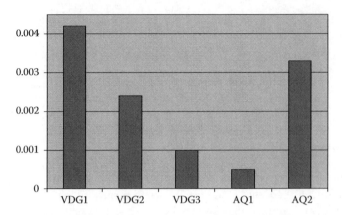

FIGURE 8.1 Nonrenewable resource depletion (kg/year). Results of life cycle assessment (per square meter of metal part cleaned). Scenarios: VDG1, open-top degreaser without NESHAP-compliant controls; VDG2, NESHAP-compliant degreaser with on-site distillation; VDG3, NESHAP-compliant degreaser with on-site distillation and carbon adsorption; AQ1, aqueous cleaning equipment with primary wastewater treatment; AQ2, aqueous cleaning equipment with primary and secondary wastewater treatment and drying. (Two additional aqueous scenarios are not included in the graphs.) (From European Chlorinated Solvents Association, CA Comparison of Metallic Parts Degreasing with trichloroethylene and aqueous solutions, Ecobilan, December 1996.)

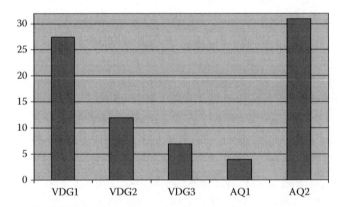

FIGURE 8.2 Total energy use (MJ). Results of life cycle assessment (per square meter of metal part cleaned).

However, air acidification can be higher due to solvent releases. In the ECSA investigation, the air pollution impact of solvent cleaning varied from 8.5 times greater to 5 times less than that for aqueous cleaning with a drying step.

Generally, a given solvent machine can treat a wider range of metal parts than a given detergent installation. This occurs because a detergent is often specific to a kind of contamination and a shape of metal part. A solvent technology without carbon recovery is competitive with an aqueous technology from an environmental viewpoint, provided that solvent emissions are limited through other means.

Cleaning Processes

The chlorinated solvents have been used traditionally in both vapor degreasing and cold cleaning applications. Recent advances in vapor degreasing help to ensure that worker and environmental regulations and concerns can be effectively addressed.

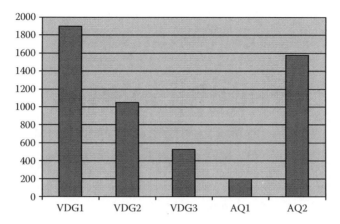

FIGURE 8.3 Greenhouse effect (gram equivalent CO_2). Results of life cycle assessment (per square meter of metal part cleaned).

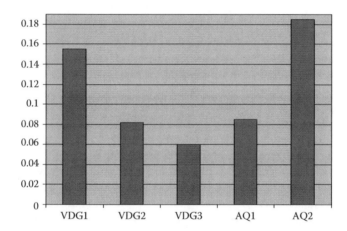

FIGURE 8.4 Solid waste (kg). Results of life cycle assessment (per square meter of metal part cleaned).

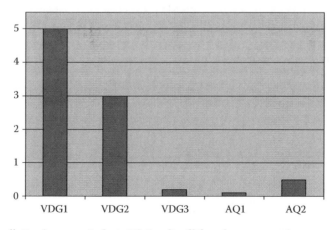

FIGURE 8.5 Air pollution (gram equivalents H^+). Results of life cycle assessment (per square meter of metal part cleaned).

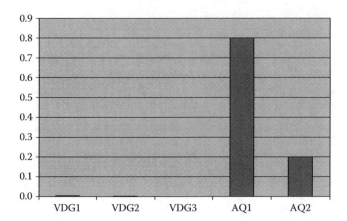

FIGURE 8.6 Results of life cycle assessment (per square meter of metal part cleaned). Water pollution (gram equivalents PO_4^{3-}).

Vapor Degreasing

The vapor degreasing process is the ideal technology for high-quality cleaning of parts. It is able to remove the most stubborn soils. It reaches into small crevices in parts with convoluted shapes. Parts degreased in chlorinated solvent vapors come out of the process dry, with no need for an additional drying stage.

Vapor degreasing is particularly effective with parts that contain recesses, blind holes, perforations, crevices, and welded seams. Chlorinated solvent vapors readily penetrate complicated assemblies as well. Solid particles such as buffing compounds, metal dust, chips, or inorganic salts contained in the soils are effectively removed by the washing action of the solvent vapor.

Vapor degreasing can be carried out in either a batch or an in-line degreaser. The traditional batch degreaser is a covered tank, with cooling coils at the top, into which the dirty parts are lowered. Solvent in the bottom of the tank is heated to produce vapor. On contacting the cooler work, the vapor condenses into pure liquid solvent. The condensation of solvent dissolves the grease and carries off the soil as it drains from the parts into the solvent reservoir below. This process continues until the parts reach the temperature of the vapor, at which point condensation ceases and the parts are lifted out of the vapor, clean and dry.

Many degreasers contain one or several immersion tanks below the vapor zone, so that parts can be lowered into liquid solvent—often in a tumbling basket—before being raised into the vapor for final rinsing. Ultrasonic cleaning can be added to remove heavy oil deposits and solid soils by installing transducers in the degreaser. When ultrasonic energy is transmitted to a solution, it imparts a scrubbing action to the surface of soiled parts through cavitation—the rapid buildup and collapse of thousands of tiny bubbles.

Several types of conveyorized equipment provide in-line vapor degreasing. These large, automatic units, which can handle a volume of work and are enclosed to provide minimal solvent loss, include monorail and cross-rod degreasers. They are particularly valuable when production rates are high.

Although conveyorized degreasers are enclosed, there is still some solvent loss through the openings where work enters and leaves the equipment. Consequently, some companies have found it cost effective to install one of the advanced types of degreasers that have no air/vapor interface. These sealed units were first introduced in Europe, but have become available in the United States in recent years.

Typically, these degreasers perform the cleaning operation in a sealed chamber into which solvent is introduced after the chamber is closed. Solvent vapor then performs the final drying stage, and all vapors are exhausted after each cycle and passed into a solvent recovery system. With the sealed chamber, control of solvent loss exceeds 90%. Operation is programmed and automated, permitting a variety of cleaning programs, including hot solvent spray.

Although these sealed units can be costly and may not be effective for some cleaning jobs, a few U.S. plants have installed them to ensure compliance with safety and environmental regulations.

Cold Cleaning

The manufacturers of the TCE, PCE, and MC generally do not recommend the use of these solvents in hand wipe and other cold (room temperature) cleaning applications. In circumstances where workplace exposure, NESHAP, and other requirements can be met, these solvents may provide a viable option for companies searching for an effective cold cleaning solvent.

Summary

Among the surface cleaning options available, one process stands out for its ability to produce a clean, dry part at a reasonable price—vapor degreasing with the chlorinated solvents. TCE, PCE, and MC are clear, heavy liquids with excellent solvency. All are virtually nonflammable, and can effectively remove the oils, greases, waxes, tars, lubricants, and coolants generally found in the metal processing industries.

The potential health effects of TCE, PCE, and MC have been very well studied. Environmental, health, and safety regulations governing these chlorinated solvents are strict, but manageable. While cleaning with these solvents is often perceived as having greater environmental impacts, LCA suggests that TCE, PCE, and MC may have a lower overall environmental impact in many situations.

The chlorinated solvents have been used traditionally in both vapor degreasing and cold cleaning applications. Recent advances in vapor degreasing help to ensure that worker and environmental regulations and concerns can be effectively addressed.

9

D-Limonene: A Safe and Versatile Naturally Occurring Alternative Solvent

Ross Gustafson
Suncor Energy

Introduction

With the production phaseout of chlorofluorocarbons (CFCs) and other ozone-depleting chemicals and increased awareness of workplace safety, many different cleaning solvents have been introduced. D-Limonene, a naturally occurring substance extracted from citrus rind during the juicing process, has shown great effectiveness in the cleaning market and is experiencing a growing acceptance as the solvent of choice in a number of different applications. D-Limonene is a non-water-soluble solvent. It can be used straight or blended with an emulsification system to produce a water dilutable/rinsable product. It is capable of effectively removing organic dirt loads ranging from light cutting oils and lubricants to heavy greases, such as cosmoline.

D-Limonene is in the chemical family of terpenes. These are products, produced by all plant life, based on isoprene. All terpenes are combinations of two or more isoprene molecules. The D-limonene structure is shown in Figure 9.1 as the addition of two isoprene molecules. Some common terpenes besides D-limonene are pinene, menthol, camphene, and β-carotene.

D-Limonene is extracted at two different points in the citrus-juicing process. During the pressing of the fruit to remove the juice, a significant quantity of the peel oil is also pressed out. This floats on top

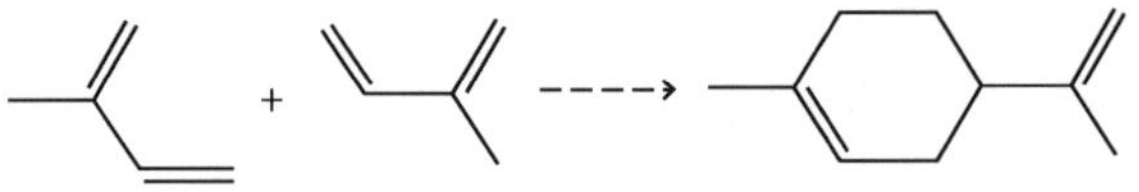

D-Limonene

FIGURE 9.1   Structure of D-limonene.

of the juice, and is separated by decanting. This fraction is called cold-pressed oil, and contains many flavor and fragrance compounds along with the D-limonene. The cold-pressed oil can be distilled to separate the D-limonene fraction from the flavor and fragrance compounds. The separated D-limonene is termed food-grade

D-limonene in the industry. After the fruit has been juiced, the peels are sent to a steam extraction step where more D-limonene is recovered. During this process, the peels are exposed to steam, which carries the D-limonene to a condenser. Here the water and D-limonene separate into aqueous and oil phase, and the D-limonene is recovered. This product is termed technical-grade D-limonene.

Historically, orange oil and D-limonene have been used extensively as flavor and fragrance ingredients in a wide variety of products, including perfumes, soaps, and beverages. D-Limonene has also been used to make paint solids through a polymerization process. The first uses of D-limonene as a cleaning solvent began in the 1970s, but with the increased environmental awareness in the 1980s and 1990s, use of D-limonene as a cleaner has shown a tremendous increase.

Typical Properties

D-Limonene is a thin, relatively colorless liquid. The typical physical and chemical properties for technical grade are:

Color	Slight yellow to water white
Odor	Orange aroma
Specific gravity (25°C)	0.83800.843
Refractive index (20°C)	1.4710 to 1.4740
Optical rotation (25°C)	+96° to +104°
Flash point	115°F
Boiling point	178°C (310°F)
Freezing point	−96°C (−140°F)
Evaporation rate	0.05 vs. butyl acetate
Water solubility	Insoluble
Vapor pressure (20°C)	1.4 mmHg

Safety and Environmental Concerns

From a personal-safety standpoint, D-limonene is a much safer product for use than most other solvents. The oral LD_{50} of D-limonene is greater than 5000 mg/kg body weight. The product also is classified as a food additive and has been granted the Food and Drug Administration GRAS (Generally Recognized as Safe) status. For comparison, the typical mineral spirit LD_{50} is around 2000 mg/kg body weight. A formal threshold limit value (TLV) or permissible exposure limit (PEL) has never been established for D-limonene.

D-Limonene is also noncaustic and nonreactive to metal surfaces. It has been classified as a slight skin irritant, because it can remove the naturally occurring oils from skin, but has not been shown to cause lasting damage. It is not carcinogenic or mutagenic, and is currently being evaluated for its chemopreventative and chemotherapeutic properties.

D-Limonene is not itself and does not contain any ozone-depleting chemicals. It is currently regulated as a volatile organic compound (VOC). The evaporation rate of D-limonene is relatively low, so the actual VOC emissions are small. D-Limonene is not considered an air toxic or hazardous air pollutant (HAP), and is not regulated under the Clean Air Act or SARA Title III. It is an approved solvent substitute under SNAP (Significant New Alternatives Program).

The issue of global warming as it pertains to the recovery and use of D-limonene is difficult, and no reliable estimate has been completed. When plants create D-limonene, or any terpene, they use carbon dioxide and water. When a terpene is destroyed or degrades, carbon dioxide and water are produced. So the creation and destruction of D-limonene would result in a net zero global-warming effect. Generally, D-limonene solutions are not heated when used. In those applications there would be essentially no global-warming effect. At some point in a cleaning process, a small amount of energy may be used for heating drying air or rinse tanks, or at least circulation of fluids. This will have a global-warming impact, but in general is not significant. The real difficulty lies in estimating the impact of producing the fruit and extracting the product from the rind. Since D-limonene is truly a by-product of the juicing industry, very little or none of the impact from growing and processing the fruit should be attributed to D-limonene production. Citrus growers do not grow fruit for the express purpose of producing D-limonene. They are in the business of producing fruit and juice. So the global-warming impacts in the form of fertilizers, irrigation, transportation, and other energy uses associated with agriculture are attributable to fruit production, but not to D-limonene production. The major source of global-warming potential attributable to D-limonene production is the energy required to perform the extraction. Although this has not been formally calculated or estimated, one would not expect to see a significant amount of energy used and one could therefore consider production of D-limonene to have a very small global-warming potential.

The closed cup flash point for D-limonene has been established at 117°F. This makes D-limonene a Class III combustible under Department of Transportation (DOT) regulations. For transportation purposes this means that when shipping D-limonene by ground, the vehicle must be placarded as hazardous if a single container contains over 110 gal of D-limonene. For storage purposes, sprinkler systems in the storage area are required. It is also recommended to use explosion-proof pumps and wiring.

When D-limonene/surfactant/water systems are made, the closed cup flash point will generally rise to about 130°F. These solutions will have a very high open cup flash point, and will not support a flame at any temperature below boiling.

Under RCRA regulations, D-limonene is classified as a characteristic ignitable hazard, and must be disposed of as a hazardous waste. However, water/D-limonene emulsions will have no flash point at less than 1.5% D-limonene concentration. So when using a water/D-limonene system and rinsing the cleaned materials, the rinse water can generally be disposed of as a nonhazardous material if the concentration is below this amount.

Effectiveness

In many cleaning applications, the soils to be removed are organic oils and greases. Because of the inherent chemical differences between organic and inorganic materials, such as polarity and ionic effects, organic solvents tend to perform much better for cleaning these types of soils than water-based solutions. To compare organic solvent strengths, the kauri-butanol (Kb) value, an ASTM method (D1133–97), has been established. The more toxic chlorinated solvents and benzene and its related compounds are all extremely effective cleaning solvents and have high Kb values. The Kb value of D-limonene is a bit lower, but higher than that of petroleum-derived products (Table 9.1). It is not possible to perform the Kb test on oxygenated compounds, so there is no listed value for methyl ethyl ketone (MEK) or acetone.

Stand-Alone Parts Washers

The typical tub-and-basin parts washers found in various mechanical maintenance shops can be adapted to use D-limonene as the washing solvent, and in most cases cleaning effectiveness will be increased and the time necessary for cleaning will be reduced. Typically the parts washer holds 18–20 gal of solvent and the solvent is pumped through a nozzle or brush into the tub for cleaning the parts. As the solvent is used and soil loading increases, the effectiveness of the solvent diminishes. The solvent is replaced at some predetermined point. When using D-limonene as the solvent, an in-line cartridge filter can be installed. A wound cotton filter preferentially absorbs the oils and greases, allowing the D-limonene to maintain its cleaning effectiveness for extended periods of time. The filters are changed out every 1–3 weeks, depending on workload, and approximately 1 gal of fresh D-limonene is added to the parts washer monthly to make up for dragout, which is solution lost by adhesion to the parts being cleaned when they are removed from the bath. While a petroleum-based product may need replacement every 1–2 weeks, parts washers using D-limonene as the solvent have stayed in service for up to 3 years without a complete replacement of the solvent. The used filters are incinerated by a waste disposal company. Because these filters are the sole waste to dispose of, a company's waste status may change to a small-quantity generator.

TABLE 9.1 Kb Values, Comparative Strength (Solvency) of Industrial Solvents (Higher Values 5 Higher Dissolving Power)

Solvent	Kb Value
Methylene Chloride	136
Trichlorethylene	129
Benzene	107
Toluene	105
Xylene	98
Perchloroethylene	92
D-limonene	67
Mineral Spirits	37
Naphtha	34
Kerosene	34
Stoddard Solvent	33
MEK	N/A
Acetone	N/A

Automatic Parts Washers

In much the same manner, D-limonene can be used in batch-loaded spray and immersion systems. In many of these systems, a water-soluble detergent is used as the cleaning agent. Typically, these detergents have an elevated pH and are used at temperatures well above 120°F. This combination of temperature and pH can have adverse effects on metal parts, such as corrosion. Depending on the temperature and cleaning system, appropriate controls may be required in consideration of the flash point. In addition, a hot caustic solution can potentially cause injuries to personnel. Water usage is also a concern. D-Limonene can be used in automatic parts washers in much the same manner as in stand-alone parts washers. By inserting a filter in-line on the system, the solvent can be used for extended periods of time. It can also be used effectively at room temperature, reducing costs associated with heating the solution. If water rinsing of the parts is necessary to remove residue of soil or cleaning agent, an emulsifier can be added to the D-limonene, which will allow the parts to be water-rinsed.

Conversion of Vapor Degreasers

The easiest conversion is the one most likely to be used. In some applications, D-limonene can be substituted directly into the line as a replacement solvent. One issue is that D-limonene will not work as a vapor-phase cleaner. The solution must be brought into direct contact with the parts being cleaned. Such conversions are suited to applications where rinsing to remove D-limonene is not required.

The approach to conversion depends partly on the design of the original cleaning system. In systems where a conveyor carries parts into the vapor cleaning zone, it can be modified so that parts are lowered into the liquid instead of into the vapor. The dwell time for the parts in contact with the cleaner will be about the same, or a little less. The parts can be air-dried just as they are with a vapor cleaning solvent. The existing equipment can probably be used with only minor modifications, as long as the parts conveyor system can be fitted to allow direct contact with the solvent and the pumps and piping are checked for material compatibility with the solvent.

Another method of conversion for a straight solvent substitution is to build a spray system into the cleaning line. This requires installing an enclosed booth with spray headers directed at the parts. The parts can then be air-dried. Some new equipment and line modifications are needed (the spraying time is determined by testing), but it is not an extensive change.

In both these cases, the D-limonene is used neat, or undiluted. It is not water soluble; it is not rinsed; it will not cause rusting or oxidation of any materials. As the D-limonene is used, soil loading of oils and greases from the parts being cleaned will increase. Also, due to dragout, D-limonene will need to be periodically replaced to maintain the initial volume.

Once the D-limonene has reached the limits of its effective lifetime, it must be disposed of properly. Ideally, waste D-limonene is incinerated. The D-limonene has a fuel value of 18,000 BTU/lb, and is valuable as a secondary fuel. Alternatively, the D-limonene can be distilled for reuse.

D-Limonene blended with surfactants can be used in aqueous emulsion. In diluting the D-limonene with water, the cost of cleaning agent is reduced. Usually these solutions in use are 10%–25% D-limonene/surfactant and the remainder is water. This requires more extensive equipment changes when converting from a vapor-type cleaning system. Because of the nature of the mixture of D-limonene, surfactants and water, the parts must be water-rinsed after cleaning to remove any residue that may remain on the parts. In general, most vapor cleaning equipment can be converted to a wash tank, as suggested above, for using the water emulsion. Again, the line must be altered to allow direct contact of the parts with the cleaning solution, but no additional equipment is necessary at the wash step. A new piece of equipment needs to be inserted after the wash tank to rinse the parts, usually a water-spray system. After rinsing, the parts can then be dried or go on for further processing. Periodic addition of cleaning agent is needed to replace the dragout, and filtration can be used to keep the solution usable. When this solution can no longer be used, again it is best to have it incinerated. A number of incinerators are equipped to deal with the presence of water in the solution. Because of the mixture created by the surfactants and water, this mixture cannot be distilled and regenerated.

Circuit Board Cleaning

D-Limonene has been used for flux removal on circuit boards. The D-limonene can be substituted directly into the washing equipment and used either in the spray or flushing method. D-Limonene does not dry as quickly as the CFCs that have historically been used for this process. But, if a cleaning system designed for use with low-flash point solvents is being used, by following the D-limonene washing stage with a rinse of either acetone or isopropyl alcohol (IPA), drying times and residue levels are comparable. In such applications, low flash point cleaning systems must be adopted. High-purity/low-residue grades of D-limonene are being introduced for printed circuit board applications with some success, although cost of this material may be twice that of regular D-limonene.

Aerosol Applications

D-Limonene has been used as an aerosol for various cleaning applications, including electronics and parts cleaning. It can be sprayed on motor windings and allowed to dry by evaporation and sprayed on contacts, switches, and connections and wiped or let air-dry. It can also be sprayed into tight areas or onto bolts to facilitate removal. In many cases the residue left after evaporation does not interfere with the process. D-Limonene can also be combined with 10% IPA or acetone to promote quicker drying times and lower residue. When packaging D-limonene in an aerosol container, carbon dioxide at 15%–20% can be used as the propellant.

A water-rinsable product can be made by combining D-limonene in a 90/10 ratio of D-limonene to emulsifier. Various types of emulsifiers can be used for this, including ethoxylated alcohols, alkylamine dodecylbenzene sulfonates, coconut diethanolamides, and sodium xylene sulfonates. The emulsified product can then be sprayed on the surface to be cleaned and rinsed off with water. Again, carbon dioxide is the propellant of choice. It is also possible to make aqueous dilutions of the D-limonene/ emulsifier blends or foaming products for aerosol packaging. These types of products require propane/ isobutane propellant.

D-Limonene can be combined in aerosol formulations to impart a pleasant citrus odor. Care must be taken to evaluate proper packaging materials. D-Limonene may cause swelling of gaskets and valves of some conventional dispensers. Viton and neoprene may be some of the best choices for aerosol stem gaskets (better than butyl or buna). Valves and cans should have an epoxy coating. Aerosol packagers and gasket suppliers should be consulted as to materials recommended for D-limonene.

Sample Formulations

A few examples of possible formulations for various applications for critical cleaning are presented here. The emulsifier in the formulations can be a wide variety of products, or a blend of several different products, depending on the properties desired.

A general-purpose concentrate for water dilution:

85%–90% D-limonene
10%–15% emulsifier

A water dilutable/rinsable printing press cleaner:

90% D-limonene
7% emulsifier
3% tripropylene methyl glycol ether

This product can be diluted to 70%–75% water to clean soy- and water-based inks.

The following guidelines can be used to formulate products for various applications. In this case the D-limonene and emulsifier or surfactant blend should be mixed in equal proportions.

Application	% D-Limonene/ Emulsifier	% Water
Engine degreaser	25	75
Tar/asphalt remova	50	50
Adhesive removal	15	85
Marine vessel cleaner	10	90
Industrial metal cleaner	20	80

These ratios are meant as guidelines.

Case Studies

The following case studies show a variety of uses and benefits that have been found from switching to D-limonene-based cleaning systems.

Martin Marietta Astronautics has replaced 1,1,1-trichloroethane (TCA) and MEK with a terpene cleaner for hand-wiping operations.[1] The terpene cleaner was selected after 16 months of extensive testing of citrus- and alkaline-based compounds. Workers prefer the citrus-based cleaner because it is more efficient. The terpene cleaner leaves less residue resulting in higher coating bond strength. Martin Marietta estimates the change has reduced toxic emissions by thousands of pounds per year. Research costs were $350,000 to find a suitable replacement for MEK and TCA. Estimated savings are $250,000/year.

In a joint research effort, the U.S. EPA and APS Materials, Inc., have investigated the use of a limonene cleaner to replace TCA and methanol.[2] APS Materials, Inc. is a metal-finishing company that plasma-coats parts for use in hostile environments. In the biomedical parts division, cobalt/molybdenum and titanium parts are coated with a porous titanium layer for use as orthopedic implants. APS Materials has converted to the terpene cleaner as a result of the investigation. Cleaning efficacy is excellent with a slight increase in bonding strength for the limonene-cleaned parts. Changing to the aqueous system required the addition of rinse and dry stations. The new system cost $1800 to install with annual operating expenses of $850. Net savings are $4800/year.

GE Medical Systems of Waukesha, Wisconsin, is a manufacturer of medical diagnostic equipment. Spray cleaning (degreasing) of parts using TCA resulted in fugitive air emissions. GE Medical Systems eliminated fugitive TCA emissions by converting to a terpene cleaner.[3] With TCA, 800 gal of solvent were purchased annually, all of which was lost to atmosphere. Because terpene cleaner is much less volatile, only 30 gal are purchased per year. In addition, terpene cleaner is recycled. No capital expenditure was required.

Northern Precision Casting of Lake Geneva, Wisconsin, switched to a citrus-based solvent for cleaning the wax patterns used in making molds.[4] Previously, they used TCA. TCA fugitive emissions amounted to 18,000 lb in 1988. The terpene solvent is water soluble and is discharged to a publicly owned treatment works. No capital costs were incurred for the change. Maintenance and operating costs are equivalent.

The Marine Corps Air Station Naval Aviation Depot, Cherry Point, North Carolina, is responsible for the complete maintenance/rebuilding of naval aircraft. In 1990 the depot used 8000 gal of CFC-113 and 15,600 gal of 1,1,1-TCA. By the end of 1992, CFC-113 usage had been reduced to 500 gal annually and TCA usage had been cut to about 4800 gal annually. Terpene cleaners were used as one of a number of approaches.[5] Approaches included soap bubbles for leak checks; aqueous power washers for electronics, motor, and engine shop use; terpene cleaners for hand wiping; steam cleaning or wet sodium bicarbonate blasting for soil and carbon removal; and plastic media blasting for paint removal.

AT&T has reduced usage of CFC-113 by converting to a semiaqueous chemistry for cleaning surface mount assemblies.[6] Parts are carried by conveyor into a power washer consisting of wash and rinse/dry modules. Low- and high-pressure sprays of a terpene cleaner are followed by nitrogen knives to reduce cleaning solution dragout and blanket the washer with an inert atmosphere to prevent fire. In the second module, the parts are rinsed with low-pressure, then high-pressure water sprays to remove the terpene cleaner. Rinsing is followed by water removal by air knives within the same module. Care must be taken in selecting materials of construction in the surface-mount components because the terpene cleaner swells some plastics and elastomers. AT&T has found that the new cleaning method is more economical than the previous CFC-113 method.

In 1988, the Motorola Corporation had 29 flux-removal cleaning systems using 250,000 lb of CFC-113 annually. By August 1991, Motorola had eliminated CFC-113 usage. Many of the printed circuit board assemblies are now assembled using a no-clean flux. Assemblies that require cleaning now use terpenes and water.[7] Benefits reported include cleaner assemblies, lower production downtime, and decreased cleaning cost. Cleaning costs are now about $8/h using the terpene/water vs. $38/h for CFC-113.

Crown Equipment Corporation, New Bremen, Ohio, manufactures electric lift trucks and television antenna rotors. Parts cleaning involves mild steel, aluminum, cast iron, and copper. In 1988 Crown used 208,000 lb of TCA in cold-cleaning (immersion) and vapor degreasing operations. Hand dipping now uses a water-based cleaner with rust inhibitor added for corrosion resistance, and 100% D-limonene spray cleaner has replaced TCA for hand-wiped parts.[8] An alkaline aqueous immersion cleaner has replaced one degreaser (with inhibitor added for ferrous parts). The other degreaser was replaced with an aqueous power washer that uses heat, agitation, and forced air drying to produce clean parts. The payback period for capital expenses was 10 months. In 1989, Crown saved $100,000 in chemical costs. Crown Equipment has switched to water-based cleaning with no decrease in production. Employees prefer the water-based cleaner for hand dipping.

The Bureau of Engraving, Industrial Division, manufactures printed circuit boards. In 1990, it decided to eliminate the use of methylene chloride and TCA, which were being used at the rate of 681,000 lb/year. Several changes in the manufacturing process were necessary to accomplish this goal, including the use of water-based and terpene-based cleaners. The Bureau of Engraving, Industrial Division, saves $250,000 annually in purchase cost and $20,000 in maintenance, energy, and disposal costs.[9]

Other Uses

Several other cleaning applications exist for D-limonene. D-Limonene is also used extensively in the asphalt industry for clean up and aggregate analysis, as an ink cleaner in printing operations, in the oil and gas fields for maintenance and as a well recovery solvent. It is used in hand cleaners and a wide variety of other general cleaning and household applications. In general, if an organic soil is to be removed, D-limonene may perform as well or better than other solvents.

In addition to cleaning applications, D-limonene has been found to have thermodynamic properties that make it a very good heat transfer fluid, especially for cryogenic applications, below 2100°C. It has also been used as a carrier solvent in paints and similar coatings, as well as adhesives. Studies have also shown D-limonene to be an effective pesticide and bactericide, although these are not yet approved uses. These are in addition to the best-known uses of D-limonene and other citrus derivatives as flavor and fragrance components.

Other Concerns

A "perfect" solvent probably does not exist, and D-limonene is no exception.

D-Limonene can cause a certain amount of swelling of polymers, so the plastic materials used with a D-limonene system must be chosen with caution. Viton is the best seal to use in joints, and nylon-braided PVC seems to be the most acceptable material for hoses. As with any solvent, gloves should be worn when working directly in the solution. The best material for these is nitrile latex.

Conclusion

D-Limonene is an extremely effective and relatively safe cleaner and solvent for use in many industries. It can be used in a wide variety of applications and in most cases will perform better and longer than the classic solvents. Although it is not perfect, it is a good option to be considered when choosing a cleaning system or looking for an effective solvent replacement.

References

1. Dykema, K.J. and G.R. Larsen. 1993. Shifting the environmental paradigm at Martin Marietta Astronautics, *Pollut. Prev. Rev.*, Spring, 205.
2. Brown, L.M., J. Springer, and M. Bower. 1992. Chemical substitution for 1,1-trichloroethane and methanol in an industrial cleaning operation, *J. Haz. Mater.*, 29:179–188.
3. Wisconsin Department of Natural Resources, Case Study: GE Medical Systems; Replacing 1,1,1-Trichloroethane with citrus-based solvents, PUBL-SW-168 92, Hazardous Waste Minimization Program (SW/3), Madison, WI.
4. Wisconsin Department of Natural Resources, Case Study: Northern Precision Casting; Replacing 1,1,1-trichloroethane (TCA) with citrus-based solvents, PUBL-SW-161 92, Hazardous Waste Minimization Program (SW/3), Madison, WI.
5. Fennell, M.B. and J.M. Roberts. Naval Aviation Depot: Hazardous minimization—saving time, money, and the environment, in *Proceedings of the Aerospace Symposium*, Lake Buena Vista, FL, 1993, pp. 39–46.

6. Terpene cleaning of surface mount assemblies, aqueous and semi-aqueous alternatives for CFC-113 and methyl chloroform cleaning of printed circuit board assemblies, EPA/400/1-91/016, June 1991, pp. 51–60.

7. Terpene cleaning of printed circuit board assemblies, aqueous and semi-aqueous alternatives for CFC-113 and methyl chloroform cleaning of printed circuit board assemblies, EPA/400/1-91/016, June 1991, pp. 61–62.

8. Kohler, K. and A. Sasson, Case studies: Multi-industry success stories to reduce TCA use in Ohio, *Pollut. Prev. Rev.,* Autumn, 407–409, 1993.

9. Currie, W.T., Vice President, Facilities and Environmental Affairs, Bureau of Engraving, Inc., Industrial Division, 500 South Fourth Street, Minneapolis, MN, 55415, MnTAP, 1993, Governor's Awards for Excellence in Pollution Prevention.

Benzotrifluorides

P. Daniel Skelly
Riverside Chemicals

Editor's note: Many factors impact the cleaning options available to components manufacturers. The situation with the benzotrifluorides illustrates the impact of overall business plans of chemical producers. As the First Edition of *Handbook for Critical Cleaning* was going to press, Occidental Chemical, the U.S. producer of benzotrifluorides, announced its intention to exit the market, ceasing production of parachlorobenzotrifluoride (PCBTF) and offering the business for sale. In 2002, the Israeli company, Makhteshim Agan Industries, purchased the entire OXSOL business from Occidental Chemical Corporation. PCBTF from China and Brazil is still available. While the current uses in critical cleaning are relatively limited, PCBTF remains a valid option for cleaning applications. For that reason, we are re-printing Mr. Skelly's chapter in this book.

Overview

Three commercially available benzotrifluorides, benzotrifluoride (BTF), parachlorobenzotrifluoride (PCBTF) and 3,4-dichlorobenzotrifluoride (DCBTF), have potential as replacements for ozone-depleting compounds and other organic solvents. Because it is exempt as a volatile organic compound (VOC), PCBTF is used in cleaning applications, particularly in areas with high regulatory constraints.

"Likes dissolves like." This is why, over the years, people have used organic solvents to clean oils, greases, and other organic contaminants off their parts and assemblies. However, most organic solvents used in cleaning operations are VOCs, which contribute to ground-level ozone or smog formation. As a result, restrictions have become burdensome and permits difficult to obtain, especially in metropolitan areas where air pollution is a significant problem. Although the three benzotrifluorides have been commercially produced in the United States since the 1960s, until the mid-1990s, their use was limited to chemical intermediates for the agricultural and pharmaceutical industries. Interest in their use as cleaning agents developed when it was discovered that the atmospheric lifetimes of the benzotrifluorides are short enough to avoid ozone depletion. The compounds are not listed as hazardous air pollutants (HAPs). The low tropospheric reactivity of PCBTF led to VOC exemption by the U.S. EPA.

Properties of the Benzotrifluorides

The structures of BTF, PCBTF, and DCBTF are indicated in Figure 10.1. Occidental Chemical Corporation (OxyChem®) is currently the only U.S. producer of these compounds, and they are currently sold under the OXSOL® 2000, OXSOL 100 and OXSOL 1000 trademarks, respectively.* Some physicochemical properties are summarized in Table 10.1; their regulatory status is indicated in Table 10.2.

In addition, properties of the benzotrifluorides obviate many of the issues associated with aqueous cleaning. Water pretreatment and wastewater disposal are not issues, nor are problems of flash rusting. Because the benzotrifluorides are organic, they have an affinity for a wide range of organic contaminants. The fluorine content provides a low surface tension, which allows penetration into blind holes and crevices. Although the efficiency of most processes tends to increase with use of multiple tanks, multiple cleaning and rinsing steps are not required. While parts will still come out wet after cold cleaning, drying is relatively fast. Cycle times are comparable with many traditional organic solvent systems.

Solvent Toxicity

Because no official exposure levels have been established by either OSHA or the ACGIH, Occidental Chemical Corporation has set a CEL (corporate exposure limit), 8-h TWA (time-weighted average) for each of the three benzotrifluoride molecules. These values, which appear in Table 10.2, are as follows: BTF 5 100 ppm, PBCTF 5 25 ppm, and DCBTF 5 5 ppm. These values were established as a safe level of exposure for an average 8-h workday. Additional studies have been completed and are available.

Worker Protection

Benzotrifluorides can be used with appropriate consideration given to worker protection. The cleaning system must be designed to minimize hazards to the worker and the environment. This may include mechanical controls such as tank covers and auxiliary cooling coils to condense solvent vapors, or

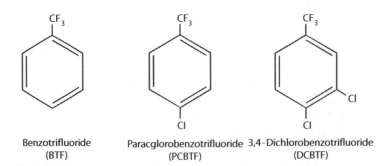

FIGURE 10.1   Structures of the benzotrifluorides.

---

* OXSOL and OxyChem are registered trademarks of Occidental Chemical Corporation.

TABLE 10.1 Physical Properties of the Benzotrifluorides

Physical Properties	BTF	PCBTF	DCBTF
CAS number	98-08-8	98-56-6	328-84-7
Chemical formula	$C_7H_5F_3$	$C_7H_4F_3Cl$	$C_7H_3F_3Cl_2$
Molecular weight	146.11	180.56	215.00
Boiling point (BP) at 760 mmHg (°C)	102	139	173.5
Density, liquid at 25°C (lb/gal)	9.88	11.16	12.3
Density vapor (air 5 1)	5.0	6.2	7.4
Dielectric breakdown (kV)	—	49	—
Dielectric constant at 25°C	11.5	—	—
Electrical resistivity M Ω (M Ω-cm)	2.5 (330)	>20 (>2640)[a]	>20 (>2640)[a]
Evaporation rate at 25°C (*n*-BuAc 5 1)	2.8	0.9	0.2
Flammability			
Flash point, tag closed cup, °C (°F)	12 (54)	43 (109)	77 (170)
Fire point, tag open cup, °C (°F)	23 (74)	97 (207)	>114 (>238)
Autoignition temperature (°C)	—	>500[a]	>500[a]
Upper flammability limit (% vol)	—	10.5	7.8
Lower flammability limit (% vol)	—	0.9	2.9
Limited oxygen index (LOI) (% vol)	18.2	26.2	~29
Freeze point (°C)	−29	−36	−12.4
Hansen solubility parameter (MPa)$^{1/2}$	16.1	17.7	18.2
Nonpolar	13.0	13.9	14.1
Polar	9.5	9.9	10.0
Hydrogen bonding	0	4.7	5.7
Kauri-butanol value (Kb)	49	64	69
Latent heat of vaporization at BP (cal/g)	53.9	49.6	46.6
Refractive index at 25°C	1.4131	1.4444	1.4736
Solubility			
Solvent in water at 20°C (ppm)	250	29	12
Water in solvent at 25°C (ppm)	290	240	153
Specific gravity at 25/15.5°C	1.185	1.3380	1.478
Specific gravity correction factor (per °C)	−0.0013	−0.00146	−0.00146
Specific heat, liquid at 20°C (cal/g/°C)	0.31	0.27	—
Surface tension in air at 25°C (dynes/cm)	23	25	29
Vapor pressure at 20°C (mmHg)	30	5.3	1.6
Viscosity, liquid at 25°C (cps)	0.53	0.79	1.52
Volatiles (% wt.)	100	100	100

[a] Limitation of test equipment.

fans and exhaust hoods to remove solvent vapors from the workstation. The flash point must also be considered in appropriate equipment design.

Cleaning Systems for Benzotrifluorides

General Considerations[1]

No single cleaning system can be used universally, and what works for a neighbor or a sister plant may be ill-fitted for your requirements. Educate yourself on the options, then choose the solvent(s) and equipment that best meet your needs. The system should be capable not only of removing your soils

TABLE 10.2 Regulatory Summary of the Benzotrifluorides

Regulatory Issue	BTF	PCBTF	DCBTF
Regulated as an ozone depleter	No	No	No
Regulated as a VOC	Yes[a]	No	Yes
Regulated as an HAP	No	No	No
SARA Title III, Sect. 313 Reportable	No	No	No
Global warming potential	Low	Low	Low
Suspected carcinogen	No	No	No
Corporate exposure limit, 8-h TWA (ppm)	100	25	5
DOT hazard class	Flammable	Not regulated	Not regulated
OSHA hazard class	Flammable	Combustible	Combustible
RCRA hazardous waste number	D001	D001	Not regulated

[a] Petitioned the Federal EPA for VOC exemption on November 3, 1997. Still pending at the time of this writing.

efficiently and economically, but also of meeting the regulatory and safety requirements of today and the future. A critical first step in selecting the cleaning system is to address, "How clean is clean?" Only you, the owner of the parts that need cleaning, can properly evaluate if a given system will meet the majority of your needs and expectations.

Size and Shape of Parts

Where there are blind holes, lap joints, or other small crevices, the solvent must not only have a good solubility for the soil, but also a low viscosity and low surface tension. Unlike water, the benzotrifluorides have a naturally low surface tension allowing them to penetrate and clean these hard-to-reach areas.

Materials of Construction

The selection of a strong solvent may be good at removing the soil, but you must also be sure that it does not damage the substrate or your equipment. Based on the Kauri-butanol (Kb) scale, all three benzotrifluorides have a moderately high solvency power. They have about twice the cleaning power of mineral spirits and half that of toluene. This makes them effective for a wide range of organic soils, but not so strong that they damage most plastics or elastomers.

If a heated process is planned, the substrate must be able to withstand the prescribed temperature. For vapor degreasing, keep in mind that the boiling points of PCBTF and DCBTF are 139°C and 173.5°C, respectively. These relatively high temperatures could damage some parts.

Volume of Parts to Be Cleaned and Amount of Soil They Contain

A process involving occasional cleaning of a few small parts is very different from a critical process in a high-volume assembly line operation. Do not shortchange yourself. You do not want to pay for excess equipment capacity, but make sure to account for future growth potential. Fabricating additional sheet metal tanks does not generally significantly increase capital costs, and additional and/or slightly larger tanks can make a big difference in the capability and capacity of the equipment.

Soil Loading

You cannot clean parts with dirty solvent. If only a small amount of soil needs to be removed, your solvent will remain relatively clean for a long time. However, if there is a large amount of contaminant, you might want to consider multiple dips in progressively cleaner solvent, and/or on-site distillation. With solvent recovery, a stable, single-component cleaner (such as PCBTF) is more easily maintained and recycled than is a complex proprietary blend.

In most applications, once the soil loading exceeds 30%, it is recommended that it be reclaimed or disposed. Soil loading may often be approximated by measuring its specific gravity at a given temperature. In the case of PCBTF, a simple graph was developed at three different temperatures and various levels of mineral oil. Once such a chart is developed for your "typical" contaminant, you can then approximate the soil loading with a quick comparison to a chart similar to Figure 10.2. It will be up to end users to determine what soil loading level is tolerable in their system for adequate cleaning performance.

Cold Cleaning

Recent regulations have focused on the VOC content of mineral spirit solvents traditionally used in these systems. PCBTF, alone or in blends, may be an appropriate replacement in recirculating overspray applications where aqueous cleaning is ineffective.

In some cases, single or multiple tank immersion units can be adapted for use with PCBTF and DCBTF with only minor modification, with the proviso that all equipment must be designed for use with low flash point solvents. However, it is important to assure that top enclosures and side workload entry be included to minimize worker exposure and to eradicate possible solvent odors. Because of its relatively high vapor pressure and the resulting solvent losses, BTF would not generally be used in immersion dip tanks.

Mechanical Agitation

The cleaning operation is generally enhanced with some level of mechanical agitation to help carry the soil away from the surface of the part. Where ultrasonics are used with the benzotrifluorides, the possible elevation of the solvent bath temperature must be taken into consideration.

Heated Cleaning

With waxes, buffing compounds, and similar soils, heating (from slightly above ambient temperature to the boiling point) may be required. Unless the equipment is properly designed for elevated temperatures,

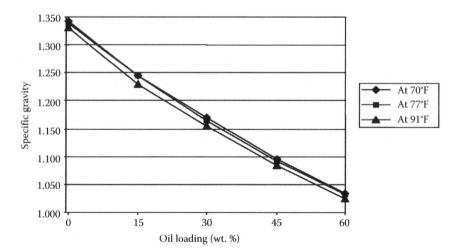

FIGURE 10.2 Specific gravity vs. oil loading in PCBTF. (From OXSOL 100—Used as a solvent flush in refrigeration conversions, pending bulletin BCG-OX-A/C, 8/95, Occidental Chemical Corp., Niagara Falls, NY.)

solvent losses will increase. Further, given the flash points, cleaning systems designed for flammable solvents must be used.

Because of its volatility and flammability, BTF is ill-suited for this cleaning method, and both DCBTF and PCBTF use is limited in heated dip tanks. Benzotrifluorides must not be used in a standard, unmodified vapor degreaser. PCBTF and DCBTF could be used in approved, specifically designed low flash point systems with sufficient containment and automation to maintain employee exposure well below the recommended limit (see Chapter 2.12, Bartell, first edition). Initial capital expenditure will be significant for such systems.

Automation is desirable to distance the worker from the cleaning operation, thereby minimizing unnecessary solvent exposure.

Solvent Blends

Although a single-solvent system is generally cheaper and easier to maintain, in some applications a blend of the benzotrifluorides with other compounds is desirable. A system containing a mixture of solvents can have advantages in areas such as solubility parameters, flash point suppression, odor reduction, and evaporation rates. However, depending on the additive selected, you may add to the regulatory requirements, complexity of solvent recovery or disposal, and total system cost.

Compatibility

Compatibility with Metals

In one study run for the U.S. Navy, PCBTF passed the ASTM F 483 total immersion corrosion test with no visible corrosion on any of the following alloys: AMS 4037, 4041, and 4049 aluminum, AMS 5046 grade 1020 steel, cadmium-plated 1020 steel, AMS 4911 titanium, and AMS 4377 magnesium.

Compatibility with Polymers and Elastomers

The solvency power of the benzotrifluorides ranges between that of mineral spirits and perchloroethylene (dry cleaning solvent). As a result, these moderately strong solvents have the ability to dissolve a wide range of uncured resins (of interest to the paint and adhesive markets) and selectively remove many organic soils from the workload. At the same time, they are generally mild enough to preserve the integrity of the articles being cleaned. A representative sample of a study of commercially available polymers and elastomers exposed to PCBTF is summarized in Table 10.3. Similar data are available for the other benzotrifluorides. In the study, test strips were weighed before and after submersion in PCBTF at room temperature for 24 h. They were removed, blotted dry, and reweighed. The differences in the weights are reported in Table 10.3 as "Δwt. % Initial" (% gain or loss). These same test strips were then allowed to air-dry at room temperature for 4 days and were again reweighed and the results reported as "Δwt. % Overall" (% gain or loss).

Approved Military and Aerospace Applications

In a study contracted by Occidental Chemical Corporation for the U.S. Navy, PCBTF passed or conformed to the following corrosion/compatibility tests:

- ASTM F 483—Total immersion corrosion (see section on Compatibility)
- ASTM F 945—Stress corrosion of titanium alloys by aircraft maintenance materials–method A
- ASTM F 519—Test method for mechanical hydrogen embrittlement testing of plating processes and aircraft maintenance chemicals

TABLE 10.3   Examples of PCBTF Compatibility with Polymers and Elastomers

Material Tested	Brand or Common Name	PCBTF (D wt. %) Initial[a]	Overall[b]
Acrylic	Lucite®	−16.5	−16.6
Butyl rubber	Bucar	83.1	9.0
Carboxylic rubber	NBR	120.4	−0.5
Chloroprene rubber	Neoprene W	73.6	−1.8
Chlorosulfonyl PE	Hypalon® 40	74.5	3.1
EPDM	Nordel® 6962	69.7	−6.6
EPD terpolymer	Nordel (Elastomer)	61.2	−5.1
Fluorocarbons			
PTFE	Teflon®	0.1	0.0
PVDF	Kynar®	−0.1	−0.5
VF-HFP copolymer	Viton® A	30.6	6.5
Ionomer	Surlyn®	1.8	0.6
Nitrile butadiene rubber	Buna N	79.9	−10.4
Polyamide	Nylon 6/6	0.0	0.0
Polycarbonate	PC	[c]	12.4
Polyester			
PBT	Valox®	0.1	0.0
PET	Rynite®	0.1	−0.1
Polyethylene, high density	HDPE	1.8	0.6
Polyphenylene sulfide	Ryton®	0.0	0.0
Polypropylene	PP	17.5	0.2
Polysulfone	Polysulfone	0.1	−0.1
Polyurethane	Adiprene® L	29.1	6.3
Polyvinyl chloride			
CPVC	—	0.0	0.0
PVC	—	0.1	0.0
Silicone	Silicone	77.3	27.2
Styrene	—	[d]	[d]
Styrene butadiene rubber	Buna S	124.3	23.7

Source: Modified from OXSOL Solvents, Compatibility with polymers and elastomers, BCG-OX-20 5197, Occidental Chemical Corp., Dallas, TX, 1997, pp. 2–4.

Note: Adiprene® L is a registered trademark of Uniroyal Chemical Corporation; Delrin®, Hypalon®, Lucite®, Nordel®, Rynite®, Surlyn®, Teflon®, Tefzel®, and Viton® are registered trademarks of I.E. du Pont de Nemours and Company; Kynar® is a registered trademark of Pennwalt Corporation; Noryl®, Ultem®, and Valox® are registered trademarks of General Electric Company; Panacea® is a registered trademark of Prince Rubber & Plastics Co., Inc.; Ryton® is a registered trademark of Phillips Petroleum Company; Thiokol® FA is a registered trademark of Morton International, Inc.

[a] Initial 5 initial weight change after 24-h solvent soak.
[b] Overall 5 overall weight change after a 4-day drying period.
[c] Attacked and cracked.
[d] Test strips dissolved.

As a result of these favorable test results, both the U.S. Navy and Air Force have approved the use of PCBTF in several protective coatings for military aircraft applications. Parachlorobenzotrifluoride has also been approved by the U.S. Air Force under T.O. 1-1-8 for use as a cleaning solvent on aircraft, as a wipe solvent for primer reactivation and as a paint thinner. After an extensive study to replace 1,1,1-trichloroethane (TCA), Alliant TechSystems (formerly Hercules Aerospace) has approved the use of PCBTF as a wipe cleaning solvent on the Navy's Trident II D/5 first- and second-stage rocket motors.

Reclamation and Disposal

Recovery of the benzotrifluoride containing waste stream[4] may be done by a competent and properly permitted contractor, or by investment in on-site explosion-proof distillation equipment. Most stills available today have efficiencies of 90% or greater. Recovery increases to 95% or more with a high-efficiency, thin-film evaporation still. To increase the amount of reclaimed solvent further, and reduce the volume of waste that must be disposed, stills can be equipped to employ steam stripping or vacuum. Because additive stabilizer packages are not required, the distillate can easily be returned into the cleaning tank for reuse; recycling equipment may have a short payback period.

The still bottoms from a typical on-site distillation operation contain between 1% and 10% solvent and must be disposed of according to proper Resource Conservation and Recovery Act (RCRA) hazard classifications. Although DCBTF is not regulated under RCRA, pure BTF and PCBTF have flash points of less than 140°F, which qualifies them as D001 "Characteristic Hazardous Wastes." Because they do not contain any listed concentrations of compounds recognized by RCRA as hazardous wastes, many state and local regulations allow these still bottoms to be added to other combustible products and incinerated as fuel oils. This can minimize costly hazardous waste disposal fees. Fuels blending programs for cement kilns are preferred as the flames are generally hotter and the alkalinity of the cement will neutralize the acid gases that will be generated.

An alternative disposal method is hazardous waste incineration in licensed equipment capable of handling HCl and HF. The heat of combustion values for BTF, PCBTF, and DCBTF are 8060, 7700, and 4830 BTU/lb, respectively. Carefully analyze the characteristics of your particular waste prior to selecting any waste disposal option.

Recycling, disposal. Is the spent cleaning solution classified as a hazardous waste? This needs to be determined on an individual basis, taking into consideration the components and characteristics of the waste stream.

References

Information was taken from the following technical bulletins produced by Occidental Chemical Corporation:

1. OXSOLs for metal cleaning, BCG-OX-19 1/95.
2. OXSOL 100—Used as a solvent flush in HFC refrigeration conversions, pending bulletin BCG-OX-A/C 8/95.
3. OXSOL Solvents—Compatibility with polymers and elastomers, BCG-OX-20 5/97.
4. Disposal of OXSOL degreasing wastes, BCG-OX-11 6/97.

II

Cleaning Systems

11

Cleaning Equipment: Overview

Barbara Kanegsberg
BFK Solutions

Cleaning Is a Process

The first section of *Handbook for Critical Cleaning: Cleaning Agents and Systems* is devoted to cleaning chemistries. However, cleaning is not just a chemical. In some cases, as with plasma cleaning, the chemical is generated in situ, as part of the process. Cleaning is not just equipment; there is an essential interaction between the cleaning chemistry and the cleaning equipment.

Why Do We Need Cleaning Equipment?

Anyone tasked with choosing new cleaning equipment asks this question, at least facetiously. Cleaning equipment often requires a substantial capital investment, and process change may be accompanied by comments and input from company management, the insurance company, the fire department, governmental regulatory agencies, the company facilities/maintenance group, the in-house environmental health and safety department, and last but not least from the technicians and assemblers who have to watch over the process.

So why do we need cleaning equipment? There are a number of reasons the most important of which should be maximizing cleaning performance. Other reasons for choosing a particular type of equipment include decreasing process time, protecting the environment, and protecting the individual worker.

In any cleaning system, it is important to consider both the cleaning agent and the cleaning action. A cleaning agent may have very high solvency and may be effective in dissolving the soil. However, it is important to have appropriate cleaning action to assure that the cleaning and rinsing agents reach all surfaces and to assure that the soil is carried away from the surface. As a chemist, my first thought in terms of cleaning action is plopping a magnetic stirrer into a beaker of cleaning agent and cranking up the rheostat. This is not practical for large-scale processes. Similarly, pilot or specialized operations may involve hand-wiping parts with a soft cloth, hand-dipping each part, or scrubbing individual portions with a cotton swab. For more systematic cleaning, ultrasonics, megasonics, spray in air, spray under immersion, and turbulation all add effectiveness to the process. In addition, weirs and filters prolong the life of the cleaning agent in general industrial processes. In high-precision applications, filtration with in-line particle monitoring may be required for adequate contamination control.

Performance

Setting aside for a moment worker safety, environmental regulatory issues, and minimization of cleaning agent loss (note that we said "for a moment," not permanently), performance involves matching the cleaning agent to the cleaning equipment, cleaning action, drying action, materials compatibility, and fixturing (including sample handling and automation). In addition, one must consider a host of additional site specifications including process costs, safety, and regulatory issues.

The Chicken or the Egg?

Integrating the cleaning system, the drying system, and overall sample handling can be crucial in maintaining process control and process efficiency. For this reason, it is important to match the cleaning agent or agents with the cleaning equipment. The following question is often asked: Should I first look at the cleaning agent or at the cleaning equipment? Cleaning agent manufacturers typically say, "Look first at the cleaning agent." Cleaning equipment manufacturers, on the other hand, often say, "Don't worry; just buy the cleaning system, you can use many different sorts of cleaning agents in it." In the opinion of this author, the most productive approach is to consider both factors at the same time.

Looking at the cleaning agent without considering the cleaning system can lead us to rule out what could be your optimal approach to cleaning. Pragmatic experience leads this author to conclude that attempting to emulate a sophisticated cleaning system at the benchtop level, by cleaning coupons or scrap parts in beakers is all too often not informative and can lead the user to discard as unworkable what could be an effective, economical cleaning process. At the same time, some engineers have been known to contact cleaning equipment manufacturers, demand cleaning equipment be fabricated to exacting specifications, and then, only after the equipment has been built and delivered, inquire as to the appropriate cleaning solution to be used. Such scenarios have lead one colleague to abandon a career as manager of an applications laboratory.

Wash, Rinse, and Dry

On the surface, it seems simplistic and obvious; the three steps of the critical cleaning process are washing, rinsing, and drying (Kanegsberg, 2005). Each step has a separate function and may require separate equipment and/or chemistries. Each step typically employs elevated temperature and/or a variety of forces. For a successful cleaning process, each step needs to be budgeted separately in terms of engineering resources and investment in capital equipment.

Wash

The wash step is also referred to as the cleaning step (after all, it involves a cleaning agent), the residue removal step, the soil removal step, or even the deblocking step. The primary purpose is to remove soils, including thin films, particles, and fingerprints, from the surface of the product. Soil removal means removal from the vicinity of the part and prevention of soil redeposition. Successful washing includes avoiding gross product deformation/damage and avoiding undesirable surface modification or corrosion. It may involve conservation or recovery of the cleaning agent using filtration or onboard recycling.

Rinse

The primary purpose of the rinse step is to remove residual cleaning agent from the surface of the part and keep it away from the part. The rinse step also removes particles and residual soils from the manufacturing process. Rinsing is not required if cleaning agent residue is acceptable, but it is typically needed for both aqueous and solvent processes. Vapor degreasing processes may involve a single, self-rinsing solvent. The rinse step must not damage the part or result in unacceptable surface modification. Several rinse steps may be required, but the rinse should not be a substitute for effective washing (residue removal).

Dry

Drying is the removal of soils that may have been deposited by the wash and rinse steps. The soils, the matter out of place, include water and adsorbed organic solvents while minimizing soil redeposition, without damage to the parts and without promoting surface modification or corrosion. Drying tends to be the overlooked and an underfunded part of the process; so read the section of this handbook devoted to drying; and plan to invest in appropriate drying equipment. Drying is often the rate-limiting step in the process. Consider cooldown time as part of the process time. In considering possible drying techniques, it must be emphasized that the drying technique must be integrated with the cleaning system. Off-the-shelf equipment may not meet your needs; you may need to specify filtration and pump design that minimizes soil deposition during the drying step.

Drying is most often considered in terms of aqueous systems. In solvent systems, drying is assumed to be an inherent part of the process. However, particularly with complex, ornate components, capability of the system to remove residual solvent rapidly and effectively without damage to the component must be considered.

We have observed instances where excessively long bake-out or even heat treatment is assumed to be a good substitute for cleaning. This is not the case; soil may become baked on, compromising the surface quality.

Materials Compatibility

Typically, cleaning agent compatibility is considered in static systems, i.e., in beakers that are perhaps heated. However, as is pointed out in the Eichenger chapter covering compatibility (refer to Chapter 14 in the second book, *Handbook for Critical Cleaning: Applications, Processes, and Controls*), the interaction of cleaning agent(s) with materials of construction can be sharply influenced by cleaning action. Ultrasonic cleaning can produce a sonochemistry effect, which may both enhance cleaning and adversely affect materials of construction. Such effects are reportedly particularly pronounced with some aqueous surfactant packages.

It should be pointed out that other types of cleaning action, such as forceful sprays, also have the potential to damage parts by deformation and erosion. Such erosion may not be visible but may show up in altered tolerances or gravimetric changes. For example, in testing alternative defluxing systems, electronics component simulators were designed with brass coupons used to simulate components raised to varying heights. Effectiveness of removal of rosin flux from under the components was determined gravimetrically. In a few cases, over 100% apparent removal of flux was observed; we finally figured out that this was due to erosion of the coupons (Kanegsberg, 1998).

Process equipment is not immune to compatibility issues; repeated contact with the cleaning agent can result in process problems. If elastomers used in seals are degraded by the cleaning agent, leaks will occur. If worn or damaged fixtures corrode or oxidize or shed particles, soil can be inadvertently introduced during the process.

Fixturing, Parts Handling, and Automation

Fixturing

Choosing the appropriate fixturing and sample handing techniques are important to maximize contact with the cleaning agent, assure that cleaning chemicals are not trapped, and avoid parts deformation or damage.

For example, even in solvent-based systems, parts rotation is often used in conjunction with ultrasonics cleaning to boost cleaning effectiveness. Inattention to fixturing and parts handling can result in inadequate cleaning and drying as well as in parts damage. It must be continuously emphasized that the use of fixture and appropriate fixture design is important for all cleaning systems, including solvent and aqueous systems. Some of the factors involved in appropriate fixture design are discussed elsewhere and are summarized in the overview of drying.

Automation

Automation requires exquisite process knowledge (Bohn, 1994). We have adapted Bohn's approach to the cleaning process (Kanegsberg, 2007). While automation can improve performance and consistency, minimize loss of cleaning agent, minimize worker exposure to chemicals, and may have desirable environmental impacts, adding parts handling equipment as an afterthought can be counterproductive.

The concept of automation is a matter of degree. Certainly, even a process consisting of a simple open-top degreaser with a manual hoist could be said to be more automated than one where assemblers hand scrub each part or component individually. However, we generally think of automation as being a bit more sophisticated. In general, parts handling is either batch or in-line. In classic in-line processes, samples are carried along a conveyor belt through various cleaning and rinsing solutions; in such cases, drying is typically through air or nitrogen nozzles. In batch automated cleaning, overhead hoists or robotics are used to carry parts to various cleaning, rinsing, and drying chambers.

A classic batch automation system generally consists of a mechanical superstructure, a drive system, a control package, and an operator interface (Aries, 1998). Engineers accustomed to in-line processes may be reluctant to try batch cleaning, in part, because they may associate batch cleaning with inconsistent, nonautomated processes. Batch cleaning actually can provide more control and more flexibility than could in-line in that with in-line cleaning, one can either vary the length of the process chambers (and this is predetermined during equipment design) or the speed of the conveyor belt; each step of the process is, therefore, inherently tied to the next. Because the duration of each phase of the process can be varied separately, batch automation can actually provide much more process flexibility. By using several robotic arms, loads can be processed sequentially to maintain process flow (i.e., one load of parts can be at the drying stage while another is in the wash stage). Cleaning is also typically automated in more sophisticated single-chamber tanks. In such batch processes, the part remains in the chamber for all cleaning steps. Such processes can be very flexible, but, since only one load can be processed per chamber, throughput of parts is limited by the size of the chamber and by the number of chambers.

An automated system may or may not provide a more rapid cycle time. Often, assemblers are accustomed to speeding up the process, by quickly submerging and then removing the part to be cleaned. However, there are benefits to automation such as

- Improved process control
- Improved cleaning consistency
- Regulatory, safety, or quality compliance
- Lower consumption of cleaning agent
- Lower solvent emissions
- Lower exposure of employees to potentially toxic cleaning agents

In designing the automation process, it is important to consider not only the immediate cleaning process but also those before and after. In a number of manufacturing, the cleaning process itself is automated, but other aspects such as assembly are carried out by hand. Or, the fixturing and sample handling for the cleaning process may not mesh well with the surrounding processes. One then observes technicians spending appreciable time reracking and refixturing samples to go from one piece of equipment to the next. A bit of advance planning in designing the entire build process can alleviate the problem.

Even in automation, the human factor is critical. Automated systems can reduce exposure of employees to cleaning agents in terms of both inhalation and skin adsorption. There are other safety issues. One of my colleagues consistently showed up at meetings with a bump on his head with ever more interesting stories of attempts to install an overhead hoist. The systems must be designed to prevent employee injury during operation, and they are best designed and installed by experts in the field.

The other aspect of the human factor is employee education and training. All of the thoughtful planning and programming can be undone if the technicians accelerate the process speed to undesirable levels. This is a real problem, particularly where costs and production pressure have built up.

Process Efficiency, Process Costs, and Environmental and Safety Concerns

Overall process costs and efficiency are very difficult to determine. It may be necessary to take educated guesses. In terms of costs, one must consider such factors as initial capital costs, costs of disposables, cleaning agents costs (concentrated versus effective dilution), bath life, loss of cleaning agent through evaporation/drag-out/drag-in, disposal costs for the spent cleaning agent, costs of safety and environmental controls, regulatory costs, energy efficiency, and rework costs.

Process costs are site specific. While efficiency claims are prevalent, hard data is not. Further, studies of process costs, whether by governmental agencies, cleaning agent, or cleaning equipment suppliers, are inevitably influenced by the economic interests and political agendas of those sponsoring the studies. All studies have validity; all must be looked at in context. A summary of factors in process costs has been reported; studies are ongoing.

Environmental requirements often inherently determine the menu of available process options in a given area. In addition, one must consider the impact not only of the cleaning agents but also of the process on worker safety. The same increased cleaning forces that boost cleaning efficiency can magnify materials compatibility concerns—including materials that make up not only the part to be cleaned but also the workers in the area. Safety has to be considered in terms of the chemical, the process, and the interaction with other processes conducted in the workplace.

Plant Facilities, Floorspace, Growth

Atop all the other considerations, one must not forget the physical limitations of the production plant, anticipated growth, and flexibility. In evaluating cleaning systems, it is important to consider utilities (e.g., water, electricity, nitrogen lines). In addition, one must consider floor space. It is important to look at overall equipment dimensions. Many of us are careful to look at length and width; fewer consider height, and still fewer consider the additional three-dimensional space needed for the robotic arm. One must also consider wall and door dimensions relative to cleaning equipment as well as total weight. Placement is important in terms of maintaining and cleaning the cleaning equipment itself. Based on a number of process remodeling anecdotes (e.g., "skylights" to accommodate the unanticipated height of robotic arms), which are amusing only in retrospect, the most realistic advice is to involve the facilities/maintenance department at the beginning of the anticipated process change.

One must also consider growth and flexibility. Purchasing a marginally sized cleaning system is a false economy. A cleaning system that just barely manages current throughput will not be effective if business improves. All estimates of process throughput with a given system, particularly vendor-generated estimates, must be critically examined to see if they are overly optimistic. In addition, while it is unrealistic to expect a cleaning system to handle all possible chemistries, one should avoid a cleaning system designed for only a single cleaning chemistry. The situation may change, and it may be necessary to use another cleaning agent. Factors that may trigger a cleaning process changes include modifications or updates to product design or materials, cleaning standards, costs, composition or availability of cleaning agents, and new safety/environmental regulations. If several cleaning systems are essentially equivalent in performance, in general, it is better to select the more flexible system.

Employee Involvement and Employee Education

The newer, sophisticated, automated cleaning systems provide very thorough, consistent cleaning. However, employee education (as opposed to attempted rote training) is necessary to maintain process quality and to assure employee safety. Equipment maintenance is also more complex with sophisticated equipment. As indicated in the Plasma Technology case study (refer Chapter 15, Kanegsberg, in the second book *Handbook for Critical Cleaning: Applications, Proccesses, and Controls*), a move to more complex equipment can be successful with careful process evaluation and thoughtful employee education.

On the other hand, sophisticated computer programming or at least reasoned following of prearranged steps can degenerate into a bravado of desperate button pushing by a terrified employee—this can be disastrous.

In addition, in this author's observation, the dedicated and resourceful employee intent on cutting corners and speeding up the process can override even the most sophisticated interlock system. Ongoing monitoring of employee performance and behavior is essential.

In-House Equipment Design

In the course of evaluating sophisticated but deceptively simple equipment, many engineers, in an attempt to control costs and/or achieve customization, are tempted to design the equipment themselves. The general advice in such cases is that, if the equipment is similar in design to the product being manufactured, there is a chance of success. Occasionally, in-house equipment design can be very successful (refer Chapter 21, Petrulio).

In most cases, however, the advice is: "don't design your own equipment." You don't have to be a rocket scientist to design a quality vapor degreaser, but you do need practical experience in the area. Most rocket scientists probably do not have this experience. In looking at a half-million dollar plastic or metal container with robotics, you may be tempted to design one. However, consider your time and research effort. Each commercially produced cleaning system had better be profitable for the supplier to product; but the initial engineering effort to produce the initial model is typically significant. If the equipment breaks, who is going to fix it? If you designed it, and if you break it, you have to fix it. Often, regulators and fire inspectors are more comfortable with standard equipment that is certified to meet design standards.

Ultrasonic Cleaning

We have devoted a number of chapters to the theory of ultrasonic cleaning, the selection and performance of ultrasonics systems, and to newer approaches to ultrasonics metrics. Some manufacturers consider the choice of ultrasonic equipment to be generic.

There are several reasons why ultrasonic cleaning should be emphasized. For one thing, ultrasonic cleaning is useful over a very wide range of applications. It is a technique that is nearly certain to become increasingly important both to extend the range of applications amenable to aqueous cleaning and to clean product with increasing miniaturization, complexity, and close tolerances. In addition, despite the advances in theoretical understanding of ultrasonics, much needs to be learned about the mechanism of action. Ultrasonic techniques are controversial; opinions differ markedly and are also influenced by the expert's view of the relative efficacy of aqueous versus solvent cleaning. Therefore, some of the statements in various chapters may be contradictory. If this makes you, the reader, somewhat uneasy, your unease reflects the fact that ultrasonic and megasonic cleaning is an exciting, dynamic, and sometimes contentious area of technology.

Ultrasonic Metrics

Lack of standardization is a real issue. One reason that ultrasonic systems are too often thought of as generic is the problem of any standardized way of comparing performance of various systems. There is always a concern with transducer degradation over time as well as a lingering questions of comparisons of systems produced by different vendors or even of two identical models produced by a given equipment supplier.

The simplest, classic approach to ultrasonics metrics is cavitation erosion of aluminum foil. A slightly non-smooth coupon of standard aluminum foil is submerged in the utlrasonics tank for a standard time (usually 30 s). A characteristic orange peel erosion pattern indicates reasonable cavitation. Extensive erosion of the foil could indicate a hot spot or a potentially undesirable cleaning solution. The test is rapid, and inexpensive; and it can be useful for quick troubleshooting. However, results are technician dependent. Unless you weigh the foil before and after exposure to the ultrasonic system, the results are not quantitative. For extensive and repeated measurements, such exercises can provide hours of play value. And, of course, for production situations, it is important to submerge the foil in a small beaker held within the larger ultrasonic tanks so as not to contaminate the system with bits of foil "glitter." Quantitative probes to estimate ultrasonic performance within the tank have been developed, as discussed in Chapters 16 (Hodnett) and 17 (Azar).

Ultrasonics Damage

Cleaning is always a balance between soil removal and substrate damage. Ultrasonic cleaning has been used successfully for decades in many applications. However, ultrasonic action has the potential to damage parts (Kanegsberg, 2007). Selecting the appropriate frequency and amplitude is helpful. Adjust the process parameters, including time, temperature, and cleaning chemistry. Tossing a delicate component into an ultrasonic tank and walking away is asking for product failure. Some products, particularly modified or filled composites, are readily damaged ultrasonic cleaning, so plan the process carefully.

Other Cleaning Systems

Cleaning systems undergo constant changes, refinement, and development. This book does not cover all cleaning systems available. Instead, some key cleaning techniques have been highlighted. It is anticipated that from the examples, principles, and reasoning processes, the practicing engineer will be able to extrapolate to evaluate other cleaning systems. A few additional cleaning systems are summarized.

Contained Spray Cleaning Chambers*

The spray cleaning process has been in use for many years with high reliability, small mechanics, optics, microelectronics, and other extremely sensitive and demanding components. Most often, it is used in those areas of production where components and subassemblies must be totally clean of all organic and inorganic contaminants such as military, aerospace, and medical applications, which meet exacting specifications for cleanliness. Because of its special ability to handle exacting cleaning requirements for delicate wire bonds without damage, clean both blind and through holes of some incredibly small diameters; e.g., 0.0005, and the removal of both organic and inorganic contaminants by utilizing a gentle agitation via the fine, usually venturi type, spray and the molecular weight of the chosen solvent, it has been a successful partner to virtually every other type of cleaning process currently in use.

Often in this type of process, when more traditional types of solvents are utilized, there is little or no waste product and virtually no effluent to handle. The solvents are generally being used at a rate of 0.2 fluid ounces per second, a minimal amount at best, and because the cleaning cycles tend to be very short in duration, an average of 5 s of solvent spray, the solvent is evaporated when these types of systems are properly used. This provides several advantages, the most important of which may be the ability of the user to use a carbon absorption process in gas phase to handle the minuscule amounts of waste. This fume and solvent extraction process can be totally unvented, therefore eliminating the exhausting of waste to the atmosphere through existing fume exhausting systems, or the expense of cutting additional holes in the physical plant. It allows the user to have a fully portable waste plant for this simple and yet incredibly effective precision cleaning partner.

The spraying process can totally replace all other processes only in small R&D-type facilities and is generally not intended for high production requirement facilities. Where requirements are less than 100 components per day, it is often the ideal solution to an otherwise highly expensive problem.

Reel-to-Reel or Continuous Web Cleaning

Reel-to-reel systems are used for such diverse processes as cleaning metal wire, strips of lead frames, combination medical devices and motion picture film. In such systems, it is helpful to picture giant spools of thread (with the thread being made of the material to be cleaned). The part is a continuous filament running from one spool, through a cleaning solution, through a drier, and then onto another spool. Ultrasonics or other agitation may assist in cleaning; rollers or brushes may assist in removing excess cleaning agent prior to drying. Such cleaning systems can be aqueous or solvent based.

Even in the age of electronics storage data, motion picture film remains a medium of choice, at least for master copies (RTI). Film cleaning is a very specialized, fascinating challenge in reel-to-reel cleaning, involving cleaning performance, speed, materials compatibility, and long term storage. The substrate is complex, composed of multiple layers of plastics and other synthetics. Before the phaseout of ozone depleting compounds (ODCs), motion picture film was cleaned with 1,1,1-trichloroethane (TCA).

After the phaseout of ODCs, most film was cleaned with perchloroethylene. Regulatory and technical issues have generated interest in other cleaning alternatives, such as hydrofluorocarbons (HFCs)

---

* The description of this technology was contributed by Rebecca Overton, who had many years of experience in this technology, most recently at Cobehn.

with ultrasonics. The delicacy and complexity of materials of construction, and long-term storage issues makes acceptance of appropriate cleaning systems for motion picture a difficult task. It is notable that because of its unusual refractive index, perfluoroethylene (PCE) is also used in a related reel-to-reel process, liquid gate processing to improve print quality; it seems plausible that PCE also has a cleaning function. In this situation, the need for most extensive film cleaning was reduced by a move to other media.

Centrifugal Cleaner

In such systems, parts are cleaned in a single centrifuge chamber filled with cleaning agent. Cleaning at low centrifugal force promotes agitation due to Coreolis mixing. Initial costs may be high; and throughput is often lower than conventional methods, but equipment is often designed for use with many different cleaning agents. Centrifugal cleaners may use water or solvent-based cleaning agents.

Spinners

Spinners are used in cleaning and surface preparation of wafers and optics. Various cleaning and rinsing solutions are applied to the surface to be cleaned; the component is on a rapidly moving turntable. Sequential cleaning with various cleaning agents as well as rinsing and spot-free drying is carried out in place. Cleaning solutions are sprayed onto the parts. These systems are typically designed to be used in clean rooms where contamination is an issue. Avoiding particles, residue from additive packages, and excess foaming are considerations. Spinners may use water or solvent-based cleaning agents; or, within a single process, both may be used.

Microclusters

Microcluster cleaning is a specialized line-of-site technique, which has been demonstrated to remove submicron particles. Microclusters are produced by atomizing a conducting liquid (typically a mixture of *n*-methyl pyrollidone and water), which is then exposed to high electric fields. The microcluster dimensions are in the range of that of the contaminants to be removed. The technique has been proposed for wafer cleaning, but may have other high-end applications (Perel et al., 1999).

Industrial Cleaners/Cabinet Washers, Dishwashers, Spray Cabinets

This category includes cleaning systems most often associated with general industrial cleaning (Peterson, 1997). Except where indicated, they are primarily used with aqueous cleaning agents. Some resemble consumer variety dishwashers, perhaps with stainless steel interiors.

Sink-on-a-drum cleaning systems are used in general metals cleaning applications with mineral spirits, or, particularly in areas that are heavily regulated due to poor air quality, with an aqueous-based cleaning agent or a volatile organic compound (VOC)-exempt solvent. They can be constructed of metal or plastic. Some solvent-based systems have been adapted to provide on-board recycling of the cleaning agent (see Chapter 20, Skelly).

Cabinet washers are typically tall, cylindrical systems. The parts (engines, etc.) are typically placed on a turntable and sprayed with hot surfactant solution. In spray cabinets, the part to be cleaned is placed inside a box-like container and sprayed with typically an aqueous-based cleaning chemistry. Spray cabinets may be manual or automated. The manual models resemble a glove box. Automated models, such as those used by the automotive industry, can be quite large and sophisticated and may include in-line (conveyor belt) or overhead robotics, with automated monitoring of the cleaning solution. Some aqueous systems have even been fitted with high-pressure spray for tube cleaning (Adam, 1995).

As with other cleaning systems, the cleaning chemistry, temperature, force of cleaning action, filtration, rinsing, and, in some cases, drying all impact cleaning quality. It is important to match the equipment to the application. In many general metals, cleaning applications, a single cleaning tank with either solvent or aqueous-based material is sufficient. The features, capabilities, prices, and quality of construction vary by orders of magnitude. Particularly because such systems will be subjected to extremes of temperature and of cleaning chemistry, purchasing a well-designed system of high quality can result in long-term benefits.

The user must consider initial costs, ongoing costs (including disposables such as filters), and labor and rework costs. Some of the more sophisticated systems may require a high level of ongoing maintenance.

Bioremediation Parts Washer

This aqueous-based technology uses bacteria in the system, but bacteria do not chomp on the oil that adheres to the part. Instead, the bacteria are part of an onboard remediation system, allowing longer cleaning agent use. While such a process would not be favored where bio-contamination is an issue, it has been used industrially and by the military (Government Sales and Service Company, 2010).

Semi-Aqueous, Co-Solvent Systems, and Bi-Solvent Processes

Semi-aqueous and co-solvent systems are related. In semi-aqueous systems, a high boiling solvent-based cleaning agent, is used for primary removal of soils. This step is followed by several rinsing and drying steps. Co-solvent cleaning can imply two or more solvents, used in a single tank or sequentially. Technically, any solvent blend or azeotrope could be considered co-solvent cleaning. In this discussion, however, we will consider co-solvent cleaning to be a process in which a high-boiling solvent blend is rinsed in a second solvent, solvent blend, or azeotrope. In co-solvent cleaning, typically a lower boiling solvent is chosen to serve as a rinsing and vapor phase drying agent. In some cases, a supplier may offer two very similar blends based on hydrocarbon, D-limonene, or ester, which differ in subtle changes in the additive packages to make them more readily removable with water or with solvent. Other products are based on complex, modified alcohols. Many high-boiling solvent blends are considered competition sensitive by the manufacturers and, therefore, unfortunately, shrouded in mystery. This, of course, makes rational process design a challenge. In such situations, the end user would be well advised to set up comprehensive product support arrangements.

While water is the rinsing agent in semi-aqueous cleaning, the rinse/vapor phase/drying agent can be any of a number of lower-boiling solvents such as HFC, HFE, isopropyl alcohol, and even isopropyl alcohol/cyclohexane azeotrope (which would, one supposes, constitute a co-solvent process).

Both semi-aqueous and co-solvent systems have potential advantages and drawbacks. In both cases, because compounds and mixtures with widely different solvency ranges are used for cleaning and rinsing (we are considering rinse water as a cleaning agent), the rinse phase can, in a sense, be considered part of the cleaning phase. Both types of systems can extend the range of soils that can be removed; both can allow the use of high-boiling cleaning agents, which may themselves leave residue on the part and/ or may not dry sufficiently rapidly.

Both types of systems can use agitation, including ultrasonic cleaning, to improve performance. In some cases, the cleaning agent is designed to form an emulsion with the rinsing agent, either for initial cleaning or as final rinsing. The emulsion can be stable or transient (i.e., the emulsion exists only during agitation). Alternatively, the cleaning and rinsing agents may be miscible. For stable emulsions as well as with miscible cleaning and rinsing agents, the issue of recovering the cleaning and rinsing agents as well as that of waste stream management becomes more complex. Depending on the situation, reverse osmosis may be needed for recovery. Multiple filtration of the waste stream is typically needed in semi-aqueous systems.

Initially semi-aqueous and co-solvent systems were widely and enthusiastically adopted. However, these systems are relatively complex and sophisticated. Like other newer cleaning systems, they require maintenance, employee education, and process monitoring. In both cases, proper fixturing of the product is crucial to assure optimal cleaning and to avoid excessive carryover of cleaning agent into the rinse tank. This author recalls implementing a semi-aqueous cleaning system with in-line automation, which initially performed very acceptably. Then came the phone call: the system does not work; the cleaning agent is gone; all of the filters for the waste water are "dead." The reason turned out to be cultural and historical, and involved the legendary third shift. You see, the facility was located in a town that held monthly auto racing. Over the years, the third shift had become accustomed to cleaning their carburetors in the vapor degreaser. With vapor degreasing, because final cleaning is in freshly distilled solvent, the extraneous soils did not present a problem. A semi-aqueous system is not as forgiving. When carburetors were placed on the conveyor belt of the semi-aqueous system, there was carryover of excessive amounts cleaning agent into the rinse tank, resulting in system failure. The scope of cleaning allowed in semi-aqueous, co-solvent, or any other cleaning system is the prerogative of management. The point is that semi-aqueous and co-solvent systems are not forgiving of mediocre process control.

It should also be noted that the physical properties, including flammability, have to be carefully considered in designing a system. A semi-aqueous cleaning system requires multiple rinse tanks (for in-line systems, a fairly long portion of the conveyor belt should be devoted to rinsing). Then, depending on the product and the next step in the process, drying will be required. Even though the primary cleaning occurs below the boiling point of the cleaning agent, a typical co-solvent system is designed more like a multi-tank degreaser, to allow for vapor phase rinsing and drying. In some cases, vapor degreasers have been converted to co-solvent systems. However, it should be remembered that if a low-flash-point solvent is employed, a standard vapor degreaser would not be suitable and would pose a fire hazard.

Wet Benches

Biomedical devices, optics, semiconductors, and microelectronics are often processed and cleaned in wet benches. This is a broad, generic term for a series of cleaning tanks, which may contain aqueous, semi-aqueous, or solvent-based cleaners.

In addition, etching with strong acids and bases may take place. While such processes might better be considered as surface modification, they are certainly related to cleaning. It should be noted that appropriate process controls, such as titrators, are desirable, and may be crucial to maintain process control. In addition, with strong acids such as hydrofluoric acid, process automation, vapor monitors, and controlled bath neutralization may be needed for adequate employee protection.

Impingement Cleaning

Impingement cleaning covers an array of processes, including line-of-sight high-force solvent and aqueous sprays, CO_2 snow (see Chapter 28, Sherman), and dry media (see Chapter 27, Swan). Impingement cleaning has been used in critical applications for many years.

In metal finishing and deburring, a variety of non-solvent, non-aqueous media are used in finishing. The equipment manufacturer often considers this equipment as separate from cleaning, but there is overlap. For example, when faced with the choice of using chemical stripping or mechanical stripping of paint, a facility that repairs pumps chose media blast for a variety of safety, environmental, and economic considerations (Maluso and Kanegsberg, 1999).

Increasingly, other forms of impingement cleaning are being adapted from general cleaning to meet critical-cleaning and surface-finishing requirements. Examples include plastic pellets, walnut shells, sand, diamond dust, nails, aluminum oxide, garnet, small nails—the list goes on and on. The impingement or rubbing action of the media itself may be achieved with blast, agitation, centrifugation, or ultrasonics; and cleaning may be dry or in an aqueous or aqueous surfactant media (Kanegsberg and Kanegsberg, 2000).

While some forms of impingement cleaning are becoming widely adopted, other media cleaning may become more important in the future for high-precision, critical applications. For example, selective removal of high-value coatings bears some relationship to selective removal of paint. As with other cleaning approaches, there will be provisos in expanding media blast. Speed, potential part damage, and preventing residual blast media from recontaminating the part are among considerations.

Point-of-Use Cleaning: Benchtop Cleaning

Point-of-use or benchtop cleaning is widely used, especially for rework and small scale or small through-put processes. Frequently, it is used in the developmental stage of a product and replaced by an automated or semi-automated process when the product goes into production.

Techniques used in point-of-use cleaning include wiping, spraying, or dipping (Kanegsberg and Kanegsberg, 2005). Wiping may use rags dipped into a cleaning agent, pre-saturated wipes and swabs. Spraying can be from lab dispensers, aerosols, or spraying equipment, including CO_2 snow sprayers (see Chapter 28, Sherman), steam (see Chapter 31, Friedheim and Gonzalez), and dry media (see Chapter 27, Swan).

Both IPA and acetone are popular chemicals for hand wipe applications. They are low cost and have relatively low toxicity. They are fairly aggressive solvents and dry rapidly. However, both are flammable and extreme care needs to be taken to avoid use near ignition sources. Acetone is popular partly because it is a VOC-exempt chemical.

Other chemicals with point-of-use applications include MEK, Methylene chloride, Perchlorethylene, and paint thinners. These generally have toxicity concerns for worker exposure and must be used with caution.

Commercial hardware store or "big-box" store products frequently find their way into benchtop cleaning situations (see "How Not to Clean Critically with Household Products" section in Chapter 1, Kanegsberg). Product variability can be an issue since products may have formulation changes without notice.

Point-of-use operations are performed by employees. This means there is the potential for human error, operator-to-operator variability or day-to-day variability with an operator.

Residues can be of concern. Many of the cleaning agents used for point-of-use cleaning need rinsing. Some of the biobased products may leave residues unless extensively rinsed. Recontamination can occur from rags, wipes, or swabs used for cleaning, including particles and fibers or films. Cleaning chemicals stored in plastic dispensers may contain plasticizers that contaminate the surface being wiped. It is important to use an appropriate chemical dispenser and to change dispensers periodically.

Conclusion

Cleaning applications and requirements are exceedingly diverse. To meet these needs, an array of cleaning agents and cleaning equipment has been developed. This chapter summarizes a few aspects of cleaning systems; other chapters discuss the specifics of cleaning systems in much greater detail.

In evaluating various cleaning processes, the reader must remember that, inherently, cleaning involves a melding of cleaning agent, cleaning action, and overall process equipment. In developing processes, some argue that the cleaning agent should be selected prior to the cleaning equipment or vice versa. There is some validity to either approach, as long as it moves us along the path to ADS (actually doing something). However, given the complexity of performance, economics, and environmental requirements, it is often more productive to consider the cleaning agent and the cleaning system in parallel, then making the choices.

In addition, even though some cleaning processes may not be widely used in your particular industry, the reader is urged to at least skim through all of the available process choices. By looking at processes creatively and in a more encompassing manner, it may be possible to adapt processes from one area of manufacturing to another. This kind of adaptation fosters overall progress, and, often, specific competitive advantage. For more information about specific applications, the reader is invited to peruse Book 2, "*Handbook for Critical Cleaning: Applications, Processes, and Controls.*"

References

Adam, S.J. Aqueous tube cleaning advances at McDonnell Douglas aerospace. In: *Proceedings of the 2nd Aerospace Environmental Technology Conference*, Huntsville, AL, August 6–8, 1995. NASA Conference Publication 3349, p. 145.

Aries, J. Moves toward automation. *Parts Cleaning Magazine*, February 1998.

Bohn, R.E. Measuring and managing technological knowledge. *Sloan Management Review*, 36(1), 61–73, 1994.

Government Sales and Service Company. Chemfree Smart Washer. http://www.gsaservice.com/chemfree_smartwasher.htm, March 31, 2010.

Kanegsberg, B. Successful cleaning/assembly processes for small to medium electronics manufacturers. Tutorial, *Nepcon West'98*, Anaheim, CA, March 3, 1998.

Kanegsberg, B. Washing, rinsing, and drying: Items to consider for the optimization of your cleaning process. *Metal Finishing Magazine*, September 2005.

Kanegsberg, B. and Kanegsberg, E. Rinsing and drying and point of use cleaning, wash, rinse, dry. In: *Cleantech05*, Chicago, IL, March 7, 2005.

Kanegsberg, B. and Kanegsberg, E. The winding road to automation. *Controlled Environments Magazine*, April 2007.

Kanegsberg, E. and Kanegsberg, B. Cleaning by abrasive impact. *A2C2 Magazine*, May 2000.

Maluso, P. and Kanegsberg, B. Hydrostatic pump rebuild: Implementing aqueous, steam and solvent free processes, presentation and proceedings. In: *Tenth Annual International Workshop on solvent Substitution and the Elimination of Toxic Substances & Emissions*, Scottsdale, AZ, September 13–16, 1999.

Perel, J., Sujo, C., and Mahoney, J.F. Microclusters make and impact on wafer cleaning. *Precision Cleaning Magazine*, 7, 18–24, 1999.

Peterson, D.S., *Practical Guide to Industrial Metal Cleaning*. Cincinnati, OH: Hanser Gardner Publications, 1997.

RTI, Research Technology International. http://www.rti-us.com/ (accessed March 31, 2010).

12

The Fundamental Theory and Application of Ultrasonics for Cleaning

F. John Fuchs

Introduction

Cleaning technology is in a state of change. Vapor degreasing using chlorinated and fluorinated solvents, long the standard for most of industry, is being subjected to increased regulatory requirements in the interest of the ecology of our planet. At the same time, cleaning requirements are continually increasing. Cleanliness has become an important issue in many industries where it never was in the past. In industries such as electronics where cleanliness was always important, it has become more critical in support of growing technology. It seems that each advance in technology demands greater and greater attention to cleanliness for its success. As a result, the cleaning industry has been challenged to deliver the needed cleanliness and has done so through rapid innovation over the past several years. Many of these advances have involved the use of ultrasonic technology.

The cleaning industry is currently in a struggle to replace solvent degreasing with alternative "environmentally friendly" means of cleaning. Although substitute water-based, semiaqueous, and petroleum-based chemistries are available, they are often somewhat less effective as cleaners than the solvents and may not perform adequately in some applications unless a mechanical energy boost is added to assure the required levels of cleanliness. Ultrasonic energy is now used extensively in critical cleaning applications to both speed and enhance the cleaning effect of the alternative chemistries. This chapter is intended to familiarize the reader with the basic theory of ultrasonics and how ultrasonic energy can be most effectively applied to enhance a variety of cleaning processes.

What Is Ultrasonics?

Ultrasonics is the science of sound waves above the limits of human audibility. The frequency of a sound wave determines its tone or pitch. Low frequencies produce low or bass tones. High frequencies produce high or treble tones. Ultrasound is a sound with a pitch so high that it cannot be heard by the human ear. Frequencies above 18 kHz are usually considered to be ultrasonic. The frequencies used for ultrasonic cleaning range from 20,000 cycles per second or 20 to over 100 kHz. The most commonly used frequencies for industrial cleaning are those between 20 and 50 kHz. Frequencies above 50 kHz are more commonly used in small tabletop ultrasonic cleaners, such as those found in jewelry stores and dental offices.

The Theory of Sound Waves

To understand the mechanics of ultrasonics, it is necessary first to have a basic understanding of sound waves, how they are generated and how they travel through a conducting medium. The dictionary defines sound as the transmission of vibration through an elastic medium which may be a solid, liquid, or a gas.

Sound Wave Generation

A sound wave is produced when a solitary or repeating displacement is generated in a sound-conducting medium, such as by a "shock" event or "vibratory" movement (Figure 12.1). The displacement of air by the cone of a radio speaker is a good example of vibratory sound waves generated by mechanical movement. As the speaker cone moves back and forth, the air in front of the cone is alternately compressed and rarefied to produce sound waves, which travel through the air until they are finally dissipated. We are probably most familiar with sound waves generated by alternating mechanical motion. There are also sound waves that are created by a single "shock" event. An example is thunder, which is generated as air instantaneously changes volume as a result of an electrical discharge (lightning). Another example of a shock event might be the sound created as a wooden board falls with its face against a cement floor. Shock events are sources of a single compression wave that radiates from the source.

The Nature of Sound Waves

Figure 12.2 uses the coils of a spring similar to a Slinky® toy to represent individual molecules of a sound-conducting medium. The molecules in the medium are influenced by adjacent molecules in much the same way that the coils of the spring influence one another. The source of the sound in the model is at the left. The compression generated by the sound source as it moves propagates down the length of the spring as each adjacent coil of the spring pushes against its neighbor. It is important to note that, although the wave travels from one end of the spring to the other, the individual coils remain in their same relative positions, being displaced first one way and then the other as the sound wave passes. As a result, each coil is first part of a compression as it is pushed toward the next coil and then part of a rarefaction as it recedes from the adjacent coil. In much the same way, any point in a sound conduction medium is alternately subjected to compression and then rarefaction. At a point in the area of a compression, the pressure in the medium is positive. At a point in the area of a rarefaction, the pressure in the medium is negative.

"Vibratory"

"Shock" events

FIGURE 12.1 Vibratory and shock events.

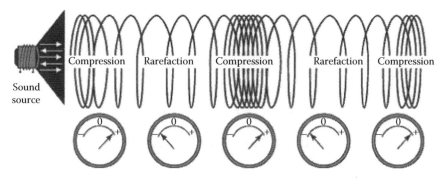

FIGURE 12.2 Coils of a spring representing individual molecules of a sound-conducting medium.

In elastic media such as air and most solids, there is a continuous transition as a sound wave is transmitted. In nonelastic media such as water and most liquids, there is continuous transition as long as the amplitude or "loudness" of the sound is relatively low. As amplitude is increased, however, the magnitude of the negative pressure in the areas of rarefaction eventually becomes sufficient to cause the liquid to fracture because of the negative pressure, causing a phenomenon known as cavitation.

As shown in Figure 12.3, cavitation "bubbles" are created at sites of rarefaction as the liquid fractures or tears because of the negative pressure of the sound wave in the liquid. As the wave fronts pass, the cavitation "bubbles" oscillate under the influence of positive pressure, eventually growing to an unstable size. Finally, the violent collapse of the cavitation bubbles results in implosions, which cause shock waves to be radiated from the sites of the collapse.

The collapse and implosion of myriad cavitation bubbles throughout an ultrasonically activated liquid result in the effect commonly associated with ultrasonics. It has been calculated that

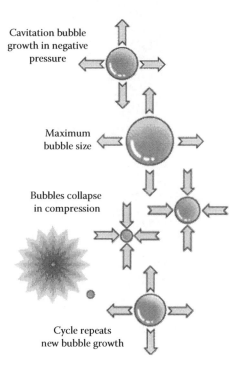

FIGURE 12.3 Cavitation and implosion.

temperatures in excess of 10,000°F and pressures in excess of 10,000 psi are generated at the implosion sites of cavitation bubbles.

Benefits of Ultrasonics in the Cleaning and Rinsing Processes

Cleaning in most instances requires that a contaminant be dissolved (as in the case of a soluble soil), displaced (as in the case of a nonsoluble soil), or both dissolved and displaced (as in the case of insoluble particles being held by a soluble binder such as oil or grease). The mechanical effect of ultrasonic energy can be helpful in both speeding dissolution and displacing particles. Just as it is beneficial in cleaning, ultrasonics is also beneficial in the rinsing process. Residual cleaning chemicals are removed quickly and completely by ultrasonic rinsing.

Ultrasonics Speeds Cleaning by Dissolution

In removing a contaminant by dissolution it is necessary for the solvent to come into contact with and dissolve the contaminant. The cleaning activity takes place only at the interface between the cleaning chemistry and the contaminant. (See Figure 12.4.) As the cleaning chemistry dissolves the contaminant, a saturated layer develops at the interface between the fresh cleaning chemistry and the contaminant. Once this has happened, cleaning action stops as the saturated chemistry can no longer attack the contaminant. Fresh chemistry cannot reach the contaminant (Figure 12.5). Ultrasonic cavitation and implosion effectively displace the saturated layer to allow fresh chemistry to come into contact with the contaminant (Figure 12.6) remaining to be removed. This is especially beneficial when irregular surfaces or internal passageways are to be cleaned.

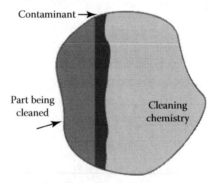

FIGURE 12.4

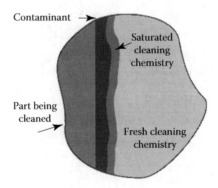

FIGURE 12.5

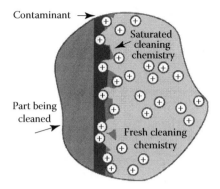

FIGURE 12.6

Ultrasonic Activity Displaces Particles

Some contaminants comprise insoluble particles loosely attached and held in place by ionic or cohesive forces. These particles need only be displaced sufficiently to break the attractive forces to be removed. (Figure 12.7).

Cavitation and implosion as a result of ultrasonic activity displace and remove loosely held contaminants such as dust from surfaces. For this to be effective, it is necessary that the coupling medium be capable of wetting the particles to be removed (Figure 12.8).

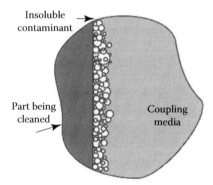

FIGURE 12.7

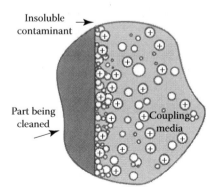

FIGURE 12.8

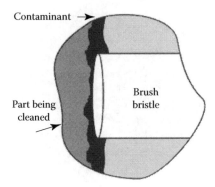

Contaminant →

Part being
cleaned

Brush
bristle

FIGURE 12.9

Complex Contaminants

Contaminants can also, of course, be more complex in nature, consisting of combination soils made up of both soluble and insoluble components. The effect of ultrasonics is substantially the same in these cases, as the mechanical microagitation helps speed both the dissolution of soluble contaminants and the displacement of insoluble particles.

Ultrasonic activity has also been demonstrated to speed or enhance the effect of many chemical reactions. This is probably caused mostly by the high energy levels created as high pressures and temperatures are created at the implosion sites. It is likely that the superior results achieved in many ultrasonic cleaning operations may be at least partially attributed to the sonochemistry effect.

A Superior Process

In the above illustrations, the surface of the part being cleaned has been represented as flat. In reality, surfaces are seldom flat, instead, they comprise hills, valleys, and convolutions of all description. Figure 12.9 shows why ultrasonic energy has proved to be more effective at enhancing cleaning than other alternatives, including spray washing, brushing, turbulation, air agitation, and even electro-cleaning in many applications. The ability of ultrasonic activity to penetrate and assist the cleaning of interior surfaces of complex parts is also especially noteworthy.

Ultrasonic Equipment

To introduce ultrasonic energy into a cleaning system requires an ultrasonic transducer and an ultrasonic power supply or "generator." The generator supplies electrical energy at the desired ultrasonic frequency. The ultrasonic transducer converts the electrical energy from the ultrasonic generator into mechanical vibrations.

Ultrasonic Generator

The ultrasonic generator converts electrical energy from the line which is typically alternating current at 50 or 60 Hz to electrical energy at the ultrasonic frequency. This is accomplished in a number of ways by various equipment manufacturers. Current ultrasonic generators nearly all use solid-state technology (Figure 12.10).

There have been several relatively recent innovations in ultrasonic generator technology which may enhance the effectiveness of ultrasonic cleaning equipment. These include square wave outputs, slowly or rapidly pulsing the ultrasonic energy on and off, and modulating or "sweeping" the frequency of the generator output around the central operating frequency. The most-advanced ultrasonic generators

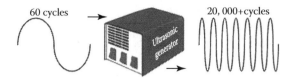

FIGURE 12.10   Generation of ultrasonics.

have provisions for adjusting a variety of output parameters to customize the ultrasonic energy output for the task.

Square Wave Output

Applying a square wave signal to an ultrasonic transducer results in an acoustic output rich in harmonics. The result is a multifrequency cleaning system that vibrates simultaneously at several frequencies which are harmonics of the fundamental frequency. Multifrequency operation offers the benefits of all frequencies combined in a single ultrasonic cleaning tank.

Pulse

In pulse operation, the ultrasonic energy is turned on and off at a rate that may vary from once every several seconds to several hundred times per second. The percentage of time that the ultrasonic energy is on may also be changed to produce varied results. At slower pulse rates, more rapid degassing of liquids occurs as coalescing bubbles of air are given an opportunity to rise to the surface of the liquid during the time the ultrasonic energy is off. At more rapid pulse rates the cleaning process may be enhanced as repeated high energy "bursts" of ultrasonic energy occur each time the energy source is turned on (Figure 12.11).

Frequency Sweep

In sweep operation, the frequency of the output of the ultrasonic generator is modulated around a central frequency, which may itself be adjustable (Figure 12.12).

Various effects are produced by changing the speed and magnitude of the frequency modulation. The frequency may be modulated from once every several seconds to several hundred times per second with the magnitude of variation ranging from several Hz to several kHz. Sweep may be used to prevent

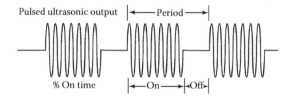

FIGURE 12.11   Pulse operation.

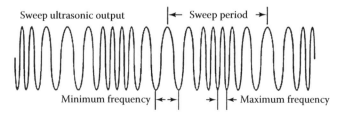

FIGURE 12.12   Frequency sweep.

damage to extremely delicate parts or to reduce the effects of standing waves in cleaning tanks. The frequency of sweep may be varied randomly to prevent damage to parts susceptible to resonating at or near the sweep rate frequency. Sweep operation may also be found especially useful in facilitating the cavitation of terpenes and petroleum-based chemistries. A combination of pulse and sweep operation may provide even better results when the cavitation of terpenes and petroleum-based chemistries is required.

Frequency and Amplitude

Frequency and amplitude are properties of sound waves. Figure 12.13a through c demonstrate frequency and amplitude using the spring model introduced earlier. If Figure 12.13a is the base sound wave, Figure 12.13b with less displacement of the media (less intense compression and rarefaction) as the wave front passes represents a sound wave of less amplitude or "loudness." Figure 12.13c represents a sound wave of higher frequency indicated by more wave fronts passing a given point within a given period of time.

Ultrasonic Transducers

There are two general types of ultrasonic transducers in use today: magnetostrictive and piezoelectric. Both accomplish the same task of converting alternating electrical energy to vibratory mechanical energy but do it using different means.

Magnetostrictive

Magnetostrictive transducers utilize the principle of magnetostriction in which certain materials expand and contract when placed in an alternating magnetic field.

Alternating electrical energy from the ultrasonic generator is first converted into an alternating magnetic field through the use of a coil of wire. The alternating magnetic field is then used to induce mechanical vibrations at the ultrasonic frequency in resonant strips of nickel or other magnetostrictive material that are attached to the surface to be vibrated. Because magnetostrictive materials behave identically to a magnetic field of either polarity, the frequency of the electrical energy applied

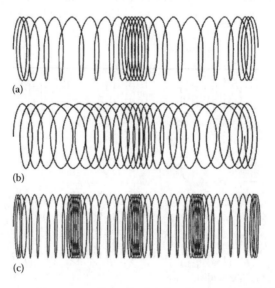

(a)

(b)

(c)

FIGURE 12.13 Demonstration of frequency and amplitude using the spring model: (a) base, (b) lower amplitude, and (c) higher frequency.

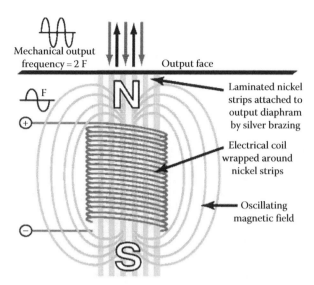

FIGURE 12.14 Magnetostrictive transducer.

to transducer is one half of the desired output frequency. Magnetostrictive transducers were the first to supply a robust source of ultrasonic vibrations for high-power applications, such as ultrasonic cleaning (Figure 12.14).

Because of inherent mechanical constraints on the physical size of the hardware as well as electrical and magnetic complications, high-power magnetostrictive transducers seldom operate at frequencies much above 30 kHz. Piezoelectric transducers, on the other hand, can easily operate well into the megahertz range.

Magnetostrictive transducers are generally less efficient than their piezoelectric counterparts. This is due primarily to the fact that the magnetostrictive transducer requires a dual energy conversion from electrical to magnetic and then from magnetic to mechanical. Some efficiency is lost in each conversion. Magnetic hysteresis effects also detract from the efficiency of the magnetostrictive transducer.

Piezoelectric

Piezoelectric transducers (Figure 12.15) convert alternating electrical energy directly to mechanical energy through use of the piezoelectric effect in which certain materials change dimension when an electrical charge is applied to them.

Electrical energy at the ultrasonic frequency is supplied to the transducer by the ultrasonic generator. This electrical energy is applied to piezoelectric element(s) in the transducer, which vibrate. These vibrations are amplified by the resonant masses of the transducer and directed into the liquid through the radiating plate.

Early piezoelectric transducers utilized such piezoelectric materials as naturally occurring quartz crystals and barium titanate, which were fragile and unstable. Early piezoelectric transducers were, therefore, unreliable. Today's transducers incorporate stronger, more efficient, and highly stable ceramic piezoelectric materials, which were developed as a result of the efforts of the U.S. Navy and its research to develop advanced sonar transponders in the 1940s. The vast majority of transducers used today for ultrasonic cleaning utilize the piezoelectric effect.

Ultrasonic Cleaning Equipment

Ultrasonic cleaning equipment ranges from the small tabletop units often found in dental offices or jewelry stores (Figure 12.16) to huge systems with capacities of several thousand gallons used in a

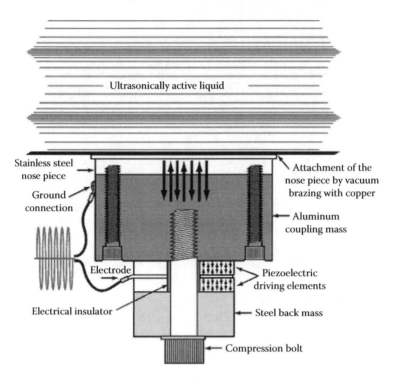

FIGURE 12.15   Piezoelectric transducer.

variety of industrial applications. Selection or design of the proper equipment is paramount in the success of any ultrasonic cleaning application.

The simplest application may require only a small heated tank cleaner with rinsing to be done in a separate container. More sophisticated cleaning systems include one or more rinses, added process tanks and hot air dryers. Automation is often added to reduce labor and guarantee process consistency.

The largest installations utilize immersible ultrasonic transducers that can be mounted on the sides or bottom of cleaning tanks of nearly any size. Immersible ultrasonic transducers offer maximum flexibility and ease of installation and service.

Small, self-contained cleaners are used in doctor's offices and jewelry stores. Heated tank cleaning systems are used in laboratories and for small batch cleaning needs (Figure 12.17). Console cleaning systems integrate ultrasonic cleaning tank(s), rinse tank(s), and a dryer for batch cleaning (Figure 12.18). Systems can be automated through the use of a PLC-controlled material handling system. A wide range of options may be offered in custom-designed systems.

Large-scale installations or retrofitting of existing tanks in plating lines, etc., can be achieved by using modular immersible ultrasonic transducers. Ultrasonic generators are often housed in climate-controlled enclosures (Figure 12.19).

FIGURE 12.16   Small and self-contained cleaner.

FIGURE 12.17   Heated tank cleaning system.

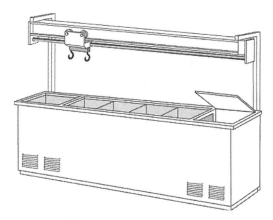

FIGURE 12.18 Console cleaning system.

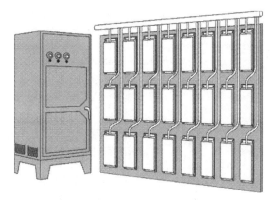

FIGURE 12.19 Large-scale installation.

Maximizing the Ultrasonic Cleaning Process Parameters

Effective application of the ultrasonic cleaning process requires consideration of a number of parameters. While time, temperature, and chemical remain important in ultrasonic cleaning as they are in other cleaning technologies, there are additional factors that must be considered to maximize the effectiveness of the process. Especially important are those variables that affect the intensity of ultrasonic cavitation in the liquid.

Maximizing Cavitation

Maximizing cavitation of the cleaning liquid is obviously very important to the success of the ultrasonic cleaning process. Several variables affect cavitation intensity.

Temperature is the most important single parameter to be considered in maximizing cavitation intensity. This is because so many liquid properties affecting cavitation intensity are related to temperature. Changes in temperature result in changes in viscosity, the solubility of gas in the liquid, the diffusion rate of dissolved gasses in the liquid, and vapor pressure, all of which affect cavitation intensity. In pure water, the cavitation effect is maximized at approximately 160°F.

The *viscosity* of a liquid must be minimized for maximum cavitation effect. Viscous liquids are sluggish and cannot respond quickly enough to form cavitation bubbles and violent implosion. The viscosity of most liquids is reduced as temperature is increased.

For most effective cavitation, the cleaning liquid must contain as little *dissolved gas* as possible. Gas dissolved in the liquid is released during the bubble growth phase of cavitation and prevents its violent implosion, which is required for the desired ultrasonic effect. The amount of dissolved gas in a liquid is reduced as the liquid temperature is increased.

Importance of Minimizing Dissolved Gas

During the negative-pressure portion of the sound wave, the liquid is torn apart and cavitation bubbles start to form. As negative pressure develops within the bubble, gases dissolved in the cavitating liquid start to diffuse across the boundary into the bubble. As negative pressure is reduced by the passing of the rarefaction portion of the sound wave and atmospheric pressure is reached, the cavitation bubble starts to collapse because of its own surface tension. During the compression portion of the sound wave, any gas that diffused into the bubble is compressed and finally starts to diffuse across the boundary again to reenter the liquid. This process, however, is never complete as long as the bubble contains gas since the diffusion out of the bubble does not start until the bubble is compressed. And once the bubble is compressed, the boundary surface available for diffusion is reduced. As a result, cavitation bubbles formed in liquids containing gas do not collapse all the way to implosion but rather result in a small pocket of compressed gas in the liquid. This phenomenon can be useful in degassing liquids. The small gas bubbles group together until they finally become sufficiently buoyant to come to the surface of the liquid (Figure 12.20).

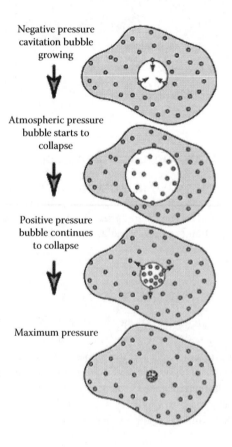

Negative pressure
cavitation bubble
growing

Atmospheric pressure
bubble starts to
collapse

Positive pressure
bubble continues
to collapse

Maximum pressure

FIGURE 12.20

The diffusion rate of dissolved gases in a liquid is increased at higher temperatures. This means that liquids at higher temperatures give up dissolved gases more rapidly than those at lower temperatures, which aids in minimizing the amount of dissolved gas in the liquid.

A moderate increase in the temperature of a liquid brings it closer to its *vapor pressure,* meaning that vaporous cavitation is more easily achieved. Vaporous cavitation, in which the cavitation bubbles are filled with the vapor of the cavitating liquid, is the most effective form of cavitation. As the boiling temperature is approached, however, the cavitation intensity is reduced as the liquid starts to boil at the cavitation sites.

Ultrasonic Power

Cavitation intensity is directly related to ultrasonic power at the power levels generally used in ultrasonic cleaning systems. As power is increased substantially above the cavitation threshold, cavitation intensity levels off and can only be further increased by using focusing techniques.

Ultrasonic Frequency

Cavitation intensity is inversely related to ultrasonic frequency. As the ultrasonic frequency is increased, cavitation intensity is reduced because of the smaller size of the cavitation bubbles and their resultant, less violent implosion. Higher frequencies are used to eliminate cavitation erosion on delicate parts.

Ultrasonic frequencies above the traditional 25 and 40 kHz have also been demonstrated more effective at removing submicron-sized particles from silicon wafers and coated optics. Other applications may also benefit from the use of higher frequencies.

Maximizing Overall Cleaning Effect

Cleaning chemical selection is extremely important to the overall success of the ultrasonic cleaning process. The selected chemical must be compatible with the base metal being cleaned and have the ability to remove the soils that are present. It must also cavitate well. Most cleaning chemicals can be used satisfactorily with ultrasonics. Some are formulated especially for use with ultrasonics. However, the nonfoaming formulations normally used in spray washing applications should be avoided. Highly wetted formulations are preferred. Many of the new petroleum-cleaners, as well as petroleum and terpene-based semiaqueous cleaners, are compatible with ultrasonics. Use of these formulations may require some special equipment considerations, including increased ultrasonic power, to be effective.

Temperature was mentioned earlier as being important to achieving maximum cavitation. The effectiveness of the cleaning chemical is also related to temperature. Although the cavitation effect is maximized in pure water at a temperature of approximately 160°F, optimum cleaning is often seen at higher or lower temperatures because of the effect that temperature has on the cleaning chemical. As a general rule, each chemical will perform best at its recommended process temperature regardless of the temperature effect on the ultrasonics. For example, although the maximum ultrasonic effect is achieved at 160°F, most highly caustic cleaners are used at a temperature of 180°F–190°F because the chemical effect is greatly enhanced by the added temperature. Other cleaners may be found to break down and lose their effectiveness at these high temperatures; for example, some should not be used above 140°F. The best practice is to use a chemical at its maximum recommended temperature, but not exceeding 190°F (Figure 12.21).

Degassing of cleaning solutions is extremely important in achieving satisfactory cleaning results. Fresh solutions or solutions that have cooled must be degassed before proceeding with cleaning. Degassing is done after the chemical is added and is accomplished by operating the ultrasonic energy and raising the solution temperature. The time required for degassing varies considerably, based on tank capacity and solution temperature, and may range from several minutes for a small tank to an hour or more for

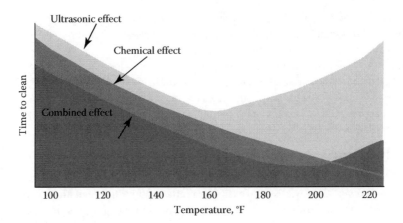

FIGURE 12.21   Temperature effect.

a large tank. An unheated tank may require several hours to degas. Degassing is complete when small bubbles of gas cannot be seen rising to the surface of the liquid and a pattern of ripples can be seen.

Ultrasonic Power

The ultrasonic power delivered to the cleaning tank must be adequate to cavitate that entire volume of liquid with the workload in place. Watts per gallon is a unit of measure often used to measure the level of ultrasonic power in a cleaning tank. As tank volume is increased, the number of watts per gallon required to achieve the required performance is reduced. Cleaning parts that are very massive or that have a high ratio of surface to mass may require additional ultrasonic power. Excessive power may cause cavitation erosion or "burning" on soft metal parts. If a wide variety of parts is to be cleaned in a single cleaning system, an ultrasonic power control is recommended to allow the power to be adjusted as required for various cleaning needs (Figure 12.22).

Part exposure to both the cleaning chemical and ultrasonic energy is important for effective cleaning. Care must be taken to ensure that all areas of the parts being cleaned are flooded with the cleaning liquid. Parts baskets and fixtures must be designed to allow penetration of ultrasonic energy and to position the parts to assure that they are exposed to the ultrasonic energy. It is often necessary to

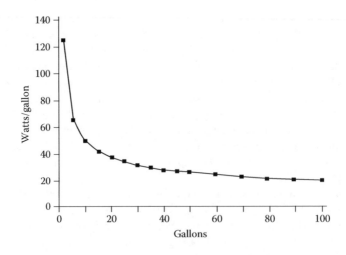

FIGURE 12.22

individually rack parts on a specific orientation or rotate them during the cleaning process to clean internal passages and blind holes thoroughly.

Conclusion

Properly utilized, ultrasonic energy can contribute significantly to the speed and effectiveness of many immersion cleaning and rinsing processes. It is especially beneficial in increasing the effectiveness of today's preferred aqueous cleaning chemistries and, in fact, is necessary in many application to achieve the desired level of cleanliness. With ultrasonics, aqueous chemistries can often give results surpassing those previously achieved using solvents. Ultrasonics is not a technology of the future—it is very much a technology of today.

13

Ultrasonic Cleaning Mechanism

Sami B. Awad
Ultrasonic Apps., LLC

Introduction

Ultrasonic cleaning is used in such diverse applications as automotive components, optics, disk drives, semiconductors, electronics, medical/pharmaceutical products, surface preparation for plating and precision coating, aerospace, general metals cleaning, precision bearings, and a variety of consumer products from jewelry to guns.

To understand the power and utility of ultrasonics, it is important to understand cavitation implosion.[1] This unique phenomenon occurs when high-energy ultrasonic waves (20 kHz to about 500 kHz, at about 0.3–1 W/cm^2) travel in a liquid or a solution. Ultrasonic waves interact with the liquid media to generate a highly dynamic agitated solution, producing microvapor/vacuum bubbles. The bubbles grow to maximum sizes inversely proportional to the applied ultrasonic frequency and then implode, releasing energy. The higher the frequency, the smaller the cavitation size and the lower the implosion energy.

Cavitation Formation Mechanism

The ultrasonic cleaning model (Figure 13.1) illustrates generation of cavitation through nucleation, growth, and violent collapse or implosion. Transient cavities (also referred to as vacuum bubbles or vapor voids), ranging from 50 to 150 μm in diameter at 25 kHz, are produced during the half cycles of the sound waves. During the rarefaction phase of the sound wave, the liquid molecules are extended outward against and beyond the liquid natural physical elasticity/bonding/attraction forces, generating vacuum nuclei, which continue to grow. A violent collapse occurs during the compression phase. It is

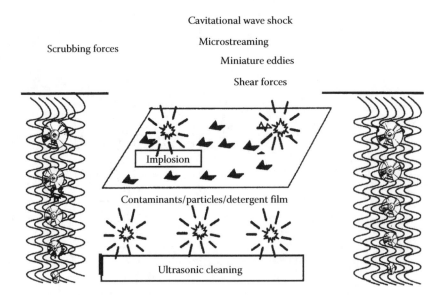

FIGURE 13.1 Scrubbing forces. (Reproduced from Awad, S.B., *Precision Cleaning*, 12, November 1996. With permission.)

believed that the compression phase is augmented by the enthalpy of the medium, the degree of mobility of the molecules, and the hydrostatic pressure of the medium.

Cavitation generates high forces in very brief bursts. Generation time of cavitation is in the order of microseconds. At 20 kHz, pressure is estimated at approximately 35–70 kPa, transient localized temperatures are about 5000°C, with the velocity of microstreaming around 400 km/h (Figure 13.2). A number of factors influence the intensity and abundance of cavitation in a given medium The ultrasonic waveform, frequency, and the power amplitude are important. Other influential factors include physical properties of the liquid medium (viscosity, surface tension, density, and vapor pressure); temperature; and liquid flow (static, dynamic, or laminar); and dissolved gases.

High-intensity ultrasonics can grow cavities to the maximum diameter prior to implosion in the course of a single cycle. At 20 kHz the bubble size is roughly 170 μm in diameter (see Figure 13.2). The vacuum bubble size becomes smaller at higher frequencies as a function of the wavelength. For example at 132 kHz it is estimated to be about half the size of cavitations generated at 68 kHz. At 68 kHz, the total time from nucleation to implosion is estimated to be about one third of that at 25 kHz.

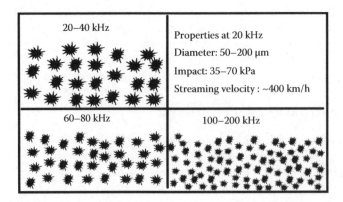

FIGURE 13.2 Ultrasonic frequency and cavitation size and population. (Reproduced from Awad, S.B., *Precision Cleaning*, 12, November 1996. With permission.)

At higher frequencies, the minimum amount of energy required to produce ultrasonic cavities is higher and must be above the cavitation threshold. In other words, the ultrasonic waves must have enough pressure amplitude to overcome the natural molecular bonding forces and the natural elasticity of the liquid medium in order to grow the cavities. For water at ambient temperature the minimum amount of energy needed to be above the threshold was found to be about 0.3 and 0.5 W/cm² of the transducer radiating surface for 20 and 40 kHz, respectively.

Matching the Frequency to the Process

Selecting the proper frequency for a particular application is critical. Estimates of cavitation abundance at various ultrasonic frequencies have shown that the number of cavitation sites is directly proportional to the ultrasonic frequency. For example, about 60%–70% more cavitation sites per unit volume of liquid are generated at 68 kHz than at 40 kHz. The average size of cavities is inversely proportional to the ultrasonic frequency. Therefore, one would expect that at the higher frequency, at a given energy level, the scrubbing intensity would be milder, particularly on soft and thin or delicate surfaces.

Because a lower number of cavitations of larger size and higher energy are generated at frequencies of 20–35 kHz, systems with lower frequencies are appropriate for cleaning large or heavy components. As the frequency increases, denser cavitation with moderate or low energies is formed. Therefore, frequencies of 60–80 kHz are recommended for delicate surfaces; frequencies of 132 and 200 kHz are recommended for cleaning ultradelicate and tiny components. The guidelines hold for both cleaning and rinsing.

Transducers

The transducers most commonly used for generating ultrasonic vibrations are piezoelectric, magnetostrictive, electromagnetic, pneumatic, and other mechanical devices. The piezoelectric transducer (PZT) is the most widely used technology in cleaning and welding applications. It offers a wide range of frequencies from about 20 kHz to the megasonic range.

PZT transducers are typically mounted on the bottom and/or sides of the cleaning tanks. The transducers can be mounted in various designs and sizes of sealed stainless steel containers or immersed in the cleaning solution/liquid (immersibles). Ultrasonic transducers should be placed on the longer sides and/or on the bottom of the tank, to provide maximum distribution of the sonic energy through the cleaning solution.

A new transducer design developed by Crest Ultrasonics[2] provides greater sound energy transmission with very low acoustic impedance at high frequencies. Benefits include high-quality surface cleanliness and efficient submicron particle removal.

Another recent design, the push–pull transducer rod, is an immersible transducer. The push–pull is made of two PZT transducers mounted on the ends of a titanium rod. The generated ultrasonic waves propagate perpendicularly to the resonating surface. The waves interact with liquid media to generate cavitation implosions.

Enhanced Transducers

Since its inception about 40 years ago, the conventional PZT transducer assembly has consisted of sandwiching a PZT crystal under compression between two metals. A newer design[2] was recently developed in which one or both metals are replaced with a ceramic material having twice or higher acoustic inductance. One important benefit is that the new transducer assembly produces sharply defined primary and tertiary resonant frequencies, including new ones not available using the classic design. A second improvement is a higher transmission coefficient of ultrasonic waves into liquid, estimated at 20%–30%.

The enhanced transducer design has been shown to improve cleaning efficiency. A study at Clarkson University (New York) by A. Busnaina et al.[3] compared one system with the conventional transducer design with a second with the enhanced transducer. Results indicate that, at 68 kHz, efficiency of removal of small particles from wafer substrates increases from 84% to 93% with the enhanced transducer system. Efficiency was also influenced by frequency; higher efficiency of particle removal (>97%) was observed at 132 kHz.

Precision Cleaning

Precision or critical cleaning of components or substrates is the *complete removal of undesirable contaminants to a preset level, without introducing new contaminants in the process.*[1] This preset level is typically the minimum level at which no adverse effects take place in a subsequent operation. In attempting to clean, it is critical not to introduce new contaminant(s). For example, in an aqueous cleaning process, it is important to have high-quality rinse water and a minimum of two rinse steps. Otherwise, new contaminants will be introduced by residual detergent and/or ionics in the rinse water. Recontamination of cleaned parts with outgassed residues produced from packaging or storing materials is another source of contamination.[1]

To meet production and quality demands, choosing the appropriate cleaning chemistry and process is essential. Rejected parts are the curse of the assembly line and improper cleaning methods are often to blame. Even beyond the factory floor, improper or inadequate cleaning of a component could directly affect warranty claims.[4,5]

Ultrasonic Cavitation and Surface Cleaning

The energy released from an implosion in close proximity to the surface collides with and fragments or disintegrates the contaminants, allowing the detergent or the cleaning solvent to displace them at a very fast rate. The implosion also produces dynamic pressure waves, which carry the fragments away from the surface. The implosion is also accompanied by high-speed microstreaming currents of the liquid molecules. The cumulative effect of millions of continuous tiny implosions in a liquid medium is what provides the necessary mechanical energy to break physically bonded contaminants, speed up the hydrolysis of chemically bonded ones, and enhance the solublization of ionic contaminants. The chemical composition of the medium is an important factor in speeding the removal rate of various contaminants.

Cleaning with ultrasonics offers several advantages over other conventional methods. Ultrasonic waves generate and evenly distribute cavitation implosions in a liquid medium. The released energies reach and penetrate crevices, blind holes, and areas that are inaccessible to other cleaning methods.[6,7] The removal of contaminants is consistent and uniform, regardless of the complexity and the geometry of the substrate.

Ultrasonic Cleaning Equipment

Ultrasonic aqueous batch cleaning equipment consists of at least four steps: ultrasonic wash, a minimum of two ultrasonic separate (or reverse cascading) water rinse tanks, and heated recirculated clean air for drying. The last drying step is not included if the postcleaning operation includes an aqueous process, as in electroplating or electroless plating. Ultrasonic transducers are bonded to the outside bottom surface, or to the outside of the sidewalls, or they are provided as immersibles and placed inside the tanks. Immersibles are usually the preferred method for large tanks. Two types of immersibles are commercially available in various sizes and frequencies. The first is the traditional sealed metal box containing a multitransducer system. The second is the cylindrical push–pull immersible, powered by two main transducers, one at each end.

Prior to selecting equipment, it is imperative that an effective cleaning process be developed. Then the number and size of the stations are determined based on required yield, total process time, and space limitation.

Typical tank size ranges from 10 to 2500 L based on the size of the parts, production throughput, and the required drying time. The tanks are typically constructed of corrosion resistant stainless steel or electropolished stainless steel. Titanium nitride or a similar coating such as hard chrome or zirconium is used to extend the lifetime of the radiating surface in the tanks or the immersible transducers.

Advantages of automation are numerous, including consistency, achieving desired throughput, and full control of process parameters.[8] Automation includes a computerized transport system able to run different processes for various parts simultaneously as well as data monitoring and acquisition.

The entire cleaning system can be enclosed to provide a clean room environment meeting Class 10,000 down to Class 100 clean room specifications. Process control and monitoring equipment consists of flow controls, chemical feed pumps, in-line particle counter, TOC (total organic carbon) measurement, pH, turbidity, conductivity, refractive index, etc.

The power requirement for most ultrasonic cleaning applications using PZTs, expressed in terms of electrical-input wattage to the transducers, ranges from 50 to 100 W/gal of cleaning fluid, or 2.8 to 3.6 W/in.[2] of transducer radiating surface.

Cleaning Chemistry

It is important to realize that the use of ultrasonics does not eliminate the need for the proper cleaning chemicals and implementing and maintaining the proper process parameters.[9]

Cleaning fluids are selected on the basis of the chemical and physical nature of the contaminants, substrate material(s), environmental considerations, and cleanliness specifications. Aqueous and solvent cleaning have advantages and disadvantages.[10] With appropriate additives, aqueous cleaning is universal and achieves better cleaning results.

Cleaning with ultrasonics using only plain water is workable, but only for short time. The question then is how long a system will work before cleaning action stops. The chemical composition of the cleaning medium is a critical factor in achieving the complete removal of various contaminants, without inflicting any damage to the components. In fact, cleaning is more complex than just extracting the contaminants from the component and moving them away from the surface. Soil loading and encapsulation/dispersion of contaminants are determining factors in the effective lifetime of the cleaning medium and therefore in effective cleaning of the part.

Requirements for the selected chemistry are many and no one chemistry is universal. For example, solvents are appropriate for removing organic contaminants but not for removing inorganic salts.[11] The solvent must cavitate well with ultrasonics and be compatible with components to be cleaned. Other properties such as wettability, stability, soil loading, oil separation, effectiveness, dispersion or encapsulation of solid residues, ability to rinse readily, and disposal considerations must be all addressed in choosing the appropriate chemistry. With so many factors to consider, an expert in the field may be better able to make this decision.

The role of additives in aqueous chemistries is multifaceted: to displace oils, to solubilize or emulsify organic contaminants, to encapsulate particles, and to disperse and prevent redeposition of contaminants. With appropriately formulated aqueous cleaning chemistries, ferrous and nonferrous metals (for example, aluminum, copper, brass, steel, and stainless steel) can be cleaned in the same bath without interaction.

Special additives are used to assist in the process of breaking chemical bonding, removal of oxides, preventing corrosion, enhancing the physical properties of the surfactants, and enhancing the surface finish. Ultrasonic rinsing with deionized water or reverse osmosis (RO) water is important to achieve spot-free surfaces. A minimum of two rinse steps is recommended. Drying and protection of steel components are valid concerns. However, the current available technologies offer effective ways to alleviate these concerns.

Contaminants

Three general classes of common contaminants are organic, inorganic, and particulate matter (organic, inorganic, or a mixture). Contaminants of any class may be water soluble or water insoluble.

Most organic contaminants such as oils, greases, waxes, polymers, paints, print, adhesives, or coatings are hydrophobic. Organic contaminants can be classified into three general classes: long-chain, medium-chain, and short-chain molecules. The physical and chemical characteristics are related to their structure and geometry.

Insoluble particulate contaminants can be divided into two groups, hydrophilic and hydrophobic. Examples of the first group include water-wettable particles, such as metals, metal oxides, minerals, and inorganic dusts. Examples of the second include non-water-wettable particles such as plastics, smoke and carbon, graphite dust, and organic chemical dusts. Similarly, substrate surfaces can be divided into hydrophilic and hydrophobic groups.

With few exceptions, inorganic materials or salts are insoluble in water-immiscible solvents. However, water-insoluble inorganics, such as polishing compounds made of oxides of aluminum, cerium, or zirconium, require a more elaborate cleaning process.

Mechanism of Cleaning

Two main steps take place in surface cleaning. The first is contaminant removal; the second is prevention of re-adherence. The removal of various contaminants involves different mechanisms, based on the nature and/or the class of the contaminant.

Organic contaminants are removed by two primary mechanisms. The first is solublization in an organic solvent. The second is by displacement with a surfactant film followed by encapsulation and dispersion.

The mechanism of removal of organic contaminants by detergent involves wetting both contaminant and substrate. According to Young's equation, wetting increases the contact angle (θ) between the contaminant and the surface, thus decreasing the surface area wetted with the hydrophobe, and reducing the scrubbing energy needed for removal (Figure 13.3).

$$\cos\theta = \frac{\gamma_{SB} - \gamma_{SO}}{\gamma_{OB}}$$

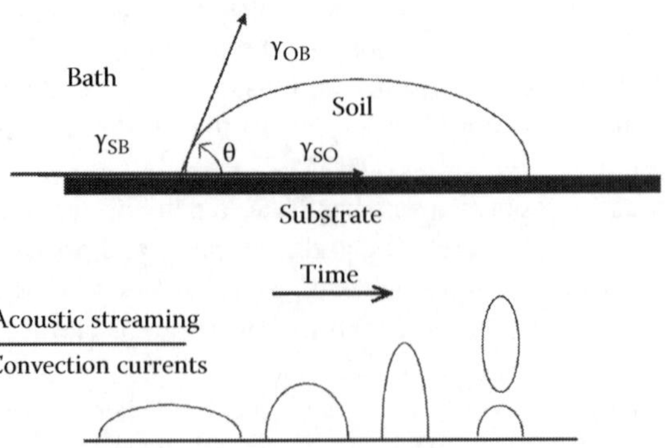

FIGURE 13.3 Liquid soil removal.

Aqueous additives contain one or more surfactants. Surfactants are long-chain organic molecules with polar and nonpolar sections. Surfactants may be ionic or nonionic. When diluted with water, surfactants form aggregates called micelles at a level above the critical micelle concentration (CMC). The micelles, composed of aggregates of hydrophilic and hydrophobic moieties, act as a solvent encapsulating the contaminants, thus preventing redeposition.

Ultrasonic cavitation plays an important role in removal of hydrophobic contaminants. The shock wave (and the microstreaming currents) greatly speed up the breaking of adhered contaminants, enhancing displacement with the detergent film. The contaminants are then encapsulated in the micellic aggregates, thus preventing redeposition. The net result is that ultrasonic cavitation accelerates displacement of contaminants from the surface of the substrate and facilitates their dispersion.

Cleaning Chemistry and Particles

Theoretically, adhesion forces, including van der Waals, electrical double layer, capillary, and electrostatic, are directly proportional to the size of the particle. One would expect the energy of detachment to decrease with the size of particles. However, smaller particles are always more difficult to detach, mainly because small particles tend to get trapped in the valleys of a rough surface.

According to the Gibbs adsorption equation, the mechanism of particle removal involves shifting the free energy of detachment to slightly above or less than zero. Surfactants play a very important role in decreasing the adsorption at particle and substrate interfaces.

Ultrasonic cavitation provides the agitation energy for detachment (i.e., the removal force). At 40 kHz, the detachment or removal efficiency of 1-μm particles is 88%. Efficiency increases to 95% at high frequencies (60–70 kHz), equaling the efficiency of megasonics of approximately 850 kHz. This is expected in light of the fact that cavitation size is smaller at higher frequencies and can reach deeper into the surface valleys. One would then anticipate that a combination of high-frequency ultrasonics at 65–70 kHz with appropriate chemistry would further improve efficiency of particle removal.

Inhibiting redeposition of contaminants involves formation of a barrier between the suspended contaminant and the cleaned surface. In solvent cleaning, a film of solvent adsorbed to both substrate and contaminant forms the barrier. In aqueous cleaning, an effective surfactant system encapsulates contaminants in the micellic structure as depicted in Figure 13.4. Redeposition of the encapsulated contaminants (soils) is prevented via stearic hindrance (nonionic surfactants) or via electrical repulsion (anionic surfactants).

Depending on the surfactant system, encapsulation can be permanent or transient. Transient encapsulation is preferable to emulsification, as it allows better filtration and/or phase separation of contaminants. Allowing soil loading to reach the saturation point significantly decreases cleaning agent efficiency; cleaning action may cease. To ensure consistent cleaning, the dispersed contaminants must be removed by continuous filtration or separation of contaminants, and the recommended concentration of the cleaning chemical must be maintained.

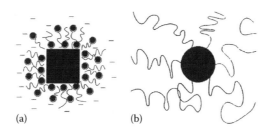

(a)   (b)

FIGURE 13.4 Antiredeposition: (a) ionic surfactant and (b) nonionic surfactant.

The physical properties of the substrate, including surface finish, are important factors in submicron particle removal.[12,13] For example, a silicon wafer surface differs from that of an aluminum disk, in their physics, topography, and finish. The inherent static charges of plastics are another challenge when dealing with submicron particles.

Conclusion

Cleaning with the assistance of ultrasonic cavitation has numerous advantages, most importantly consistency in results.

Advantages:

- Efficient cleaning in recessed areas and blind holes
- Capability of cleaning assemblies or devices
- Removal of micro- and submicrocontaminants
- The proper chemistry → exceptional and consistent cleaning
- Shorter process time
- Full automation and controls, batch and continuous processes

For best cleaning results, selection of the ultrasonic frequency or the cleaning medium (solvent or aqueous) for an application must be precise and specific.

References

1. S.B. Awad, Ultrasonic cavitations and precision cleaning, *Precision Cleaning,* November 1996, p. 12.
2. J.M. Goodson, U.S. Patent 05,748,566.
3. A. Busnaina et al., Microcontamination Research Lab, Clarkson University, Potsdam, NY, 1998, 1999, results to be published elsewhere.
4. H.A. Bhatt, How now, *Parts Cleaning,* May 1998, p. 17.
5. J.B. Durkee, *The Parts Cleaning Handbook,* Gardner Pub. Inc., Cincinnati, OH, 1994.
6. M. O'Donoghue, The ultrasonic cleaning process, *Microcontamination,* 2(5), 62–67, 1984.
7. F.J. Fuchs, Ultrasonic cleaning principles for parts cleaning potential, *Parts Cleaning Mag.,* December 1997, p. 14.
8. J. Harmon, Ultrasonic applications in the life sciences, *A^2C^2 Mag.,* March 1999, p. 7.
9. S.B. Awad, Ultrasonic cleaning of medical and pharmaceutical devices and equipment, *A^2C^2 Mag.,* February 2000.
10. S.S. Seelig, The chemical aspects of cleaning, *Precision Cleaning,* 1995, p. 33.
11. B. Kanegsberg, Aqueous cleaning for high-value processes, *A^2C^2 Mag.,* September 1999, p. 25.
12. S.B. Awad, Ultrasonic aqueous cleaning and particle removal of disk drive components, *Datatech,* 1999, p. 59.
13. K.L. Mittal, Surface contamination concepts and concerns, *Precision Cleaning,* 3(1), 17, 1995.

14

Ultrasonic Cleaning with Two Frequencies

K.R. Gopi
Crest Ultrasonics Corp.

Sami B. Awad
Ultrasonic Apps, LLC

Introduction

Ultrasonic cleaning is very efficient when compared to other conventional cleaning methods. In various industries, the method of single frequency (SF) per cleaning tank is still being used in the majority of ultrasonic applications. Typical frequencies are in the range of 20–470 kHz.

An active interest in using two frequencies or more for cleaning enhancement started in the late 1990s with a combination of two individual frequencies below 100 kHz in one tank [1]. The technology found limited use in some applications. With the recent developments of advanced higher frequencies in the range of 100–500 kHz, a novel design was commercially introduced by Crest Ultrasonics Corporation [2]. The transducers are arranged in equilateral triangular patterns along diagonal lines on a wall of the tank so that each transducer has an adjacent transducer of a different frequency. Different combinations of low/high and high/high frequencies were designed and tested. This technology was named the dual frequency (DF).

Recently, it was reported [3] that multifrequency effects are more than the sum of single-frequency effects. In theory, a combination of low/high frequencies should effectively combine the known characteristics of both frequencies, namely, cavitation and acoustic streaming. This should enhance the cleaning process. Three frequency combinations were studied. They were 58/132, 58/192, and 132/92 kHz. The study included using specific contaminants to evaluate the cleaning efficiencies of these DF systems.

Ultrasonic Cleaning Mechanism

Two main principal mechanisms are involved in the ultrasonic cleaning process:

1. Cavitation
2. Acoustic streaming

Technically, both cavitation and acoustic streaming exist side by side in all frequencies but at different ratios. Empirically, cavitation intensity is much higher at low frequencies, moderates at higher frequencies, and almost diminishes in the megasonic range.

When an ultrasonic wave is projected in liquid, negative pressure is created and causes the liquid to "fracture," leading to a phenomenon known as cavitation [4]. As the ultrasonic waves continue, micro bubbles are created and continue to grow to a critical volume and become unstable, eventually collapsing with a violent implosion. These implosions radiate high-powered shockwaves that dissipate repeatedly at 25,000–30,000 times per second. Additionally, the implosion of the cavitation bubbles creates temperature estimated to exceed 5000°C and pressure that exceeds 700 kg/cm^2. Violent bubble collapse produces a microscopic jet of liquid that can impinge on the surface of the parts to be cleaned. These high-velocity jets remove particles from surfaces and simultaneously allow cleaning chemicals to interact with organic and inorganic contaminants on the surface.

Acoustic streaming can be defined as steady flow generated by the propagation of acoustic waves in a viscous fluid. It is due to transfer of momentum and energy of the acoustic field to the medium through its attenuation coefficient [5]. In acoustic streaming, bulk movement of fluid occurs [6]. Acoustic streaming can penetrate through the boundary layer of motionless fluid that surrounds all of the surfaces in the ultrasonic tank. Particles removed by cavitation action are swept away from the surface by streaming action. At low frequencies, cavitation is dominant; and fluid motion is randomized and omnidirectional.

Cavitation and acoustic streaming work together in all forms of ultrasonic cleaning, but the relative contribution of each is a function of frequency.

Coupling of the two phenomena in a single process helps to exploit the full effect of cavitation on cleaning with no damage to delicate components. Our systematic investigation has demonstrated that the surface cleaning efficiency of DF systems is significantly better than that which can be achieved with a SF. This superior efficiency has been demonstrated in applications ranging from removal of submicron contamination to cleaning large metal surfaces with stubborn stains.

Dual-Frequency Equipment

Ultrasonic cleaning systems consist of four fundamental components: transducer, generator, tank, and liquid medium. Most cleaning tanks are made of stainless steel. The piezo-ceramic transducers are typically bonded to the bottom or the side wall. SF systems consist of one type of transducer bonded to the bottom of the tank. In the DF systems, two specific transducer designs are diagonally placed as shown in Figure 14.1. Each shade indicates one frequency. A repeated equilateral triangularity is achieved, with both specific frequencies in every triangle. Each set of transducers is powered by a separate generator. This design provides maximum efficiency, equilateral distribution, and minimum interferences. This design provides the options of using the ultrasonics tank in three ways, either in two individual frequencies or the two combined.

Cleaning Comparisons

The objective of this study is to evaluate the efficiency of the DF system (58/132 kHz) versus SF system (e.g., 40, 58, 68, 192 kHz) as indicated in experiments below. Three different substrates with different contaminants were evaluated.

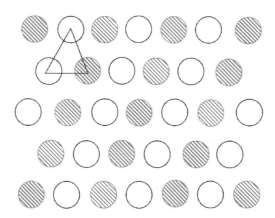

FIGURE 14.1 Displays the transducer stacking pattern of a DF system.

High-Precision Stainless Steel, Particles, and Fingerprints

Voice coil parts made of high-precision stainless steel and contaminated with particles and fingerprints were chosen for this study. The contamination level of the particle ranges from 0.2 to 1 μm. Prior to cleaning process evaluations, the particle contamination on parts, as received, is quantified by the liquid particle counter (LPC). A flow type liquid particle counter (LPC) LS-200™, which senses the size range from 0.2 to 2.0 μm, was used for measurement. Deionized water (16 MΩ resistance) was used as cleaning medium. The test pieces were stacked in a washing jig and subjected to two different cleaning processes. Process 1 consists of a two-stage ultrasonic cleaning process with SF system in the first stage (40 kHz, 1000 W) and a DF system in the second stage (58/132 kHz, 1000 W). Process 2 consists of a three-stage ultrasonic cleaning process with a SF system in each of the three stages (40 kHz, 1000 W; 68 kHz, 1000 W; and 192 kHz, 1000 W). For both processes, ultrasonic cleaning was followed by spray rinse and drying steps. In this instance, efficacy of cleaning was judged based on the decrease in particle counts before and after cleaning.

Epoxy Contamination of Aluminum Parts

A second process comparison utilized aluminum parts with epoxy contamination. The test piece was stacked in a washing jig and subjected to two different cleaning processes. Chem Crest 121 (NMP [*n*-methyl pyrrolidone] based) 100% solution was used as cleaning medium. In Process 1, ultrasonic immersion cleaning was performed using a SF system (58 kHz, 1000 W); Process 2 utilized a DF system (58/132 kHz, 1000 W) as the wash step. Both processes 1 and 2 utilized deionized (DI) water rinse and drying. Efficacy of cleaning was judged based on visual examination, before and after cleaning.

Adherent Carbon Stains on Large Titanium Plates

A third cleaning comparison involved removal of stubborn baked carbon stains from large titanium plates (8 in. × 22 in. × 0.79 in.). The frequency protocols were 40 kHz, 1000 W; 58 kHz, 1000 W; versus 58/132 kHz, 1000 W in a single tank. The cleaning medium was 275™ aqueous detergent (30% concentration). The ultrasonic cleaning processes were followed by identical rinsing and drying operations. Here, cleaning efficiency was judged visually based on removal of carbon particles from the substrate.

Results and Discussion

High-Precision Stainless Steel, Particles, and Fingerprints

It is evident from the result displayed in the Figure 14.2 that there was a significantly higher drop in the level of particle contamination level after cleaning by the process involving DF system when compared to process involving SF system.

In this case, use of the DF turns out to be an advantage in terms of shortening the process time and optimizing floor space. The time consumed for the two-stage ultrasonic cleaning process involving DF system in the second stage is 18 min. By contrast, the time consumed for the process involving a SF system in all three stages is 25 min.

Epoxy Contamination of Aluminum Parts

A visual comparison of Figure 14.3a and b illustrates the effective cleaning efficiency of DF system as compared with the SF under identical test conditions. The parts cleaned using the DF system show better contamination removal and enhanced surface finish as compared with the SF system. The total process time consumed for the both process is 1 h 6 min.

Adherent Carbon Stains on Large Titanium Plates

Three titanium plates with stubborn carbon stains were chosen for ultrasonic cleaning evaluations. DF ultrasonic cleaning has successfully replaced manual cleaning. The manual cleaning method proved to be unproductive because it did not remove undesirable stains and because the substrates became scarred. Here, cleanliness means perfectly free from carbon particles. Cleanliness was judged visually before and after cleaning process as shown in Figure 14.4a and b.

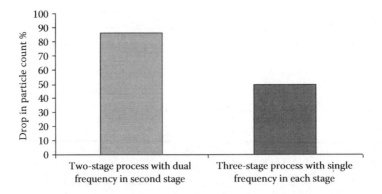

FIGURE 14.2   Liquid particle drop versus process involving DF ultrasonic system and SF ultrasonic system.

FIGURE 14.3   Aluminum parts cleaned using (a) SF (58 kHz) and (b) DF (58/132 kHz).

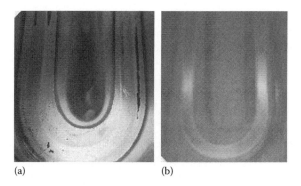

FIGURE 14.4 (a) Before ultrasonic cleaning. (b) After cleaning using DF ultrasonic system.

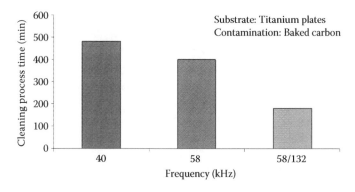

FIGURE 14.5 Comparison of cleaning time required to remove baked carbon from titanium using two SFs or DF in a single tank.

Good surface cleaning was achieved with a significantly shorter process time; when cleaning was performed using DF ultrasonic cleaner (58/132 kHz) compared with cleaning at a SF at either 40 or 58 kHz under identical test conditions (see Figure 14.5).

Conclusion

It is evident that the DF system (58/132 kHz) displays significantly higher cleaning efficiency for the tested substrates/contaminants. Advantages of the combined frequencies include a decrease in the process time, enhancement of the cleanliness level, and optimizing the equipment configuration by reducing the number of the cleaning steps.

Glossary

CDA   Compressed discharged air
DI   Deionized
HDD   Hard disk drive
kg   kilogram
kHz   kilohertz
LPC   Liquid particle counter
MΩ   Mega ohm
μm   Micrometer
W   Watts

References

1. M. Pedziwiatr and E. Pedziwiatr, Ultrasonic Cleaning Apparatus, U.S. Patent number 5,865,199 (1999).
2. M. Goodson, Ultrasonic Processing Method and Apparatus with Multiple Frequency Transducers, U.S. Patent number 7,247,977 (2007).
3. R. Feng, Y. Zhao, C. Zhu, and T.J. Manson. Enhancement of ultrasonics cavitation yield by multi-frequency sonication. *Ultrasonics Sonochemistry,* 9, 2002, 231–236.
4. K.S. Suslick, Sounding out new chemistry. *New Scientist,* 1702, 1990, 50–53.
5. G. Madelin, D. Grucker, J.-M. Franconi, and E. Thiaudiere, Magnetic resonance imaging of acoustic streaming: Absorption coefficient and acoustic field shape estimation. *Ultrasonics,* 44, 2006, 272–278.
6. J. Lighthill, Acoustic streaming. *Journal of Sound and Vibration,* 61(3), 1978, 391–418.

15

Megasonic Cleaning Action

Mark Beck
Product Systems Inc.

Introduction

Megasonics has been a widely accepted cleaning method for contamination-sensitive products for nearly 20 years. Megasonics was initially developed in the early 1940s as a result of U.S. Navy research into advanced sonar instrumentation for antisubmarine warfare. In the late 1970s, RCA adapted this technology for wafer cleaning, and by 1982 commercial megasonic cleaning equipment was being delivered to the semiconductor industry.

More recently, advances have been made in this acoustic cleaning technology, through a better understanding of high-frequency acoustic streaming, and controlled acoustic cavitation, the megasonic cleaning technique has proved effective for removing submicron particles from silicon and other substrates without damage. As a result, growing numbers of manufacturers in the integrated circuit, hard drive, raw silicon, mask, flat panel display, and other industries have been turning to megasonic cleaning to help meet stringent cleaning requirements. Megasonic cleaning is now increasingly accepted by industry as a cost-effective, efficient, and safe method for the removal of nanoscale particles from contamination-sensitive products.

Overview of Megasonic Cleaning

Megasonics utilizes the piezoelectric effect at high frequencies to generate controlled acoustic waves in a liquid bath to enable removal of submicron particles from substrates.

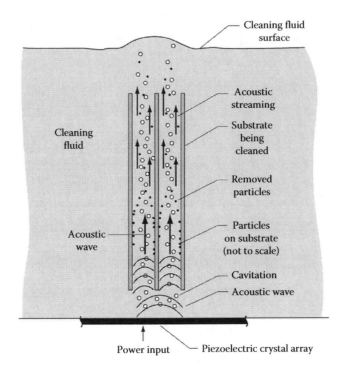

FIGURE 15.1 Megasonic cleaning uses the piezoelectric effect to produce acoustic waves that move through the cleaning liquid.

In megasonic cleaning (Figure 15.1), a piezoelectric crystal array transducer converts alternating electrical energy directly to mechanical energy using the piezoelectric effect, in which certain materials change dimension when an electrical charge is applied. A ceramic piezoelectric crystal is excited by high-frequency AC voltage, between 500 and 2000 kHz, causing the ceramic material to change dimension rapidly, or vibrate. These vibrations are transmitted by the resonant masses of the transducer, and directed into the liquid through a resonating plate, producing acoustic waves in the cleaning fluid. Acoustic cavitation, produced by pressure variations in the sound waves moving through the liquid, and the effects of acoustic streaming cause particles to be removed from the material being cleaned.

Megasonic Cleaning Compared with Ultrasonic Cleaning

There are two types of acoustic cavitation: transient cavitation and stable, or controlled, cavitation. Ultrasonic cleaning frequencies, between 20 and 350 kHz, produce transient acoustic cavitation. Transient cavitation is characterized by transient bubbles that exist for only a few acoustic cycles, after which they collapse violently, producing very high local temperature and pressure. Transient acoustic cavitation generates shock waves that are powerful enough to erode solid surfaces nearby,[1] and to damage some substrate surfaces.

Megasonic cleaning operates at much higher frequencies, 500–2000 kHz, which produce controlled acoustic cavitation. Controlled cavitation is characterized by stable bubbles that are relatively permanent, can exist for many acoustic cycles, and do not cause damage to substrate surfaces,[2] because the cavitation radii are much smaller at higher frequencies and have less energy upon collapse. Thus, megasonic-controlled acoustic cavitation is best suited for sensitive substrate surfaces that cannot withstand the heat and pressure of transient cavitation.

In addition, ultrasonics simultaneously cleans all surfaces of a submerged object. This means that ultrasonic cleaning subjects all areas of the substrate, including areas that may not need to be cleaned to the previously described effects of transient acoustic cavitation. Megasonics accomplishes line-of-sight cleaning; it affects only those surfaces of the object that are in the path of the acoustic wave.

Applications of Megasonics and Ultrasonics

The mechanical effects of both ultrasonic and megasonic cleaning can be helpful in speeding particle dissolution and in displacing particles. Both ultrasonics and megasonics have also been demonstrated to speed or enhance the effect of many chemical reactions. In addition, residual cleaning chemicals can be removed quickly and completely by either ultrasonic or megasonic rinsing. However, there are applications for which megasonic cleaning clearly would be favored.

The effects of ultrasonics and megasonics on substrate surfaces and particle removal results provide the basis for identifying the best applications for each process. Ultrasonic cleaning is most appropriate for strong, heat-tolerant substrate materials requiring multisurface cleaning. Ultrasonics is also well suited for the removal and/or dissolution of large particles from chemically tolerant substrates.

Megasonics is most appropriate for heat- or chemical-sensitive substrates that cannot withstand the heat and pressure of transient cavitation and for applications requiring line-of-sight-dependent cleaning. Parts that cannot be cleaned with ultrasonics, because they are sensitive to the frequency or transient cavitation effects can often be cleaned with megasonics. Megasonics cleaning is also the application of choice for the removal and/or dissolution of small particles (less than 0.3 µm; Figure 15.2). For example, this cleaning technique has been proved effective for removing 0.15 µm particles from silicon wafers and other cavitation-sensitive products, without causing substrate damage.

Table 15.1 summarizes the relative strengths of megasonic and ultrasonic cleaning.

Positive Environmental Effects of Megasonics

The use of megasonic cleaning yields several positive environmental results. The high pressure and temperatures produced by ultrasonic cleaning result in the evaporation of large volumes of both chemicals

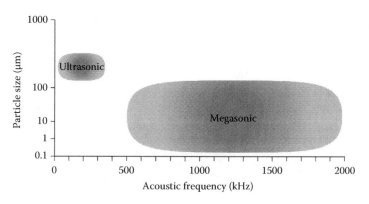

FIGURE 15.2   Particle size vs. frequency for megasonic and ultrasonic cleaning.

TABLE 15.1   Strengths of Megasonics and Ultrasonics

Applications for Megasonic Cleaning	Applications for Ultrasonic Cleaning
Cavitation-sensitive substrates	Strong substrates
Small particle removal/dissolution (>0.3 µm)	Larger particle removal/dissolution
Chemically sensitive substrates	Chemically tolerant substrates
Line-of-sight-dependent cleaning	Multisurface cleaning
Heat-sensitive material	Heat-tolerant material

and ultrapure water. This has two negative effects. First, the chemical compositions of cleaning solutions cannot be maintained at constant levels. Second, large amounts of chemical vapors are released, increasing the loads on clean-air exhaust systems.

The lower pressures and temperatures produced by megasonic cleaning enable processes that drastically reduce both chemical vapor evaporation and the load on air exhaust and replacement systems.

In addition to their environmental benefits, megasonic cleaning methods optimize the use of cleaning fluids and reduce the costs associated with the acquisition and disposal of toxic substances.

Discussion of Underlying Physics

Properties of Piezoelectric Transducers

A basic characteristic of the piezoelectric crystal is that when a sine wave is applied to it, through the application of AC voltage, it expands. For the purpose of megasonic cleaning, the molecules in the piezoelectric crystal have been aligned, or poled, in the thickness extension mode. Upon application of the sine wave, the first expansion takes place to the side of the crystal, a second expansion takes place to the end of the crystal, and the third expansion takes place in the thickness of the crystal (Figure 15.3).

The frequency at which this third expansion, which is the first thickness expansion, takes place is known as the fundamental frequency. The fundamental frequency occurs at approximately 1000 kHz (Figure 15.4), with harmonic frequencies at 3 and 5 MHz.

The most efficient transfer of the energy generated by crystal expansion would occur from direct contact between the piezoelectric crystal and the liquid bath. However, cleaning solutions can damage the piezoelectric crystal, and the piezoelectric crystal is not pure and can add impurities to the cleaning solution. To prevent this, a resonator is adhered to the top of the crystal, between the crystal and the fluid.

Ideally, the resonator should have no effect on the energy being transferred, that is, there would be no energy loss or frequency distortion. The velocity of sound in the resonator is an important factor in

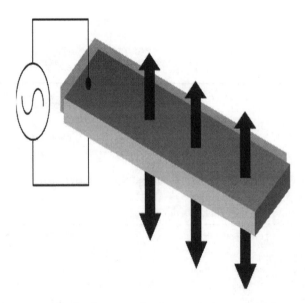

FIGURE 15.3 Piezoelectric crystal transducer rapidly changes dimensions or vibrates with the application of high-frequency AC voltage.

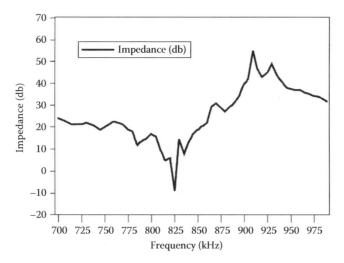

FIGURE 15.4 Piezoelectric transducer characteristic of impedance as a function of frequency. Impedance changes near fundamental frequencies.

approaching this ideal. Based on acoustic KLM transmission theory, the resonator should be designed to be half the wavelength thickness and should operate optimally at the fundamental frequency.

$$\lambda = \frac{V_L}{f}$$

where
   λ is the wavelength of sound in resonator
   V_L is the velocity of sound in the resonator, mm/s
   f is the frequency of sound, MHz

Particle Attraction and Removal Forces

Megasonics cleaning is able to overcome the attraction forces that hold very small particles to a surface. Particle adhesion force is a function of the type of medium surrounding the particle and the surface. In general, it is weaker in liquid media than in gas media,[3] and it increases linearly with an increase in particle diameter[1] (Figure 15.5).

Although adhesion force is lower at smaller particle diameters, small particles are more difficult to remove. The weight of the particle decreases as a function of the diameter cubed, and for small particles the adhesion force can easily exceed the gravitational force by a factor of 10^3 or more.[1,3] In addition, van der Waals attractive forces (Fvw) vary depending upon the composition of the particle. They are about ten times larger for silicon particles than for polystyrene latex (PSL) particles, for example, indicating that some particles may require much greater removal forces than others.[1]

Principal Mechanisms of Megasonic Cleaning

An understanding of exactly how particles are removed when megasonic cleaning techniques are used has been the subject of increased investigation during the past decade. To date, researchers have not been able to explain precisely why megasonics works, and there is disagreement on whether controlled acoustic cavitation or acoustic streaming is the more effective mechanism in removing particles. The

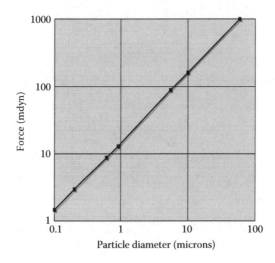

FIGURE 15.5   van der Waals forces vs. particle size.

effect of various parameters on particle removal can be determined, but whether removal is the result of acoustic cavitation or acoustic streaming, or both, is not always clear.[2]

However, it is accepted that controlled acoustic cavitation, acoustic streaming, and reduction of the boundary layer are the principal particle removal mechanisms in megasonic cleaning.

Acoustic Cavitation

Acoustic cavitation is the generation and action of cavities, or bubbles, in a liquid. Acoustic waves moving through a liquid produce variations in the liquid pressure. When the liquid pressure drops momentarily below the vapor pressure during the low-pressure portion of the acoustic wave, small evacuated areas, or cavities, are formed that quickly become filled with gas (a foreign contaminant such as dissolved oxygen or air) and/or vapor (a gaseous form of the surrounding liquid).[4] These tiny bubbles are set in motion by the acoustic wave. The bubbles may be suspended in the liquid medium, or they may become trapped in voids either in the boundary surface of the liquid or in solid particles suspended in the liquid.

The tiny bubbles can expand and contract in the liquid. Bubble expansion can be caused by reducing the ambient pressure in the liquid, either by static or dynamic means. The bubbles can then become large enough to be seen by the unaided eye. The bubbles may contain gas or vapor or a mixture of both. If the bubbles contain gas, then their expansion can be caused by rectified diffusion, pressure reduction, or an increase in temperature.[3]

Rectified diffusion is the diffusion of dissolved gas from the liquid into the bubble, and vice versa, with the pressure oscillations resulting in a net diffusion into the bubble. This net inward diffusion occurs because the bubble surface area increases during inward diffusion and decreases during outward diffusion; a higher surface area leads to more diffusion.[2] If the ambient liquid is not saturated with gas, then rectified diffusion must compete with ordinary diffusion from the bubble to the liquid. In that case, the sound pressure amplitude must exceed a certain value in order for the bubbles to increase significantly in size.[2]

The pressure oscillations that created the bubbles can also cause them to expand and contract. If the pressure variation is great enough to reduce the local liquid pressure down to, or below, the vapor pressure in the negative parts of the acoustic cycle moving through the liquid, any minute cavities or bubbles that are present will grow larger. If the range of the pressure variation is increased to produce zero and then negative pressures locally in the liquid, then bubble growth is increased. Gas from the liquid diffuses into a bubble during expansion, and leaves the bubble during contraction.

When the bubble reaches a size that can no longer be sustained by its surface tension, the bubble will expand and then collapse, or implode, which is an important action of the cavitation phenomenon. The bubble action of cavitation has sufficient energy to overcome particle adhesion forces and to dislodge particulates attached to substrates in the stream of bubbles. Essentially, imploding cavitation bubbles generate shock waves that dislodge particles from substrate surfaces. Cavitation breaks down the molecular force by which a particle is held to a surface either by direct impact from bubble implosion or by the fatiguing action caused by repeated bombardment.[3]

Cavitation implosion force varies with the size and contents of the bubble. Larger bubbles are unstable and implode with larger force; smaller bubbles are stable and collapse with less force. Vapor collapses more quickly, resulting in larger implosion force, whereas gas cushions and slows the collapse, resulting in smaller implosion force.

Cavitation does not occur until a specific threshold is reached.[3] The cavitation threshold is defined as the minimum pressure amplitude required to induce cavitation.[2]

A number of methods have been developed for detecting cavitation, including acoustic emissions, visual observations, sonoluminescence (SL), and surface erosion. Of these, SL is believed to be the most suitable method for characterizing cavitation in a megasonic tank,[1] because it is related to the cavitation collapse of bubbles.

The intensity and effect of cavitation on materials being cleaned are related to the type of acoustic cavitation produced. Two types of acoustic cavitation have been identified and studied: transient cavitation and stable cavitation. Transient acoustic cavitation is produced by ultrasonic cleaning frequencies, between 20 and 350 kHz, which transform low-energy-density sound waves into high-energy-density collapsing bubbles. In transient cavitation, the mostly vapor-filled bubbles exist for only a few acoustic cycles, followed by a rapid and violent collapse. This type of cavitation is likely to produce violent events in the acoustic field, such as radiation of light (SL) and shock waves. The level of violence produced is believed to be dependent on the maximum size of transient bubbles, which is related to the acoustic frequency.[1] Because transient cavitation concentrates energy into very small volumes and tends to produce very high local temperatures and pressure, it can cause surface erosion and damage to sensitive substrates.

Bubble size decreases as acoustic frequency increases, and the smaller the maximum bubble size, the less violent the cavitation produced.[1] The high frequencies used in megasonic cleaning, 500 to 2000 kHz, produce controlled acoustic cavitation, which is characterized by mostly small, gas-filled cavities. Unlike the violent implosion associated with vapor-filled cavities in transient cavitation, controlled cavitation bubbles exhibit less violent collapse,[4] producing lower temperatures and pressure. As a result, megasonic cleaning substantially minimizes surface erosion and damage to substrates being cleaned. Stable cavitation produces light in the visible range (violet), while the light produced by transient cavitation is primarily in the ultraviolet range (with a peak at 270–290 nm).[4]

The bubble action of controlled acoustic cavitation is believed to be a primary particle removal mechanism in megasonic cleaning.

Acoustic Streaming

Acoustic streaming is considered another primary particle removal mechanism of megasonic cleaning. Acoustic streaming is time-independent fluid motion generated by a sound field. This motion is caused by the loss of acoustic momentum by attenuation or absorption of a sound beam. Acoustic streaming enhances particle dissolution and the transport of detached particles away from surfaces,[4] thereby decreasing particle redeposition. It also produces a much thinner boundary layer (less than 1 μm) than would be found in a cleaning tank without megasonics.

Acoustic streaming velocity is a function of energy intensity, geometry, energy absorption, liquid density and viscosity, and sound speed in the liquid. Streaming velocity has been found to increase linearly with acoustic intensity (power). Velocity also increases linearly with frequency. Streaming velocity also decreases with distance from the source, due to attenuation.[2]

Acoustic streaming comprises several important effects: (1) bulk motion of the liquid, (2) microstreaming, and (3) streaming inside the boundary layer.

The primary effect of acoustic streaming is bulk motion of the liquid, the strong localized flow of cleaning solution. The shear force of the bulk liquid motion is the primary particle removal agent. In a closed tank, forces due to sound pressure variation create this bulk fluid motion, which carries particles away from the substrate once the molecular attraction of the particle to the surface is broken and the particle is dislodged. Bulk fluid motion increases linearly with acoustic intensity. The bulk fluid motion shear force combines with the other effects of acoustic streaming to increase particle removal.

A second effect of acoustic streaming is microstreaming. Microstreaming, also known as Eckart streaming, occurs near oscillating bubbles, or any compressible substance in the liquid. Microstreaming occurs at the substrate surface, outside the boundary layer, because of the action of bubbles as acoustic lenses that focus sound power in the immediate vicinity of the bubble. This is a powerful type of streaming, in which the bubbles scatter sound waves and generate remarkably swift currents in localized regions. The currents are most pronounced near bubbles that are undergoing volume resonance and are located along solid boundaries. Microstreaming aids in dislodging particles and contributes to megasonic cleaning.[4]

Most of the flow induced by acoustic streaming occurs in the bulk liquid outside the boundary layer. However, there is a third effect of acoustic streaming, called Schlichting streaming, which is associated with cavitation collapse and is believed to assist in the removal of small particles and their transport away from surfaces. Schlichting streaming occurs outside the boundary layer and is characterized by very high local velocity and vortex (rotational) motion. The vortices are of a scale much smaller than the wavelength. Schlichting streaming results from interactions with a solid boundary. Steady viscous stresses are exerted on the boundaries where this type of rotational motion occurs, and these stresses may contribute significantly to removal of surface layers.[2]

The combined effects of acoustic streaming produced in megasonic cleaning may slide, roll, or lift a particle from its initial position on a substrate, depending on the size and shape of the particle, as well as the nature of the hydrodynamic force being applied. Acoustic streaming, both inside and outside the boundary layer, clearly enhances cleaning and other chemical reactions. Particle transport is aided significantly by the strong currents and small boundary layer thicknesses that result from acoustic streaming.[2]

Significance of the Boundary Layer

During megasonic cleaning, the cleaning solution flows swiftly past the substrate being cleaned, forcing chemistry into contact with contaminant particles, removing them from the surface, and carrying them away. On a microscopic scale, during acoustic cleaning, fluid friction at the surface of the substrate being cleaned causes a thin layer of solution to move more slowly than the bulk solution. This layer of slow-moving fluid at the surface is called the boundary layer (Figure 15.6). The boundary layer effectively shields the substrate surface from fresh chemistry and shields contaminant particles from the removal forces of the bulk fluid.

Within the boundary layer, van der Waals attractive forces have been shown to be substantially stronger than the removal forces that result from acoustic pressure oscillations, acoustic velocity oscillations, or bulk fluid motion associated with acoustic streaming.

Megasonic cleaning has proved especially effective at removing submicron particles in part, because it reduces the boundary layer. The higher frequencies of megasonic cleaning reduce the boundary layer to less than 0.5 μm, compared to the boundary layer of 2.5 μm produced by ultrasonic cleaning frequencies. The primary effect of acoustic streaming is the bulk fluid motion of the cleaning solution. The thickness of the boundary layer decreases as the velocity of bulk fluid motion increases.

Reduction of the boundary layer yields several benefits. It allows fresh chemistry to come closer to the substrate,[2] and come into contact with smaller particles. This higher chemistry refresh rate results in

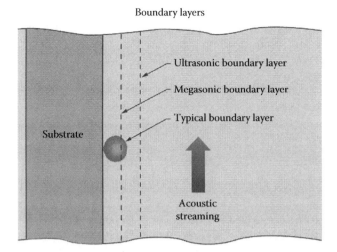

Boundary layers

Ultrasonic boundary layer

Megasonic boundary layer

Typical boundary layer

Substrate

Acoustic streaming

FIGURE 15.6 Comparison of boundary layers in ultrasonic and megasonic cleaning.

faster cleaning. Boundary layer reduction increases the effectiveness of the acoustic streaming removal forces by allowing the cleaning solution to rush past the substrate closer to the substrate surface, forcing chemistry onto particles, removing them from the surface, and carrying them away. The small, controlled cavitation bubbles generated by megasonics are able to remove contaminants within the thinner boundary layer. This effect is especially important in removing small particles and accessing small surface features. Reducing the boundary layer results in increased removal of submicron particles, particles that were previously protected by the boundary layer, as well as increased particle removal overall.

In megasonic cleaning, the combined results of boundary layer reduction, acoustic streaming, and controlled acoustic cavitation are very effective at enabling smaller particles to be removed.

Cleaning Chemistry and Other Factors

Several additional factors contribute to the effectiveness of megasonic cleaning. These include cleaning chemistries, fluid temperature, process time, and power.

Cleaning Chemistries

Megasonics cleaning may be used with a variety of chemistries, including water, neutral aqueous solutions, alkaline aqueous solutions, acidic aqueous solutions, ethyl lactate, alcohol, acetone, *N*-methyl pyrollidone, dibasic esters, and glycol ethers. Although megasonic cleaning is used primarily for particle removal, it can also be used to increase the efficiency of chemical cleaning with surfactants or detergents. Efficacy of removal of other contaminants depends on the solution in the tank.

Cleaning chemistries play a significant role in megasonic cleaning, because the chemical composition of the cleaning solution may affect how quickly the cavitation threshold is reached. In megasonic cleaning (as contrasted with ultrasonic cleaning), it is believed that operating at or below the cavitation threshold produces better cleaning results.

The cavitation threshold, defined as the minimum pressure amplitude required to induce cavitation, has been found to increase with increasing hydrostatic pressure (under most conditions) and to decrease with increasing surface tension, with increasing temperature, and with an increasing number of solid contaminants. A reduction in the number of hydrophobic ions (such as C^2 and F^2) will also decrease cavitation threshold, since these ions collect at bubble surfaces and prevent cavitation bubbles from dissolving.[2]

A lower cavitation threshold allows cavitation to occur more readily. This suggests that cavitation could be mitigated under the following conditions: low surface tension, high hydrostatic pressure, low temperature, and the presence of as few solid surfaces and contaminants as possible.[2]

The original RCA Standard Clean consists of sequential immersion in two chemicals: Standard Clean 1 (SC-1) and Standard Clean 2 (SC-2). The formula for SC-1 is one part H_2O_2, one part NH_34OH, and five parts H_2O. The formula for SC-2 is one part H_2O_2, one part HCl, and five parts H_2O. The addition of megasonic cleaning to the SC-1 solution substantially enhances particle removal.[5] Chemists have succeeded in getting very dilute solutions to clean effectively with the addition of megasonics to the cleaning process.

For example, in statistically designed experiments on semiconductor wafer cleaning, megasonic power was observed to be the dominant factor for particle removal using SC-1 type chemistries. Both the bath temperature and the ratio of ammonium hydroxide to hydrogen peroxide were found to modify the effect of megasonic power on particle removal. Using substantially diluted chemistries, together with high megasonic input power and moderate to elevated temperatures, resulted in very high cleaning efficiencies for small particle removal.[6]

Table 15.2 presents typical chemicals used in the wet cleaning of silicon wafers.

TABLE 15.2 Typical Chemicals for Wet Cleaning of Silicon Wafers

Contaminants	Chemicals
Organics	SPM (H_2SO_4/H_2O_2)
	APM (NH_4OH/H_2O_2) 5 SC-1
Particles	APM (NH_4OH/H_2O_2) 5 SC-1
Metallics	HPM ($HCl/H_2O_2/H_2O$) 5 SC-2
	SPM (H_2SO_4/H_2O_2)
	DHF (HF/H_2O)
Native oxides	DHF (HF/H_2O)
	BHF ($NH_4F/HF/H_2O$)

Other Cleaning Factors

Cleaning fluid temperature, process time, and power are additional factors that can affect megasonic cleaning results. In general, sound speed decreases with increasing temperature. The optimum temperature for the cleaning fluid will vary with the type of substrate being cleaned and with the type of particle that must be removed. The choice of temperature will also depend on the specific cleaning solution being used and how effective it is to begin with at room temperature. In megasonics cleaning, exposure time and megasonic power are the most significant variables. The combination of megasonic controlled cavitation and acoustic streaming enables typical substrate exposure times of 1–30 min. with most exposure times between 10 and 30 min. As megasonic power or exposure time increases, particle redeposition decreases. Increasing the power level directly affects bulk streaming. Higher power levels increase the microstreaming component of megasonic cleaning, reducing the boundary layer, and can shorten the process time required. Megasonic power is affected by array geometry, manufacturing method, and bath geometry.

Design Considerations for Megasonic Systems

Initial megasonic systems developed for general industry use had transducer array lifetimes of a few months. Current technology has increased reliabilities to tens of thousands of hours (years).

For overall megasonic cleaning effectiveness, one must take tank design, fluid circulation and filtration, and system electronics into consideration. The tank size should be small, to minimize the amount of cleaning fluid used. Additional fixturing may be required to position the substrate accurately in the bath; only those surfaces located within the acoustic stream will be cleaned.

The system should incorporate efficient fluid circulation and filtration, to assist in final particle removal from the fluid. Acoustic cavitation dislodges the particles, and acoustic streaming carries them away, but they must be removed from the fluid to prevent their redeposition on the surface being cleaned.

The electronics that drive the resonator are crucial. Piezoelectric impedance is very dynamic over frequency, temperature, and age. If computer-controlled electronics with positive feedback are not used to supply the RF power source to the piezoelectric material, reliability is seriously impaired.

Additional important considerations when choosing a megasonic cleaning system are choosing the appropriate power level and resonator for the fluid type to be used and the type of particle to be removed.

Conclusion

Megasonics provides several advantages over ultrasonics for damage-sensitive substrates. Megasonic-controlled cavitation and high-power acoustic streaming enable substrate exposure times of 1–30 min. and provide effective submicron particle removal without the substrate damage typically associated with ultrasonics. The lower pressures and temperatures produced in megasonic cleaning reduce substrate surface erosion while also providing significant environmental benefits.

References

1. Gouk, R., Experimental study of acoustic pressure and cavitation fields in a megasonic tank, MS thesis, University of Minnesota, Minneapolis, MN, 1996, p. 47.
2. Gale, G., Physical and chemical effects of high frequency ultrasound (megasonics) on liquid based cleaning of Si 100. Surfaces, PhD thesis, 1995, p. 4.
3. Zhang, D., Fundamental study of megasonic cleaning, PhD thesis, University of Minnesota, Minneapolis, MN, 1993, p. 18.
4. Gale, G., Busnaina, A., Dai, F., and Kashkoush, I., How to accomplish effective megasonic particle removal, *Semiconductor Int.*, 19, 133, August 1996.
5. Hottori, T., Trends in wafer cleaning technology, in *Solid State Technology*, Penwell Publishing, Nashua, NH, May 1995, p. S8.
6. Resnick, P.J., Adkins, C.L.J., Clews, P.J., Thomas, E.V., and Korbe, N.C., A study of cleaning performance and mechanisms in dilute SC-1 processing, in *Ultraclean Semiconductor Processing Technology and Surface Chemical Cleaning and Passivation*, Liehr, M., Heyns, M., Hirose, M., and Parks, H., Eds., Materials Research Society, Pittsburgh, PA, 1995, p. 21.

16

Snap, Crackle, or Pop: How Do Bubbles Sound?

Mark Hodnett
National Physical Laboratory

Ultrasonic cleaning is a widely applied technology and has been used in a variety of marketplaces for many decades. A typical ultrasonic cleaning vessel is essentially a simple device, consisting of a stainless steel tank to which are coupled one or more ultrasound sources, operating in the kHz to MHz frequency ranges. When energized, these transducers generate sufficient acoustic pressures to cause cavitation to occur in the fluid contained within the tank, and it is this phenomenon that is responsible for the cleaning process.

Acoustic cavitation [1–3] may be defined as the growth, oscillation, and collapse of microbubbles within a fluid driven by a sound field (Figure 16.1), and it plays a key role in ultrasonic cleaning and other processes [4–6].

Whether cavitation occurs or not is strongly dependent on the fluid and specifically on the availability of sites for *cavitation nucleation*: in a typical sample of tap water, there may be upwards of 100,000 sites per cc where this may occur, in the form of dust particles, on which microscale pockets of air may be trapped. Irregular surfaces on any fluid container offer additional nucleation locations. Cavitation is hence a stochastic phenomenon, and so developing techniques to quantify its occurrence and extent has been challenging.

The detectable secondary effects of cavitation are clearly application specific. For example, it is estimated that the temperatures inside a collapsing bubble may reach 15,000 K, with pressures in excess of 1,000 atm, and the microjets generated during asymmetric collapse may reach speeds of Mach 1, impacting with pressures of 1,000 MPa. These extraordinary effects result in the well known-phenomena of sonochemistry, light emission, free radical generation, and surface erosion, which not only occur in many applications, but also afford the opportunity for measurement.

To try and gain an understanding of the cleaning process, and provide tools for manufacturers and users of such systems, there has been a long-standing requirement to develop real-time measurement methods for quantifying acoustic cavitation [4,6,7]. This is particularly important for safety-related applications, such as the cleaning of surgical instruments [8], where it is vital to ensure that the vessel cleans effectively throughout as much of its volume as possible. However, although a range of techniques have been attempted, none has been found suitable for standardization [9], and the best way of assessing cavitation in a cleaning vessel remains the use of thin aluminum foils, which can record the occurrence of cavitation through erosion of the material and surface indentations.

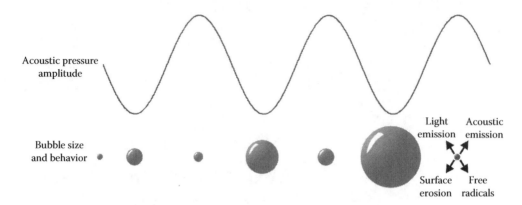

FIGURE 16.1 Bubble size and behavior in response to acoustic pressure. A bubble initially at rest can grow during the rarefactional half-cycles of an applied acoustic field. Through rectified diffusion, gas and vapor are transported into the bubble, unit it reaches a critical size and collapses. The bubble contents are compressed rapidly, resulting in extreme local conditions and a range of secondary effects that drive processes, and may also be measured.

There are many reasons for this deficit of traceable measurement methods: cavitation is a destructive process, so any measurement devices must be rugged enough to withstand hostile conditions, yet maintain refinement to extract crucial information. Cavitation has different meanings in different applications, and so measurements are in some ways subjective, and cavitation itself cannot be measured directly—it is only its secondary effects that can be quantified.

High-power ultrasonic systems such as cleaning vessels are complex acoustic environments as, at the frequencies of main interest (20–200 kHz), the sound fields generated are strongly reverberant, and the development of cavitating bubbles will scatter and absorb the applied sound. Further, high applied powers result in heating and degassing of the liquid, such that the cavitation produced in the vessel is strongly time dependent. All of these factors increase the complexity in developing standardized measurement methods.

Cavitation Sensor

Comprehensive review articles and reports have been published on how cavitation could be quantified [10,11], examining which of the secondary effects might best be utilized for measurement, and so such a discussion is beyond the scope of this chapter. The approach adopted at National Physics Laboratory (NPL) for cavitation characterization is to detect the broadband acoustic signals known to be emitted by bubble clouds from shockwaves generated during the collapse process [10]. Such an approach was first conceptualized many years ago by Neppiras [7], who suggested

- The use of underwater microphones to monitor "white noise" emitted from inertial cavitation, which has been shown to be linearly related to effects such as erosion, dispersion, and surface cleaning
- The need for a device that has a very high frequency response (15–20 MHz), to produce spatial resolution, due to the tendency of low-frequency receivers to detect signals over large distances

The availability of wideband piezoelectric polymers, alongside impedance-matched acoustical absorbers has made such a device feasible and realizable. The constructional and performance details of the NPL sensor have been described extensively [12,13] and so are summarized and illustrated (Figure 16.2) as follows:

- Open-ended hollow right-circular cylindrical construction
- 34 mm high, 40 mm diameter; polyurethane rubber construction minimizing perturbation

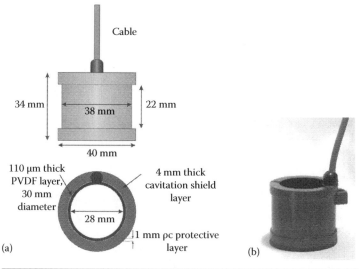

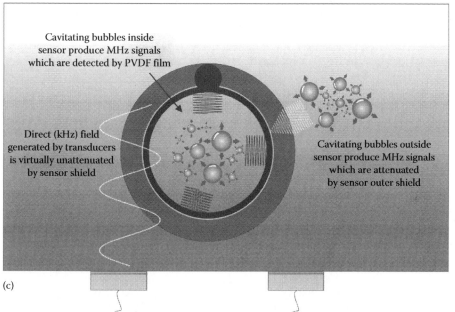

FIGURE 16.2 Schematic (a), photograph (b), and operational concept (c) of cavitation sensor.

- 110 µm thick polyvinylidene fluoride (PVDF) layer, providing a passive detection bandwidth beyond 10 MHz
- Ruggedized inner layer for protecting the sensing element from erosion

In use, the sensor is immersed vertically in the liquid medium under test, which then fills the hollow cylinder. The PVDF layer senses acoustic signals generated by cavitation occurring within the volume enclosed by the cylinder. The PVDF is bonded to a 4 mm thick polyurethane-based rubber material, which forms a rigid outer coating for the sensor and which is highly attenuating at MHz acoustic frequencies. It therefore acts as a "cavitation barrier," shielding the inner PVDF film from acoustic signals originating from cavitation throughout the remainder of the fluid system under test.

The sensor exploits the highly nonlinear aspects of cavitation, in which input acoustic drive frequencies, typically in the tens of kilohertz range, result in acoustic emissions from bubbles well into the megahertz ranges. NPL's cavitation detection research has confirmed the existence of these high-frequency components. The outer material construction attenuates these signals at high frequencies, such that signals detected beyond 1.5 MHz must originate from collapsing bubbles and bubble clouds inside the hollow sensor enclosed volume. In this way, the passive sensor, when connected to appropriate monitoring electronics, produces a spatially averaged signal that corresponds to the high-frequency white noise emitted by cavitation activity in a specific region of fluid. Such a capability is unique, and the cavitation sensor has been patented in the major worldwide territories.

Detection Electronics

To detect, display, and analyze the signals picked up by the cavitation sensor, monitoring electronics are required. Emissions from cavitation contain a wealth of information about the dynamics of the cavitation process, and so using a low-Q sensor means that broad bandwidth instrumentation is needed. "Reference" type measurements can be made using a spectrum analyzer, which reveals the full detail of detected signals, and allows data to be analyzed in specific bands and ranges. However, such an approach is time-consuming, and frequently requires data to be analyzed offline, and so to provide a more rapid indicator of cavitation-related signals, bespoke electronics have been designed and built. These have evolved over several iterations, but with the basic philosophy of processing and displaying information in discrete channels to provide

- A frequency band corresponding to the direct field
- A lower frequency band corresponding to the primary subharmonic
- A high-pass frequency range corresponding to the white noise emitted by bubble collapse

The intention of the latter two channels was to provide information on two particular aspects: the appearance of the subharmonic has been reported previously as a precursor to inertial cavitation, thereby acting as a threshold measure; and the broadband high-frequency white noise as an indicator of the "energy" or "violence" of inertial cavitation. The current realization of the multichannel processing unit [14] is shown in Figure 16.3 with two cavitation sensors.

In combination with the sensor, the detection electronics have been utilized in a broad range of research studies [15–18], and in particular, to examine in detail the cavitation produced by commercial cleaning vessels. Measurements of this type are considered here, in the form of two case studies.

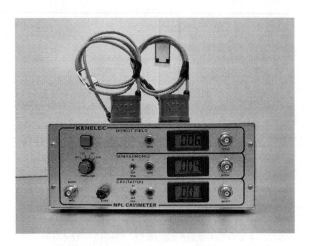

FIGURE 16.3 NPL CaviMeter™ and sensors.

CASE STUDY 1: COMPARING CAVITATION DETERMINED VIA EROSIVE AND ACOUSTICAL TECHNIQUES

Aspects of the work described here are described in more detail in Ref. [15]. The system used for study was a prototype Olympus cleaning vessel, similar to systems used during sterilization processes for surgical devices such as endoscopes. The vessel, of dimensions 330 mm × 300 mm × 130 mm (depth), consisted of four transducers operating at 40 kHz, with a sweep system varying the frequency over a +3, −2 kHz range. For all measurements, the vessel was filled to the manufacturers' operating specification which required 7 L of water, to an average depth of 80 ± 2 mm. Using a mains controller, the input voltage to the vessel was varied over the range 30%–100%, corresponding to an applied electrical power range of 75–250 W. To maintain stability in temperature and gas content, most measurements were carried out at a drive level corresponding to an output power of 140 W.

The vessel was filled with deionized, filtered water that was initially saturated in terms of its gas content (fluid gas content has a strong effect on cavitation and hence cleaning behavior, so much so that without gas, it is acknowledged that little or no cleaning is produced [19]. To this well-defined and repeatable medium, 1% surfactant by volume was added, according to the manufacturer's recommendations. This also had the effect of pre-wetting the sensor.

At a given power setting, and in specific measurement planes, the spatial variation of acoustic cavitation was estimated using aluminum foil erosion. The aluminum foil used was Bacofoil, of thickness around 23 μm. The foil was deployed in sheets suspended over a frame constructed from 2 mm diameter wire, in vertical and horizontal orientations. Measurements were then made of the spatial variation of cavitation activity when using the cavitation sensor, scanning over the same plane as that used for the foil studies with a computer-controlled positioning system. For these acquisitions, the sensor was oriented with its line focus located vertically in the vessel. The sensor output was connected to a variant of the monitoring electronics described above, in which signals over the range 1–7 MHz were measured. The unit provides a visual display of the *rms* signal level over this frequency band and for the comparison study, was connected to an outboard *rms* voltmeter, allowing cavitation sensor signals to be acquired under PC control.

The photograph (Figure 16.4) shows the erosion patterns produced by the whole cleaning vessel when using a horizontally mounted piece of Bacofoil (of dimensions 210 mm × 210 mm) designed to cover all four transducers.

The foil was positioned at a distance of 9 mm below the water surface, and left for 30 s at the operating power of 140 W. The approximate positioning of the four transducers is clearly identifiable, through the areas of greater erosion.

The map (Figure 16.5) shows a two-dimensional cavitation sensor scan of the entire vessel, with data acquired at 10 mm intervals, with the average water temperature being 31°C. The structure of the four transducers is again clearly seen, with transducer B generating the strongest cavitation activity and transducer C the lowest. Comparing the two techniques shows a good correlation between the spatial variation of cavitation determined using the new sensor and erosion

FIGURE 16.4 Two-dimensional variation in inertial cavitation estimated by aluminum foil placed in the 40 kHz vessel.

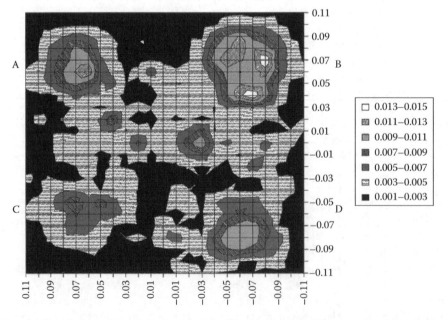

FIGURE 16.5 Two-dimensional variation in inertial cavitation determined by scanning the cavitation sensor throughout the 40 kHz vessel.

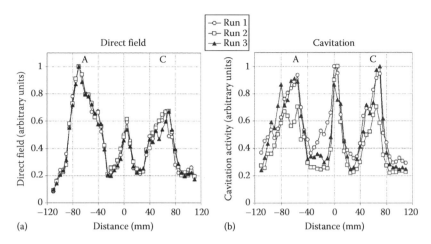

FIGURE 16.6 Variation in (a) direct field and (b) inertial cavitation determined by scanning the cavitation sensor over the left side of the 40 kHz vessel.

tests carried out using thin aluminum foils, confirming other literature findings [20–22]. The expected peak activity locations over the transducers are clearly seen in both cases, and there are more subtle structures due to interference effects between transducers that are also discernible.

A more detailed acoustic scan is shown in Figure 16.6, carried out using the cavitation sensor over the center line of the left-side transducers, with the sensor center located 22 mm below the water surface. In this case, the low-frequency driving signal (determined using the low-frequency band-pass channel of the monitoring electronics), and the resulting cavitation (as above) were monitored.

There is an encouraging and intuitive general correlation between the two data sets: three distinct peaks as a function of position are observed in each plot (Figure 16.6). Strong maxima are seen over the transducer locations (the labels A and C in both cases correspond to the labels in the map above), with an additional peak in the field overlap region. Closer inspection shows that the three measurement runs in the direct field graph show good agreement, whereas this is not the case for the cavitation graph. This is an important outcome, and demonstrates the random nature of cavitation. Looking at both graphs around location "A," the driving acoustic pressure represented by the direct field is similar, but the resulting cavitation in the three runs varies by 50%. It is, therefore, clearly important to use detection techniques that have sufficient capability and fidelity to characterize a direct consequence of cavitation itself, rather than only measuring the driving field, in order to truly understand the spatial and temporal variation of the phenomenon.

CASE STUDY 2: DUAL FREQUENCY CLEANING SYSTEM

To increase the versatility and applicability of commercial products, more cleaning vessel manufacturers are now producing baths that are able to operate at more than one frequency, with the aim of giving the user the choice between different "strengths" of

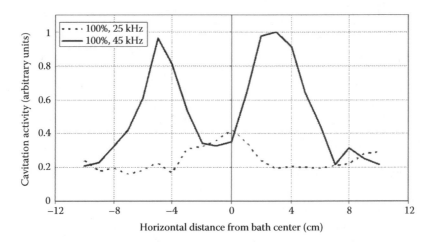

FIGURE 16.7   Variation in normalized inertial cavitation activity as a function of position and frequency.

cleaning performance, for removing a range of contaminants. These normally use the same transducers, driven at different excitation frequencies, and so the resulting cavitation distributions may differ significantly.

Measurements were made on a commercially available Elma Ultrasonic TI-H-55 bath, operable at frequencies of 25 and 45 kHz, with a maximum electrical power input of 600 W available, and with temperature control over the range 30°C–80°C. For a bath operating temperature of 30°C, and with the sensor positioned at half-depth in the 5 L vessel volume, filled with deionized water with 2% surfactant, measurements of cavitation were made using the cavitation sensor and trademarked meter, plotting the spatial variation in the broadband cavitation signal produced by the 25 and 45 kHz drive frequencies (maximum output power setting).

The normalized values are calculated from the full distributions for both frequencies, and so in this way, the inertial cavitation variation as a function of position and frequency is considered. The plot (Figure 16.7) shows that at the operating frequency of 45 kHz, there are two distinct peaks, corresponding approximately to the spatial locations of the transducers, whereas at 25 kHz, the overall cavitation level is reduced to less than half of what is measured at 45 kHz, and the spatial peak occurs over the center of the vessel. This is likely to correspond to a region of constructive interference for the particular scanned height in the vessel. This significant difference has implications for users, who may assume that in switching the operating frequency between the two available settings, the location and "strength" of a given cleaning effect does not change: this clearly is not the case, and again reinforces the benefits of using quantitative detection techniques with spatial resolution.

The first case study presented suggests that a high-frequency, spatially sensitive cavitation sensor can be used as a replacement for the current state of the art but purely qualitative aluminum foil method as an objective means of determining the spatial distribution of cavitation activity within a medium.

Also, both studies show the nonuniformity of cavitation activity in commercially available cleaning systems, even in situations when sweep circuitry is present, and reinforce the benefit to manufacturers and users of employing spatially sensitive quantitative methods.

The proliferation of ultrasonic cleaning and quality assurance requirements means that more and more manufacturers and users are demanding characterization techniques that can provide numerical information on vessel output. The cavitation sensor discussed here is one way in which this can be provided, and the extensive tests carried out with it demonstrate that it is perhaps the most versatile and widely applicable method. Current studies are examining the performance of the sensor and electronics in a range of industrial environments, including elevated temperatures, and in non-aqueous media, to increase the understanding and applicability of the technique.

The ultimate goal of developing cavitation measurement devices and techniques is to provide standardization and traceability, and achieving this aim will prove a challenging, but reachable target in the coming years.

References

1. B.E. Noltingk and E.A. Neppiras, Cavitation produced by ultrasonics, *Proc. Phys. Soc. London B* 63, 1950, 674–685.
2. T.G. Leighton, *The Acoustic Bubble*, London, U.K.: Academic Press, 1994.
3. R. Apfel, Sonic effervescence: A tutorial on acoustic cavitation, *J. Acoust. Soc. Am.* 101, 1997, 1227–1237.
4. B. Zeqiri, M. Hodnett, and T.G. Leighton, A strategy for the development and standardisation of measurement methods for high power/cavitating ultrasonic fields: Final Project report, Teddington, United Kingdom: National Physical Laboratory, NPL Report CIRA(EXT)16, 1997.
5. J.R. Frederick, *Ultrasonic Engineering*, New York: Wiley, 1965.
6. A.E. Crawford, The measurement of cavitation, *Ultrasonics* 2, 1964, 120–123.
7. E.A. Neppiras, Measurement of acoustic cavitation, *IEEE Trans. Sonics Ultrason.* SU-15, 1968, 81–89.
8. I.P. Marangopoulos, C.J. Martin, and J.M.S. Hutchinson, Measurement of field distributions in ultrasonic cleaning vessels: Implications for cleaning efficiency, *Phys. Med. Biol.* 40, 1995, 1897–1908.
9. IEC, Investigations on test procedures for ultrasonic cleaners, Geneva, Switzerland: International Electrotechnical Commission, IEC Report 60886, 1987.
10. T.G. Leighton, A strategy for the development and standardisation of measurement methods for high power/cavitating ultrasonic fields: Review of cavitation monitoring techniques, ISVR Technical Report No. 263, 1997.
11. ANSI Technical Report, Bubble detection and cavitation monitoring, Acoustical Society of America, Report ANSI S1.24 TR-2002, 2002.
12. B. Zeqiri, P.N. Gélat, M. Hodnett, and N.D. Lee, A novel sensor for monitoring acoustic cavitation, Part I: Concept, theory and prototype development, *IEEE Trans. Ultrason. Ferroelectr. Freq. Control* 50, 2003, 1342–1350.
13. B. Zeqiri, N.D. Lee, M. Hodnett, and P.N. Gélat, A novel sensor for monitoring acoustic cavitation, Part II: Prototype performance evaluation, *IEEE Trans. Ultrason. Ferroelectr. Freq. Control* 50, 2003, 1351–1362.
14. M. Hodnett and B. Zeqiri, Towards a reference ultrasonic cavitation vessel: Part 2-investigating the spatial variation and acoustic pressure threshold of inertial cavitation in a 25 kHz ultrasound field, *IEEE Trans. Ultrason. Ferroelectr. Freq. Control* 55, 2008, 1809–1822.
15. B. Zeqiri, M. Hodnett, and A.J. Carroll, Studies of a novel sensor for assessing the spatial distribution of cavitation activity within ultrasonic cleaning vessels, *Ultrasonics* 44, 2006, 73–82.
16. M. Hodnett, R. Chow, and B. Zeqiri, High frequency acoustic emissions generated by a 20 kHz sonochemical horn processor detected using a novel broadband acoustic sensor: A preliminary study, *Ultrason. Sonochem.* 11, 2004, 441–454.
17. M. Ashokkumar, M. Hodnett, B. Zeqiri, F. Grieser, and G.J. Price, Acoustic emission spectra from 515 kHz cavitation in aqueous solutions containing surface-active solutes, *J. Am. Chem. Soc.* 129, 2007, 2250–2258.
18. B. Felver, D. King, S. Lea, G.J. Price, and A.D. Walmsley, Cavitation occurrence around ultrasonic dental scalers, *Ultrason. Sonochem.* 16, 2009, 692–697.

19. O.A. Antony, Technical aspects of ultrasonic cleaning, *Ultrasonics*, 1(4), Oct.–Dec. 1963, 194–198.
20. A.S. Bebchuk, On the problem of cavitation erosion, *Sov. Phys. Acoust.* 4, 1958, 372–373.
21. L. Gaete-Garréton, Y. Vargas-Hernandez, R. Vargas-Herrera, J.A. Gallego-Juarez, and F. Montoy-Vitini, On the onset of transient cavitation in gassy liquids, Part 1, *J. Acoust. Soc. Am.* 101, 1997, 2536–2540.
22. L.D. Rozenberg, On the physics of ultrasonic cleaning, *Ultrason. News* 4, 1960, 16–20.

17

Principles and Quantitative Measurements of Cavitation

Lawrence Azar
PPB Megasonics

Introduction*

A wide variety of substrate surfaces are cleaned in ultrasonic or megasonics cleaners, whereby ultrasound emanates into a fluid to generate cavitation. The fluid will typically contain an aqueous solution, which includes additives such as surfactants and detergents that enhance the cleaning performance of the system. The standard approach is to have a bath of the fluid with bottom-mounted or side-mounted transducers. Recently, more specialized means of delivery have been employed, most notably for single-wafer applications. One application has the megasonics diverted into a stream of fluid that impacts the substrate surface. In another, megasonics imparts directly on a film of fluid no more than a few millimeters thick on the wafer surface. Ultrasonics can also be delivered via an ultrasonic horn, which is a popular method not for cleaning, but for cell disruption, emulsification, and homogenizing of biological matter. In both cleaning and cell disruption applications, it is the occurrence of cavitation that drives the actions.

Background of Ultrasound

The term "ultrasonic" represents sonic waves having a wave frequency above approximately 20 kHz and includes both the traditional ultrasonic cleaning spectrum, which extends in frequency from approximately 20–500 kHz, and the more recently used megasonic cleaning spectrum, which extends in

---

* Photos and charts are courtesy of PPB Megasonics, unless otherwise indicated.

frequency from about 0.5 to about 5 MHz. The device used for cell disruption has traditionally been the ultrasonic horn. This device works at a fixed frequency, normally between 20 and 50 kHz, and is designed to be resonant in the longitudinal mode of vibration. In a typical ultrasonic cleaner, a transducer generates high-frequency vibrations in the cleaning medium in response to an electrical signal input. Once generated, the transducer vibrations propagate through the fluid medium in the cleaner until they reach the substrate to be cleaned.

TABLE 17.1   Acoustic Impedances for Various Materials

Material	Z (g cm^{-2} s^{-1})
Air	0.431
Water	1.5×10^5
Quartz	14.5×10^5
Steel	39×10^5

Acoustic Impedance

Ultrasonics used in cleaning is typically a continuous cycle of compression and rarefaction of pressure waves. The velocity of the ultrasound can be expressed by the density of the material (ρ), the modulus of elasticity (E), and Poisson's ratio (v):

$$c_L = \sqrt{\frac{E}{\rho} \cdot \frac{1-v}{(1+v)(1-2v)}}$$

These longitudinal (compressive, dilatational) waves are bulk waves, where the direction of particle motion coincides with the direction of wave propagation. A relationship can be established between the particle velocity (v) and the propagation velocity as follows:

$$\frac{\sigma}{v} = \rho c$$

where
  σ is the wave stress
  ρ is the density of the material

The term ρc is the acoustic impedance (Z) of the material, which varies from one type of material to another.

Table 17.1 gives approximate values for the acoustic impedance for various materials.

Transmission and Reflection Coefficients

If an interface between different materials exists (see Figure 17.1), some of the acoustic energy will continue to transmit through, while some will be reflected back. The product of the transmission coefficient

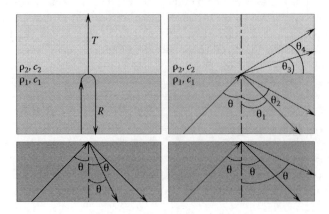

FIGURE 17.1   Reflection and transmission coefficients at media interface.

(*T*) or the reflection coefficient (*R*), with the magnitude (or amplitude) of the original pulse, gives the value of the transmitted and reflected pulses. These coefficients, for normal incidents, are based on the ratio of acoustic impedances of the two materials (z_1 and z_2)[1]:

$$\beta = \frac{z_1}{z_2} = \frac{\rho_1 c_1}{\rho_2 c_2}$$

$$R = \frac{1-\beta}{1+\beta}$$

$$T = \frac{2}{1+\beta}$$

The coefficient of reflection at a water–air interface is high because of the comparatively much greater acoustic impedance of water with that of air. As such, most of the ultrasound is reflected back to the water medium. A negative value for *R* or *T* implies a phase change.

As Figure 17.2 shows, stainless steel will absorb the energy of the acoustics and release very little. This has the effect of attenuating the acoustic signature that tries to pass through it. Careful consideration should be taken when deciding upon how loaded your cleaner will be with the substrates to be cleaned, as well as the design of the part carrier. If the wavelength is considerably greater than the dimension of this interface, the interface will be "invisible" to the passing wave. It is therefore better to use a carrier that has low impedance (i.e., quartz) or very small geometries relative to the wavelength for high-impedance material (i.e., stainless steel). Heavily loaded cleaners will have significant issues cleaning away from the ultrasound source, adequate spacing should be utilized to allow more uniform acoustic distributions.

The waves that transmit energy from one position to another are called "traveling" or "progressive" waves. However, if two identical traveling waves, traveling in opposite directions, are superimposed, there is clearly no net flow of energy in any direction. Such a field is then called a "standing wave." Normal reflection at plane rigid boundary causes a reversal of the particle velocity and of the wave

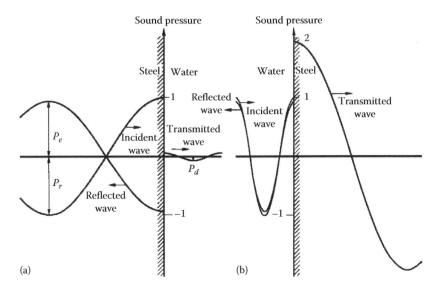

FIGURE 17.2 Acoustic reflection and transmission between water and steel. (From Krautkrämer, J. and Krautkrämer, H., *Ultrasonic Testing of Materials*, 4th edn., Springer-Verlag, New York, 1990.)

velocity. The pressure amplitude is a maximum at the boundary. Reflection from a free interface causes a reversal in normal wave velocity, and in pressure (i.e., a compression is reflected as rarefaction). The pressure is zero at the interface.

Constructive/Destructive Interference

The generating ultrasound field at any one location and at any one moment in time is the culmination of both the amplitude, in either compression or rarefaction, and the phase, thereby generating constructive/destructive pattern. Depending on the location within the bath, you may have ultrasound arriving in phase, generating constructive interference, and in another location, it will be out of phase, generating destructive interference. Tank manufacturers change the pattern by sweeping the frequencies, thereby improving the uniformity within the bath. As a consequence, portions of a substrate that are located at different locations within the bath will experience different levels of cavitation energy. It has been somewhat challenging to uniformly clean a substrate.

The following simulation is based upon Huyghen's principle, which states that wave interactions can be analyzed by summing the phases and amplitudes contributed by a number of simple sources. The pressure at a given distance from the source is computed as follows:

$$p(r,t) = \frac{p_o}{r} \exp\left[j(\omega t - kr) - \alpha r \right]$$

where
  p_o is the initial pressure (Pa)
  α is the attenuation coefficient (Np/m)
  r is the radial distance from the source (m)

Our simulation routine is similar to that utilized by Buchanan and Hynynen,[2] in that they modeled the transducer as an evenly spaced array of simple sources ($a \ll \lambda$), whereas our model treats it as an ensemble of elements of finite width. Figure 17.3 demonstrates how the pressure at any given point can then be attained by adding the contributions of a discrete number of simple sources, which make up a given element. The contributions of all the elements are then added up.

Figure 17.4 shows a numerical simulation of a 1 MHz megasonic bath with a square plate transducer on the bottom. The constructive/destructive pattern is evident, even though only one transducer is shown. The energy is directed primarily over the transducer because the overall dimension of the transducer is much larger than the wavelength of the ultrasound.

Fundamentals of Cavitation

A cavitation bubble will form and grow when a liquid is put into a significant state of tension. Liquids, though unable to support shear stresses, can support compressive stresses, and for short periods, tensile stresses.[3] The acoustic pressure wave undergoes a compression and rarefaction cycle, and the pressure in the liquid becomes a negative during the rarefaction portion of the cycle. When the negative pressure falls below the vapor pressure of the fluid medium, the ultrasonic wave can cause voids or cavitation bubbles to form in the fluid medium.

These cavitation bubbles subsequently collapse (i.e., implode) as the site transitions from the pressure minimum to the pressure maximum, although they can also grow larger and large over successive cycles. Coleman et al. showed a remarkable photograph, as shown in Figure 17.5, of a bubble collapsing near a boundary.[4] Once the cavitation is generated, a cavitation bubble may undergo two different kinds of radial oscillations. One may oscillate nonlinearly during many cycles of the acoustic wave, termed

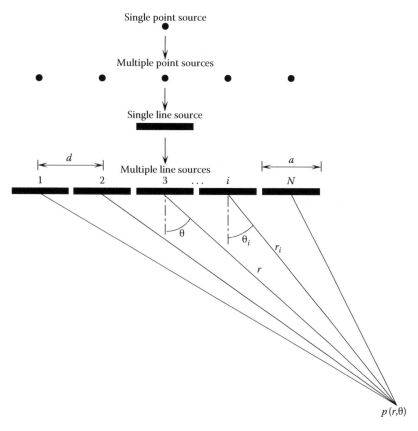

FIGURE 17.3 Approach to modeling the numerical acoustic pressure distribution: the transducers are an ensemble of multiple line sources, each of which is composed of an infinite number of point sources.

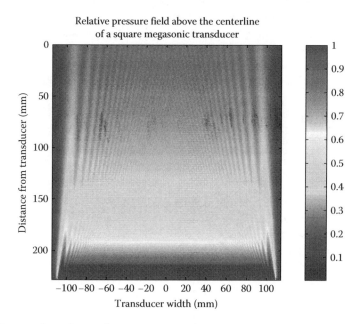

FIGURE 17.4 Numerical simulation of a megasonic transducer.

FIGURE 17.5 Photograph of liquid jet formation during cavitation bubble collapse. (From Coleman, A.J. et al., *Ultrasound Med. Biol.*, 69, 1987.)

"stable cavitation." The other may grow rapidly and collapse violently in one or two acoustic cycles, termed "transient cavitation."[5]

During bubble implosion, surrounding fluid quickly flows to fill the void created by the collapsing bubble and this flow results in an intense shock wave, which is uniquely suited to substrate surface cleaning. Specifically, bubble implosions that occur near or at the substrate surface will generate shock waves that can dislodge contaminants and other soils from the substrate surface. When the bubble collapses, pressure up to 20,000 psi and a "high local temperature, possibly in the order of 5000 K,"[6] are achieved.

Cavitation Equation

In almost all cleaning applications, it is important to control the cavitation energy. When an insufficient amount of cavitation energy is provided, undesirably long process time may be required to obtain a desired level of cleaning, or in some cases, a desired level of surface cleaning may not be achievable. On the other hand, excessive cavitation energy near a substrate having delicate surfaces or components can cause substrate damage. The levels of cavitation energy are also critical in assuring complete and rapid cell disruption. The presence of solid impurities and dissolved gas determines the threshold of cavitation. Many use tap water, which varies widely in solid and gas content. A simple way of ensuring more uniform results is to use distilled water and then degas it. The liquid can now be engassed by bubbling the desired gas through it, which will ensure optimal cavitation.[3]

The bubble dynamics in the acoustic field is described by the well-known Rayleigh–Plesset equation as follows:

$$R\frac{d^2R}{dt^2} + \frac{3}{2}\left(\frac{dR}{dt}\right)^2 = \frac{1}{\rho}\left[P_i - P_\infty - \frac{2\sigma}{R} - \frac{4\mu}{R}\left(\frac{dR}{dt}\right)\right]$$

where
    R is the radius (m) of cavitation bubble at any time
    μ is the viscosity of the liquid medium (N s/m^2)
    σ is the surface tension (N/m)
    P_i is the pressure inside the bubble (N/m^2)
    P_∞ is the pressure in the liquid far from the bubble (N/m^2)[7]

Studies have shown that high density, low viscosity, and middle range surface tensions and vapor pressure are the ideal conditions for most intense cavitation.[8] There are significant temperature effects on these properties, and cavitation itself will be dramatically affected with increasing temperature.

Influence of Frequency and Amplitude on Cavitation

The major factor that affects the size of the cavitation bubbles and the corresponding cavitation energy is the frequency of the ultrasonic wave. Specifically, at higher wave frequencies, there is less time for the bubble to grow. The result is smaller bubbles and a corresponding reduction in cavitation energy. Low-frequency ultrasound has superior particle removal efficiencies for large particles, and that high-frequency ultrasound is best suited for submicron particles.[9] Another factor that affects cavitation energy is the intensity of the ultrasonic wave (i.e., wave amplitude) produced by the transducers. In greater detail, higher wave intensities cause each point along the wave to oscillate over a larger pressure range (between rarefaction and compression), which in turn, produces larger cavitation bubbles and larger cavitation energy. Thus, there is a direct correlation between the intensity of the ultrasonic wave, the pressure range that the fluid medium oscillates between, and cavitation energy.

Microstreaming

In additional to cavitation, there is another effect from ultrasound, acoustic streaming, which occurs when the momentum absorbed from the acoustic field manifests itself as a flow of the liquid in the direction of the sound field. There is a second type of streaming associated that occurs near small obstacles placed within a sound field, called "microstreaming," generated by the oscillations of an acoustically driven bubble.[10] This can lead to additional biological effects and enhanced cleaning effects. The microstreaming associated with bubble motion can be very significant in biological systems. Suspended cells or macromolecules are carried in streaming orbits and may be brought momentarily into the boundary layer near a bubble once during each traverse of an orbit. When in this boundary layer, they may be distorted or fragmented by the high shearing stresses.[3]

Recirculation Flow

The flow rates of fluid within a bath can have serious influence on the cavitation energy present. Chart 17.1 shows the cavitation energy being measured by a cavitation meter immediately after the recirculation pumps were turned off. It is clear that the initial energy with the flow present was at 40% and that it

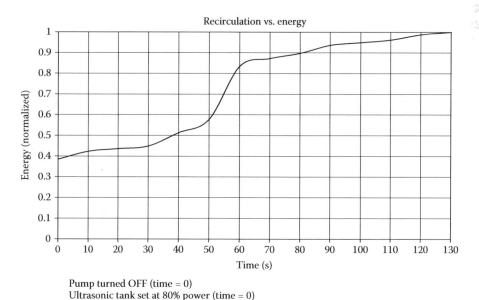

Pump turned OFF (time = 0)
Ultrasonic tank set at 80% power (time = 0)

CHART 17.1   Cavitation meter readings versus time for recirculation flow recovery.

took nearly 80 s to return to 90% levels and more than 2 min to return to 100% efficiency. It is clear that recirculation plays a critical role in cavitation bubble production. Recirculation can entrain air into the fluid medium, raising the attenuation of the ultrasound and minimizing cavitation bubble growth.

Some have thought to increase the effectiveness of their ultrasonic bath by incorporating spray jets into the cleaning equipment. The addition of the added flow into the system brought the cavitation energy nearly to zero, and they lost all benefits the ultrasonic cleaning could provide. Many changes to parameters in the bath, whether it is the solution type, transducer design and location, frequency, flow rate, temperature, level of loading, location of the substrate, can all have significant influence on the cavitation energy at any one location. Only a cavitation measurement at each specific location will provide the necessary information to properly quantify the uniformity and cleaning capacity of the ultrasonic or megasonic cleaner.

Cavitation Meters

Measurement Principle

Ultrasonic cavitation can produce a cavitation noise spectrum including harmonics, sub-harmonics and continuous noise. Further, continuous monitoring of the driving frequency and the relative intensity of ultrasonic cavitation can be acquired by analyzing the cavitation noise.[5] PPB Megasonics (Lake Oswego, Oregon) has developed a method to analyze the cavitation fields and measure the cavitation intensity by separating the acoustic energy intensity at a particular location in the ultrasonic fields into two components: the energy intensity due to the ultrasonic itself and the cavitation activity. The cavitation meter then outputs the RMS of the cavitation energy intensity (units of watts per square inch) and the frequency of the ultrasound. These meters are not hydrophones, which typically filter out the higher frequency cavitation signatures and inherently spatially average compression and rarefaction resulting in what can be misleading data. The cavitation signature is superimposed on the acoustic measure, but has no spacial average component. The acoustic signature—compression and rarefaction—is removed, leaving behind only the cavitation implosion forces.

Comparison with Existing Testing Methods

The aluminum foil test is a rudimentary method sometimes used to evaluate ultrasonic baths. There are three fundamental drawbacks to the foil test. The first being that it may contaminate any bath, as pieces of the foil may break off. The second is the relativistic nature of the foil test, where one is measuring the level of bumps and wrinkles in a very unscientific way. Mishandling of the foil could also lead to wrinkles and folds that could throw off the relative measures. The time-consuming and relativistic nature of the foil test may work for some with a lot of time on their hands and only one or two baths, but a more quantitative process is required in the real world.

John Kolyer from Boeing has written articles on this topic, including one that compared the use of the foil test as compared to a cavitation meter. He states in the article: "A Meter That Works" that "… the Ultrasonic Energy Meter will take over the task of monitoring tank performance."[11] His data showed a dramatic improvement in accuracy and time to evaluate a bath using the cavitation meter. His article also charts how the energy measure of the unit changes with varying power settings from a standard 40 kHz bath, which are shown in Chart 17.2. It is clear that the measurements are directly proportional to the energy present within the bath.

Figure 17.6 shows a detailed mapping of a megasonic bath at mid-depth using the cavitation probe. The constructive/destructive patterns are quite evident, as are the subsequent peaks and valleys present within the bath. Figure 17.7 shows a comparison of two different megasonic bath designs; the data obtained by a leading cleaner manufacturer using a cavitation meter.[12] Figures 17.8 and 17.9 show a comparison of contamination and cavitation meter measures, which demonstrate very good correlation in this single-wafer cleaner.[13]

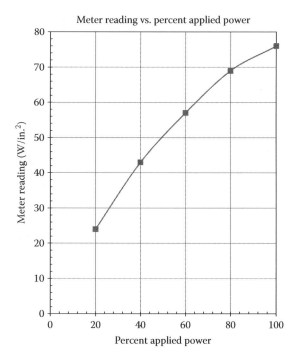

CHART 17.2 Cavitation meter readings versus applied generator power. (Chart courtesy of J.M. Kolyer, Boeing.)

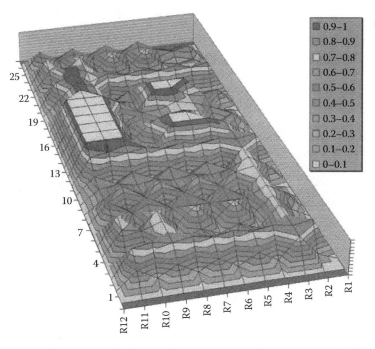

FIGURE 17.6 Mapping of a megasonic bath using the cavitation meter. (Image courtesy of Sebastian Barth.)

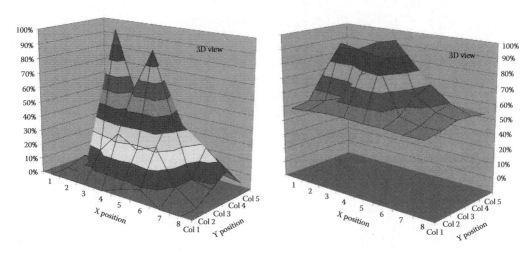

FIGURE 17.7 Traditional and sweeping megasonic power distribution pattern in a cleaning vessel. (From Buchanan, M.T. and Hynynen, K., *IEEE Trans. Biomed. Eng.*, 41(12), 1178, 1994.)

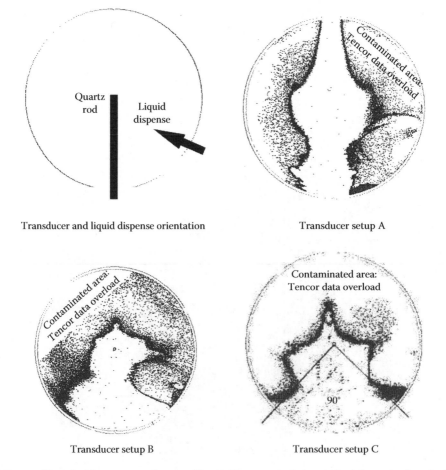

FIGURE 17.8 "Static" wafer test—transducer and liquid dispense orientation; wafer defect maps for a single-wafer cleaner. (From Wu, Y. et al., Acoustic property characterization of a single wafer megasonic cleaner. *Presentation and Proceedings, Electrochemical Society*, Honolulu, HI, October 1999.)

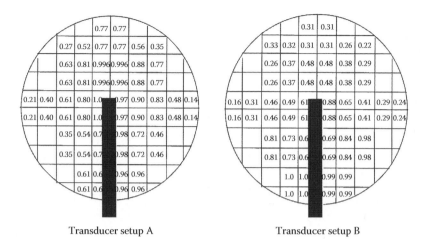

Transducer setup A   Transducer setup B

FIGURE 17.9 Cavitation probe intensity measurements—normalized. (From Wu, Y. et al., Acoustic property characterization of a single wafer megasonic cleaner. *Presentation and Proceedings, Electrochemical Society,* Honolulu, HI, October 1999.)

Cavitation Meter Designs

PPB Megasonics manufactures a cavitation meter that provides the RMS of the cavitation energy intensity, as well as the frequency of the driving ultrasound (see Figure 17.10). The meters allow quantitative measures of critical data, including the average cavitation energy, maximum, minimum, standard deviation, and frequency measure. The unit allows the data to be stored and later downloaded to a PC for mapping of the energy distribution within the bath.

The probes for ultrasonic cleaning applications have been specially designed to isolate any resonant affects by utilizing a chemically resistant polymer to house a sensor mounted on an acoustically matched quartz lens. All-quartz probes are used for megasonic applications, and quartz probes with side-mounted sensors have been developed for measurements within megasonic nozzle streams. A newly released "beaker" style probe has been introduced to quantify the energy emanating from ultrasonic sonicator®/sonifier® horns used for cell disruption.

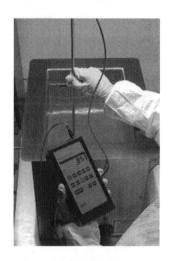

FIGURE 17.10 Ultrasonic Energy Meter.

Custom probes that are a representation of the substrate being cleaned have been manufactured, which include a prototype probe to represent a 300 mm wafer, as shown in Figure 17.11, another for

FIGURE 17.11 Multichannel cavitation probe: wafer representation prototype.

photomask applications, and others. We are also introducing patented multisensor probes that provide real-time energy distribution profiles across wafer or other substrate surfaces being cleaned, vital in minimizing damage but also assuring effective cleaning. An early prototype is already being used by a leading megasonic manufacturer on a single-wafer cleaner. The first-generation model will include up to 64 channels that are displayed simultaneously.

Conclusion

The cavitation meter has proven to be an invaluable metrology tool to quantify the energy within the ultrasonic and megasonic cleaners and those emanating from ultrasonic horns. Using the cavitation meter as a process control tool will improve both the yield and throughput of the cleaning and cell disruptor operations. NIST traceable calibration certificates are available.

References

1. J. Krautkrämer and H. Krautkrämer. *Ultrasonic Testing of Materials*, 4th edn., Springer-Verlag, New York, 1990.
2. M.T. Buchanan and K. Hynynen. Design and experimental evaluation of an intracavity ultrasound phased array system for hyperthermia. *IEEE Transaction on Biomedical Engineering* **41**(12), 1178–1187, December 1994.
3. F.R. Young. *Cavitation*, Imperial College Press, London, U.K., 1999.
4. A.J. Coleman, J.E. Sauders, L.A. Crum, and M. Dyson. Acoustic cavitation generated by an extracorporeal shockwave lithotripter. *Ultrasound in Medicine & Biology* **13**(2), 69–76, 1987.
5. Z. Liang, G. Zhou, S. Lin, Y. Zhang, and H. Yang. Study of low-frequency ultrasonic cavitation fields based on spectral analysis technique. *Ultrasonics* **44**, 1, 2006.
6. A. Henglein and M.J. Gutierrez. Sonochemistry and sonoluminescence: effects of external pressure. *Journal of Physical Chemistry* **97**, 158, 1993.
7. P. Kanthale, M. Ashokkumar, and F. Grieser. Sonoluminescence, sonochemistry and bubble dynamics. *Ultrasonics Sonochemistry* **15**, 2, 2008.
8. C. Simpson. Bearing the load. *CleanTech* **2**, 2, 2002.
9. R. Rodrigo and T. Piazza. Intelligent ultrasonics for the disk drive industry. A^2C^2 **3**, 8, 2000.
10. T.G. Leighton. *The Acoustic Bubble*, Academic Press, London, U.K., 1994.
11. J. M. Kolyer, A.A. Passchier, and L. Lau. New wrinkles in evaluating ultrasonic tanks. *Precision Cleaning Magazine* May/June, 2000.
12. J.M. Goodson and R. Nagarajan. Megasonic sweeping and silicon wafer cleaning. *Solid State Phenomena* **145–146**, 27–30, 2009.
13. Y. Wu, C. Franklin, M. Bran, and B. Fraser. Acoustic property characterization of a single wafer megasonic cleaner. *Presentation and Proceedings, Electrochemical Society*, Honolulu, HI, October 1999.

18

Equipment Design

Edward W. Lamm
*Branson Ultrasonics
Corporation*

Introduction

In the late 1980s as the concern about the effects of CFCs on the ozone layer came to a head, solvent cleaning was subjected to severe scrutiny. During the next few years, most cleaning applications were evaluated regarding their ability to be converted to aqueous or semiaqueous. With this new activity, the aqueous marketplace began to explode with new manufacturers eager to gain a piece of the market that always seemed solvents were granted by royal decree. The new regulations now offered a slice of the fief that was previously out of reach. The manufacturers of chemicals that could be rinsed with water (semi-aqueous) and had minimal emissions also began to see a crack in the solvent armor. They began probing into this new potential growth market. Those applications that could not undergo the conversion to aqueous or semiaqueous were tested with new solvents that had almost no ozone depletion potential, but emissions continued to be an issue. As this market began to grow, new equipment was required to meet the tighter environmental regulations that were instituted to assure solvent loss be kept to a minimum.

All these events changed the distribution of the cleaning market and the available equipment. The new designs that were born will be explored along with the original designs and the ancillary support equipment that enhances their performance.

Application

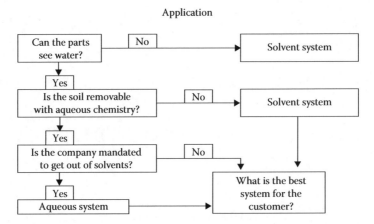

FIGURE 18.1 Equipment path questions. (From Genet, C., *CleanRooms East '99*, Philadelphia, PA, March 1999, Penn Well Publishing, Nashua, NH, pp. 125–143. With permission.)

Choosing the Correct Equipment Path

To look at the types of equipment, it is first necessary to understand how the decision to select a cleaning process is reached. Essentially there are a few questions that must be answered (Figure 18.1). The questions basically help the engineer decide the approach to take in regard to the cleaning agent to be used. This will then determine the type of equipment to be investigated, solvent or aqueous.

Once the selection of the cleaning agent category has been completed, the field of available designs is significantly reduced. This is a great start because there are over 100 equipment companies that provide in excess of 200 products.[1]

Solvent Equipment

Equipment designed for cleaning with solvents is divided into two simple groups, cold batch cleaning and vapor degreasing. The first group, cold batch, is somewhat a throwback to the paintbrush and coffee can era with a bit of scale-up and sophistication. A sink on a barrel is a good example. This system is quite labor intensive, but simple to operate. With the addition of an agitation lift, the process is enhanced; however, it is still relegated to fairly noncomplex parts and far from the tight tolerances of precision cleaning.

Vapor degreasers on the other hand can run the gamut of low-end open-top to extremely sophisticated closed designs depending on the cleanliness required. The basics of the concept are the same for all operations. Vapor degreasers are designed not only to vaporize solvent for cleaning and drying the parts, but also to confine and recycle the solvent and solvent vapor to maintain a healthful environment and to keep cleaning costs low. Vapor degreasing equipment must provide for:

- Cleaning to remove soluble and particulate soil
- Recovery of solvent by distillation for repeated use
- Concentration of the soils

Open-Top Vapor Degreasers

These requirements are met by the basic degreasing unit, which is an open rectangular tank with a pool of solvent in the bottom (Figure 18.2). The solvent is heated in the boil sump and vaporized into a dense vapor layer (specific gravity greater than 1, heavier than air) that lies above the liquid and constitutes the

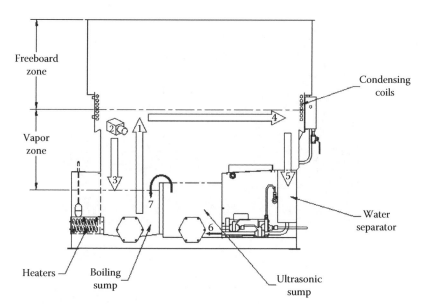

FIGURE 18.2 Solvent cycle. (1) Heaters boil solvent to make vapors; (2) vapors condense on parts; (3) condensate drips into boil sump; (4) excess vapors condense on coils; (5) condensate flows to water separator where water is removed; (6) dry condensate overflows to ultrasonic sump; (7) solvent overflows from ultrasonic sump to boiling sump.

vapor cleaning zone. The condensing coils, which are installed high on the inside periphery of the tank, condense the vapor reaching that level. The solvent condensate is returned to the solvent boil sump via the condenser trough and water separator.

The very important vertical extension of the degreaser wall is the freeboard. The freeboard provides a stationary air zone above the vapor level. It shields the normal vapor zone from moderate drafts, which could carry the vapors into the work environment. In addition, a very thin solvent film evaporates from the work, as it is slowly withdrawn from the vapor zone, and the freeboard area confines these vapors.

The degreaser work opening must be adequate to handle the work dimensions and cleaning cycle, but should be kept to a minimum to maintain economical operation and acceptable working conditions. Even when vapor loss is controlled and no work is passed through, every square foot of exposed surface permits loss of a given amount of solvent related to the equipment design and the solvent used. Since vapor in the center of the tank must be passed to the walls for condensation, extending the tank width increases turbulence, which causes entrainment of air resulting in vapor loss. As narrow a tank as feasible is recommended.

Freeboard is the distance from the top of the vapor line to the top of the confining side wall at the top of the tank. The freeboard zone reduces vapor disturbance caused by air motion in the work area. The freeboard zone also permits drainage of the work being removed, evaporation of residual solvent, and drying of the part with a minimum of solvent loss as well as reduced solvent emissions into the air. In degreasers of extreme length, the height of the freeboard is increased. Generally speaking, the higher the freeboard, the lower the solvent consumption.

The basis of effective vapor degreaser design is control of the vapor level. The control of the vapor zone provides for cleaning in freshly distilled solvent and also helps minimize solvent loss. This control is best done by use of condensing coils. Condensing coils are located within the degreaser tank at a height above the boiling solvent equal to the work height plus allowances for clearance below the work and a 3–9 in. vapor layer above the work (Figure 18.3). The usual design of a degreaser provides for the normal vapor level to be at the midpoint of the vertical span of these coils. Thus, the positioning of the condensing coils also established the freeboard height in a given tank.

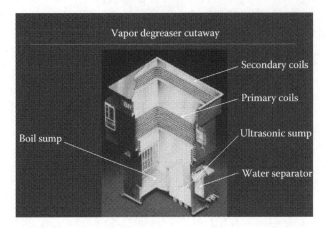

FIGURE 18.3   Vapor degreaser.

To prevent excessive condensation of atmospheric moisture on the coil surfaces above the vapor line, the temperature of the water leaving the coils should be above the dew point of the ambient air. To accomplish this, the water should flow into the lowest coil and out the top coil.

The refrigerated freeboard chiller is designed to reduce solvent emissions at the solvent vapor–air interface by placing a cool, dry layer of air above the vapor zone. This cool air blanket assists in confining the solvent vapors. The refrigerated freeboard chiller consists of a coil placed on the inside perimeter of the unit, immediately above the primary condensing coils. An external refrigeration unit supplies the coils with the necessary cooling.

The chiller unit will condense, and in some cases freeze, atmospheric moisture onto the coils. The additional water from these coils should be handled by placing a separate trough under the coils, draining to its own water separator for a holding tank. If an additional trough is not placed on the equipment, the coil should drip into the solvent condensate trough, but a larger separator would then be considered for effectively separating the water from the solvent. Additional water in the solvent may cause corrosion and shorten equipment life.

Water enters a degreaser from several sources:

- Condensation of atmospheric moisture on the condenser coils
- Moisture on the work being cleaned
- Steam or cooling water leaks
- Water-soluble cutting oil

Water can form a boiling mixture with the solvent (an azeotrope) that is vaporized, causing equipment corrosion, decreased solvent life, and increased vapor losses. All degreasers should be equipped with a properly sized water separator.

In the water separator, the condensed solvent–water mixture drops into a trough below the condenser coils and flows by gravity to the separator. The mixture enters the separator below the solvent level. The water with a lower specific gravity and insolubility rises to the top and is discharged through a water drain. Relatively moisture-free solvent is then discharged through the solvent return line to the degreaser. This separation requires time. Since 5 min is a practical minimum, the separation chamber should have a capacity of at least $\frac{1}{12}$ the hourly solvent condensing rate.

A deeper separator is more efficient to operate than a shallow one of equal volume, because the solvent–water interface area is smaller in the deeper design.

To minimize solvent loss induced by air turbulence over the vapor zone, a degreaser should be placed away from excessive air currents, open windows or doors, heating and ventilating equipment, and any device causing rapid, uncontrolled air displacement. Typically, the usual air circulation is sufficient to dilute small quantities of vapor that normally escape from the degreaser. When the degreaser must be placed in an unfavorable location, a baffle on the windward side will divert drafts and protect the vapor level.

Closed-System Degreasing

The contained or closed degreaser (Chapter 24, Gray and Durkee) offers all the benefits of the open-top design, but enhances the process by eliminating the solvent–air interface. This is accomplished by conducting the cleaning in a sealed chamber, thereby preventing emission. Another benefit of the closed chamber is the addition of a vacuum step during processing. This feature aids in soil displacement from blind holes and complete removal of solvent during drying. The major concern when evaluating this technology is the associated cost and low throughput as compared with open-top systems.

The major difference in design between the open and closed systems is the parts movement. Unlike the open-top degreaser, when items are placed in the closed chamber to be cleaned in closed-system degreasing, the parts never leave the chamber until the process is complete. The cleaning solvent is brought to the parts where they are cleaned, rinsed, and dried. This eliminates the possibility of a disturbed or collapsed vapor zone, which contributes to solvent consumption and emissions. In addition, a number of agitation options, which include spray, ultrasonics, and rotation, can be employed during any of the process steps (Figure 18.4).

These systems typically include multiple feed tanks for a variety of solvent cleanliness levels. For reclamation of the solvent and to provide a fresh uncontaminated rinse source, a still is an integral part

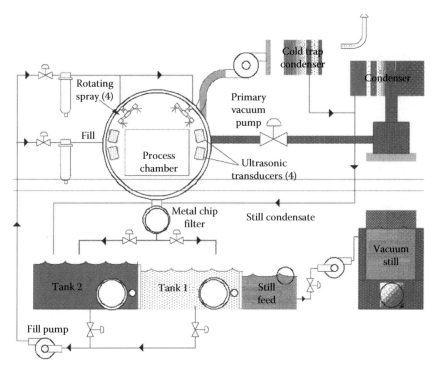

FIGURE 18.4 Schematic of a closed degreaser system.

of the design. As elimination of emissions is a key principle, heat exchangers are installed to remove solvent from the discharge of the vacuum system.

Semiaqueous and Aqueous Similarities

In cleaning applications where water does not have a negative impact, both aqueous and semiaqueous systems have been given substantial consideration. There are a number of similarities between semiaqueous and aqueous equipment. The major one is that they both use water as the medium to remove the wash medium, which provides for a huge similarity in the last two thirds of the process. With both using the same rinse design, the drying options for elimination of moisture are identical for these processes. Obviously, the ancillary equipment is also similar. Pumping, filtration, and water purification are all handled by equipment of identical design.

Aqueous Equipment

The process of aqueous cleaning can be divided into three specific components: cleaning, rinsing, and drying. This is no different from solvent cleaning except that each process component is conducted in a different piece of equipment, each section is generally more sophisticated than a section of the vapor degreaser. Of course, the cost associated in providing the added detail is significant for aqueous design.

When washing a part, the contaminant is often removed through the introduction of cleaning chemistry and mechanical force. Rinsing involves the removal of any residual soil and chemistry that remain after washing. It is important to perform this task without introducing new contaminants, such as dust or impurities in the water. Drying is the process by which residual rinse liquid is removed without introducing any new contaminants.

In the cleaning step, a detergent that is typically diluted in water actually bonds to the soil (oil, grease, or particulate). To be effective, the detergent requires temperature and mechanical activity to loosen the dirt. Both of these are important when evaluating the design of the equipment.

Mechanical force is typically used in both the cleaning and rinsing stages. There are a number of options available (Table 18.1). Spraying is a fairly effective, low-cost method for large parts without intricate details or holes. However, for smaller components or parts with blind holes, immersion with an additional source of agitation is required. All the options provide a relative level of removal of both soluble and particulate soils and must be evaluated for the type of contaminant to be removed. Of course, all come at a price and must be judged on their need and effectiveness.

TABLE 18.1　Mechanical Forces—Separating the Soil from the Substrate

Method	Relative Energy	Solubles Removal	Particle Removal	Relative Cost
Spray	High	Good	OK	Low
Immersion	Low	OK	Poor	Low
Agitations				
Bubbler	Low	OK	Poor	Low
Lift	Med	Good	Good	Med
Propeller	High	Good	Good	Med
Ultrasonics	High	Excel	Excel	High

Source: Genet, C., *CleanRooms East '99,* Philadelphia, PA, March 1999, Penn Well Publishing, Nashua, NH, pp. 125–143.

When people think of cleaning applications, the focus is typically placed on the washing stage of the operation. In precision cleaning, however, rinsing becomes a much more important step. The allowable contamination levels are lower, and spot-free drying is almost always a requirement.

Water Rinsing

Rinsing is a technology, just as washing is. It is measurable, controllable, and directly contributes to the effectiveness of the cleaning process. Effective rinsing can improve yield, reliability, and appearance; it is also an important factor in containing operation costs.

Rinsing removes two basic types of soils: (1) solubles, which encompass washing chemistries and other soils that dissolve in the cleaning media, and (2) insolubles, consisting of particulate dispersed throughout the cleaning media. Rinsing is based on the principle of dilution. To develop an effective rinsing process, three questions must be answered:

1. What soils are present?
2. How much soil is there?
3. How much residue is acceptable?

What soils and how much soil there is can be determined with analytical testing. How much residue is acceptable is a more difficult question—one that must often be answered empirically by the end user.

Often, acceptable residue levels are defined by testing a cleaned part for acceptable performance in its next operation or use. If the part performs acceptably after being put through the cleaning process, the cleanliness level is assumed to be acceptable.

It is important to minimize contamination in the rinsing steps and to allow the use of less rinse water. Two rinsing techniques commonly used to minimize rinse water volume are spray rinsing and countercurrent immersion flow rinsing. Spray rinsing, as the name implies, uses spray nozzles to direct the flow of rinse water over the parts. This type of rinsing can be very effective, using much less water than a typical flowing rinse. Proper application of spray rinses is necessary to ensure that all areas of the parts can be rinsed and also that the spray is only activated when the part is present to be rinsed. Effective immersion flow rinsing is based on the successful completion of two tasks: first, the soils must be separated from the part; then, the soils must be prevented from redepositing onto the part. This can be accomplished by several means—for example, sparging the surface to remove buoyant soils, filtering the solution for particulate, and maintaining continuous dilution of solubles and fine particulate.

Separating the soil often requires mechanical energy, especially with parts having complex shapes, or those that are "nested" in blind holes and/or crevices. Table 18.1 lists several options to accomplish this.

If ultrasonic agitation is used in the wash, it might also be helpful in the rinse. Many times, a higher frequency is used in the rinse than has been used in the wash. This facilitates removal of smaller particles and reduces the potential for part damage.

Continuous filtration of the rinse baths is very important in precision rinsing. The level of retention of the filter should reflect the level of cleanliness required. In systems with multiple rinse tanks, the filter retention level is often reduced with each succeeding bath. Continuous dilution is also a method of preventing redisposition and involves four key elements:

1. Concentration of tank chemistry in dragout (C), which is measured in parts per million (ppm) (1 oz/gal 5 approximately 7500 ppm).
2. Volume of dragout (V), which is the volume of water/chemistry moved (with the parts and carrier) from the wash to the rinse stage.
3. Flow rate of rinse water (F), which is measured in gallons per hour (gph).
4. Rinse tank equilibrium concentration (E), which is a function of flow rate and dragout, to the point at which incoming and outgoing chemistry levels are equal.

These four rinsing factors are related by the following formula:

$$C \times V = F \times E$$

$C \times V$ defines the amount of chemistry entering the rinses. Precision rinsing generally requires low E values, therefore, high F values (or overflow rates) are required. Figure 18.5 illustrates the process of continuous dilution.

One method for improving rinsing is the use of several rinse tanks in a series (Figure 18.6).The rinse formula applies to each successive tank. This allows a significant reduction in equilibrium concentration with a fixed overflow rate (F). This arrangement increases the capital costs of achieving a particular cleanliness level by requiring more rinse tanks, but it reduces operating costs by lowering the required overflow rate. It is important to note that the water flow is in the opposite direction of the work flow. This is called "counter cascade" rinsing. In applications with high dragout, a spray rinse may be used to remove the gross chemistry before the first immersion rinse, further increasing efficiency.

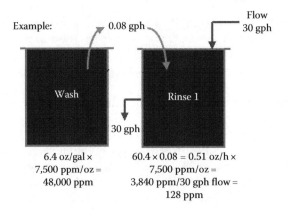

FIGURE 18.5 Continuous dilution. (From Genet, C., *CleanRooms East '99*, Philadelphia, PA, March 1999, Penn Well Publishing, Nashua, NH, pp. 125–143.)

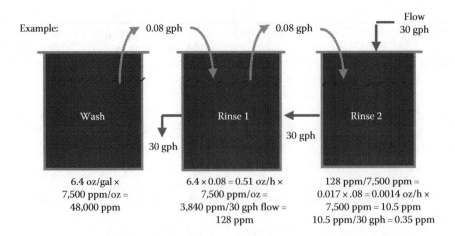

FIGURE 18.6 Continuous dilution—several rinse tanks. (From Genet, C., *CleanRooms East '99*, Philadelphia, PA, March 1999, Penn Well Publishing, Nashua, NH, pp. 125–143. With permission.)

TABLE 18.2 Deionized Water Quality

Resistance (Megohms)	Conductance (Microsiemens)	Total Dissolved Solids (ppm)
18.2	0.055	None (higher quality)
10.0	0.100	0.115
4.0	0.250	0.288
1.0	1.000	1.150
0.4	2.500	2.875 (lower quality)

Source: Genet, C., *CleanRooms East '99,* Philadelphia, PA, March 1999, Penn Well Publishing, Nashua, NH, pp. 125–143.

In precision applications, the quality of the rinse water itself can be a factor in the effectiveness of the rinsing stage. In most cases, deionized (DI) water is required. In the deionization process, organics are removed by carbon, and special functional exchange resins remove the ions. Biological growth is controlled with ultraviolet lights and special filtration. One method of measuring rinse water quality is through resistivity or conductivity (Table 18.2). This is a measure of the electrical insulation properties of the water. Dirty water and tap water may contain many ions that conduct electricity, lowering the resistivity.

Rinse Tank Design

In addition to the process variables, rinse tank design can impact the effectiveness of rinsing. The flow pattern of the water can be important in rinsing, and this pattern is a function of the tank design. The most common design for rinse tanks is the single-sided, overflow weir design. This design depends on dilution for effectiveness and has "dead spots" in the corners where mixing does not take place, thereby reducing its effectiveness.

Another, more effective rinse tank design, which has become popular in precision rinsing, is the four-sided overflow model (Figure 18.7). This design utilizes a laminar upflow of water, which improves mixing and eliminates dead spots. This design is also very effective at sweeping fine particulate off the surface, preventing redeposition on the parts. The overflow design is often augmented by the use of high-flow recirculation filtration, which further increases the sweeping action. The return of the filtered water typically enters the bottom of the rinse tank. Return manifolds that are specifically configured for the fixture are often used.

One of the important variables in rinsing is the cleanup rate. This is defined as "the time it takes for the contamination in the rinse tank to return to a steady level, after the parts enter the bath." An

FIGURE 18.7 Four-sided overflow weir with 360° saw-tooth weir design.

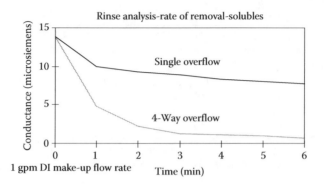

FIGURE 18.8   Equipment considerations. Cleanup rate using solubles.

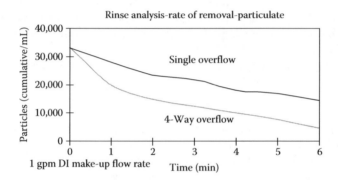

FIGURE 18.9   Equipment considerations. Cleanup rate using particulate.

experiment was run to determine the cleanup rate for a four-sided overflow with recirculation/filtration and a single-sided rinse with recirculation/filtration and a sparger. The four-sided overflow rinse had a faster cleanup rate than the single-sided overflow rinse. There are several factors that contribute to the improved efficiency. The high internal flow and mixing in the four-sided design enhances solubility. The high-volume laminar flow is efficient for particle removal and minimizes redeposition. In the four-sided design, the distance to the overflow is minimized, improving the sweeping action.

The four-sided overflow design can provide up to 60% advantage in rinsing over the conventional single-sided design. It has improved efficiency for both soluble soils and particulate. For fixed overflow rate, process throughput, and cleanliness level, fewer rinse tanks may be required with a four-sided design. Alternately, for a fixed number of rinse tanks, a higher throughput may be possible. Figures 18.8 and 18.9 show the cleanup rate for the two designs, using soluble soils and particulate.

Drying

The drying stage of aqueous cleaning has the objective of removing the residual water that is carried over from the last rinse. Removal of the remaining water can be quite difficult depending on the geometry of the part being processed. The two basic additives required to increase water evaporation are temperature and flow of air. The mechanical methods used to produce these effects include[3]:

Compressed air blow-off	Vacuum oven
Infrared lamp bank	Centrifuge
Recirculating air oven	Solvent displacement

If moisture can be tolerated, an air blow-off or centrifugation of the parts may be all that is needed. If the parts need to be dry to the touch, infrared lamps or oven drying may be required. If a higher level of dryness is needed prior to subsequent processing, a combination of drying steps may be used.

Parts configuration, the substrate involved, and the degree of dryness will dictate which drying method or methods are most suitable for the majority of parts involved. Plastics, copper, and aluminum containing alloys may have temperature restrictions that need to be considered. Parts with blind holes, threads, depressions, and narrow cavities may require special handling. Small parts tightly nested together also offer special drying challenges when considering the best or most efficient design.

Compressed air is economical and can be used directly over a process tank to minimize dragout. It is especially effective on large flat surfaces. This method is ideal if some moisture can be tolerated or if used in conjunction with another drying process. Air velocity dictates the percentage of moisture removed as droplets and the percentage removed by evaporation. A lower velocity results in a greater amount of evaporated moisture.

Infrared heat lamps are designed to focus heat where needed. A full line of area and chamber heated lamps are available. Infrared heat is clean, fast, and controllable. Sample part configurations of delicate construction respond best to this type of drying.

Recirculating hot air ovens are commonly used by industry to dry parts cleaned with water. Drying times are directly related to air velocity and temperature. These ovens are ideal if a high degree of dryness is required.

Vacuum ovens should be used only as a polishing step as required to get the last little bit of moisture off the parts. Vacuum of 1 T or greater is used at temperatures of 120°F or greater.

Centrifugal dryers are used for small parts with simple configuration. Parts are spun at speeds approaching 1000 rpm for up to 10 min. The liquid can be recovered and returned to the process or rinse tank as desired. This type of drying process requires little space, and operation costs are relatively low. This process is, however, for small parts and it is not adequate if a high level of dryness is required.

Parts with complicated internal components or blind cavities may require final moisture removal using a water-displacing solvent. Any solvent immiscible with water can be used for this process. Molecular structure and physical characteristics all must be considered carefully.

The primary cost of parts drying is energy. For that reason it is prudent not to dry parts any more than is essential for subsequent processing. Less energy is required to run a centrifuge than an air compressor needed for forced air blow-off. If heat is added to the drying process, the cost increases as the temperature rises. In addition, depending on circumstances, manual labor could be the most costly part of the drying equation.

Semiaqueous

In semiaqueous cleaning, there are basically two distinct categories of agents. They differ based on the miscibility of the cleaning agent in water or the boiling points, i.e., the method used to separate the cleaning agent from the rinse water.[4] This is conducted either by gravity or by difference in boiling point. Separation by gravity is based on the immiscibility of the solvent and rinse water. Boiling point differences are separated by distillation of the water-soluble solvents and rinse water.

As in any industry, there are a number of semiaqueous formulations created from different solvent bases. Typically, the semiaqueous cleaning agent suppliers are the manufacturers of the main components included in their products.

In the fundamental semiaqueous process, parts are cleaned of soil with a suitable solvent that often may contain a detergent. The solvent is then removed from the parts by washing with progressively cleaner water. The parts are dried with hot forced air. To be economical, the cleaning agent must be separated from the rinse water by gravity or distillation. The rinse water may be purified further for recycle with membranes that reject organic materials.

A major advantage of the semiaqueous process is the high degree of waste recovery—the only direct waste is a concentrate of the soil in the cleaning agent. A major disadvantage is equipment complexity. Relative to a vapor degreaser, semiaqueous equipment is expensive.

The cleaning tank is designed similarly to those for other cleaning agent systems. The operating temperature is from ambient to as high as 180°F, because of the high flash point of semiaqueous cleaning agents. Soil concentration at equilibrium should be no more than 5–10 wt%. Cleaning time typically runs from 30 s to 5 min. Ultrasonic cleaning is often used for removal of particulates.

The next stage is called emulsion cleaning. The parts are removed from the tank and contacted with a rapidly moving stream of air (air knife) to blow off liquid cleaning agent from the parts. This is done for two reasons: (1) to adjust the soil concentration in the cleaning stage and (2) to adjust the cleaning agent concentration in the next emulsion cleaning stage. Too little blow-off will harm cleaning performance by raising the soil concentration in the cleaning stage and reducing the cleaning agent concentration in the next emulsion cleaning stage.

Water is sprayed onto the parts in the emulsion cleaning stage. Again, this is done for two reasons: (1) to remove cleaning agent and (2) to continue the cleaning process with a water emulsion of the cleaning agent. The water emulsion is often a better cleaner than the concentrated semiaqueous chemistry used in the cleaning stage because little soil is present in the emulsion cleaning stage. The temperature is increased slightly. Cleaning time is in the same range.

Cleaning agent concentration in water is from 1% to 10%. This is deliberately low to minimize organic cleaning agent flow to the final rinsing stages.

Separation Stage

The separation stage is not part of cleaning per se, but refers to recovery of the semiaqueous cleaning agent or to removal of oils from aqueous cleaning agents. Since the separation stage is the keystone of a semiaqueous process, the opportunity to avoid problems in that stage is worthwhile.

The term *gravity separation* refers to the driving force that controls the rate of separation. That is the density difference between water and the cleaning agent, and is typically 0.15–0.2 g/cc.

The emulsion is fed to a decanter for separation (in the gravity-separation process) and to a distillation column (in the distillation-separation process). Conditions in the decanter are deliberately different from those in the cleaning and rinse tanks; usually the temperature in the decanter is higher by 20°F–40°F. The separation should take place in between 5 and 30 min. An interface monitor in the decanter is used to activate pumps that withdraw the top organic phase and the bottom water phase. Removal is usually done in batch mode to maintain the organic/water interface between prescribed levels.

Problems occur in a decanter system when the withdrawal of one phase becomes contaminated with the other phase. A change in soil chemistry is a major potential cause of contamination. Another potential problem is foaming in the rinse tank, which can occur if spray nozzles are not correctly sized and positioned.

Distillation separates chemicals based on differences in their boiling points. For most solvents of interest, the difference between the boiling point of the solvent and of water is more than 70°C. That is well above the minimum of the 10°C–15°C acceptable for good operation. Further, boiling points of soil are typically 200°C above the boiling point of water.

The key advantages of a distillation separation system are reproducible and forgiving separation of soil from the rinse water and of water from the cleaning solvent. Operation could be with batch or continuous mode, depending on cleaning load. Batch distillation systems probably are less expensive.

Both types of separation schemes have been used in a variety of industrial situations. Decanters and distillation columns commonly are used in chemical plants and refineries. If the successful cleaning situation is one in which two solvents can be used—one of each separation type—the distillation option will work best. Distillation requires more capital ($5000 vs. $2000) and consumes more energy than does operation of a decanter. However, distillation is a more positive separation approach than decantation. It can be more easily monitored, and is less affected by changes in soil chemistry.

Ancillary Equipment

Cleaning solution chemistry can be as benign as hot water or can be a mixture of water and cleaning chemicals.[5] Cleaning chemicals are typically used where heavier soils such as oils need to be removed. Hot water is used where water-soluble contaminants (such as water-soluble fluxes) need to be removed from the part. The recovery and reuse techniques described apply to chemical-based cleaning solutions. Those cleaning solutions comprising water only can be dealt with using the techniques applicable to the recovery and recycling of rinse water.

The key to minimizing the disposal of cleaning solutions lies in extending their useful life. At some point, the cleaning solution becomes too concentrated in contaminants for the cleaner to perform adequately. The contaminants that cause a cleaning solution to become spent include both organic compounds such as free and emulsified oils and inorganic components such as dissolved metal, which are introduced into the solution as part of the process. They may also be components inherent in the cleaning chemistry or makeup water, which build up over time. Processes that are used for recovering aqueous cleaning solutions include oil skimming, media/membrane filtration, and coalescing.

Oil Skimming and Filtration

Oils removed from parts during cleaning can either be emulsified or "free," depending upon the cleaning chemical formulation. Some cleaners are formulated to reject soils, which allows the soils (typically oils) to float on the surface of the solution. Skimmers are used to remove these free oil layers. For those cleaners that are formulated to emulsify oils, the oil can be removed via a coalescing-type filter or membrane filtration.

Media Filtration

Media filtration (e.g., cartridges, bags, and sand) is used to remove suspended solids from cleaning solutions and associated wastewater. No dissolved materials are removed and these total dissolved solids (TDS) remain in the water.

Membrane Filtration

Membrane filtration processes are pressure driven and are used for various aqueous separations. Several types of membrane processes are used (microfiltration, ultrafiltration, nanofiltration, and reverse osmosis) depending upon the size of the contaminant to remove. The two most important membrane separation processes used in the recovery and reuse of aqueous cleaning solutions are microfiltration and ultrafiltration. The limitations on these processes are those created by the presence of material that can foul, scale, or damage the membrane.

Physical Principles of Coalescing

Liquid/liquid coalescing technology is used to accelerate separation of an emulsion. The principal driving force for coalescing action in either a gas or liquid stream is the interfacial tension of the droplets. Interfacial tension is the excess free energy due to the existence of an interface at the surface of a droplet, arising from unbalanced molecular forces. A relatively small interfacial tension value is typically required to obtain a coalescence rate low enough for practical application.

In a carrier stream of dispersed liquid droplets, the total interfering effect of surface active agents, particulate masking, or electrical charge is not great enough to render the dispersion permanent. The interfacial tension value between the two liquids is neither drastically reduced nor destroyed. Therefore, the dispersed droplets can be physically induced to agglomerate and the natural process of fluid coalescing can be mechanically accelerated to separate economically the liquids making up the emulsion. This provides the basis for liquid/liquid coalescing technology.

There are several different methods available to promote coalescence in an industrial process. Three primary mechanisms of coalescence are generally observed: impaction, Brownian diffusion, and turbulent field coalescence. Impaction occurs when the momentum of a droplet in the carrier stream causes it to collide with a droplet attached to a fiber or surface media, resulting in coalescence. The second mechanism occurs when the Brownian motion of a droplet in the carrier stream causes it to collide either with another droplet in the carrier stream or with a droplet attached to a fiber or surface media. In turbulent field coalescence, drops that have associated in pairs are pushed through the small capillary passage of the bed or barrier, resulting in turbulence in the carrier stream. The associated droplets eventually coalesce as a result of their relative motion when passing through the capillary.

Coalescing Equipment

Industry uses a variety of mechanical means to effect fluid coalescing. A settling tank reduces the velocity of a liquid emulsion and provides a quiescent zone. At low velocity, the dispersed droplets agglomerate and form a second continuous phase because of differences in specific gravity.

Additional techniques are used to improve the coalescing rate in settling tanks, including directional flow inducers and baffles. System modifications may include recycling the excess dispersed phase and flowing the emulsion through beds of coarse, porous media, such as wire mesh or fiberglass.

Similar methods are used to effect gas/liquid coalescing. Surge tanks are used to reduce the velocity of the gas stream, encouraging the agglomeration of liquid droplets. After the droplets settle, they are removed from the system. In many instances, vessels use devices to induce centrifugal flow and create abrupt changes in the direction of flow.

Coalescing Elements

Using elements with a medium of engineered surface and pore-size characteristics can augment coalescing of fluids. Several factors need to be considered when selecting the most effective fluid coalescing element.

1. The size and range of the openings (pores) in the porous material.
2. The relative surface tension value of the fluids.
3. The degree of wetting of the porous material exhibited by the fluid. (This is related to the surface tension value between the liquid and porous media.)
4. The fluid pressure drop across the coalescing media.
5. The chemical compatibility of the fluid system and the coalescing element.

Liquid/Liquid Systems

A liquid/liquid system that is a candidate for coalescing is generally in the form of an unstable emulsion. An emulsion is a dispersion of fine droplets of one liquid in a second in which the first liquid is completely immiscible or incompletely miscible. Generally, emulsions are formed by the mixing or mechanical agitation of liquids.

The dispersed fine droplets will rise or fall in the continuous liquid column as a result of differences in liquid densities. The droplets may impact other droplets, agglomerate, and become larger (coalesce). However, interfering factors usually retard or prevent natural coalescing at an acceptable rate.

Water Quality

Water of defined quality is needed for controlled cleaning. High-purity water is usually needed for precision cleaning; 18.3 MV-cm is considered the measure of perfection most commonly sought the world over when talking about water purity. The only commonly available way to achieve this resistivity level is by use of deionization. To appreciate fully what deionization is and how it works, one must first look at the contaminants found in water and what purification processes are needed, in addition to deionization, to provide water purity for a specific application.

Because pure water is the "supreme" solvent, it actively gathers contaminants from everything it passes over or through, including, potentially, the parts that are trying to be cleaned. Dissolved ionized solids such as sodium (Na), calcium (Ca), and chloride (Cl) are stripped from rock and soil. Organic molecules are gathered from decaying debris and environmental pollutants. Particulates include organic debris, dirt and rust from soil and piping; bacteria and microbials (including pyrogens) from normal growth in water; dissolved gases such as chlorine (Cl) and carbon dioxide (CO_2) from water treatment and organic decay, and colloids from rock and sand. All these contaminants are present in varying concentrations in water. Each presents different problems depending upon the application.

Deionization alone can allow achievement of 18.3 MV-cm resistivity, guaranteeing water free of ionic contaminants, but it does not remove organics, particulate, bacteria, or microbials. To remove these contaminants, other types of purification are used in conjunction with deionization. Activated carbon is used to remove organics and chlorine gas. Filtration is used to remove particulate and bacteria. Ultrafiltration is used to remove microbials, including pyrogens.

Resistivity is the measure of how much electrical current will pass between two electrodes at a specific distance. When an electrical current is passed through a solution such as water, ionic molecules are used as stepping stones by the electrical current. The fewer stepping stones, the more difficult the passage becomes, and the higher the resistivity reading. Most organic and bacteria are not adequate stepping stones to change the resistivity of water appreciably.

The temperature of water will also have an impact on its resistance. For this reason most water systems incorporate a meter that will automatically compensate temperatures to 25°C, the standard for water purification. The maximum achievable resistivity reading of water at 25°C is 18.3 MV-cm.

Ionic contaminants exist dissolved within the chemical structure of water. Dissolved ionized solids and dissolved ionized gases are removed using ion-exchange resins, which act like tiny magnets stripping ions from water, replacing them with H and OH ions, which ultimately join to form water (H_2O).

Ion-exchange resins are for the most part synthetic polymers with several ion-exchange sites attached to the surface. Two basic types of ion-exchange resins are used. Cation removal resins have several hydrogen ions (H^+) attached to their surface, capable of exchanging for positively charged ions. Anion resins have several hydroxyl groups (OH^-) attached to their surface, each capable of exchanging for negatively charged ions.

In a two-bed cartridge, these reactions occur separately with the cation removal resin being used first, followed by anion removal resin. A two-bed cartridge is used to remove the bulk of ionic contaminants, because when the two resins are separated, the cartridge has higher effective capacity for ionic molecules. However, a two-bed system cartridge cannot fully remove all the ionic contaminants because the reaction is never completed.

To achieve totally deionized water, a mixed-bed cartridge is required. The mixed-bed cartridge is configured so that the cation and anion resin are mixed. When a reaction takes place in a mixed-bed cartridge, the by-products of one reaction are picked up by the corresponding reaction, thus taking it to its completion.

As previously explained, deionization alone may not be enough for a specific application. This is the reason a system should incorporate more than one method of purification to deliver water free of any and all contaminants. The system should employ a pretreatment cartridge that utilizes a combination of macroreticular resin and carbon to prepare the water for the deionization that takes place in the following steps. The feed water first passes through the carbon to remove organics and chlorine. These components could potentially reduce the effectiveness of the ion-exchange resin. From the carbon, the water passes through a layer of macroreticular colloids. Colloids are very slightly ionized, extremely small particles that both clog conventional filtration and reduce the ability of the resin to produce high-purity water. This would be followed by a two-bed high-capacity cartridge to remove the majority of ionic contaminants as a preparation for the ultrapure mixed-bed cartridge.

An ultrapure mixed-bed cartridge is then employed to remove all remaining ionic contaminants yielding up to 18.3 MV water. Organics, which are still present after initial carbon adsorption and deionization, are removed now using high-efficiency synthetic carbon. Membrane filtration is used as the final treatment to remove bacteria and particulate, which have passed through the previous steps. A 0.2-mm hollow fiber filter attached to the faucet block performs the final filtration. For most applications, water after this step is sufficiently pure for use.

Automation

Whether the cleaning system is aqueous, semiaqueous, or solvent, automated parts handling can add enormous value to the process in terms of throughput, total output, and ease of equipment operation. The obvious requirement for a mechanical assist to moving parts through a system is the sheer weight of the load. However, there are other benefits that automation provides. In addition to eliminating the labor cost required if the unit were to be operated manually, automation increases consistency in the process, provides a traceable process, and permits the use of static process control.[6]

Automated systems are composed of four main components: the mechanical superstructure, the drive systems, the control package, and the operator interface. All these subsystems need to mesh with the entire tank line, which includes the tanks themselves, and environmental equipment. Consideration should also be given to up- and downstream production.

The most visible feature that differentiates the various automation options of a system is the mechanical superstructure. A number of standard designs are available and described below.

Mechanical Superstructure

The basic objective for horizontal and vertical travel should be a clean design with minimal moving parts, especially over the tank line. The automation system needs to be rigid, durable, adaptable to available footprint, and compatible with the chemistry in use. Concerns include overhead clearance and accessibility for the tank line to operators.

Overhead conveyors are chain or belt pulley systems mounted laterally over the centerline of the tanks. Stated tank lengths are typically exaggerated to allow for the transitions for vertical travel. There is no flexibility in altering processing and the only variability in throughput is by altering the speed of the conveyor.

Tank level conveyors use powered rollers to move payloads between stations and vertical movement is implemented by lifts in each tank. This can be an efficient approach to automation. However, processing flexibility is limited, tanks are significantly oversized, and it may not be appropriate for delicate parts.

A walking beam is typically a top- or side-mounted fixture that indexes payloads simultaneously. It can be advantageous in single-recipe, high-volume applications. It has the same limitations in flexibility and throughput described above. Additionally, these systems limit tank design in that all stations must be the same distance apart and all station process times are identical.

I beam or cable systems employ suspended independent head(s), which travel over the centerline of the tanks. The only advantage to these systems is where ceiling clearances are an issue. By design, the moving parts of the heads inherently create potential for contamination of the payload. Alternatives for low-ceiling applications include motion multipliers, or where footprint constraints require front-to-back tank layout, three-axis automation. However, given the potential for contamination, caution is required in clean room installations.

Cantilevered design has one horizontal frame mounted behind the tank line along which one or more heads travel and execute vertical movement. Properly designed, this concept is considered optimal for general applications since it creates the least contamination, uses the smallest footprint, and affords unimpeded operator access to the front of the tank line. Multiple heads, which overlap travel zones, can be an efficient way to increase throughput, especially during "dead travel" with no payload. Any head can lift more than one payload at a time for simple high-throughput applications (use of a "gang fixture"), although as in a walking beam the distance and processing time between stations must be equal.

Gantry/rim runners are two horizontal frames, one along each long axis of the tank line. From here the system can be essentially two I beam systems with associated contamination concerns, or mated cantilevered heads sharing weight distribution of the payload. The main disadvantage to this concept is that access to the front of the tank line is limited.

References

1. Reynolds, R., Cleaning equipment directory, *Precision Cleaning Magazine,* Witter Publishing, Flemington, NJ, February 1997.
2. Genet, C., Key requirements for proper rinsing in precision applications, *CleanRooms East '99,* Philadelphia, PA, March 1999, Penn Well Publishing, Nashua, NH, pp. 125–143.
3. Quitmeyer, J.A., Aqueous cleaning process challenges, in *Precision Cleaning '96 Proceedings,* Anaheim, CA, May 1996, Witter Publishing, Flemington, NJ, pp. 275–284.
4. Durkee, J.B., *The Parts Cleaning Handbook: How to Manage the Challenge without CFCs,* Section II, Semi Aqueous Cleaning, Gardner Publications, Cincinnati, OH, 1994, pp. 36–42.
5. Riley, C.T., Reduction/recycle/reuse concepts for aqueous cleaning process, in *CleanTech '98 Proceedings,* Rosemont, IL, May 1998, Witter Publishing, Flemington, NJ, pp. 128–136.
6. Aries, J., Automation: designing the right system for your cleaning equipment and production integration, *Precision Cleaning '97 Proceedings,* Cincinnati, OH, April 1997, Witter Publishing, Flemington, NJ, pp. 296–305.

19

From Laboratory Cleaning to Production Cleaning

Ronald Baldwin
Branson Ultrasonics
Corporation

Introduction

All aqueous cleaning processes involve three steps: washing, rinsing, and drying. The focus of most laboratory tests is to determine what must be done in the wash phase to actually remove contamination from the part while rinsing is often done in the most expeditious manner possible. Equipment constraints and material availability may dictate that only a few parts are cleaned using whatever equipment is available in the laboratory. Sometimes only one or two parts are cleaned using temporarily available production equipment from another application. On the surface, this might seem to be a path leading toward failure, but, just by applying a few well-known tools, the results of the cleaning tests can be scaled up to satisfactorily handle the full, ongoing production requirements. The purpose of this chapter is to provide these tools. While the discussion focuses on scaling up and automating a batch aqueous immersion washing and rinsing process, many of the principles can be applied to other aqueous and solvent cleaning processes.

Laboratory Cleaning: An Overview

What is the laboratory? It is any place where tests are performed to determine cleaning procedures. It may be a traditional laboratory filled with high-tech equipment, a small tabletop cleaner on a workbench or even a production cleaning line co-opted to test clean a different product. The persons performing the tests may be application chemists, process engineers, or just you, working by yourself. The laboratory may be located in your factory, or laboratory services may be provided by a private contractor or a chemical or equipment supplier. The end result of the laboratory test should be a clean part with a clear definition of the essential cleaning parameters.

Cleaning methods in the laboratory differ from production cleaning in several important ways. The first is in the number of parts that are cleaned. The laboratory will have a limited number of dirty parts available; those parts will be divided into batches to be cleaned using different chemistries or other process variables. Often, only one or two parts are cleaned with the final approved process parameters.

The second major difference is in how the parts are handled. The evaluators in the laboratory should have examined the parts to determine the best way to position them so that all cavities fill and drain completely. They then held the parts in some sort of fixture to keep them in the correct orientation. For many parts, this fixture may be as simple as a wire mesh basket. Others may require an elaborate clamp or they may be suspended by a wire. Still other parts may have to be rotated to promote efficient filling and draining or to be sure that the parts have adequate contact with the washing and rinse chemistries and forces. It is rare that the arrangement of parts in the cleaning tank matches the production configuration, especially since the laboratory cleaning tank is seldom the same size as a production tank and production baskets are not available.

Rinsing in the laboratory may be a one-step process since the amount of dragout from a small sample of parts is dwarfed by the volume of the rinse tank. At most, one additional rinse may be required to achieve the required cleanliness. Laboratory rinse water is usually high quality and low processing rates mean that the water does not become overly contaminated with use.

Sometimes, a laboratory may allow parts to air dry naturally or they may blow the water off with a handheld clean air nozzle. More rigorous drying tests may be performed with a hot air dryer similar to one that would be used in production. Since drying is often the limiting factor on production throughput, it is worth the laboratory's effort to duplicate the production process as closely as possible.

Determining Production Requirements

The throughput requirements must be determined before a production cleaning system can be specified. Generally, the production rate is known before the specification process starts. Make a preliminary basket or fixture layout to determine how many parts can be cleaned in one load. Verify that the size and weight of the basket are reasonable from ergonomic and automation standpoints. Divide the production rate by the basket loading to determine the total number of baskets that must be processed in a given time period. At this point, decisions must be made about how the parts will be processed through the cleaning line. These decisions may have to be revisited as the specification process progresses.

Make an estimate of the total number of stations in the cleaning line. The chemistry station types and dwell times may have been developed in the laboratory and can be used as specified. Estimate the production rinse requirements using the following guidelines. If the concentration of the last, or only, chemistry station is 10% or less, two final rinse stations are likely to be needed for spot-free drying. Higher chemistry concentrations will probably require three rinse tanks. If the final station contains a rust inhibitor, only one rinse station may be needed prior to applying the rust inhibitor. Use a single rinse tank between any two different type chemistry tanks.

Determine the dwell times for each of the tanks. In a batch immersion system, the drip time is the time the part is held after removal from one tank and before being placed in the next tank. Typically, the longest drip time occurs between the wash tank and the first rinse, to avoid dragout of cleaning chemistry. The chemistry station and drying dwell times are usually specified in the laboratory report. Rinse dwell times are somewhat flexible and may range from several seconds up to the same duration as the chemistry station dwell time. An exception to the rinse dwell flexibility rule occurs when the parts are prone to flash rusting. In those cases, the laboratory report should specify the dwell.

Determine how the baskets will be processed through the cleaning system. There are two methods of processing parts through a cleaning line, sequential and indexed. In sequential processing, one basket at a time is moved through the entire cleaning line. In indexed processing, multiple baskets are processed by advancing each basket one tank each cycle. If material handling is automated, one or the other method must usually be chosen, although more sophisticated systems allow for indexed processing. If the baskets are moved manually, the movements may be indexed but the maximum throughput may be reduced if the individual dwell times are short.

To estimate the processing time in a sequential system, add the individual tank dwell times and add a transfer factor, usually about 30 s including a 10 s drip or dwell time, for each transfer. Multiply this time by the number of baskets to be processed to determine if the entire production can be processed in the time available. If it cannot, re-evaluate the process as an indexed system or change the basket size or the tank size.

To estimate the processing time for an indexed system, first look at the times for each individual step. The maximum throughput time is the dwell for the longest tank plus the transfer time, usually between 45 s and 1 min. If one step requires two or more times the dwell of the other steps, consider adding a second tank and either processing one half the dwell time in each tank or alternating baskets between the tanks on successive cycles. The number of loads that may be processed in a given time period is the time period divided by the cycle time minus the number of stations in the cleaning line. This reduction is to account for the time that it takes to fully charge the system with baskets.

Now that the rate at which baskets enter a cleaning tank has been determined, the soil loading, dragout, chemistry concentrations, and water makeup rates can be used to estimate the conditions in each tank to make sure that parts are consistently cleaned to the same standards as achieved in the laboratory tests.

Determining Dragout and Soil Content

In order to effectively scale up from laboratory tests to production requirements, the types and amounts of soils that will be removed and the dragout, the amount of water or cleaning chemistry that will be carried from tank to tank by the parts and baskets, must be determined. The amount of soil dictates the frequency of solution or filter changes. The type of soil determines the auxiliary equipment needed to keep the cleaning solution clean. The amount of water carried from tank to tank, known as dragout, determines the ultimate soil concentration in a chemistry tank and how many rinses are needed to ensure spot-free drying.

Estimating Dragout

The easiest method to estimate dragout is to weigh a clean, dry, loaded basket, dip it into the wash solution at the operating temperature, remove it from the solution and reweigh it. The difference in weight is the dragout. The scale should have a resolution of at least 0.2% of the loaded basket weight. High dragout rates deplete the chemicals in chemistry tanks, increase the number of rinses needed to get spot-free drying and increase the time required to dry the parts. Some dragout is unavoidable since parts are wet when they are removed from a tank, but the biggest cause of excess dragout is improper basket and fixture design.

Basket Design

In the laboratory, parts may have been cleaned one at a time or in small batches in generic fixtures. When developing production processes, you have to pay attention to how parts are oriented and how they relate to one another and the cleaning system.

Mesh Baskets

Mesh baskets are well suited for holding parts where orientation is not an issue or where the parts will be naturally stabile when they are placed in the basket in the proper orientation.

Basket meshes can interfere with ultrasonic cleaning. Meshes with openings larger than ¼″ are good since they appear open to ultrasonic waves. Meshes with greater than 50 wires per inch are also good since they act as a solid surface and transmit ultrasonic waves with minimal loss. In between these limits, there will be losses caused by individual mesh wires scattering the ultrasonic waves. The amount of loss may be minor or severe depending upon the relationship between the mesh and the ultrasonic frequency.

Part Orientation

Many parts must be properly oriented so that they can fill and drain and also to expose critical surfaces to cleaning chemistries and to ultrasonics or sprays. Blind holes present particular orientation problems. If parts are positioned with the opening downward, air will be trapped and the hole will not be cleaned. If they are positioned with the opening upward, the hole will not drain, so soils and chemistry will remain after cleaning. It is usually best to orient blind holes horizontally. The part may have to be rotated if it contains blind holes in multiple orientations or if it contains threads or other features that would prevent static draining.

Some parts have critical surfaces that must be accurately positioned with respect to ultrasonics or sprays, while others have critical surfaces that cannot be allowed to touch anything. These parts must have fixtures to keep them in the proper orientation. Parts with specific gravities less than one must be restrained or contained.

Shadowing

Shadowing occurs when exposure to an energy source is blocked by another part or object resulting in a reduction in cleaning effectiveness. When spraying, a major cleaning effect comes as a result of the impact of the spray on a surface. Anything that prevents the spray from hitting a surface will reduce the cleaning effectiveness. Careful positioning is needed when spraying baskets containing multiple layers of parts since the upper layers will shadow the lower layers.

Ultrasonics are omnidirectional, so they are not as sensitive to part orientation as are line-of-sight sprays. However, ultrasonics are also subject to shadowing. Provide pathways for ultrasonic energy to get between layers or rows of parts. When cleaning large quantities of loose parts, use multiple trays with single layers, side-mounted ultrasonics and space the trays so that energy can travel to the farthest point between trays.

Dripping

Avoid placing parts so that they drip on one another when removed from a tank. Dripping may significantly increase drying time.

Contact Points

It is often difficult to clean and dry the points where parts contact the fixture or each other. Avoid surface-to-surface contact. Edge to crossing edge contact or wire to an edge contact is preferred. Provide room for part movement so that liquid can get under the contact points.

Coatings

Coatings on baskets or fixtures may be used to prevent damage to part surfaces. Avoid soft coatings on ultrasonic systems since the coating may absorb ultrasonic energy and reduce the energy available for cleaning parts. Hard nylon and teflon are commonly used coatings that have minimal effect on ultrasonics. All coatings are vulnerable to flaking and peeling.

Wash Tank Considerations: Optimizing the Cleaning Force

The goal in specifying production cleaning equipment is for the parts to see the same conditions as were experienced in the laboratory tests. In laboratory tests, it is rare that the soil loading capacity of the solution is approached or that production size tanks are used. Ultrasonic activity may be different because the ratio of the mass of production parts will be higher in relationship to the tank size and to the ultrasonic power.

Ultrasonics: Power and Location

The position of the ultrasonic source in the tank must be determined. In laboratory cleaning of a single part or of a small sample of parts, it often makes little difference whether the ultrasonic source is located on the bottom of the tank or on one or two sides. This is because the ultrasonics are omnidirectional and an open tank fills with cavitation bubbles that emit shock waves in all directions as they implode. However, when production loads are put into the tank, layers or rows of parts may restrict the passage of the ultrasonic waves resulting in areas of low activity. Use bottom-mounted ultrasonics when cleaning a single layer of parts or if the parts are racked with clear paths for the ultrasonics to travel between parts. If multiple layers of parts are to be cleaned, it may be best to use side-mounted ultrasonics so that the sound energy can travel in the space between layers to reach the farthest points. The separation should be proportional to the distance the sound must travel. Use ultrasonics on opposite sides of a tank if the distance is too great to clean with single-sided ultrasonics or to clean both sides of massive parts such as injection molds.

Parts and baskets absorb and reflect ultrasonics so that, as the parts loading increases, the amount of energy available for cavitation decreases. To more closely emulate production conditions, it is best to run laboratory trials on sample parts at 70%–80% of the power intensity that will be available for production.

Sprays—Impingement and Coverage

Sprays can have multiple uses in a precision cleaning system. They can remove bulk contamination in a preclean tank so that precision cleaning tanks do not have to handle large soil loads. They can remove excess cleaning solution prior to the first immersion rinse. They can dilute the dragout when transferring between rinse tanks, or, they can be used as a precision cleaning step if care is taken to ensure that impingement angles and forces are carefully controlled.

In dilution and bulk soil removal applications, the emphasis is on maintaining the cleanliness of the subsequent tank, not necessarily on achieving a specified cleanliness level for individual parts. For this reason, the spray system for bulk soil removal may not be subject to the same design considerations that would be required to achieve appropriate coverage in precision washing or rinsing applications.

The effectiveness of spray cleaning is very dependent upon the spray angle and force at the impact point. Use great care to ensure that each production part sees the same conditions as experienced in the laboratory tests. In the laboratory, spray angles and distances should be varied to simulate the expected conditions at each basket location. Part fixtures are usually required in precision spray cleaning to keep parts in their required positions and orientations.

Agitation

In immersion cleaning systems, moving parts within the cleaning tank or moving the wash and/or rinse solution past the parts can improve cleaning. While agitation in immersion cleaning seldom matches the force provided by a spray system, agitation enhances the cleaning process by repeatedly exposing soils to fresh cleaning solution and by flushing away contaminants that are loosened by ultrasonics or chemical action.

There are numerous ways to agitate cleaning baths such as mechanical oscillation, spray under immersion, gas bubbling, ultrasonics, or propellers. Generally, the laboratory report will specify a particular type of agitation if it is required.

Maintaining the Cleaning Solution

Concentration Control

Cleaning solutions change over time as soils are introduced and water content changes due to evaporation. Some chemistries, such as surfactants, change the character of the water but do not act directly on the soil and are not consumed in the cleaning process. Other chemistries bind with soils and must be replenished as they are used up. Many cleaning products are combinations of the two types and have to be adjusted or replaced periodically. Fortunately, most cleaning chemistries are effective over a broad range of concentrations, so there is seldom a need to have extremely precise concentration control. A good laboratory test will establish the upper and lower concentration limits required for effective cleaning and the system designer must determine how to operate within those limits.

Knowing how much water will be lost from a tank through evaporation allows a tank to be designed so that water may be added at convenient intervals. It also allows facility designers to determine how much water the ventilation system must handle. Water evaporation rates can be substantial in a heated tank and are a function of the surface area of the tank, water temperature, and the air velocity across the water surface.

Use Table 19.1 to estimate how much water will evaporate per hour for each square meter of tank surface area. Four kilometers per hour represents a shop area with a heater or air conditioning outlet blowing on the water surface. Twenty-five kilometers per hour is typical for a tank equipped with a lip vent exhaust plenum. Multiply the tank surface area by the value from the table to determine the total evaporation rate.

When adding liquid to a tank, it is important to note that evaporation removes water, not chemistry. If no parts are being cleaned, then makeup should be pure water. Dragout, on the other hand, removes

TABLE 19.1　Typical Evaporation Losses

Temperature (°C)	Air Velocity		
	Calm (L/h-m²)	4 kph (L/h-m²)	25 kph (L/h-m²)
50	2.8	4.1	7.0
60	5.1	7.5	13
70	8.8	13	22
80	14	20	35

TABLE 19.2 Makeup Solution Concentration

Tank Concentration (%)	Ratio = Evaporation/Dragout				
	20 (%)	10 (%)	5 (%)	3 (%)	2 (%)
1	0.0	0.1	0.2	0.3	0.3
2	0.1	0.2	0.3	0.5	0.7
5	0.2	0.5	0.8	1.3	1.7
10	0.5	0.9	1.7	2.6	3.4
20	1.0	1.9	3.4	5.3	7.1
30	1.4	2.8	5.3	8.1	11.1

chemistry as well as water, so the makeup should include chemistry. Since evaporation is always present, the makeup concentration will always be less than the basic tank concentration. Table 19.2 shows the required concentration of makeup solution as a function of the intended concentration in the tank and the ratio between the evaporation rate and the dragout rate. To use the table, first calculate the ratio of evaporation to dragout. Be certain that the units are the same, such as liters per hour.

Add solution at the proper concentration to restore the tank to its normal operating level after each operating session or whenever necessary. To replace losses that occur during periods of shutdown, use only water since there will have been no loss of chemistry due to dragout.

Automated Concentration Control

The simplest form of automated concentration control adds premixed solution as required by level sensors located in the tank. Use the preceding table to determine the concentration of the premixed solution. This method relies on predictable, steady production because it cannot compensate for variations in dragout due to changing production rates. Check the concentration periodically and adjust it as required.

The concentration of solution in a tank can also be controlled automatically (or manually) by monitoring a parameter such as pH, conductivity, or specific gravity that has been shown to have a clear relationship to concentration. Water or cleaning chemistry is added as appropriate, based on changes in that parameter. There are several factors that should be taken into consideration when using automated concentration control.

- Use a sensor that has good linearity in the controlled region. For example, do not use a conductivity meter for controlling a strong acid since the conductivity changes very little with concentration once a threshold value has been exceeded.
- Select injection devices for water and chemistry with flow rates that will not cause overshooting of the desired concentration.
- Avoid placing sensors in stagnant areas of the tank where they cannot rapidly sense the effects of adding chemistry or water.
- Inject water and chemistry where they can mix well with the tank solution before concentration is sensed.
- Install sensors so that they can be easily removed for calibration.

Particles and Filter Selection

Particulate filtering may not have been included in the laboratory tests because the number of particles dispersed into a bath when cleaning a few parts is often insignificant compared to the volume of the tank. But, if not filtered out, the concentration of particles will increase during production cleaning. In extreme situations, there may be nearly as many particles being carried in the dragout as are brought in on the dirty parts. The parts are just as dirty when leaving the tank as when they entered.

Select a filter with a retention rating that is appropriate to achieve the cleanliness needed for the part. Over-filtration simply adds expense due to increased maintenance and filter cost. While the product engineer must determine the largest particles that the parts can tolerate, the following examples are included to show typical filtration ranges. If the parts need to be visually clean, draw on the fact that the unaided eye can only see particles larger than 35–45 μm and select a 20–30 μm filter. Loose-fitting mechanical components may need 3–10 μm ratings while close-fitting bearings or flow valves may require 1–3 μm retention. Medical applications may need 0.45 or 0.22 μm filters to remove bacteria. Ultraprecision devices such as disk drives are forcing filtration limits to ever smaller ratings. A micron rating is a general term to indicate the ability of a filtration media to remove contaminants of various sizes.

Filters may have nominal or absolute ratings. An absolute filter will remove at least 98.7% of the particles of the rated size or larger. A nominally rated filter will typically remove about 90%; however, the ratings may vary considerably between types of filters and manufacturers. Because absolute rated filters are more expensive, most system designers use them only where particle counts are a major concern.

One problem is that it is difficult to analytically determine the number of particles that are brought in on the parts. Weighing is usually not effective since the weight of the soil is usually a very small percentage of the weight of the parts and baskets. A laboratory particle counter may be used for high-end applications. It provides tabular data on the number and size distribution of particles on the part. Measurements may be made directly on the surface or after extraction.

When dirty parts are cleaned, soils are removed from their surfaces and distributed into the volume of the tank. The filter system removes particles and returns clean, filtered water or cleaning agent to the tank. It therefore acts as a dilution system where the tank concentration is the ratio of the number of particles brought into the tank divided by the flow rate of the filtration system.

$$\text{Concentration} = \frac{\text{Particle input rate}}{\text{Filtered flow rate}}$$

The result has units of particles per unit volume. Since the volume of the dragout is known, the number of particles carried into the next tank, or remaining on the part after the final tank, is the number of particles per unit volume times the dragout volume. These calculations assume that all particles are removed by the cleaning process. This is not always true since there is a correlation between ultrasonic frequency and particle size removal. Frequently, a cleaning process will start with relatively low frequency ultrasonics and then use progressively higher frequencies as the parts progress through the system. The filter ratings should get progressively finer to match the higher frequencies. If this approach is to be used, it will normally be part of the laboratory specification.

Another way of looking at filtration is to determine how fast a filtered tank recovers after particles are introduced. The rate at which this occurs can be expressed in turnovers where one turnover is defined as flowing one tank volume through the filter. Table 19.3 shows the idealized relationship between initial soil concentration and the number of turnovers in a completely mixed tank.

The turnover rate is how many turnovers a tank experiences in an hour. Typical recirculation turnover rates vary between 3 and 12 turnovers/h. Twelve turnovers per hour is the maximum rate recommended in ultrasonic tanks because high flows can interfere with ultrasonic action and lead to uneven cleaning.

There are numerous other factors involved in selecting a filtration system such as chemical compatibility, pressure drop, and soil capacity.

TABLE 19.3 Relationship between Turnovers and Initial Concentration

Turnovers	Concentration (%)
0	100
0.5	61
1	37
2	14
3	5
4	2
5	1

Consult your local filter supplier for assistance in specifying filters for a particular application. Most suppliers are happy to lend assistance in return for continuing filter sales.

Oil Separation: Emulsifying and Splitting Cleaning Chemistries

Cleaning systems are frequently required to remove machining and process oils from parts; emulsifying or splitting chemistries are used. Oils present unique problems in cleaning systems because free oil floats on water and could form a layer that would recontaminate the parts as they are removed from the tank.

Emulsifying Chemistries

Emulsifying chemistries prevent this by dispersing the oil as tiny stable droplets throughout the bath. The stable droplets are held away from the part; however, excessive soil loading may result in redeposition of soils. Replace emulsifying chemistries when they become loaded with oil and start to lose their cleaning effectiveness. Sometimes, simply adding solution to make up for dragout losses is sufficient to maintain the cleaning bath.

Splitting Chemistries, Control, and Regeneration

Splitting chemistries initially emulsify the oil but then allow it to coagulate and float to the surface of the wash tank. The splitting chemistries are regenerated; however, the oil must be removed from the surface. When using splitting chemistries, one of the most effective ways to prevent recontamination from a floating oil layer is to sweep the tank surface with a stream of liquid (sparging) and force the oil over a weir into a reservoir where it cannot contact the parts. A sparged cleaning system generally consists of a tank equipped with a manifold containing closely spaced holes or nozzles located slightly below the tank surface. Opposite the manifold is an overflow weir and a reservoir with sufficient volume to allow the oil and water or cleaning agent to sit quietly so that the oil may completely separate from the water before the water is pumped back to the sparging manifold. The reservoir is often separate from the cleaning tank. Excess floating oil may be automatically removed from the reservoir with a skimmer.

If there is insufficient time available in the reservoir to allow complete separation, the water leaving the reservoir will have to pass through a coalescing filter before being returned to the sparger. When passing through a coalescing filter, the tiny oil droplets clump together until their combined buoyancy is large enough to float them to the water surface where they can be drained.

Temperature Control

Most processes have optimum operating temperatures that should be determined in the laboratory tests. The production cleaning system must be able to maintain the correct temperatures while meeting all production requirements. To maintain temperature, the heater must be able to compensate for the total of all heat losses.

Surface Heat Losses

Heat is lost through a liquid surface as water evaporates into the air. Use Table 19.4 to estimate heat losses from the liquid surface. Multiply the table value by the surface area of the tank to get the heat required to make up for surface losses. Four kilometers per hour represents a shop area with a heater or air conditioning outlet blowing on the water surface. Twenty-five kilometers per hour is typical for a tank equipped with a lip vent exhaust plenum.

Tank Wall Losses

For insulated tank walls, the heat losses will be much lower than water surface losses and can generally be ignored.

TABLE 19.4   Surface Heat Losses

| | Air Velocity | | |
Temperature (°C)	Calm (kW/m²)	4 kph (kW/m²)	25 kph (kW/m²)
50	1.7	2.5	4.4
60	3.2	4.6	8.1
70	5.5	8.0	14
80	8.6	12	22

For non-insulated tank walls, the heat losses will vary according to water temperature. To determine the tank wall heat loss, multiply the sum of the surface areas of the side walls and tank bottom, in square meters (Table 19.5).

TABLE 19.5   Tank Wall Heat Losses

Temperature (°C)	Heat Loss (W/m²)
50	297
60	410
70	524
80	638

Heat Lost to Parts

Submerging cold parts in a hot cleaning tank cools the water. This heat must be replaced to maintain the tank temperature. To determine the amount of heat required to bring parts up to operating temperature, multiply the total weight of parts processed in 1 h by the factor from Table 19.6. Do not forget to include the weight of the baskets. The table assumes that parts are 25°C when they are put into the tank.

Total Heat Loss

The total heat loss is the sum of surface losses, wall losses, and part losses. This is the minimum heat required to maintain a given tank temperature and can be used to estimate average operating energy usage. Typically, a tank will require additional heat to bring it up to its operating temperature.

Start-Up Heat Requirement

The heat required to bring a tank from room temperature to operating temperature usually exceeds the operating heat loss. If the tank is covered during start-up, the heat required is the amount of heat to bring the liquid up to operating temperature plus the amount of heat lost through the tank walls while heating. The heater size is the determined by the total heat requirement divided by the length of time allowed for start-up. Typical start-up heat requirements for a non-insulated tank are shown in the following Table 19.7.

As indicated in Table 19.8, additional heat is required if the tank is uncovered due to evaporative losses at the liquid surface.

TABLE 19.6   Heat Lost to Parts

Temperature (°C)	Heat Loss for Steel or Stainless Steel Parts (W/kg)	Heat Loss for Aluminum Parts (W/kg)
50	3.3	6.1
60	4.6	8.4
70	5.9	11
80	7.2	13

TABLE 19.7 Start-Up Heat for Covered Non-Insulated Tank

Temperature (°C)	1 H Startup (W/L)	2 H Startup (W/L)	3 H Startup (W/L)
50	33	18	13
60	45	24	17
70	58	31	22
80	70	38	27

TABLE 19.8 Start-Up Heat for Uncovered Tank

Temperature	1 H Startup (W/L)	2 H Startup (W/L)	3 H Startup (W/L)
50	39	23	18
60	57	36	29
70	79	53	44
80	106	73	62

Note that the heat requirements indicated in the above tables are approximations based on normal tank configurations where the tank width and depth are approximately equal and the length is between one and two times the width. The approximations may not hold for exceptionally shallow or deep tanks.

Use a programmable timer to start the heat before the first shift if start-up times are unacceptably long. Be certain to protect the heaters with a low liquid level safety switch when using programmable timers.

Rinsing

Rinsing removes the cleaning chemistry and any residual soils, and, because of dragout, removing the cleaning chemistry can be a significant part of the process. During the cleaning process, soils are removed in the wash tank and distributed throughout the wash solution. When the parts are removed, dragout can carry wash solution and residual soils into the rinse tank. In poorly controlled processes, the amount of cleaning chemistry carried out is often much greater than the amount of the original soil on the parts. This can result in parts that are actually dirtier than when they started. Fortunately, cleaning chemistries are usually much easier to remove than the original soils.

Reverse Cascade Rinsing in Production

Rinsing is often a one-step process in the laboratory since the amount of cleaning chemistry carried out with the parts is small compared to the volume of the rinse tank. In contrast, in production, multiple stages of cascading flowing rinses may be needed to reach the desired level of cleanliness. The system specifier must determine how many rinses are required to achieve the desired cleanliness level.

Water Quality and Spot-Free Drying

Any residues in the water dragged out of the rinse tank will remain on the part when the water evaporates leaving a stain on the surface. There are several sources for these residues. The first is the original soil that was on the part. A fraction of these soils will be carried from tank to tank with the dragout. Normally, this is a very small quantity of residue. A second much more important contributor to spots and stains is the chemistry used to clean the part because the chemistry is typically present in much higher concentration in the dragout than is the original soil. The amount of chemistry in the dragout is reduced at each successive rinse station.

The third source of spots and stains is the incoming rinse water, and the quality of the final rinse water is the key to spot-free drying. The quality of the rinse can never be better than the quality of the incoming water. Tap water frequently contains large amounts of dissolved impurities. Use reverse osmosis (RO) or deionized (DI) water to prevent introduction of contaminants that could cause spotting.

Resistivity is often used to measure rinse water cleanliness. $18.3\,M\Omega$ cm water is considered to be pure water. 2–$3\,M\Omega$ cm water will generally insure spot-free drying. While a well-maintained DI water system can deliver nearly pure water to a rinse tank, the resistivity reading within the tank will quickly drop to 3–$7\,M\Omega$ cm as gases such as carbon dioxide dissolve in the water upon contact with air.

Static Rinsing

Some processes use static rinsing, where there is no continuously flowing water. The water starts out clean and then becomes dirtier as each successive basket of parts is rinsed. Eventually, in uncontrolled processes, a situation may be approached where the amount of contamination in the dragout from the rinse equals the amount of contamination introduced to the rinse tank. Fortunately, in most cases, the water in static rinses is changed frequently, often once per shift or more, depending on the application. In some critical applications, wash and rinse baths may be changed after each basket.

Cleanup Ratio: Dragout versus Flow Rate

When production rates are higher than would be practical with static rinses, overflowing rinse tanks may be used in which water is added either continuously or intermittently and the excess water overflows to a drain.

The cleanup ratio is defined as the ratio between the average makeup water flow rate and the dragout rate. Typical cleanup ratios vary between 30 and 100. As an example, if there is one liter per hour of dragout and the makeup water flows at 1 L/min (60 L/h), the cleanup ratio would be 60 and the concentration of cleaning agent in the rinse tank would be 1/60th of the wash tank soap concentration. If the concentration of cleaning agent additives in the wash tank is 2% (20,000 ppm), the concentration of residual cleaning agent in the rinse tank would be 20,000/60 = 333 ppm (parts per million). While the rinse tank is far cleaner than the wash tank, the water in the rinse tank is not clean enough to prevent spotting. To put the situation in perspective, compared with 333 ppm in the rinse tank, $1\,M\Omega$ cm DI water has the equivalent of a 0.2 ppm sodium hydroxide concentration.

The previous discussion provides estimates of the average contamination observed in a single overflowing rinse tank after a large number of baskets of parts have been washed and rinsed. It is assumed that the dragout and rinse water mix well. The reality is that there will be a spike in concentration immediately after a basket is introduced to the rinse tank. The concentration will then decrease as makeup water dilutes the tank contents. The minimum concentration is achieved just before the next basket is inserted. Initially, this minimum concentration will be very low. However, it will increase as additional baskets are processed until the concentration calculated from the cleanup ratio is approached.

Cascade Rinsing: Number of Tanks

Why are multiple cascading rinse tanks so effective?

As seen in the previous example, the reduction in cleaning agent concentration in a single rinse tank may not be sufficient for spot-free drying. If the flow rate were doubled, the concentration would be cut in half, and the DI water cost and heating costs would double. However, the concentration of residue would still be too high to assure spot-free drying.

It would be much more effective to add another rinse. While the flow and heat would double, the parts would be 60 times cleaner. In addition, if the overflow on the final rinse were to supply the makeup water to the previous rinse, the required water flow and heat would be the same as for a single rinse,

TABLE 19.9 Recommended Number of Rinses (ppm)

Source Concentration (%)	Ratio Makeup/ Dragout 30	Ratio Makeup/ Dragout 60	Ratio Makeup/ Dragout 100	Ratio Makeup/ Dragout 150
1	3 (0.4 ppm)	2 (2.8 ppm)	2 (1.0 ppm)	2 (0.4 ppm)
2	3 (0.7 ppm)	3 (0.1 ppm)	2 (2.0 ppm)	2 (0.9 ppm)
3	3 (1.1 ppm)	3 (0.1 ppm)	2 (3.0 ppm)	2 (1.3 ppm)
5	3 (1.9 ppm)	3 (0.2 ppm)	2 (5.0 ppm)	2 (2.2 ppm)
7	3 (2.6 ppm)	3 (0.3 ppm)	3 (0.1 ppm)	2 (3.1 ppm)
10	3 (3.7 ppm)	3 (0.5 ppm)	3 (0.1 ppm)	2 (4.4 ppm)
20	4 (0.2 ppm)	3 (0.9 ppm)	3 (0.2 ppm)	3 (0.1 ppm)
30	4 (0.4 ppm)	3 (1.4 ppm)	3 (0.3 ppm)	3 (0.1 ppm)

but the cleanliness would be 60 times better than with a single overflow rinse. In the example, the concentration after a second rinse would be 5.5 ppm. Adding a third rinse would bring the concentration down to 0.09 ppm, well within the levels achieved for in-process DI water.

Use Table 19.9 to determine the number of rinse tanks required to keep the ppm of the cleaning agent in the final rinse below 5 ppm. The calculated ppm for the final rinse is also shown.

Adding multiple rinse tanks is the operationally least expensive way to improve rinsing. Constraints include increased floor space and higher initial investment costs.

Rinse Filtration

Rinse filtration is required if particles on the surface of the part must be controlled. While the basic principles are the same as described previously for a wash tank, controlling particle levels in the rinse tank is easier. You only have to control the particles remaining in the dragout from the previous tank, not the higher level particles present initially on the part. In addition, the micron rating of rinse tank filters is often smaller than for the rating for the chemistry tanks since the lower total particle count means that there is less danger of clogging a fine filter.

When a tank has a smaller filter rating than the previous tanks, assume that 100% of the particles smaller than the previous tanks' filter ratings are passed through to this tank in the dragout.

In calculating turnovers and cleanup ratios, add the amount of makeup water to the filtration rate to estimate the total volume of clean water. This means that even if an overflowing rinse tank is not filtered, the particle count will be reduced by the same makeup-to-dragout ratio as the chemistry dragout is diluted.

Spray Rinsing

Spray rinses are frequently used as initial rinses after the wash step to remove excess foam and to dilute the dragout. Spray rinsing alone may be sufficient in non-precision applications. The efficacy of spray rinsing is difficult to quantify due to variables including coverage and shadowing.

Sprays are sometimes located over immersion rinse tanks to provide an additional dilution of the dragout as parts are removed from the tank. Runoff and excess sprays drop into the rinse tank where they act as additional makeup water for calculation purposes. The results of spray rinsing are difficult to quantify due to coverage and shadowing issues. Air that becomes dissolved in the spray effluent may briefly reduce ultrasonic activity while it degasses.

Ultrasonics

In general, if a part is cleaned in ultrasonics, it is recommended that at least one rinse be ultrasonic. This is because, compared with spray or static rinsing, the ultrasonics are more likely to get into the same nooks and crannies that were cleaned in the chemistry tank and remove the residual cleaning solution. Ultrasonic rinsing may not be required for simple parts with easily accessed surfaces.

One technique that is becoming more common recently is to rinse with higher-frequency ultrasonics than were used to clean the parts. This is because higher-frequency ultrasonics can remove smaller particles. Since the soils that frequently prevent access to the small particles are removed by the lower-frequency ultrasonics in the wash tank, the high-frequency rinse ultrasonics have good access to the particles and can effectively remove them. Given that smaller particles are removed in the rinse than in the wash, a filter with a smaller filter rating should be used.

Resistivity Monitoring and Makeup Water Flow

Monitoring the resistivity of the final rinse can be used to ensure that the water quality in the dragout is adequate for spot-free drying. One method is to continuously flow makeup water into the rinses at a rate that should ensure good water quality and simply sound an alarm if the quality falls out of specification. The alarm is typically set to sound at a water quality level that is better than that required for satisfactory rinsing. This means that parts currently in the system can complete processing without exceeding the water quality limits. Because this method has the disadvantage of using makeup water even if no parts are being processed through the rinses, it is often used when the makeup water is processed in a closed-loop system.

Another method is to use the resistivity monitor to control the flow of makeup water between a high- and a low-resistivity setpoint. Makeup water starts to flow when the resistivity falls below the "add water" setpoint. The water stops flowing when the resistivity increases a fixed amount, called the hysteresis, above the add water setpoint. The add water setpoint resistivity is set to a higher resistivity than the warning or alarm setpoints. The advantage of this system is that it only uses water when it is needed and is well adapted to single-pass water systems where waste water goes to drain.

Temperature Control of Rinse Tanks

Static rinses have similar heat requirements to those of a wash tank.

Overflowing rinses require additional heat to bring the makeup water up to operating temperature. This is often accomplished by preheating the water with a standalone hot water heater. The amount of heat required is a function of the desired operating temperature, the incoming water temperature, and the water flow rate. Heater manufacturers rate the temperature difference as the temperature rise. Table 19.10 indicates heater requirements for typical makeup water flow rates and temperature rises.

The heat losses in a flowing tank are the same as in a static tank. Most overflowing rinse tanks are provided with heaters to improve initial startup times and make up for normal heat losses.

TABLE 19.10　Makeup Water Heat Requirements

Flow Rate (L/min)	Temperature Rise					
	10°C	20°C	30°C	40°C	50°C	60°C
2	1.4 kW	2.8 kW	4.2 kW	5.6 kW	7.8 kW	8.3 kW
5	3.5 kW	6.9 kW	10 kW	14 kW	19 kW	21 kW
10	6.9 kW	14 kW	21 kW	28 kW	39 kW	42 kW
15	10 kW	21 kW	31 kW	42 kW	58 kW	63 kW
20	14 kW	28 kW	42 kW	56 kW	78 kW	83 kW
25	17 kW	35 kW	52 kW	70 kW	97 kW	104 kW

Drying

There are many techniques available to dry parts after rinsing. The laboratory report will detail any specialized drying processes that may be required such as slow pull, air knife, vacuum, infrared, or water displacing solvents. For hot air drying, the laboratory report should specify the drying temperature, dwell time, and whether HEPA filtration or prior air blow-off is required. Verify that the dryer manufacturer's offering will meet the laboratory specifications.

System Design Verification

Recheck the initial system assumptions after determining the number of rinses required and selecting a dryer. The throughput calculations may have to be adjusted if the number of tanks has changed. Determine the system utility and space requirements.

Conclusions

Scaling up from laboratory tests to production equipment requires a basic understanding of cleaning principles. It starts with analyzing the throughput requirements and determining how the parts will be positioned and handled. The chemistry tanks must then be specified so that they duplicate the laboratory cleaning process as closely as possible with particular attention being paid to chemical concentration, temperature control, filtration, agitation, and other critical process enhancements. Proper rinsing techniques must be followed so that the required cleanliness level is achieved. Following these basic principles will assist you in specifying a system that will achieve the same level of cleanliness as was experienced in the laboratory.

20

Cold and Heated Batch: Solvent Cleaning Systems

P. Daniel Skelly
Riverside Chemicals

Introduction

In light of current and expected regulations, the trend of the 1990s has been to adopt aqueous (water-based) cleaning systems. In some applications, this may be the best choice. However, in other applications, water just does not work. Some considerations and problems include requirements for pretreatment of water supply, waste stream handling requirements and costs, limited efficacy of cleaning due to low solvency for many soils of interest and high surface tension, energy costs of heating and drying, requirements for rinsing and drying, high total cycle time, compatibility/flash rusting, complicated bath maintenance, high capital equipment costs, high maintenance costs, and large equipment footprint.

The Ideal Solvent

When evaluating a new or replacement cleaning system, the ideal solvent would have the following properties:

1. Environmentally friendly
   a. Does not create air or water pollution
   b. Biodegradable
2. Not regulated at the federal, state, or local levels
   a. Not implicated in ozone depletion
   b. Exempt from VOC regulations
   c. Not a HAP
   d. Not implicated in global warming

    e. Not on the SARA 313 or other regulatory lists
    f. Not a RCRA Hazard
 3. Solubility parameters match those of the contaminant to be removed
 4. Works well as a single-component solution to avoid complex proprietary blends
 5. Widely available at a reasonable cost
 6. Compatible with all construction materials in the operation
 7. Stable, does not readily break down in the presence of heat, metals, or chemical contact, and does not require the addition of stabilizers to achieve this goal
 8. Nonflammable at operating and handling temperatures
 9. Easily (and inexpensively) distilled or recycled
 10. Low toxicity (a high PEL), with extensive animal testing and a long application history
 11. Low or pleasant, yet detectable odor
 12. Worker exposure easily controlled under the prescribed conditions of use
 13. Fast evaporation rate for quick dry times
 14. Low vapor pressure to minimize solvent losses

Unfortunately, no chemicals have every desirable property, and development of an ideal solvent is unlikely. Therefore, the end user must evaluate the particular cleaning requirements as well as specific regulatory constraints.

Solvents are often characterized by their degree of perceived toxicity and rated as low, moderate, or severe. However, it is possible that the largest category, especially for the new-generation products, should be "unknown" or "unsure." Classic solvents, including the aliphatic and aromatic hydrocarbons, alcohols, ketones, and chlorinated solvents have each been studied by numerous organizations and testing laboratories. Even with this sizable database, scientists, toxicologists, and regulators seldom agree on the significance of their results. It is wise to assume that all chemicals have some degree of toxicity and a priority should be to minimize emissions and worker exposure.

Once the solvent options have been reviewed, the cleaning method must be chosen. Since there are no completely nontoxic solvents available for cleaning applications, the system must be designed to minimize hazards to the worker and the environment. This may include mechanical controls such as tank covers and auxiliary cooling coils to condense solvent vapors, or fans and exhaust hoods to remove solvent vapors from the workstation. Each of the solvent alternatives can be used safely with an appropriately controlled cleaning system.

With organic solvents, the choice of cleaning methods generally falls into one of three categories: ambient temperature (cold cleaning), elevated temperature (hot liquid dip), or vapor degreasing (cleaning in boiling solvent vapors and often immersion in the liquid solvent).

Cold Cleaning

Cold cleaning with organic solvents and solvent blends is often used when water is detrimental or ineffective, when the soils are of an oily or greasy nature, or when the capital costs of vapor degreasing cannot be justified. Generally speaking, the majority of the industrial cleaning applications can be accomplished in a cold solvent system. If cold cleaning provides results that meet expectations, use it. This method will ordinarily be the simplest, most trouble-free, have the lowest utility requirements, and be the least capital intensive of the cleaning system options.

Cold cleaning methods are as varied as the solvent choices that go with them. The most significant limitations to cold cleaning are decreased cleaning efficiency as a function of workload, absence of a drying system, difficulty in controlling flammability, potential worker exposure hazards, and regulatory compliance. However, these limitations can be countered by the solvent selection and by equipment design.

Pail and Scrub Brush

This method is very basic and has a low capital investment. However, solvent losses and worker exposure may be excessive, particularly with solvents having a high vapor pressure and low allowable exposure limits. Brushing provides some abrasive action, but is generally not effective on small or intricate parts. A rinse in clean solvent is often necessary after brushing, and there is generally no means for reclaiming the solvent once it becomes contaminated.

Hand Wipe

Hand-wipe cleaning can be accomplished by carefully pouring solvent on a reusable rag, or the purchase of presaturated disposable wipers. Mechanical rubbing with the wipe provides some abrasive action, but unless the soil loading is low, it is likely to leave a thin residue film.

Aerosol Spray

Aerosol cleaning is effective for removing soluble soils and the spray action helps to flush away insoluble particulates mechanically. However, it is generally inefficient in solvent utilization and is therefore reserved for small bench-scale and precision cleaning applications. Depending on the solvent selected, there is a potential concern for flammability and/or worker exposure to high levels of the atomized solvent.

Recirculating Overspray ("Sink-on-a-Drum") Parts Washer

This is a standard method for garage and maintenance shops, and has reasonable cleaning potential until the solvent becomes dirty. The solvent (traditionally a mineral spirits blend) is often replaced under a service contract, but it is necessary to assure that the solvent will be replaced often enough to meet the soil loading requirements. In addition, the convenience of this service generally comes at a high price. These systems are not generally suitable for high-vapor-pressure, low-flash-point solvents.

It may be worth considering some of the new systems with built-in vaccum distillation for on-site solvent recovery (Figure 20.1). Such a system can reduce overall solvent usage and minimize off-site waste disposal. In addition, freshly distilled solvent is available on a regular basis and the need for frequent solvent change-out is eliminated, a particular consideration in heavy-duty operations. With solvents and solvent blends where there are concerns for worker exposure and odor, the unit should be equipped with a hood and exhaust fan for proper ventilation. In areas of poor air quality, recent regulations have focused on the VOC content of solvents traditionally used in sink-on-a-drum systems. As a result, water-based cleaners have been the suggested replacement. Where organic solvents are required for performance, using a recirculating system with a hood and exhaust fan and with exempt solvents such as parachlorobenzotrifluoride (PCBTF) or volatile methyl siloxanes (VMSs) may provide an additional option.

Immersion Cleaning, Single-Dip Tank, with Manual Parts Handling

Immersion cleaning is often the most economical cold cleaning method. These are simple cleaning systems where the workload is lowered and raised hydraulically, mechanically, or manually into liquid solvent. Agitation generally increases efficiency. Air agitation is not recommended because of high solvent losses to the atmosphere, but ultrasonic agitation is often recommended because of its powerful scrubbing action. Mechanical agitation can be supplemented with a pump and filter.

Standard single-dip cleaning systems are offered by many equipment manufacturers for aqueous cleaning. With only minor modifications, these units can sometimes be adapted for use with organic solvents (Figure 20.2). Where worker inhalation exposure and odor must be controlled, top enclosures and side workload entry can be added.

FIGURE 20.1 Recirculating overspray parts washer with vaccum distillation. (Courtesy of Machine by SystemOne, Miami, FL. With permission.)

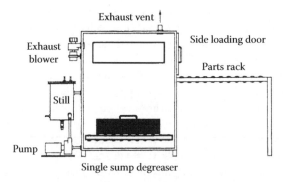

FIGURE 20.2 Single immersion dip cleaning system. (Courtesy of Machine design by Magnus Equipment, Willoughby, OH. From BCG-OX-36, Occidental Chemical Corp., May 1996. With permission.)

In Figure 20.2, the parts are manually loaded on a roller conveyor, fed through a side opening on the machine, then immersed and hydraulically agitated in solvent. At the end of the cleaning cycle, the deck is raised to the top position and the parts are allowed to dry. Drying is accomplished by passing a stream of ambient or heated air over the basket. This design is useful for light workloads and is adaptable to a wide variety of parts. Addition of a still would enhance removal of oil.

Automated Immersion Cleaning, Multiple-Dip Tanks

A multiple-dip system (typically two to four tanks) is recommended for applications having high soil loading. Sequential dipping into progressively cleaner dip tanks provides for efficient solvent usage,

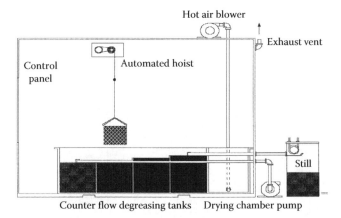

FIGURE 20.3 Automated multiple-dip cleaning system with cascade overflow. (Courtesy of Machine design by Finishing Equipment, Eagan, MN. From BCG-OX-36, Occidental Chemical Corp., May 1996. With permission.)

and the final rinse is in the cleanest solvent. Automated parts handling is recommended to maximize process control and reduce worker exposure (Figure 20.3). The system is generally unsuitable for containing solvents that have a high vapor pressure and low boiling point. Depending on the regulatory and toxicological profile of the solvent, additional controls may be needed.

In the design in Figure 20.3, an automated hoist controlled by a microprocessor picks up the workload at the operator station (far left) and processes the part(s) through a cleaning cycle, a hot air drying chamber, and then returns clean dry parts to the operator station. The operator has the option of controlling duration of immersion, number of immersions, rotation, drying time, and drying temperature. A distillation system could be added to remove oils and keep the final dip tank supplied with fresh clean solvent.

Heated Solvent Cleaning Methods

In applications where parts are not adequately cleaned with a cold solvent, a combination of temperature and solvency may be required. For example, buffing compounds, spinning compounds, and waxes are solids at room temperature and must be converted to a liquid for effective removal.

Heated Dip Tank

Although solvents are generally more effective cleaners when they are heated, there are a significant number of disadvantages. Depending on the solvent selected, flammability may be a concern, solvent losses increase, and there is an increased potential for worker exposure. To address these issues, extensive safeguards may be required, equipment design becomes more complex, and costs increase.

Vapor Degreasing

Although the capital investment can be significant, vapor degreasing (Figure 20.4) is a very effective and forgiving technology. Cleaning can be accomplished by immersion in hot solvent with agitation and ultrasonics. The final cleaning takes place in freshly distilled solvent. This vapor blanket also helps to minimize solvent loss. The most important problems relate to the additional engineering controls required to comply with environmental regulations and to control solvent loss, to minimize worker exposure, and to use specific equipment design for low-flash-point solvents. In addition, for certain solvents, buildup of water and acidity must be controlled, so the process has to be monitored.

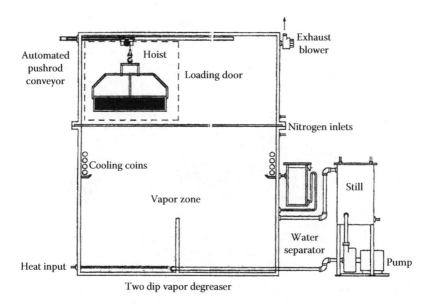

FIGURE 20.4 Open-top vapor degreaser with still, hood, automated conveyor, and inert atmosphere. (From BCG-OX-36, Occidental Chemical Corp., May 1996. With permission.)

Summary

Aqueous cleaning is not suitable for all applications; some solvent cleaning is appropriate. There is not now and there is never likely to be an ideal solvent. With appropriate controls and subject to the particular regulatory climate, solvents can be used responsibly in a variety of cold cleaning, heated cleaning, and vapor degreasing systems. The end user must consider his or her specific application to select the best option.

References

The following references were taken from technical bulletins produced by Occidental Chemical Corporation:

1. OXSOLs for Metal Cleaning, BCG-OX-19, January 1995.
2. Cleaning Systems for OXSOL 100, BCG-OX-36, May 1996.

21

Flushing: A Dynamic Learning Process in Soils, Chemistry, and Equipment

Richard Petrulio
B/E Aerospace, Inc.

Overview

The activity of cleaning the inside of items such as tubes, heat exchangers, and complex fluid passages presents a unique and difficult challenge. By using carefully chosen fluids and application-specific equipment to flush out identified soils, the desired results can be achieved. This chapter takes a look at the process of developing a flushing system in the context of compact commercial aircraft refrigeration equipment. As it turns out, the process has spanned nearly two decades and resulted in a number of iterations encompassing every aspect of the flushing activity. The lessons learned could be summed up as "no process is perfect." Although the steps discussed provide an insight as to how to approach developing a flushing process, the key is to continually question the premise and assumptions that lead to the elements of the chosen process. Given that, it can be seen how this is an iterative dynamic learning process. The first question is: what is flushing?

What Is Flushing?

What makes flushing so different from the mainstream of cleaning processes? Flushing is truly a unique process with subtleties that can make it difficult to develop. So what is flushing? Webster's says it is to "cleanse with a rush of water." From this simple definition the basic idea of a fluid moving across a surface to mechanically remove soils can be visualized. However, what is missing is the idea of cleaning enclosed, basically inaccessible surfaces. This goes beyond blind holes and slots to the concept that parts requiring flushing, to clean internal surfaces, are unable to utilize surface inspection to verify cleanliness. Thus, the process takes on a completely different context.

Why Flush?

The purpose of flushing is to remove soils from enclosed inaccessible surfaces. Soils left inside parts may cause catastrophic failure or they could simply limit performance. Long or complex passages that require fluid flow can become partially or totally blocked. Small valves with delicate or precise sealing surfaces can fail to seal. Surfaces that transfer heat may become insulated by soils causing a loss of heat transfer rate. For many products these problems will result in annoying or embarrassing performance situations. However, other products perform tasks that are safety critical. For these, performance and reliability are not just desires, they are requirements. An example of such products can be found in the liquid oxygen systems aboard the U.S. space shuttle. Of course, what drives the need for flushing may have nothing to do with safety, performance or reliability; it may hinge on aesthetics or industry regulations. The point being, any issue that forces the need to flush must be addressed in the context of its own requirements. However, if the product can operate and fully meet its intended function without flushing, then stop there. Do not flush time and money down the drain.

CASE STUDY: COMPACT AIRCRAFT REFRIGERATION

The inability to verify internal component cleanliness of a vapor cycle refrigeration system without dissecting each part is what drove the manufacturer of custom aircraft galley equipment to develop its own flushing process and equipment. Refrigeration systems such as these circulate relatively small volumes of hydrofluorocarbon refrigerant (HFC R134a) and polyolester (POE) oil through a closed-loop system. Both a precision refrigerant pump (compressor) and the electric motor that drives it are sealed within the system and thus are exposed to all that is circulated with the refrigerant. Valves with small orifices and heat exchangers with many meters of tubing join in as main parts of the fluid loop. Soils such as metal chips, moisture, acid, cleaning solvent residue, and persistent contaminants can do catastrophic damage to the internal components and materials in a refrigeration system. Unfortunately, poor cleaning in this application will not show up until the equipment has been assembled and operated (sometimes for many months). Having a compressor pump seize or an electric motor burn up is a very expensive method of verifying internal cleanliness. The question may still be asked, so why is flushing required? The answer comes in two forms; either what is inside a part is known or it is not. As such, the ability to remove both known and unknown soils from components of the refrigeration systems produced by this company set the context for developing a robust and comprehensive flushing process.

Mostly, removal of soils known or unknown from complex internal surfaces is driven by the need for product performance and reliability. As with refrigeration equipment, metal fines and solid particles can damage moving parts or foul critical passages. Other soils may drive chemical reactions within enclosed systems and subsequently degrade the materials. This is commonly seen within refrigeration systems when water or acid is left inside the parts. Water, when exposed to heat and refrigerant or oil can react to form various acids and sludge. Earlier refrigerant systems containing chlorofluorocarbon (CFC) or hydrochlorofluorocarbon (HCFC) refrigerants along with mineral-based oils would develop strong inorganic acids. Hydrochloric (HCl) and hydrofluoric (HF) acids would form due to the breakdown of refrigerant in the presence of metals, water and heat. Sludge, consisting of metal salts and deposits, would result from the corrosion of metal by the acid. In contrast, current systems using hydrofluorocarbon (HFC) refrigerant along with POE oil tend to experience decomposition of the oil into its base acid and alcohol components. In this case, the resulting acid is a weaker (less aggressive or damaging) organic material. In both cases, however, the acid is able to strip material from the walls of the commonly used copper tubing and redeposit it on the surfaces of the steel compressor parts. Deposits of copper will grow on the moving parts and violate the clearances required for proper operation. Thus, after the equipment has been in service for a short period of time, it grinds to a halt leaving the customer with unrefrigerated food and the manufacturer with a tarnished reputation.

How Clean Is Clean?

In the case above, the soils that could reside in the system were not benign. As such, the question of whether flushing was needed had already been answered. However, the next logical step before deciding on a flushing process is determining the required level of cleanliness. For each part or component that may require a flushing process, an analysis of the potential soils, sources and quantity must be performed. Based on the information compiled, the potential damage or part degradation needs to be evaluated. Potentially, a flushing process could be designed to eliminate every possible soil at its highest level of accumulation (design to the worst case scenario). This potentially points to a highly aggressive (read costly) process. However, what if the goal is to only clean as much as required to consistently meet the product (and customer) needs? This points to a more pragmatic and economics-based discussion. At what point does the level of cleaning required exceed the cost of system parts replacement? How often does the worst case scenario actually occur? Further, do other aspects of a manufacturing, refurbishment or repair process affect the need for cleaning? The example refrigeration system required flushing to remove particles that could do mechanical damage to a compressor or block critical valve passages. In addition, fluids that could form sludge or acid had to be removed. Although it is critical for these soils to be removed, is flushing the only prescription for eliminating the soil problem? Is it possible that other parts of the product design, manufacturing, and repair process play a role in reduction or removal of the identified soils?

What and When to Flush?

In order to determine if, and to what extent, flushing is required, a complete look at the parts and components is needed. In the repair process of the refrigeration equipment described previously, the components are subject to three potential soil sources: soils in new parts, soils remaining in the reused parts (repair activity), and soils from inadequate critical processes. Each of these soils presents different cleaning challenges as well as frequency of occurrence. Initially, it may be tempting to approach the development of a flushing system by trying to capture all of those soils. However, an analysis of each source can reveal other solutions thereby reducing the scope, and thus cost, of the flushing activity.

Soils remaining in newly manufactured parts can potentially be eliminated during the manufacturing process. In the case of the refrigeration equipment, the repair facility is a part of the OEM manufacturing plant. Feedback to the design, purchasing, and production groups drove the establishment of in-process cleaning and flushing of components prior to the parts being supplied for either new or repaired equipment. This reduced the need to flush parts immediately prior to installation and eliminated secondary contamination due to the effects of those soils. The secondary contamination can be chemical breakdown of the refrigerant or oil as previously described, as well as physical damage. Ultimately, soils in new parts represent a continuous potential source of contamination that if reduced or eliminated, can also reduce the overall flushing requirement.

With the reduction of soils in new parts, the quantity and frequency of occurrence of soils in reused parts begins to decline. At that point, an analysis of critical processes is in order. Aside from flushing, evacuation of refrigeration systems is one of the most critical processes in terms of eliminating contamination and avoiding system breakdown. The formation of acids in sealed refrigeration systems is directly related to the presence and quantity of water in the system. Elimination of water is primarily accomplished by evacuating the refrigeration system prior to charging it with refrigerant. Other critical aspects center on the highly hygroscopic nature of current synthetic refrigerant oils such as POE. Since acid cannot only damage parts but also remain in the system to cause future damage, maintaining the evacuation process becomes a critical part of establishing the flushing process requirements.

At some point, the reduction and control of soil sources will approach the point of diminishing returns. It is then that an analysis of flushing parts versus replacement of parts becomes viable. As it had been found with the repair of refrigeration equipment, the soils that were identified had multiple sources and varying frequency of occurrence. During the first few flushing process development and use iterations, the focus was on removal of all soils, including the most aggressive. Given the possibility of heavy contamination, it was thought that the process and solvent would need to be able to remove anything that could exist within the components.

Choosing a Solvent

Since the soil(s) to be removed drives the solvent required to achieve the desired cleaning results, all known and potential soils should be listed. Each of the soils should be evaluated to determine if it actually needs to be removed. This is another opportunity to choose a no clean option. Again, some of the soils may have no impact on the performance or reliability of a part. For those soils that do need to be removed, the process to flush them may not remove them all in one step. Aqueous flushing will require multiple steps to clean out the soils as well as rinse out the cleaning solution. In addition, the rinse water will have to be removed. Solvent flushing may also require additional steps to rinse out the main solvent. Thus, the choice of solvent will have an impact on the process time as well as the required equipment.

An early flushing technique used on the refrigeration heat exchangers employed a Stoddard-solvent-based cleaner that had been enhanced with perchlorethylene and methylene chloride. The solvent cleaned well but was incompatible with the refrigerant and oil used in the final product. Thus, it required a rinse step using HCFC-141b to flush out the Stoddard solvent. Both aqueous and solvent flushing chemistries likely require a drying step. The drying step is most difficult for aqueous and solvents that do not readily evaporate. As mentioned earlier, moisture is an enemy of refrigeration systems. Therefore, flushing with water was not a good candidate for the heat exchangers and plumbing.

As the methods and chemistry for flushing are being narrowed down, it is also important to consider material compatibility. In the same manner that soils were listed and their potential effects on the parts explored, compatibility of the cleaning compounds and process with the materials of construction must be evaluated. Substantial effort should be put forth to ensure that all materials that will be exposed to the flushing process are not degraded in either the short or long term.

Verifying the Process: Is It Really Clean?

Since the nature of parts that require flushing is that they have inaccessible internal surfaces, how can you verify that those surfaces are actually clean? Before completing the development of a flushing process, it is vital to answer that question. Two directions can be taken for flushing process verification. The first is continuous inspection of parts. The other is to develop the process so that it will yield the desired results without inspection. Although these methods could apply to almost any process, flushing presents a unique inspection challenge. To gain visibility of internal surface cleanliness one would require special equipment such as a borescope, methods such as chemical analysis of cleaning fluid, or dissection of the part.

Dissection of a part can only be used for spot or batch inspection due to its destructive nature. However, for high-volume, low-cost parts, this might be effective. Fluid sampling can give results for 100% inspection via analysis of solvent samples from each part. Such a method would be suitable for small quantity parts, which require high precision and consistency of cleanliness. The downside of this method lies in the tracking required for each sample and the potentially long turn around time for results. Immediate results can be obtained by using sophisticated equipment such as a borescope. However, such equipment is expensive and requires proper training to be used effectively. Additionally, the interiors of some parts are not conducive to accepting the borescope into its interior. For these reasons, developing the flushing process to do the job right every time becomes an attractive option. In the case of refrigeration equipment, heat exchangers, and plumbing, developing the flushing process to ensure consistent performance results proved to be the best method.

Environmental Considerations

Consideration for cleaning level, soils to remove and material compatibility has now been given. But, what thoughts have been given to the environment? Remember that cleaning is a dirty business. A manufacturing or repair facility must be able to provide a safe and appropriate environment for the flushing process to be performed. Further, the process must be designed to have a minimal impact on our surroundings. The solvent flushing process used for the heat exchangers required a location with ample space, electrical power and ventilation. Safety for the operators as well as governmental emission limits required that the process minimize release of the chosen solvent. These considerations were made a part of the design for the equipment as well as the process steps. Even so, upon initial operation of the system, air monitoring was conducted to ensure the safety of the operators and nearby employees. Long-term monitoring consists of emission logs to track any losses of solvent. Ultimately, common sense, sound ethics, and regulatory requirements will dictate what equipment and steps are needed to build a safe and environmentally sound process.

The Virtual System

Before the equipment is built and dropped on the production floor, the whole process should be put together virtually. Nail down the chemistry desired to fit with the projected requirements (look to have a backup) and envision the equipment needed to use the chosen chemistry. Next, write down each step of the process from start to finish and decide what skill level of operator is required to perform those steps. Then, project ahead to when the process has matured some and look at who will be in charge of the process and equipment. Will the equipment be reliable? Who will be responsible for maintenance? What about record keeping and follow-on training?

Building a virtual process will help shake out some of the bugs and shed light on potential pitfalls. Now it is time to make the process a reality. Equipment can be obtained outside or developed in-house. Since flushing is unique even to the cleaning industry, finding a turnkey system that is off the shelf is nearly impossible. Custom-designed systems can be fabricated but are generally very costly. In addition, the fabricator may not fully understand the process thereby making it difficult to get the desired system. For these reasons, it may be justified to develop a system in house. In support of this phase of the process, it is worth employing outside help to provide industry contacts and keep the sales glitz to a minimum. Time spent interviewing the industry with a well-versed consultant along for the ride is truly worth the cost.

CASE STUDY REVISITED: THE FIRST SYSTEM

The first flushing process developed for cleaning heat exchangers and plumbing in the galley refrigeration equipment was designed and built completely in house. The unique requirements of refrigeration eliminated the ability to comfortably use an aqueous process; not flushing was an unacceptable option. Thus, the use of a solvent-based flushing process was the only viable choice. The level of cleaning for the flushing process was difficult to determine. It was clear that moisture of any significance could not be left in the system. Past failures due to copper fines ingested by the compressor pump lead to the discovery that forming oils within heat exchangers and complex tube shapes held these fines in place. These were the main challenges for flushing of new equipment components. However, the repair side presented the challenge of cleaning out mechanical debris, acids, and overheated or burnt oil. A condition labeled as a "burn out" would periodically occur when the electric motor overheated or shorted out and caused the refrigerant and oil to burn. This generally will coat the entire refrigerant system with a tough, adherent black residue from the burnt oil. Despite the fact that such severe types of contamination did not occur frequently, it was thought that the flushing system developed should be required to handle all the soils. However, to meet these needs, a very aggressive solvent and flushing system would be required.

Fortunately, there had been a history of flushing methods used for the cleaning of the refrigeration equipment parts. The downfall of these past methods was shaped by the increased environmental awareness. Without the creation of new processes to take over for the old unacceptable methods, product performance and reliability suffered severely. One of the first methods employed for flushing heat exchangers was to blast liquid dichlorodifluoromethane (Freon R-12 or R12) through the tubes and out to the parking lot through a hole in the wall. Verification of cleanliness was performed by allowing some of the expelled R12 to flow through a white towel. If the towel remained white, the flushing was deemed complete. Although very effective, this method was environmentally unsound. The replacement method used the previously mentioned enhanced Stoddard solvent to break down the oils and loosen debris. A flush of HCFC-141b using the spray wand inside a vapor degreaser rinsed out the Stoddard solvent residue. Finally, shop air was blown through the heat exchanger to remove and evaporate the HCFC-141b. Again, the method worked but it was costly and also environmentally unsound. In fact, HCFC-141b is no longer produced. It was at this point that a systematic approach to developing a flushing process was initiated.

The understanding of soils and materials had been investigated and the time had come for professional assistance. With the help of a top consultant, numerous solvent chemistries and equipment options were investigated (Petrulio et al., 1998). As a result, n-propyl bromide was chosen to perform the cleaning task. However, that still left the equipment end open. After numerous conversations, demonstrations, and some hard-to-swallow quotes, it was clear that the only way to obtain the flushing equipment needed was to design and build it in-house. The result would be a safe, effective system with a reasonable price tag.

During the virtual process phase, a minimum of five different systems were penned out. Each focused on the need to introduce solvent into the parts, flush out the soils and then remove the solvent without exposing the operator to the chemical. The largest challenge was removal of the solvent from the part such that it was not vented to the atmosphere and yet left the part clean and dry. A sophisticated system was prototyped and given to the repair department for evaluation. Although it worked, a couple of the premises needed to be revisited. The system was modified on paper for creation of the first full-scale system. Its operation employed a vapor degreaser to provide clean hot solvent. Pumps drew the solvent out of the degreaser and directed it through the part being flushed. Solvent was passed through filters to collect particulate soils while the degreaser separated out the oils from the solvent. Automatic pump timers terminated the solvent circulation cycle. The solvent left in the part was then pushed out with a carefully controlled flow of nitrogen. The mixture of nitrogen and solvent was delivered to a separating tank to allow the solvent to be captured and the nitrogen expelled.

Each of the components needed for this process was selected from commercially available hardware to ensure reasonable cost and allow for future replacement. Electrical controls and safeties were employed to allow for simple and safe operation. As the system took physical shape, a comprehensive procedure and maintenance manual was written so as to be available as soon as the equipment was put into operation. Initial use of the system demonstrated the success of the process. The development effort had paid off. As to be expected, however, initial operation yielded a few issues; a couple of valves did not operate as expected, technicians found ways to let small parts be ingested into the pumps and the solvent separation left something to be desired. Minor changes corrected the valve and debris issues; however, the separation issue took some thought. Due to the boiling point of the solvent being close to its operating temperature, when the solvent entered the separation tank, some of it would exhaust with the nitrogen. The solution turned out to be a significant feature of the total flushing system. Refrigeration was used to sub-cool the solvent prior to reaching the separation tank. With the solvent now in a fully liquid state, it separated from the nitrogen effectively.

The Iterative Process: A Second System

Two complete original systems had been in operation with continuous successful use. Cleaning results from the flushing process had been consistent and effective. However, a new situation arose and some lurking questions crept in. Changes in environmental regulations in Southern California required substantial reduction of solvent emissions. Since the solvent used was very aggressive and under strict regulation, a complete change of flushing equipment was in order. Again the driver of the design decision rested on the soils to be removed and the solvent required to accomplish the task. At that point, the criteria for the flushing system still embraced removal of all soils that could possibly be present within the system. In order to accomplish this, the solvent still needed to be very aggressive and the equipment to contain it very sophisticated.

The change in perspective of this iteration focused on meeting emission regulations while still being able to flush out all possible soils. In order to satisfy these needs, a sophisticated "airless" machine would have to be developed by an outside vendor. As with the development of the first process, multiple designs (from different vendors) were evaluated with assistance of an experienced consultant. The chosen system was developed, fabricated and presented to the local authorities for certification. In choosing this type of solution to meet the perceived flushing need, a substantial amount of financial and human resources were required to complete the project. The result of the effort and time was both a capable flushing system but also the catalyst to approaching flushing from a different perspective.

Match the Process to the Need

Return to the question; what is the true flushing need for any particular product? In the case of the refrigeration equipment, a significant driver was extreme soils that occurred with only a single-digit percentage frequency (dubbed one-percenters). Does it make sense to develop the capability of the process based on the one-percenters? What this lurking question implies is that soils and contamination that do not occur often, or can be eliminated by addressing other areas of the manufacturing or repair process, may not need to be flushed out. It became more apparent that an economic trade-off between the cost of replacing parts and that of the aggressive equipment was in order. As such, a significant part of the flushing system development process should include a true needs assessment for flushing the product. On one side of the equation is the cost of the flushing equipment, solvent, operating expenses, and maintenance. On the other side is the cost of replacement parts when they cannot be effectively flushed. As it turns out in the case study of refrigeration units, the cost of replacing heat exchangers and plumbing parts could not justify the cost of the equipment needed to flush with highly aggressive solvent. Further, analysis and improvement of critical processes along with improved design and new part cleanliness lead to a dramatic reduction of soils needing to be flushed out. The need for flushing still existed; however, the need for aggressive solvents was nearly eliminated. This revisiting and assessment of the true flushing need uncovered another lurking question: how do we support our customer's (repair shop) flushing needs?

A significant need exists in the support of customer repair shops. In the case of aircraft galley refrigeration, the equipment is used around the globe. Many aircraft operators choose to perform their own repairs and thus require both equipment and processes that meet the OEM factory specifications. Typically, these operators have relatively small numbers of refrigeration equipment to repair and they may already have a shop and personnel who are capable of the work. In that context, it becomes even more difficult to justify expensive flushing equipment and aggressive solvents. Further, numerous operators expressed having difficulty purchasing and being allowed to use the originally specified aggressive solvent. The flushing need began to reshape itself into that of a cost-effective process requiring readily available and permissible solvent while covering 95% or more of the flushing situations.

The Final Iteration

From this account, the perspective of developing a flushing system takes on a different context; one that could only have been seen through numerous iterations of development, use, and assessment. The main steps still focus on

1. Evaluation of soils
2. How clean the product must be to perform the intended function reliably
3. Choice of solvent
4. Verification of the flushing process
5. Source of equipment (make or buy)

However, additional keys to the development of a successful process lie in matching the process to the true need and assessing the results once the process is in place.

The final flushing process chosen for cleaning out the refrigeration equipment components was quite simple. As it turns out, one of the original concepts explored in the development of the first process became the basis for the final iteration. The process removes the soils contained in over 95% of the system components. Where components contain soils that cannot be eliminated through flushing, the components are replaced. Verification of the process came by way of testing using a prototype setup, development of contamination criteria, and analysis methods for the equipment components along with tracking of equipment performance and reliability data. The solvent chosen is simply the same refrigerant used within the refrigeration equipment. As such, every repair shop around the globe is able to obtain and use the specified solvent. The process operation meets all the requirements for safe use and handling of the solvent and it was able to be put into a compact package for purchase by the airline repair shops.

Each of the iterations of flushing process development experienced by the refrigeration system manufacturer lead to further insight (this might be called wisdom). In the end, they were able to develop a process that met their needs in the context of both performance as well as economics. Given the willingness and patience, along with the guidelines and wisdom presented above, any organization has the ability to develop their own effective and economical flushing process.

A Final Thought

Although the process can be summarized in five main steps, there is an additional step that encompasses the entire activity. The step could be listed as "start over from the beginning." Each of the iterations presented provided a view into the learning dynamic that goes hand in hand with process development. As such, the true "final" step is to dig through the entire process and continually look for ways to make it better.

Reference

R. Petrulio, B. Kanegsberg, and S.-C. Chang. A practical search solves aerospace cleaning quandry. *Precision Cleaning Magazine*, 7, 22–31, August 1998.

22

Solvent Vapor Degreasing: Minimizing Waste Streams

Joe McChesney
Parts Cleaning Technologies

Introduction

With the ever-tightening rules and regulations imposed on solvent cleaning by federal, state, and other regional governing bodies, today's cleaning systems must be safe, efficient, effective, and environmentally compliant and must provide the required throughput to meet the end user's needs.

As existing solvents are regulated and new solvents emerge, it is important for those engaged in industrial and precision cleaning as part of the manufacturing or repair process to understand the guidelines for using solvents in conjunction with available cleaning systems on the market today.

One would think these statements would not indicate growth in the solvent cleaning market. But the contrary is true. Replacement with modern equipment and new processes yield a limited upward trend in solvent applications.

This chapter provides a current comprehensive look at the following aspects:

- State-of-the-art equipment and their operational processes
- Available/EPA-approved solvents review (Environmental Protection Agency)
- Current EPA/OSHA/NESHAP solvent regulations review (Occupational Safety and Health Administration; National Emissions Standards for Hazardous Air Pollutants)
- Market trends for solvent cleaning
- Gazing into the crystal ball …"What is the future of solvent cleaning from today's view?"

To understand today's use of solvent cleaning systems, one must first take a retrospective look at how our current environmental rules and regulations were developed. This brief glance backward will establish the criteria that ushered in the current regulations for solvent systems. It will also help you understand how today's companies all have common goals of meeting the environmental parameters established by governing bodies worldwide. Companies include those that use solvent cleaning systems, that manufacture and provide the equipment, and those that manufacture and provide the solvents. These goals consist of achieving even lower limits of solvent usage to better our environment, preserving a process that is currently required for specific applications, as well as providing a cost-effective process and a safe working environment.

Retrospective

For several decades, solvent degreasing was the preferred cleaning method for most manufacturers. This method had replaced aqueous processes in most factories because degreasing was a simpler, one-step process that required less floor space, less maintenance, and greater compatibility with numerous soils/substrates and was generally less costly to operate. It provided a cleaning system that generally

- Accepted any type of metal substrate
- Removed most lubricants or contaminants in a single tank or process step
- Cleaned any size/shape parts providing maximum solvent penetration into the smallest tolerance achievable
- Yielded parts that were dry upon removal. If any residual solvent remained, it soon evaporated into the atmosphere, leaving minimal residue
- Consolidated waste into one area for recycling/disposal
- Was easy to operate/easy to maintain

Pre-Montreal Protocol

As scientists studied the earth's environment, one such area of interest in the 1980s was the ozone layer. It was soon a topic of great interest and concern. The scientific community theorized that the use of these solvents for degreasing, carrier fluids, and other applications was harmful to the ozone layer around the earth.

One of the features that contributed to the success of solvent usage soon became the focal point of action. Public awareness increased, world government bodies formed investigative committees, chemical manufacturers performed tests, and environmental organizations made models of the atmosphere to analyze these theories. The results of these actions were the driving factors to create a worldwide pact to decrease and eliminate the use of certain solvents, first as aerosol carrier fluids and later as degreasing fluids as well as in other applications.

Based on worldwide environmental stewardship, a group of nations formed and sanctioned the rules and regulations concerning these solvents. This meeting and joint agreement became known as the "Montreal Protocol."

Post-Montreal Protocol

After the initial regulations were established and the final review and approval by each member nation was obtained, new environmental regulations were enacted.

The regulations prohibited the use of some of these solvents, regulated the amount of others that could be used, and establish future replacement/phaseout dates for other solvents. Warning label notices were required for products cleaned in certain solvents during the phaseout period.

Given other environmental and worker safety issues, new regulations regulated or specified the equipment that used certain solvents. Regulations limiting air emissions concerning equipment operator exposure, equipment emissions to the atmosphere, and the amount of solvent that a company could use within a defined time frame were developed and instituted.

Effects in the 1990s

The Federal Halogenated Solvents NESHAP resulted in changes in degreasers; approximately 50% of degreasers nationwide were retrofitted to meet these rules. Decisions of the EPA SNAP (Significant New Alternatives Policy Program), the group that determines acceptability of substitutes for stratospheric ozone depleters, caused some solvent users to change to different, SNAP-approved solvents. Many companies switched to alternative processes based on misinformation or fear of any future solvent regulations. Many companies switched their cleaning process based on environmental outlook/corporate direction. Many companies switched based on federal, state, or local record keeping requirements.

Sometimes the switch was too quick. Companies chose the wrong equipment, the wrong chemistry, the wrong application, and/or the wrong process. In addition, there was a learning period going to aqueous from solvent. Some companies had success. Others had partial success; they are still using limited solvent cleaning.

Other companies failed in their initial process modification and either converted back to solvent or replaced the wrong equipment with processes appropriate to the application.

Some companies are still using solvent but not in compliance with the law. Reasons include lack of awareness of requirements, avoiding learning about requirements, and the high costs of compliance. Still others may be waiting for the future to bring an inexpensive universal environmentally accepted drop-in solvent.

Today's State-of-the-Art Equipment and Their Operational Procedures

Today's regulations change quickly regarding environmental concerns for solvent degreasers. Research by the major solvent equipment manufacturers and the solvent suppliers evolves at the same expediency to match the regulatory changes. New solvent choices are constantly being developed and offered, sometimes within the regulations time frame and sometimes ahead of them. The degreasers of today meet and typically exceed all federal guidelines concerning design criteria, operator exposure, and atmospheric emissions. Just as today's automobiles are more reliable with greater fuel economy, today's degreasers clean better with less solvent consumption. Equipment manufacturers have been able to incorporate new designs and technology that solves cleaning problems quicker and with less operator involvement/exposure to the actual cleaning process.

New solvents on the market have also contributed to this equation by reacting with certain lubricants quickly and effectively. The new solvents are designed for environmental compliance in terms of VOC content, vapor pressure, lower air emissions, stability, and other concerns. When combined with current generation degreasing systems, the user realizes the full effect of solvent conservation and improved cleaning that yields a higher quality product at less cost than in the previous years. Operator exposure is minimum, and the solvent emissions from the actual equipment are the lowest in years.

Vapor Degreasing Equipment

Regardless of whether the cleaning process involves vapor only, immersion, spray-in-air, ultrasonics, or spray-under-immersion degreasers can be generally classified in three categories: traditional open-tops, totally enclosed low-emission maximum efficiency, and airless or vacuum. Current designs are configured to insure environmental compliance and maximize production while minimizing solvent consumption and/or drag-out.

Traditional Open-Top Degreasers

Features

Open-top degreasers are available in various shapes and sizes depending on the weight/size of the products and the throughput. The basic design lends itself to be loaded through the tank top opening either manually or with a conveyance device. The throughput is generally higher than for other types of degreasers based on loads per hour and size of these loads. The simple design of open-tops equates to standard equipment offerings minus complex material handling and cleaning processes. This generally means lower equipment cost.

Disadvantages, Traditional Open-Top Degreasers

Because open-tops can be used with minimal parts handling equipment, there is the potential for increased solvent drag-out based on operator control. Because there is a greater vapor–air interface, the potential exists for increased emissive losses compared with enclosed systems. There is unrestricted access to the full tank opening, so the potential exists for the operator to process larger size parts/baskets than the unit is rated for, thus increasing solvent consumption via vapor displacement, vapor migration, or carryout.

In addition, larger units require more efficient cooling for both water-cooled primary condensers and any freeboard refrigeration devices. Open-top units are more susceptible to ambient temperature and air drafts across the top of unit, thus increasing emissive losses. The operator has greater potential for exposure to solvent vapors.

Improvements in Current Open-Top Designs

More effective cooling systems use more efficient materials to control vapor levels; this equates to lower vapor emissions. Modern heat controls insure a regulated heat supply that senses actual equipment conditions and makes appropriate adjustments. Benefits include energy conservation and reduced emissive losses.

Increased freeboard ratios in combination with tight sealing covers (mostly powered) also reduce emissive losses. In addition, freeboard chillers are now designed with maximum efficiency in mind. They are installed in stratifications to achieve full coverage of the areas required. With larger units, the chiller coils are installed using alternating banks or levels that allows one part of the system to be operational (at −9°C, −20°F) while the alternate level is being defrosted. This feature insures that the chilled air blanket is maintained at all times, thus reducing emissive losses.

Uses of material handling devices are now recommended with all open-top degreasers to decrease operator interface and to increase control of process parameters. This allows control of solvent exposure time: dwell cycles, work basket tilt to increase drainage (if possible), spray application, and cover operation.

Updated water separator systems now feature increased cooling capacity to assist in removal of water or moisture from the process or air. This decreases emissive losses, affects solvent stability, and favorably impacts equipment operation and structural integrity.

Depending on the specific application parameters, other devices such as carbon adsorption systems and superheated vapors can also assist in reducing overall solvent consumption.

Summary: Current Open-Top Degreasers

Today's open-top degreasers are at the most efficient design in their history. The overall solvent idle emissions are at the lowest ratio per square foot of vapor–air interface. The units yield acceptable performance, generally at a minimum cost. They are simple to operate, maintain, and generally provide more access and flexibility to the degreaser tank than any other type system.

However, these units are not the most environmentally effective systems. The degreaser tank (in normal operation) is generally open to the ambient atmosphere and is subject to ambient air currents and temperature. There is also the human factor of an operator who has the potential to control the actual machine-operating conditions. The operator may decide what/when/how goes into the system; this results in variables that may contribute to the overall solvent efficiency and environmental impact of a particular machine.

Totally Enclosed Low-Emission, Maximum Efficiency Degreasers

Attributes and Current Design of Enclosed Degreasers

In enclosed degreasers, the design enhancements of the today's open-top units are utilized, and the degreaser system (including tank(s) and load/unload area) is enclosed in a cabinet. A material handling system to automatically convey the work in/thru/out of the degreaser is added. This yields a system that provides the required cleaning process based on set parameters and an enclosed unit that provides minimum operator interface/exposure while providing maximum environmental conservation. The system is not subject to air drafts, provides limited operator control of the actual process, reduces vapor–air interface based on largest size work basket designed for the system, reduces ambient air/moisture entry to the system, and is more energy efficient.

With automation, the work sees the same process parameters each load, with identical dwell times as established for the vapor/immersion/freeboard. The conveyor system is interfaced with any spray cycles to control exact timing and location, thus reducing emissive losses. The enclosure (typically including powered isolation doors) shields the degreasing tank(s) from the ambient plant and the operator and conversely isolates the vapor layer from the operator and the plant environment.

Various types of enhanced cleaning action such as spray-under-immersion, ultrasonics, and vertical agitation can be incorporated. Product rotation can enhance cleaning of parts that have blind holes, passageways, complex geometry, etc. Rotation also helps in draining entrapped solvent from the parts/basket as they are rotated above the liquid tank(s). This reduces solvent carryout.

The equipment solvent tanks are "right sized" for the workbasket. This generally means a lower solvent capacity in the system and less vapor–air interface, thus reducing solvent emissive losses.

Solvent Recovery

The enclosure decreases emissive losses because the tank has limited contact with the outside atmosphere. An enclosure also allows the addition of solvent recovery devices such as carbon adsorption systems that will capture solvent vapors.

When one processes parts through a traditional vapor degreaser, some solvent remains on the parts. The volume of solvent depends on the parts or even the workbasket itself may be in liquid and/or vapor form inside the parts or as a film coating the outside of the product. This solvent will eventually evaporate as the parts/basket are exposed to the atmosphere. By delaying the removal of these parts until the solvent can be extracted from them, the user achieves the maximum efficiency possible within the process time frame and equipment used. There are several methods of extraction but a carbon adsorption system provides the most efficient means to remove and recover solvent before it reaches the atmosphere. By incorporating a carbon adsorption system into the enclosure, one can isolate the work from the degreaser tank after cleaning, thus insuring that solvent is not extracted from the working tank. The clean work with entrained solvent is then exposed to an air stream that pulls across the work, extracting solvent with it as the air and solvent mixture is pulled into the carbon recovery unit's vessels. Makeup air is pulled from the ambient area outside the tank to provide a cooler temperature air, which mixes with the solvent for better adsorption efficiency.

The solvent-laden air is then pulled into the carbon vessels for future recovery and reuse in the degreaser. This results in lower solvent usage, lower emissive losses, lower plant/operator exposure limits, and maximum efficiency. In some cases, carbon recovery efficiency has been reported at 98% and more. Users report reduction of solvent purchases by 50% annually.

Disadvantages of Enclosed Degreasers

Equipment costs are higher than are traditional open-tops. Throughput is generally reduced compared with a traditional open-top system. This is based on a cycle time that includes conveying parts into the system, opening and closing the isolation doors, controlling the cycles, and conveying the finished parts out of system.

The work processed is limited to size of workbasket. There is decreased flexibility in utilizing the tank area and decreased tank access. Typically, increased plant floor area is required.

Because an enclosed degreaser is a more complex system than an open-top, it generally requires regularly scheduled thorough maintenance. Decreased solvent consumption requires attention to solvent condition inside the unit.

Summary: Enclosed Degreasers

Today's totally enclosed solvent degreasers yield very low emissions and provide maximum efficiency. The units meet the most current applicable environmental regulations. Solvent vapors within the plant are reduced and can be further reduced by adding solvent conservation devices. The units are automated and provide controlled, repeatable performance. While such degreasers generally provide lower throughput than an open-top degreaser, they may provide higher throughput than an airless system.

Airless or Vacuum Degreasers

Airless and Vacuum Degreasers

Airless and vacuum systems are discussed elsewhere in this book. Similar to the totally enclosed systems, these units yield minimum emissive losses and low actual solvent usage at a generally lower throughput. Equipment costs are generally higher than for open-tops. The user that must meet strict environmental regulations and needs solvent processing might consider these units.

Solvents

The solvents of today are a mixture of old and new. We still have some of the typical chlorinated degreasing fluids that have been used for years while others have been effectively regulated out of existence. New solvents have been formulated while others not generally thought of as degreasing fluids are now being used for that purpose.

Environmental impact is the major deciding factor for most solvents. Some current regulations regarding solvents and solvent degreasers are of particular concern. One is EPA regulations including Clean Air Act/Title V permitting hazardous waste stream management. Another is the federal National Emission Standards for Hazardous Air Pollutants (NESHAP), including both equipment and recordkeeping issues. In terms of equipment, maximum achievable control technology (MACT) and lowest achievable emission rate (LAER) control technology are perennial concerns. OSHA worker exposure limits must be achieved. And local environmental regulations, exceeding federal regulations, may be present.

Other factors in solvent selection include safety, performance compatibility with substrates, global warming potential, ozone depletion potential, EPA approval as a substitute for ozone depleters, volatile organic content level, flammability, worker exposure profile, purity, cost, availability, and technical assistance/stewardship.

Fugitive Solvent Emissions Reclamation Using Carbon Adsorption

Overview of Fugitive Solvent Emissions

Most manufactured products must be cleaned in order to remove lubricants, cutting oils, drawing compounds, miscellaneous contaminants, etc. used in the fabrication process.

When the cleaning process involves typical solvents, it is practical, efficient, and sometimes mandatory that the emissive solvent vapors be recovered and possibly reclaimed.

Carbon adsorption is one of the most efficient and cost-effective pollution control/solvent recovery processes available today. Carbon adsorption reclaims solvent vapors that would normally be dissipated to the atmosphere.

Carbon is the preferred material used in adsorption systems because it exhibits unique surface tension properties. Due to its nonpolar surface, activated carbon will preferentially attract other nonpolar materials such as organic solvents rather than polar materials like water. Carbon's granular multifaceted geometry also possesses tremendous surface area (with 1 lb having an area greater than 750,000 ft^2). This characteristic allows carbon to adsorb up to 30% of its own weight in solvent.

Solvent recovery consists of passing solvent-laden air through an activated carbon bed (Figure 22.1). The activated carbon captures the solvent molecules allowing residual denuded air to be exhausted to the atmosphere.

Process

Adsorption

Solvent-laden air is directed from the exhaust source to the activated carbon bed by a blower/fan assembly (Figure 22.2, left). The carbon adsorbs the solvent vapor and residual purified air is exhausted through the ventilation duct. This process continues until the entire carbon bed is near saturation.

Desorption

At the end of the time allowed for adsorption, the unit will automatically switch the incoming air flow from the first carbon bed to a second carbon bed. This will allow incoming solvent-laden air to flow through a fresh activated carbon bed while the first bed is desorbed or stripped (Figure 22.2, right). The first bed is now injected with steam, which passes through the carbon bed vaporizing the adsorbed

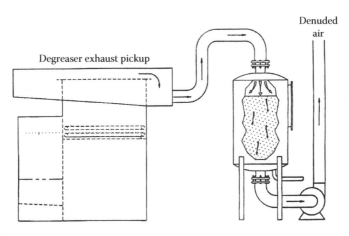

FIGURE 22.1 Solvent recovery.

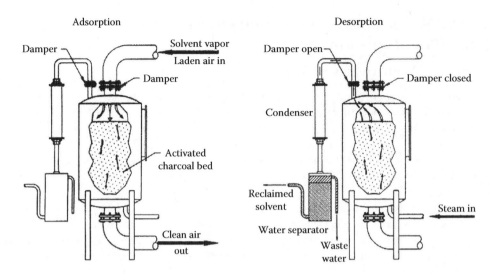

FIGURE 22.2   Solvent adsorption/desorption.

solvent. Additionally, the physical characteristics of the steam condensate passing through the carbon assists in removing solvent residue. The mixture of steam condensate and solvent then passes through a water-cooled heat exchanger, which cools the solution. This allows for gravity separation to occur in a water separator device due to the difference in specific gravity of the liquids. The reclaimed solvent is now ready for reuse or disposal. The water discharge is channeled for treatment or disposal.

Cool Down

As hot wet carbon will not readily adsorb solvent, the carbon must be dried and cooled before the next adsorption cycle. Ambient air or process air is drawn through the bed for a preset period of time, which dries and cools the carbon. At the end of this cycle, the unit shifts into a stand-by mode ready for the next adsorption cycle.

Calculations of Solvent Emissions/System Design

The proper sizing of a carbon adsorption system (CAS) for a particular solvent application depends on two main factors:

1. The volume of recoverable solvent that is to be directed to the CAS
2. The amount of air mixture that is to be directed to the CAS

With respect to the first factor, obviously, if the fugitive emissions that are lost cannot be picked up in an air stream and directed toward a carbon adsorption system, these fugitive emissions are not recoverable by normal means. The best way to determine the recoverable loss is to measure the amount of solvent in the air stream that is presently being emitted from the source, which is usually through a vent duct. This can be done with a variety of meters on the market today but is most accurately performed with a recording device connected to a properly calibrated meter. It should be emphasized that solvent losses may not be constant. This necessitates either continuous monitoring via a recorder connected to the meter or periodic sampling. For best accuracy, all factors should be gathered, starting with the total solvent purchases and then determine how much is used, disposed as waste, and lost as a fugitive emission.

With respect to second factor, accurate equipment measuring solvent/air mixture is useless without knowing the exact amount of air flow. This can be best determined by using a precise air-measuring

TABLE 22.1 Capacity of Activated Carbon—Lb Solvent per Lb Carbon at 80°F

Solvent Level (PPM)	TCE	MC	PCE	CFC 113
20	0.052	0.00413	0.069	0.033
50	0.066	0.00935	0.092	0.041
100	0.077	0.0154	0.099	0.050
200	0.091	0.022	0.112	0.060
500	0.110	0.037	0.129	0.077
1000	0.124	0.049	0.142	0.091
2000	0.138	0.0633	0.152	0.102
5000	0.153	0.08	0.167	0.125

Notes: PPM, parts per million; TCE, trichloroethylene; MC, methylene chloride; PCE, perchloroethylene.

TABLE 22.2 Loss Factors

Trichloroethylene
lb per hour loss = PPM × CFM × 0.00002175
Perchloroethylene
lb per hour loss = PPM × CFM × 0.00002756
Freon 113
lb per hour loss = PPM × CFM × 0.0000286
Methylene chloride
lb per hour loss = PPM × CFM × 0.00001414

Note: CFM, cubic feet per minute.

instrument. In most cases, lower air flow rates are preferred to reduce the amount of fugitive emissions being generated and increase concentration of the mixture. This will allow the CAS to be more efficient.

Calculation examples for determining CAS size make use of the carbon capacity with the solvent of interest (Table 22.1) and the rate of loss of the solvent (Table 22.2).

Size of CAS System

Customer vent system air flow is measured to be 2500 cubic feet per minute (CFM). The solvent is trichloroethylene. The daily stack loss average measurement is 1500 ppm for the first 8 h and for the next 16 h is 800 ppm; the operation runs 6 days a week.

Referring to Table 22.2,

$$lb/h \ Loss = PPM \times CFM \times 0.00002175$$

$$For \ first \ 8 \ h \ 1500 \times 2500 \times 0.00002175 = 81.375 \ lb/h$$

$$For \ next \ 16 \ h \ 800 \times 2500 \times 0.00002175 = 43.4 \ lb/h$$

$$Or \ for \ total \ day \ 81.375 \times 8 = 651.0$$

$$43.4 \times 16 = 694.4$$

$$= 1345.4 \ lb/day$$

The user's records indicate machines connected to this vent are using 70 drums per month and they operate it 26 days per month, which means they are losing out the stack 26 days/month × 1345.4 lb/day = 34,980.4 lb/month but are using 70 drums/month × 660 lb/drum = 46,200 lb/month. Therefore,

34,980.4/46,200 or 76% is directly available for recovery. A CAS is approximately 95%–98% efficient so we can expect to recover a maximum of.95 × 34,950.4 = 33,231 lb/month.

Referring to Table 22.1, at the rate of 1500 ppm, a unit will hold approximately 0.13 lb of solvent per lb of carbon. Therefore, with a system containing 1200 lb of carbon, 0.13 × 1200 = 156 lb of solvent can be recovered before desorption is required.

At this high rate, the CAS will require desorption in a little over 2 h so the timers can be set to desorb every 2 h during the first 8 h. At the lower rate, i.e., 800 ppm, approximately 0.12 lb per lb or 0.12 × 1200 = 144 lb of solvent can be recovered between desorbs and since only 43.6 lb/h is lost during the remainder of the day, the timers can be set for 4 h between desorption for the 16 h the CAS is on low rate, thus saving water and steam.

Thus, by calculating or obtaining the following information, a carbon adsorption system size that will safely handle the solvent emissions can be determined:

- Operational hours
- Amount of solvent to be adsorbed
- Adsorption characteristics of applied solvent
- Pounds of carbon required to adsorb incoming solvent amount
- Air stream velocity/volume
- Discharge limits to atmosphere
- Safety factor

Operational Cost Estimates

The CAS system described above, with 1200 lb of carbon, needs 500 lb/h of steam, 2400 gal/h of water, a 5 HP motor for the steam blower, and a 3 HP motor for the condenser water to recover 4.5 gal/h of TRI.

Electrical

$$\text{Steam power: } \frac{500\,\text{lb/h} \times 945\,\text{BTU/lb}}{3415\,\text{BTU/kW}} = 138\,\text{kW}$$

A 5 HP blower motor (460 V/3-phase/60 Hz) needs 6.04 kW; similarly, a 3 HP motor will have an energy use rate of 3.8 kW.

Therefore

$$138 + 6.04 + 3.8 + 52\,\text{kW (Misc.)} = 200\,\text{kW Total Loading.}$$

Using Tennessee Valley Authority General Power Rate for a 200 kW total load × 24 h operation × 31 days/month = 148,800 kWh at 100% loading.

GP-12 Rating for these conditions per TVA area is $8.708/h average costs.

Water

40 GPM at 85°F inlet = 2400 gph
At a rate of $0.0013/gal, water cost is $3.20/h

Return on Investment

Electrical costs	8.708/h
Water	3.20/h
	11.9/h operating costs

Using trichloroethylene at $5.87/gal × 4.6 gal/h (recovered) = $27.00/h (recovered) less $11.91 operation cost = 15.09/h Net payback.

Therefore

$$\frac{\$33{,}275.00 \text{ (system cost)}}{\$15.09/\text{h}} = 2205 \text{ h payback or } 138 \text{ days operating } 16 \text{ h/day}$$

The recovery system can therefore pay for itself in well under half a year.

*n*PB Solvent Recovery/Reclamation

*n*PB solvent has found general industry acceptance as a replacement fluid for other chlorinated compounds that are being phased out or face more stringent regulations by the U.S. EPA as well as OSHA.

For those solvent users that have switched to *n*PB, carbon adsorption technology allows the recovery of emissive losses as well as lowers the ambient ppm levels in the workplace.

When a carbon adsorption system is properly sized and installed, the ambient air around the degreaser and the fugitive solvent emissions are pulled into the suction duct. As long as the carbon unit is operational, any fugitive emissions are continually removed from the workplace to lower the ppm levels in the immediate area. Eventually the activated carbon in the vessel adsorbs enough solvent to reach a point near saturation. At this point, the solvent-laden vessel must be desorbed to recover the solvent.

The injection of live steam is the most common and preferred method for chlorinated and other solvents recovery in the industry when using a traditional carbon adsorption system. However, laboratory tests revealed that the usual method of solvent reclamation using a traditional carbon adsorber with *n*PB would not be advised because, when one uses live steam with *n*PB solvent, a chemical reaction occurs that removes the stabilizers from the solvent. This destabilization could create problems with the solvent and/or equipment.

A *recently awarded patent design* describes a chemical and mechanical system installed on the carbon adsorption system to insure that neither the solvent nor equipment is affected during the adsorption/desorption process. This process insures the solvent that has been reclaimed through normal steam injection will be able to be reused in the original degreaser process without chemical reaction to the solvent in the degreaser, to any other solvent it comes in contact with, any other equipment containing solvent (such as another degreaser or distillation unit), or actual products being degreased with this reclaimed solvent.

The capture and recovery of *n*PB results in operational cost savings. A typical recovery rate is 98% or better of the captured solvent laden air stream in a properly designed and installed carbon adsorption system. Typically one can reduce gross annual solvent purchases by 50% when properly operating a degreaser with a carbon adsorption system installed.

Summary: Carbon Adsorption Systems

Carbon adsorption systems are ideal for solvent recovery. Numerous systems exist in the field today reclaiming various solvents with high efficiency. Recovery efficiency as high as 95%–98% of the incoming solvent-laden air stream can be achieved or exceeded.

A new, patent-pending technology allows for the capture and recovery of *n*PB solvent vapors and chemically treats the system to insure proper solvent quality levels during normal operation.

A recovery system quickly pays for itself in solvent savings. Typical reduction in gross solvent purchases due to reclamation is 50%. In addition, carbon adsorbers help comply with safety and environmental regulations while providing a better workplace environment for your employees.

Market Trends in Solvent Cleaning

While this market will never reach the level it once was, there is a definite need and/or demand for solvent cleaning in specific applications. From the sheer size of some products and the cleanliness criteria that today's aerospace/military/medical industry demands, solvent degreasing offers a simple yet effective solution.

Based on demand for modern environmentally compliant degreasers, equipment suppliers are seeing a limited increase of solvent degreasers. Some are replacing existing dated solvent systems. Others are replacing aqueous systems that were purchased to replace old degreasers but did not perform as expected, proved to be to labor-intensive, or were operationally cost prohibitive. Still others must now institute cleaning, must meet higher production rates, and/or must meet increased cleanliness specifications. Typically solvent is an easier choice to resolve this issue.

Environmental drivers are moving equipment design toward totally enclosed systems where possible. The enclosed design, whether atmospheric shielded tank or hermetically sealed chamber, is the choice of many companies. When one looks at the cost of lower solvent usage and emissive losses combined with a controlled process that yields repeatable results with minimum operator interface, the advantage of these systems becomes readily apparent. One still has to determine the required throughput, process parameters required to yield clean parts, material handling aspects, selected fluids, and the system that best suits their needs from an overall scope. But this is getting easier in today's world. The current generations of degreasers are designed with many factors considered. Among them are performance, environmental compliance, solvent conservation, reliability, and market demands. These factors can be placed in any priority based on the needs of the individual user. There are many factors to consider when purchasing a solvent degreaser system. Please seek out professional knowledgeable assistance to provide you with information that you will need to make your decision.

Crystal Ball: "What is the Future of Solvent Cleaning?"

There are several ways to answer this question. While we manufacturers readily admit that the market has shrunk over the last decade, it is not dead. The industry as a whole is seeing new machines being ordered. The smaller units tend to be standard in design based on work chamber size versus production requirements and options desired. The larger units tend to be more custom designed based on specific applications.

We see a number of trends from our own vantage point. Large enclosed systems, for aerospace applications in particular, that most users would not have considered several years back, are being ordered based on the future environmental outlook. Manufacturers involved in small precision cleaning applications want totally enclosed systems, featuring automation that will minimize operator interface while conserving as much solvent as possible.

The adage that solvent cleans better, faster, and in more restrictive places than water can ever reach is still true. The fact that today's cleanliness specifications for critical cleaning cannot have any contaminant residue or rinse water residue on the end product drives some users toward solvent. The cost to clean is a part of this equation. The cost for environmental compliance is another part. The end user confidence is yet another. Some applications can go either solvent or aqueous. In some cases, one process is generally perceived as superior based on several different inputs. That's why there are alternate choices. This is also why the government recognized the need for solvent cleaning and enacted certain laws to preserve this process until a better method is discovered.

As a solvent equipment manufacturer, we believe that existing process will remain viable for many years into the future. We know the equipment will evolve as well as the solvent that goes into them. While the basic process will remain similar, improvements will come in material handling concepts, solvent recovery systems, and even tighter environmental compliance. Just like the automobile, today's systems are more reliable, perform better, and use less fluid.

23

Vapor Degreaser Retrofitting

Arthur Gillman
Unique Equipment Corporation

Introduction

The first question would be why? If a unit is in good working order, and there are no particular complaints, why make costly changes? There are good reasons, and they include regulation, including federal (NESHAP), state, and regional. There are also economic and safety issues.

The Halogenated Solvents NESHAP (National Emission Standard for Hazardous Air Pollutants) is a federal regulation that specifically regulates vapor degreasers using trichloroethylene, perchloroethylene, l,l,l-trichloroethane, and methylene chloride (two additional solvents not typically used in vapor degreasing are also part of the NESHAP). It is made up of a series of emission reduction choices. Assuming the vapor degreaser is in relatively constant use, retrofitting is most often the best choice.

Economics

Where low-cost chlorinated solvents either cannot be used or are not a good choice, there are new so-called exotic solvent and solvent blends to choose from. One of the common threads among many of these newer solvents is cost. They are expensive per pound, per gallon, or per drum. They are so expensive, costing perhaps \$10,000–\$15,000/drum, that unnecessary solvent losses are worth preventing. Proper operating procedure, combined with a decent retrofit, can produce operating costs on a par with the older solvents.

Safety

Each solvent has a toxicity listing called a threshold limit value (TLV). Many of the newer solvents are more toxic (lower TLV) than the solvents they replaced. Further, the chlorinated solvents are being reevaluated and may see lower limits set. One of the newer solvents, *normal*-propyl bronide (*n*PB), has a recommended exposure rate but it has not been firmly established. The government-approved rate has not been set as of this date. This all means that reducing operator exposure makes good sense.

Retrofitting

Retrofitting means making physical changes and additions to the vapor degreaser. Although there are theoretically many things to be done, here are the tried-and-true "best of the list."

Freeboard Ratio

This is defined as the distance from the point where the boiling solvent vapor idles (usually around the middle of the cooling coils or cooling jacket) to the top of the machine opening. This dimension must be at least equal to the narrowest width of the overall vapor area. *Example:* If vapor depth measures 20 in. and the tank measures 24 × 48 in., then the freeboard ratio must be raised to the narrowest dimension of 24 in. or an increase of at least 4 in. This is a standard that has been changing. The early vapor degreasers were typically manufactured with a freeboard ratio of 50%. Later, the federal government dictated that this ratio should be raised to 75%. The federal NESHAP rules now dictate a ratio of 100%. The question is often asked, "Will the ratio be moved higher?" This author's opinion is no. The reason is that a ratio above 100% does not improve idle losses by much and the continual raising of the freeboard causes operating problems, including interference with hoist mounting and ceiling heights. There remains the question of how much freeboard is "best." The answer is, the more the better, *but be practical!* Even if one is not affected by NESHAP, a freeboard ratio of 100%, or greater, is going to reduce idle solvent losses.

Freeboard ratio can be accomplished by installing a stainless steel collar of the appropriate height. Make certain that the collar is sealed and that the top is flat and sturdy enough to support a proper cover. We recommend that, where possible, the top should have a lip that is turned in horizontally toward the tank opening and then formed down toward the vapor.

Major Emission Reduction Devices

Freeboard Chiller

This consists of a second set of cooling coils, powered by a separate refrigeration compressor condenser, using an EPA-approved refrigerant such as 404A. The coils are mounted as close to the primary cooling coils, or jacket, as practical. Freeboard coils can be mounted one side of the tank wall or around all sides. The combination of refrigeration power (motor horsepower) and coil surface area must produce a temperature at the center of the tank, and center of the coil system, that does not exceed 40% of the boiling point of the solvent. The larger the opening of the vapor degreaser, and the higher the boiling point of the solvent, the more power and surface area is required to achieve the desired temperature. We recommend using finned tubing and mounting on all four sides for best results. There are situations where this is not practical because the addition of either finned tubing, or the mounting on all four sides, chokes off the tank area too severely. In this case, do what you need to get results. This might include mounting finned coils on one side only or using straight, nonfinned tubing, and raising the number of coil wraps to increase the surface area.

Carbon

Activated carbon systems have been used for decades and can be quite effective. A system consists of one or two specifically sized canister(s), a lip exhaust, a heat source to release the solvent from the carbon, and a condensing system to collect the condensate. The idea is to draw off vapor from the top of the vapor degreaser and trap (adsorb) it in the carbon canister. When loaded, the canister is desorbed by heating the carbon, causing the trapped solvent to turn to vapor where it is condensed. A two-canister system allows for continuous operation. Carbon systems typically are chosen only for very large vapor degreasers. Cost is the reason. Carbon systems can easily cost $100,000 and up. For this reason, carbon is warranted only when there is no other economic choice.

Super Heat

Superheat involves the addition of a heated surface placed in the vapor zone. By raising the temperature of the solvent vapor above the boiling point of the solvent, liquid solvent that is entrained in the parts is boiled off and the result is less solvent dragout. Heating must be carefully controlled because surface temperatures that are too high can damage the solvent. Each solvent has its own limits. The most common heating method is circulating hot oil. The problem with super heat as a retrofit device is that it significantly reduces tank area. For that reason it is not as popular as a retrofit choice and is most often considered a major emission control device on new equipment.

Additional Emission Control Devices

Cover

Here is a simple test to determine if the cover style is adequate. Can one open and close the cover quickly and not disturb the vapor? If not, then a non interfering-style cover is necessary. Cover styles that do not

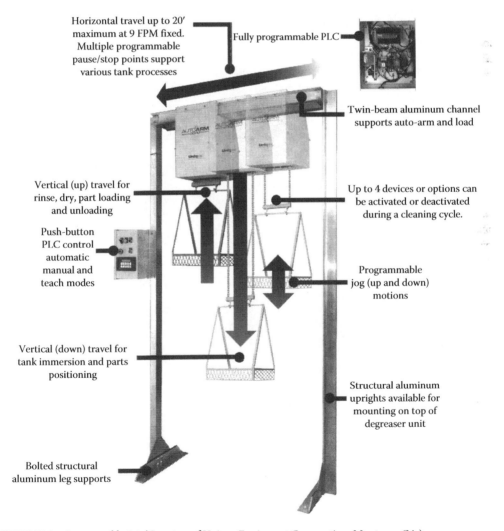

Horizontal travel up to 20′ maximum at 9 FPM fixed. Multiple programmable pause/stop points support various tank processes

Fully programmable PLC

Twin-beam aluminum channel supports auto-arm and load

Vertical (up) travel for rinse, dry, part loading and unloading

Push-button PLC control automatic manual and teach modes

Up to 4 devices or options can be activated or deactivated during a cleaning cycle.

Programmable jog (up and down) motions

Vertical (down) travel for tank immersion and parts positioning

Structural aluminum uprights available for mounting on top of degreaser unit

Bolted structural aluminum leg supports

FIGURE 23.1 Automated hoist. (Courtesy of Unique Equipment Corporation, Montrose, CA.)

disturb the vapor include roll top, sliding, and pivot. Some of these styles are available in both manual and automatic versions. Power covers are best in two situations. The first is with large vapor degreasers where reaching across the opening presents a risk and unnecessary exposure. The second situation is any vapor degreaser utilizing a programmable hoist or material handling system. The reason is because it is often possible to integrate the automatic open/close function as part of the hoist or material-handling controller. The bottom line advantage is solvent saving, knowing that the cover will be closed after each cycle and during idle periods.

Controlled Speed Hoist

The most overlooked emission control device is the controlled speed hoist. This device guarantees that the speed of the workload will always be at or below the 3.3 m/min maximum mandated by NESHAP and other regional rules. This is the speed limit determined to be necessary to prevent dragging out vapors. With large vapor degreasers and heavy loads, a hoist seems obvious. But with smaller systems it is often overlooked. The problem is that an operator has no concept of what 3.3 m/min means. Even if the operator did, there is another problem and that is that the typical basket handle puts an operator's arm at an awkward and uncomfortable angle. Going that slow is almost impossible. The result is that *most solvent losses occur during load insertion and withdrawal.* That means nothing anyone can do will save more solvent than automating the speed of the parts in and out of the vapor degreaser. Hoist systems can be as simple as pendent-controlled chain hoists, costing around $1000, to microprocessor hoists (Figure 23.1) that can automate all of the movement involved in the cleaning cycle, as well as automatically turn on/off various vapor degreaser accessories, such as automatic covers, ultrasonics, and pump/filter systems. Cost of these automated systems can range from approximately $12,000–$50,000, depending on weight and complexity of function.

Retrofit Sources

Contact the vapor degreaser supplier, as well as the solvent supplier. In addition, there are independent retrofit suppliers who specialize in this area.

24

Enclosed Cleaning Systems

Don Gray
University of Rhode Island

John Durkee
Consultant

Background and Definitions

While enclosed cleaning systems have been used for specific applications for several generations, their popularity as a solution to broader industrial cleaning problems has only emerged in the United States in the 1990s.

At the heart of every enclosed cleaning system is a cleaning process. The general purpose of the enclosure is to protect the environment from emissions arising out of the cleaning process. A second purpose is to implement some unique cleaning process within the enclosure, which could not be implemented outside the enclosure.

Enclosed cleaning systems are of three general types. They differ by the degree and method by which they are sealed from the ambient environment. As one would expect, sealing is the key issue in defining enclosed cleaning systems.

The three types are as follows:

Airtight	These systems are sealed to contain a light pressure above ambient. Typically, maximum pressure is around 0.5 psig (380 mmHg)
Airless	These systems are sealed to contain either full vacuum (~1 mmHg) or a pressure significantly elevated above ambient (~800–10,000 mmHg). The word airless has been used generically, and later as a trademark
Externally sealed	These systems are not sealed to contain either pressure or vacuum. Rather they are sealed to restrict interaction of the internal environment with the ambient environment

Another Distinction

All enclosed cleaning systems bring the value of keeping the cleaning solvent "in the tank." There are two very different methods by which this is accomplished. Usually both are incorporated in any enclosed cleaning system—however, one is the dominant method of emission control. Reliance on each method has very different consequences for users.

The first method is described by environmental engineers as "tailpipe control." Generally this means solvent vapors leave the cleaning system in a stream of air and pass through a bed of activated carbon to be adsorbed prior to discharge to the environment. Nearly every enclosed system uses carbon treatment for "tailpipe control" to meet environmental standards.

The second method would be similarly described as "pollution prevention." This means that the operating process has steps through which solvent liquid and vapor are recovered and not allowed to leave the cleaning systems. Though not necessary practical, excellent internal recovery of solvent could mean that external carbon treatment is not required.

A Dynamic Field

This subject is a "moving target." The commercial application of enclosed cleaning systems is affected by environmental regulations, investment at purchase, perception of economics in use, competitive offerings, customer needs, availability and price of solvents, and quality of design. As this is written all factors are in flux, especially environmental regulations and investment at purchase. New and existing firms are providing new offerings of enclosed cleaning systems. Existing firms, currently offering enclosed cleaning systems, are retrenching. Prices in the United States and attitudes about enclosed cleaning systems are in a state of flux.

Consequently, a comparison by supplier of offerings will be obsolete within several years. For example, such a comparison written in 1997 would not have included the impact on the marketplace of the LAER/BACT (Lowest Achievable Emission Rate/Best Achievable Control Technology) regulations and would be of little value to current readers.

So, we will focus in this chapter on basic differences among enclosed cleaning systems, general principles of operation, common process steps, and lasting disadvantages and advantages of their use.

Rationale

In a sense, the initial applications of enclosed systems were chemical reactors, autoclaves, or storage vessels. Only very seldom would a process engineer consider completion of a chemical reaction in an open system, and depend on controlling atmospheric diffusion rates to keep the feeds and products of reaction out of the ambient environment.

Similarly, when it became necessary to contain emissions from cleaning systems, it was natural for process engineers to turn to sealed vessels. In traditional, liquid/vapor degreasers, diffusion-based controls (high side walls with refrigeration) are used with open-top cleaning systems because they are low cost not because they produce high containment efficiency. At best, older open-top vapor degreasers contain emissions at a level around 80% or slightly higher; enclosed degreasers can routinely achieve emission control at a level in the high 90 percentiles. Here the establishment of effective sealing mechanisms offers much higher efficiency of containment. When this degree of containment is demanded by environmental regulations, or for other reasons, cleaning experts and users turn to enclosed systems.

Principles of System Design

Design of enclosed systems is partially based on what is known about the equivalent cleaning process in an open-top vapor degreasing system. The designer of any enclosed system must consider the following principles:

Principle I: Most of the same processes practiced in an open-top vapor cleaning system can be well-used in an enclosed cleaning system. That is, most any cleaning process (immersion in one or more stages, ultrasonics or megasonics, hot rinse in one or more stages, superheat, dry, etc.) practiced in an open-top system can be ported to an enclosed system for the purpose of emission reduction.

Principle II: In addition, other process features may be added or subtracted. For example, vapor spray onto cooler parts (which enables rinsing with pristine distilled solvent) often leads to unacceptable emissions in an open-top system, but is a normal cleaning technique readily practiced in enclosed systems. In addition, enclosed cleaning machines can operate with multiple cleaning cycles where parts are immersed in liquid-cleaning solvent, followed by drainage of that solvent to an integrated receiver, which is then followed by another contact vapor spray. In this way, multiple cycles of rinsing with pristine distilled solvent can be readily achieved. This rinsing process is essentially impossible in a conventional open-top degreaser.

Principle III: Environmental contaminants must not be allowed to enter the cleaning chamber of the enclosed system, or additional precleaning process steps will be necessary. Basically, the items of concern in the outside environment are humidity (water), noncondensables (nitrogen and possibly oxygen), and airborne particulates. Because the system is sealed, the process designer must be careful to eliminate the entry of impurities, which are not normally purged from the enclosed system. For example, retention of water and oxygen can lead to rapid deterioration of the solvent via chemical reaction. Finally, any process implemented in an enclosed system must allow for separation of the solvent from the internal environment prior to release of that environment.

Principle IV: The environment inside the chamber of the enclosed cleaning system must not be allowed to enter the ambient atmosphere. Since this environment is rich (or possibly saturated) in solvent, the result would be significant air pollution. One cannot simply "open the door" in the enclosed chamber when the cleaning cycle is complete, for the chamber is loaded with solvent.

We will discuss each of the three types of enclosed cleaning systems, and provide some examples where they have been successfully and unsuccessfully used.

Airtight Systems

Compared with other enclosed cleaning systems, airtight systems are simpler to design, cheaper to construct, and more inexpensive to operate. Usually, the cleaning cycle in an airtight system is rapid. In fact, a cleaning cycle in an airtight system may well be shorter than the same cycle in a open-top system.

Here is an example of a cleaning cycle for parts with a low thermal mass (ball point pen refills, catheter wires, electronic connectors, gold foils, battery casings, jewelry components, eyeglass frames, etc.).

Example: Airtight I

	Total Elapsed Time (min:s)
A. Parts loaded in racks; racks loaded in chamber	0:00
B. Chamber sealed; cycle selected and started	0:15
C. Hot liquid solvent sprayed on cold parts; parts heat rapidly	1:45
D. Superheated solvent vapor sprayed on hot parts to dry them	3:15
E. Solvent vapors displaced with dry forced hot air	4:30
F. Hot air displaced with cool air to cool parts	5:15
G. Cycle complete. Chamber unsealed automatically	5:30

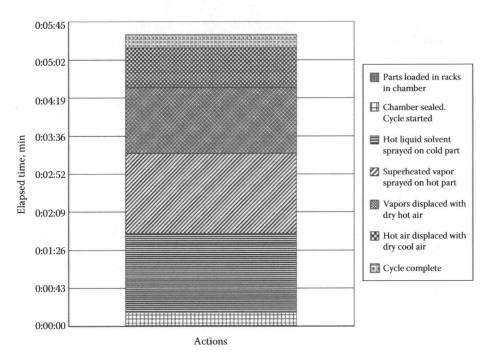

FIGURE 24.1 Time chart for airtight cleaning.

The progress of this cleaning cycle is shown in the elapsed time chart of Figure 24.1.

The equivalent process in an open-top vapor degreaser would be immersion in a single sump followed by spray drying with hot vapor.

Steps E and F are required to satisfy the fourth principle to keep solvents from escaping. The process equipment required for step F is a holdup chamber followed by a huge carbon absorption column.

The holdup chamber is mandatory because the hot air solvent mixture is forced from the cleaning chamber at a rate higher than solvent can be adsorbed by the carbon absorber. This is an important point. The presence or absence of a properly sized holdup chamber can be used to distinguish low-cost enclosed cleaning systems of poor design, from enclosed cleaning systems offering real value.

Those who would choose this process

1. Desire solvent cleaning to avoid mineral residues (water spotting)
2. Require extraordinarily low solvent emissions
3. Greatly value short and controlled cycle time
4. Have a low level of soil on their parts
5. Have a high throughput and a highly automated process
6. Should be able to rack parts for exposure to liquid, vapor, and air sprays
7. Require good-to-excellent drying

The first example illustrates a short cleaning cycle. As a second example, envision a situation where the soil is difficult to remove and the parts have a high thermal mass. Such a process is similar to the first in respect to points one, two, six, and seven, but could take nearly 50 min because long soaking in hot solvent is required to remove soil (e.g., wax/gum, soot, or buffing compound).

This is not vapor degreasing. Because of the reduced operational pressure, the parts are already at the reduced temperature of the hot solvent; and therefore there is no cold surface upon which vapor could condense.

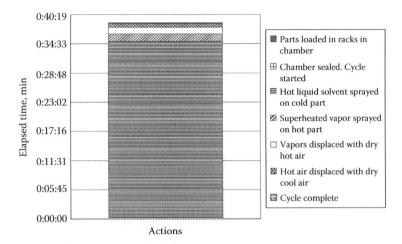

FIGURE 24.2 Elapsed time chart for airtight cleaning with difficult to remove soils.

Ultrasonic cleaning may be required during the immersion process to remove particles. In this case step C, from the previous case, is replaced by a step lasting (an assumed) time of 34.75 min where the parts are immersed in hot boiling liquid solvent.

The remainder of this process is similar to Example I except that the parts are elevated above the immersion vessel prior to the start of drying steps D and E. In step F, 45 s rather than 15 s might be needed to cool parts with a higher thermal mass.

The progress of this cleaning cycle is shown in the elapsed time chart of Figure 24.2.

These users would also need to monitor solvent quality frequently because, in contrast to Principle III, water and oxygen can enter the sealed chamber.

While process operations within an airtight system are quite flexible, there is one common operation that cannot be completed in an enclosed cleaning system: true boiling of the solvent. True boiling happens when the vapor pressure of the solvent equals the total pressure of the atmosphere. The former is a property of the solvent molecules. The latter is atmospheric pressure in open space, or some pressure in an enclosed chamber.

The sum of partial pressures of the two vapors (air and solvent) must equal the total pressure. The presence of a diluent (air) means that the partial pressure of the solvent can never equal the total pressure. Thus, the solvent can never truly "boil." If the temperature in the chamber is raised, the partial pressure of air is raised per Dalton's law and the partial pressure of solvent follows the vapor pressure curve upward.

Can one have high solvent evaporation rates without boiling? Absolutely! But the net rate of vapor generation will not be as high as that in true boiling. The vapor condensation rate on the parts is likewise impeded by the presence of air, which, with no place to go, cannot be displaced by a vapor blanket as in an open top degreaser. The solvent now must diffuse to the solid part surface through the air surrounding the part, thus adding a significant resistance to condensing heat transfer.

There are no inherent limitations as to which solvent can be used, even blends and binary azeotropes. Naturally, the solvent should be chosen to match the soil. Newer, engineered solvents (hydrofluorocarbon [HFC], hydrofluoroether [HFE], hydrochlorofluorocarbon [HCFC]) are well suited for the first application (short cycle time) because of their rapid evaporation rate. However, trichloroethylene (TCE), perchloroethylene (PCE), and normal propyl bromide (nPB) are often used to remove drawing oils and ink residues, which require additional soaking time from ball point pen components. Flammable solvents are not commonly used with air-spray processes because of the absence of the control of sparks. But they could be used readily with the second application (soaking) with appropriate secondary controls in case of system failure.

Airless Systems

Airless systems offer the most capability and power for customization of cleaning operations, albeit with the highest price tag. The capability of pressure (vacuum) greatly expands the possibilities for use of airless systems.

The initial airless systems were developed for three applications: use of highly regulated toxic solvents (such as perchloroethylene and methylene chloride), cleaning of large parts, and cleaning/drying of complex parts. Airless systems are commonly used in Europe, chiefly with hydrocarbon solvents. In addition, these authors believe there are many untapped, unrecognized applications.

Typically, an airless system is a vacuum system, although a few use methylene chloride or other solvents[1,2] under pressure in order to raise the boiling temperature. Paint stripping is one application for a pressurized systems. Temperature and pressure are linked because the boiling point of a solvent decreases as the pressure is reduced. The reverse is also true. For example, it is quite possible to use trichloroethylene (TCE) (boiling point 189°F) at the temperature at which CFC-113 would have been used (117°F). The vapor pressure curve indicates what pressure should be selected to attain that temperature.

Several commercial processes using airless equipment will be described in the following. The first[3,4] is typical of an approach to cleaning and drying highly porous or complex parts. Examples include metal structures impregnated with grease for lubrication, multiport injectors, or aluminum/alloy honeycomb structures used as panels in construction of aircraft. The process schedule is described later.

Example: Airless I

	Elapsed Time
A. Parts loaded in racks; racks loaded in chamber	0:00
B. Chamber sealed; cycle selected and started	0:15
C. Air removed by vacuum pump-down to 1 mmHg (to remove water and oxygen—see Principle III)	3:30
D. Parts sprayed with hot solvent vapor (which raises total pressure); the vapor condenses on the cold parts	5:30
E. Solvent vapors are removed by vacuum; the parts are naturally cooled	7:30
F. Step D is repeated (solvent vapor spray)	9:30
G. Step E is repeated (vacuum evacuation)	11:30
H. Step D is repeated (solvent vapor spray)	13:30
I. Step E is repeated (vacuum evacuation)	15:30
J. Chamber is filled and flushed with air (fed to carbon absorption column)	16:30
K. Step E is repeated (vacuum evacuation)	18:30
L. Chamber is filled and flushed with air (fed to carbon absorption column)	19:30
M. Cycle complete; chamber unsealed automatically	20:00

Note that step C reduced the air concentration (v/v) in the cleaning chamber to 1300 ppm (equal to 1/760 mmHg), the oxygen concentration to around 250 ppm (equal to 20% of 1300), and the water concentration in a warm, humid ambient environment (equal to 100% RH in air at 90°F) to ~50 ppm. For additional control of the process environment, the evacuated chamber could be filled with clean dry nitrogen and again vacuum-evacuated to 1 mmHg. That would reduce each concentration by a factor of 759/760.

The progress of this cleaning cycle is shown in the elapsed time chart of Figure 24.3.

The above process provides three separate stages of vapor degreasing. Parts are cooled between stages by vacuum removal of vapor. Obviously, this process could be shortened via use of fewer wash stages. A cycle time of 15:00 to 20:00 is typical. One accepts this extended cycle time in order to achieve cleaning of complex structures with multiple cycles of vapor degreasing without removing the parts from the machine.

Another example where airless systems can provide excellent value is with hydrocarbon-based solvents, which are used in Europe because of a preference against chlorinated solvents. Hydrocarbons are

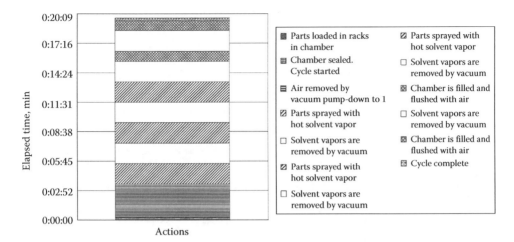

Legend (right side of chart):

- ▓ Parts loaded in racks in chamber
- ▦ Chamber sealed. Cycle started
- ▤ Air removed by vacuum pump-down to 1
- ▨ Parts sprayed with hot solvent vapor
- ☐ Solvent vapors are removed by vacuum
- ▨ Parts sprayed with hot solvent vapor
- ☐ Solvent vapors are removed by vacuum

- ▨ Parts sprayed with hot solvent vapor
- ☐ Solvent vapors are removed by vacuum
- ▨ Chamber is filled and flushed with air
- ☐ Solvent vapors are removed by vacuum
- ▨ Chamber is filled and flushed with air
- ▦ Cycle complete

FIGURE 24.3 Elapsed time chart for airless cleaning of porous or complex parts.

high boiling (some are at 400°F), can have a flash point above 200°F, are excellent solvents for hydrocarbon-based soils, have low odor, low skin irritation, etc. The drawback is evaporation/drying rates are very low—because of their high boiling point.

Airless systems overcome the drying problem as (a) nitrogen is added to the chamber atmosphere so as to increase the pressure and dilute the hydrocarbon concentration and then (b) the chamber is vacuum-evacuated. When the total pressure is quickly reduced to around 10 mmHg, the "oily" hydrocarbons "fly off" the parts. Some vendors have excellent videos showing this effect. If drying is not sufficient, the cycle may be repeated. The process schedule is shown in the following table. Steps A–F are omitted because the operations and timing are similar to airless example I. During step D, parts are sprayed, then immersed in hot solvent, and ultrasonic agitation may be used.

Example: Airless II

	Elapsed Time
A.–F. As above except step D as noted	9:30
N. Step E is repeated (vacuum evacuation); First drying	11:30
O. The chamber is filled with hot nitrogen and the pressure increased to ~500 mmHg	12:30
P. Step E is repeated (vacuum evacuation); final drying	14:30
Q. Chamber is filled and flushed with air (fed to carbon absorption column)	15:30
R. Cycle complete; chamber unsealed automatically	16:00

That this technique is only seldom practiced in the United States does not mean it is not of value in the United States. The progress of this cleaning cycle is shown in the elapsed time chart of Figure 24.4.

A recently commercialized process[5–8] provides good value for cleaning of small parts. In the previous systems (Airless I and II), air is removed by vacuum before solvent is added. This is the desirable system (according to Principle IV).

In the recently commercialized process, the vacuum step is only used for final drying. In Airless III (below), at step C, air and water are displaced after cleaning at ambient pressure by flushing with hot nitrogen.

Note that this system is not a "true" vacuum system in the sense that the cleaning is not done at reduced pressure—under vacuum. But if the cleaning process downstream of step D is appropriate, the user should receive clean dry parts.

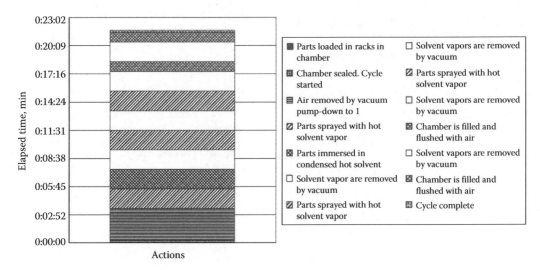

FIGURE 24.4 Elapsed time chart for airless cleaning with hydrocarbon based solvents.

Example: Airless III

	Elapsed Time
A. Parts loaded in racks; racks loaded in chamber	0:00
B. Chamber sealed; cycle already started	0:00
C. Chamber flushed with hot nitrogen (to displace oxygen)	1:00
D. Chamber filled with oxygen-free hot nitrogen after purging	1:30
E. Hot solvent liquid introduced for immersion cleaning	2:00
F. Immersion cleaning with ultrasonics	4:00
G. Drain liquid and flush with hot clean solvent	4:30
H. Second immersion cleaning with ultrasonics	6:30
I. Drain liquid and flush with hot clean solvent	7:00
J. Continuous flushing with hot clean solvent	9:00
K. Drain liquid solvent from tank	9:30
L. Blow hot nitrogen across parts to dislodge liquid	10:30
M. Evacuate chamber to 1 mmHg, to dry parts	13:30
N. Replace chamber environment with clean dry air	14:45
O. Cycle complete; chamber unsealed automatically	15:50

A final example involves an unusual solvent—water.[9] This technique used in this example could also be used with other solvents. Vacuum technology is used to overcome the major limitation of aqueous cleaning—drying. Evaporation of water at atmospheric pressure is slow; and soluble mineral salts are left behind on the parts as imperfections, stains, scars, or spots.

The progress of this cleaning cycle is shown in the elapsed time chart of Figure 24.5.

The evaporation rate is raised by the huge partial pressure difference between water on the part surface and the water in vapor space after evacuation by vacuum. Naturally, the partial pressure of water in the vapor space is low because the vacuum pump is continually removing all vapor from the chamber.

Evaporation of water under vacuum is quick, but brings an unexpected problem: ice. Remember, evaporation involves both transfer of mass (water) as well as heat (of vaporization). This is true for evaporation in a vacuum or under pressure. If water (or any other liquid) is evaporated, the heat

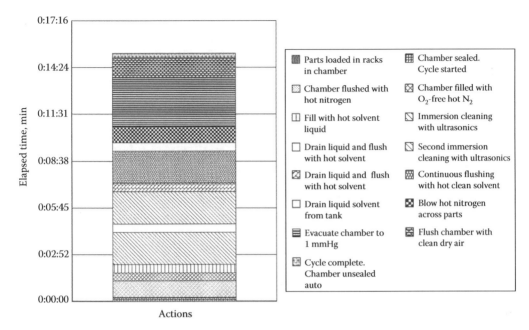

FIGURE 24.5 Elapsed time chart for airless cleaning with water.

of vaporization must be supplied. In this case, heat comes from the surroundings—the parts and the walls of the machine. Typically, without process modification, the parts become chilled and the remaining water becomes frozen. This situation is much more critical with water than with organic solvents because the heat of vaporization of water is ~1000 BTU/lb and that for organic solvents is ~200 BTU/lb.

The process modification is to add heat: hot air, hot water, or radiant heat, so that the heat of vaporization is supplied externally. Hot air is less useful because of its low capacity for holding heat. Hot water, having a high specific hear relative to most liquids, is much more useful in heating the parts (water not adhering to the parts is drained and is not removed by evaporation). Using the system for drying, with hot water, has a cycle time of about 14 min.

Water spots are usually oxides and salts of metal ions. The metal ions are soluble in water—that is why the final rinse is often with metal-free (deionized) water. The oxygen component is thought to come from the air, entrained with or absorbed on the parts. No metal salts are left on the surface since the oxygen has been removed from the chamber prior to evaporation of the water, and since metal-to-nitrogen bonds are exceedingly difficult to form.

What happens to the metal ions left on the parts by the last rinse with water? These authors do not know, but believe that the ions remain on the parts as ionic contamination.

Externally Sealed Systems

Basically, these systems are open-top systems in an isolation chamber. Thus, the solvent is effectively separated from the atmosphere. Exhaust-evaporated solvent is vented through a carbon absorption trap. In some cases, the isolation chamber is retrofitted on an existing open-top vapor degreaser.[10] Additional designs have recently become available. Externally sealed systems have been designed around traditional vapor degreasers as well as low flash point systems.

In externally sealed systems, loading and unloading of parts requires either more labor or additional capital for automation. In addition, cycle time is likely to be increased.

TABLE 24.1　Comparison of Types of Enclosed Systems

Item	Airtight	Airless	Externally Sealed
Operating pressure	Low pressure	Vacuum or pressure	Atmospheric
Operating temperature	<Normal boiling point	Any	Normal boiling point
Choice of solvents	Volatile	Any (volatile or non-volatile)	Volatile
Type of parts	Best with low thermal mass	Any	Best with low thermal mass
Best type of process	Hot soaking in liquid	Hot vapor spray	Traditional vapor degreasing
Drying quality	Normal with volatile solvents	Vacuum quality	Normal with volatile solvents
Estimated investment (smaller units)	2.0 × open-top	2.5 × open-top	1.25 × open-top
Estimated operating costs[13] (with TCE and PCE)	~10% > open-top	≅ Open-top	≅ Open-top
Suppliers	1–2	~5	2 (plus two with flammables)

Summary: Enclosed Systems

The information discussed in the previous sections is summarized in Table 24.1.

Regulation of Enclosed Systems

Two major air pollution regulations cover enclosed systems. One is federal and the other is regional. The federal regulation applies to all users in the United States. The regional regulation, which is more stringent, applies to U.S. firms in one region. If other regions are defined by the Environmental Protection Agency (EPA) as having similar characteristics, they might adopt similar regulations.

Federal

The federal regulation is one of many national emission standards for hazardous air pollutants known by its acronym NESHAP. A hazardous air pollutant is one defined by EPA as causing serious health and environmental problems. NESHAPs regulate their use.

The halogenated solvent NESHAP was published in December 1994, and took effect in December of 1997. This NESHAP covered cleaning operations using six chlorinated solvents: 1,1,1-trichloroethane (TCA), trichloroethylene (TCE), methylene chloride (MC), and perchloroethylene (PCE); and two others not used normally in cleaning operations.

The basis for this standard was maximum achievable control technology (MACT) as defined in the 1990 Clean Air Act[10] for these solvents. This was embodied in the engineering requirements for compliance as 50%–70% control efficiency. This NESHAP does not cover other halogenated such as *n*-propyl bromide and nonhalogenated solvents such as acetone or isopropanol.

The adjectives enclosed, airless, airtight, and externally sealed were not in common use when the NESHAP for chlorinated solvents was developed. These terms are not mentioned in the NESHAP. However, these three types of cleaning systems are covered by interpretation of other language. In the NESHAP, airless systems are included under "solvent systems without an air–solvent interface." Although airtight systems are full of air and solvent, as are externally sealed open-top systems, airtight systems are classified in the same manner as airless systems.

Unfortunately, when the NESHAP was written (1994), externally sealed systems were not ecognized. The EPA only considered "systems without an air–solvent interface" as being "one(s) that does not expose the cleaning solvent to the ambient air during or between the cleaning of parts.[11]" The interpretation by these authors is that this definition includes only airless (and airtight) systems, but does not include externally sealed open-top systems. The EPA has confirmed this understanding.[13]

Regional

Los Angeles, and surrounding counties, have a significant problem with smog caused by emissions of volatile organic compounds (VOC). The EPA defines this region, regulated by the South Coast Air Quality Management District (SCAQMD) as "nonattainment" for federal VOC guidelines. The concepts of lowest achievable emission rate (LAER) or best achievable control technology (BACT) apply in all nonattainment areas. These concepts are beyond the scope of this chapter. The situation is contentious, not necessarily driven by science or by a dispassionate risk–benefit analysis. As of this writing, it is subject to continuing interpretation and change.

Given these attitudes, manufacturers wishing to use nonexempt (VOC) solvents in new operations (including change of machine locations and changes of the solvent used in a given operation) might do well to consider an enclosed cleaning system with documented, demonstrated emissions values.

Other regions of the United States are nonattainment areas. If one is considering solvent cleaning in nonattainment areas, one should insist on a commercial-scale demonstration or a supplier certification (about emission levels with proper operation) prior to purchase of a new or rebuilt solvent cleaning system. In addition, the regional environmental regulatory agency should provide written approval of this evidence to satisfy compliance requirements.

In summary, while all enclosed systems should easily meet the NESHAP requirements, in areas determined by the U.S. EPA to have poor air quality, performance-based evidence (even with solvents not covered by the NESHAP) should be required from the supplier and accepted by the regional regulatory body before enclosed systems are purchased for use in nonattainment regions.

Costs

Enclosed systems, of all types, are more expensive than open-top liquid/vapor degreasers, in part because their construction, operation, and maintenance are more complex and also because there are relatively few producers of enclosed cleaning systems. Technical justification for choosing the more complex systems, beyond cleaning and emissions control, may include better drying, operation at a lower temperature with less potential damage to parts, and the capability to conduct operations not available with open-top machines.

For commonly used smaller systems, we estimate that the average ratio of initial investment costs of open-top systems is 2.5 for airless, 2.0 for airtight, and 1.3 for externally sealed systems. In general, the larger the cleaning chamber, the more costly the system. However, the differential relative to open-top systems begins to converge for large systems (at approximately 75 ft³ chamber volume). These are estimate*; additional process control and parts handling may increase the initial investment.

Analysis

While some data has been provided by equipment suppliers, the analysis and conclusions are those of the authors. Those considering purchase of an enclosed machine are concerned with answers to questions such as: Are the costs of enclosed systems justified? Do enclosed systems pay for themselves? Over what period?

Unbiased answers are difficult to obtain. Nearly all studies are based on estimates or forecasts.[12] With chlorinated solvents, whose prices are less than $1/lb, the decrease in solvent purchases with 99.X%

---

* Basis: 5-year use of capital at 8% annual interest, cost components are: capital use, solvent, operating labor, disposal, and miscellaneous which includes power.

control efficiency system over 70% control efficiency probably will not pay for the additional ~2.5 times investment needed from an enclosed system.

For more costly solvents, a stronger case may be made for enclosed cleaning systems. What if a "designer" solvent costing $15/lb is used? For HFC-4310, HFE-7100, or HCFC-225, we estimate that the enclosed airless system costs less to operate than does the open-top cleaning system.

Studies based on customer experience are difficult to apply to other situations because of the narrow focus of the customer's application, the customer's youth on the learning curve, and the small number of systems constructed in the United States.

One must be cognizant of additional cost issues which do not normally register on a cleaning cost sheet: reduction of hazardous solvent-based waste, reduced cost of obtaining and complying with an environmental permit, increased footprint, and the probably of increased maintenance cost because the unit is more complex.

One reason to purchase an enclosed system is to reduce labor costs. The main component of operating costs (approximately 80% for very large systems) for all enclosed systems is capital payback. Additional costs include, in decreasing order of significance, labor, solvent, and miscellaneous [power/waste disposal]. In contrast, for open-top systems, capital payback is much less significant, and operating labor is much more significant.

In our surveys, capital investment was found to be the main barrier to implementation of enclosed airless systems. Yet, users should be focused on total cost of ownership which is: capital payback + labor + solvent purchase + solvent disposal + miscellaneous. That total should be significant in the minds of users.

In all of our analyses, open-top cleaning systems show less total annual cost of ownership than do airless enclosed cleaning systems. We do not have adequate data of total cost ownership with airtight and externally sealed enclosed systems. However because they require less investment than do airless enclosed systems, it seems reasonable, at worst case, to assume their cost of ownership is no different than for open-top systems.

Hidden Costs

Not all costs are readily quantified. First there are the environmental/regulatory costs. For example, there are costs in time and legal fees to get a permit for an open-top system. In contrast, because normal emissions are below a de minimis value, permitting might not be required for an enclosed airtight system. There are savings associating with avoidance of the need for environmental monitoring, where it probably not be required with an externally sealed system. In addition, there may be value to being able to use PCE in your process, and only an airless system will meet the regulatory emission requirements to enable that end. Further, justification of the high operating costs of an open-top system may be if one needs a "designer" solvent costing $15/pound.

Costs of cleaning quality are also difficult to quantify, leading to difficult questions such as: what is vacuum drying worth if it comes "free" with purchase of an airless system? Will an enclosed system allow one to precisely match the solvent with the soil rather than compromise based on what can be readily contained in an open-top system? What is it worth to be able to repeatedly clean and dry a complex structure, which would retain residual solvent when processed in a nonvacuum system? These questions are the most important of all because they force understanding of the reasons for doing the cleaning operation in the first place.

Vacuum Cycling Nucleation

New technology has led to an increased interest in airless systems as a substitute for ultrasonic or agitated fluid convective aqueous cleaning. The new technology is referred to as vacuum cycling nucleation (VCN)[16] where vacuum and pressure are rapidly alternated within the cleaning vessel in order to grow and implode vapor bubbles on a part's surface. The process enables bubbles to be grown upon the part surface, especially where there is a surface imperfection that can act as a nucleation site for bubble growth. The cleaning effect is caused by the collapse of bubbles.

There must be bulk flow within the cleaning bath. It is probably generated by a pump and is used to convectively transfer material liberated by the action of the bubbles. The liberated material can be either dissolved or dislodged material. It must be removed away from the part surface into the bulk fluid to complete the cleaning process.

Unfortunately with VCN technology, at these surface sites, the transfer of fluid (or suspended solid) to or from the submerged and wetted solid surface generally encounters considerable resistance to mass transfer. It is within this region, adjacent to the solid surface, that the bulk fluid velocity is diminished and decreases rapidly as one approaches the solid surface.

The bulk velocity of even very fast-moving fluids approaches zero at a surface. So there is a region surrounding the part surface (where VCN is ongoing) in which the fluid is actually flowing much slower than the bulk fluid in a cleaning vessel. This region is called the boundary layer. It is defined as the distance from the solid surface within which the fluid velocity moves much slower than the bulk of the free stream of fluid flowing past the surface.

It is within this boundary layer that the rate of mass transfer slows due to a dependence upon molecular transfer mechanisms near the surface as opposed to the more rapid eddy transfer mechanism encountered in bulk fluids.

Increasing the fluid velocity reduces the boundary layer thickness thus enhancing the transfer rate. However, the boundary layer can never be totally eliminated. Similarly ultrasonic or megasonic processes reduce the boundary layer size with increased frequency, however sonic bubbles always form within the bulk liquid and thus a fluid boundary layer always exists.

This is basically why water-soluble cutting fluids for instance cannot be cleaned easily with water alone. Unlike hydrocarbon oil being cleaned with organic solvents, water is generally more viscous than chlorinated solvents and even though the water-based fluids are dissolved in the water, the transfer of these fluids from the part surface to the bulk fluid is highly dampened by the viscous mixture produced at the surface. When the part is drained, the viscous cutting fluid will simply redeposit upon the part.

Studies with VCN have shown successful removal of water-based cutting fluids from small tubes. The cleaning was successful on both the inside and outside surfaces for tubes as small as a ¼″ diameter. Only pure heated water was used. The pressure pulsations lasted around 1 min.

This is a significant advantage of VCN to be able to clean inside of parts and tubes. Because the bulk pressure is equal in all volumes of the cleaning bath, bubbles react the same as if there inside or outside a tube. This eliminates the need to manifold and pump fluid through the tubes as is frequently done to clean the inside of tubes.

For insoluble contaminants, generally chemistries are added to enhance the cleaning. The objective is to use very small quantities of a chemical that can be diffused to the surface of the part and rapidly increased in concentration so as to interact with the parts' surface and contaminant.

The chemical, preferably an oxidizing agent, is dissolved in the treating solution. It concentrates at the bubble–solid interface, because when a bubble expands water is vaporized but the chemical is not so.

The chemical then oxidizes the surface. Oxidative etching of the surface leads to a release of the cleaning fluid from the surface. Observers believe the vapor being formed at the surface tends to lift the residue (soil) from the surface and transport the residue to the bulk liquid.

The reacting chemical may also oxidize the residue. However the residue is not emulsified with water. Consequently, it rises to the surface where it can be physically removed from the vessel. At this point in the VCN treatment cycle, the cleaning fluid is essentially clean and can be recycled for reuse. That leads to reduced water, energy, and chemical use.

Transfer of the contaminant to the bulk fluid and transfer of chemical to the solid surface is aided by detaching or imploding bubbles. The process is thus much faster than force convection or ultrasonic systems. Hydrogen peroxide in concentrations from 0.1% to 1% have been shown to be an effective cleaner (etchant) for most water-insoluble contaminants.

The VCN technology is in its early stages and should be tested prior to use to assure cleaning and no parts damaging; however, to date, the process has shown great success.

Summary of Benefits and Disadvantages of Airless Degreasers

Low emissions of NESHAP solvents to the environment must presently well documented given the implication of the 3 month EPA reporting of NESHAP solvent usage. According to the NESHAP: Halogenated Solvent Cleaning Final Rule May 3, 2007 (which was a required report by the EPA based on their analysis of the outcome of the 1994 rule):

> it was determined that a 95 percent reduction in emissions would result from switching from an existing (open-top) solvent cleaning machine to a vacuum-to-vacuum cleaning machine.

This leads to safety in the work area. Independent testing of airless (vacuum to vacuum) degreasers in industry consistently show work areas containing less than 2 ppm—well below OSHA's requirement for most chlorinated solvents that generally are set at 10–25 ppm TWA limits.

The absence of air solvent interfaces leads to low water absorption and less acid production problems, although acid production is application-specific. Systems have been reported where solvents have been in use for over 5 years without solvent change out.

These same systems emit less than 1 gal/month.

Recycling solvent using vacuum distillation, which is an inherent process present on an airless degreaser, additionally increases solvent life. Typically waste concentrations of less than 2% solvent can be attained. This means solvent waste volume and total disposal volume are very low.

Airless degreasers expand the use of commonly used solvents as well as providing for use of other solvents. Those used in job shop environments are therefore more versatile and can be "value-added machines."

Since an airless system is free of noncondensable gases, the solvent is at its boiling point at any temperature (depending upon the controlled pressure). Thus, solvents can carry out vapor degreasing at temperatures well below their normal boiling point. *N*-propyl bromide, for instance, can be used effectively down to 45°C (normal boiling point of 161°C), and can be used as a substitute in degreasers previously using methylene chloride boiling points 40°C.

Airless degreasers are adaptable to flammable solvents as well. Since the oxygen is initially removed from the chamber prior to exposure to solvent vapors, the enclosed nature of the system provides for safe use of non-NESHAP solvents. Systems that are National Fire Protection Association (NFPA) and Factory Mutual Insurance (FM) approved using acetone (VOC exempt by EPA), IPA and cyclohexane have reportedly been operating within the SCAQMD area for many years.

The fact that air is removed prior to exposure of the component to solvent also effectively enhances cleaning of small or tight offset parts. Air pockets are eliminated and solvent now can enter into tight areas without concern of surface tension effects as encountered in open-top machines. The vacuum drying also is very effective for small parts since pressure is not sight sensitive and solvent flash from parts equally in all areas on the part.

Airless systems, however, do require more sophisticated equipment and can become maintenance intensive. Instrumentation calibrations and fluid-handling equipment require more maintenance than simple evaporation degreasers. The initial capital cost as mentioned above and the added maintenance cost often is prohibitive when simple degreasing is required.

The more flexible the units are, the less operator friendly these machines become and often operator training is required. The energy cost for airless systems is generally higher when cycling, however during idling an airless degreaser uses less energy. Generally energy costs overall are probably comparable to open top degreasers.

As a general rule, the larger the unit required, the more economical an airless system becomes. Solvent- and waste-saving costs have well surpassed the payback time for many of the larger units. Some, which have earlier repaid their capital expense, have been in operation for over 10 years.

Why Purchase an Enclosed System?

The question arises because (1) cost of ownership (with low-priced solvents) is slightly higher for an enclosed system and (2) the purchase cost is higher for an enclosed system.

Regulators easily find an acceptable answer: to control solvent loss.

However, opinions do vary. The view of the authors is that an enclosed system is preferred over open-top systems for nearly all cleaning operations for at least two reasons: (1) everywhere, allowed future emission levels will be lower than they are now and (2) the variety of available cleaning processes produces superior value.

Yet, for low-price solvents, which pose no environmental concern, the open-top system is slightly more cost effective, and for low-price solvents, with environmental issues, the open-top system is slightly cheaper to operate. In both cases, the value of recovering lost solvent within the cleaning machine is quite low. Yet, because most of these solvents dry very slowly, one might purchase an airless enclosed system, which provides vacuum drying.

For medium- and high-price solvents, enclosed systems are readily justified on the basis of reduced total operating cost due to solvent savings.

Location can be a reason to consider enclosed machines, especially if one is located in an area of poor air quality which is heavily regulated? Even where the predicted cost of ownership of an open-top system is less than that of an enclosed system, there are the hidden costs.

In summary, the strength of the case for the open-top cleaning system is diminishing relative to the strength of the case for some type of enclosed cleaning system.

Conclusions

The sole purpose of an enclosed system is to conduct some cleaning process within a contained environment. Enclosed cleaning systems are a relatively new and not commonly used commercially. They were developed in the United States for a few specific applications which were difficult to satisfactorily accomplish with open-top cleaning systems. Operation with reduced emissions is the principal motivation for purchase. Other factors include excellent drying performance, and cleaning of unusual shapes or structures. Airless systems also have significant potential to conduct additional types of cleaning processes.

The three types of systems include airtight, airless (a vacuum chamber), and externally sealed (an open-top system with an airlock). The airless system is the most costly and is suitable for certain unique applications. The airtight system operates with reduced emissions, cleans components, and is priced intermediately. The externally sealed system operates with reduced emissions and is priced lowest.

High purchase investment has restricted commercial adoption of enclosed cleaning systems. Our cost analysis indicates that total operating costs for enclosed systems are close to those for open-top systems, even with low-cost solvents. With high-cost solvents, users rapidly recover the higher capital investment.

References

1. Grant, D.C.H. Solvent recovery and reclamation system. USP 5,232,476. August 3, 1993. Assignee is Baxter International.
2. Grant, D.C.H. Method for cleaning with a volatile solvent. USP 5,304,253. April 19, 1994. Assignee is Baxter International.

3. Gray, D.J. and Gebhard, P.T.E. Cleaning method and system. US5,469,876. November 28, 1995. Assigned to Serec.

4. Gray, D.J. and Gebhard, P.T.E. Solvent cleaning system. US5,538,025. July 23, 1996. Assigned to Serec.

5. Tanaka, M. and Ichikawa, T. Cleaning system using a solvent. USP 5,193,560. March 16, 1993. Assigned to Tiyoda.

6. Tanaka, M. and Ichikawa, T. Cleaning method using a solvent while preventing discharge of solvent vapors to the environment. USP 5,051,135. September 24, 1991. Assigned to Tiyoda.

7. Grant, D.C.H. Emission control for fluid compositions having volatile constituents, and method thereof. USP 5,106,404. April 21, 1992. Assignee is Baxter International.

8. Turicco, T. Pressure controlled cleaning system, USP 5,449,010. September 12, 1995.

9. Nafzifer, C.P. Single chamber cleaning, rinsing, and drying apparatus, and method therefor. USP 5301701. April 12, 1994. Assignee is Hyperflo.

10. Grant, D.C.H. Vacuum airlock for a closed-perimeter solvent conversation system. USP 5,343,885. September 6, 1995. Assignee is Baxter International.

11. Durkee, J.B. *NESHAP Recap—"Dr. PC" Explains All: Precision Cleaning*, April 1995, 39.

12. Almodovar, P., US EPA, personal communication, August 14, 1997.

13. US EPA, Guidance Document, Part 2, Section 1.2, 1995.

14. Almodovar, P., US EPA, personal communication, April 16, 1999.

15. Office of Air Quality Planning and Standards at Research Triangle Park. Impact analysis of the halogenated solvent cleaning NESHAP. US EPA-453/D93–058. November 1993. This analysis was based on 15 year recovery of capital cost and 10% cost of capital.

16. Frederick, C. and Gray, D. Sub micron cleaning of microelectronics devices via vacuum cavitational cleaning process, *Process Cleaning Magazine*, January–February 2007.

25

Organic Solvent Cleaning: Solvent and Vapor Phase Equipment Overview

Wayne L. Mouser
Crest Ultrasonics
Corporation

What Is Vapor Phase Cleaning

Vapor phase solvent cleaning has been a mainstay in the metal processing industries since the early 1940s. Its popularity was driven by the ability to quickly remove organics such as oils, greases, lubricants, coolants, and resins in a single step. The part, at ambient temperature, is lowered into a solvent vapor. The vapor, hotter than the part, condenses on the part and dissolves the organics. In some protocols, a cleaning or washing step occurs where the part is immersed in the solvent with or without ultrasonics. This is followed by additional cleaning/rinsing in solvent vapor. The part is then withdrawn to the freeboard area where the solvent evaporates from the part, leaving it clean, spot free, and dry. A vapor degreaser as shown in Figure 25.1 was inexpensive to own and operate.

Protecting the Ozone Layer

Beginning in the early 1970s, use of the most popular chlorinated solvent, trichloroethylene (TCE), dropped from a high of 609 million pounds in 1970 to a low of 90 million pounds in 1992. TCE fell out of favor due to environmental issues, both air and water pollution. TCE is a volatile organic compound (VOC) and a hazardous air pollutant (HAP). Use of TCE was partially replaced by 1,1,1-trichloroethane (TCA) in the early 1970s and then by CFC-113 (chlorofluorocarbon 113) in the late 1970s and 1980s. Both CFC-113 and TCA were considered to be very safe in terms of the worker exposure profile. Because TCA is the more aggressive of the two solvents, TCA was favored in the metal working world while CFC-113 became the top choice in electronics, aerospace, and many other precision cleaning applications. CFC-113 was effective, low-cost, nontoxic, nonflammable, and considered to be environmentally preferable. Worldwide consumption of CFC-113 peaked at 279 million pounds in 1989.

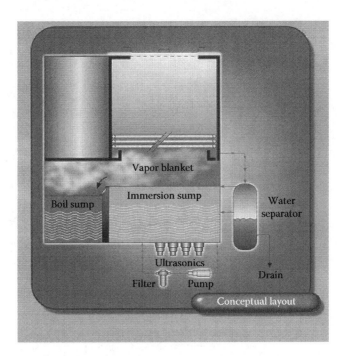

FIGURE 25.1   Conceptual layout of a vapor degreaser.

The situation changed radically with the discovery of the hole in the ozone layer. In September 1987, the international community signed the Montreal Protocol, an agreement to reduce the usage of CFC-113 by 50%. At that time, computer modeling suggested that such a reduction would be sufficient to halt the ozone depletion. It was later learned through more sophisticated examination that a complete ban not on the usage but on the production of CFC-113 and TCA was needed to protect the ozone layer. In the United States, production of both had ceased by the mid-1990s.

The impending loss of these two important solvents started a frantic stampede by users and suppliers to alternative processes such as aqueous, semi-aqueous, and alcohol-based processes. No-clean fluxes and soldering processes were developed in the electronics industry. Metal working primarily shifted to aqueous cleaning and to no-clean processes using vanishing oils. In some instances, there was a move to other chlorinated solvents. In some precision cleaning applications, water-based cleaning was found to be unacceptable due to inadequate soil removal, entrapment of cleaning agent residue, or product damage. Precision cleaning applications where water could not be tolerated were converted to vapor phase cleaning with alcohol, engineered solvents, or other chlorinated or non-chlorinated solvents.

Current Solvent Cleaning Options

Today, solvent cleaning has become an increasingly attractive option for manufacturers. Several factors have contributed to the growth in solvent cleaning. For one thing, a number of new solvent choices have come on the scene. In addition, recent advances in equipment have allowed the safe use of some older solvents that may be very effective in removing soil but may have relatively unfavorable worker safety and/or environmental characteristics. Further, in comparison with aqueous systems, solvent cleaning can be accomplished with a smaller footprint of the equipment, faster throughput time, and spot-free finish produced by vapor phase drying as opposed to mechanical drying utilized in aqueous processes.

The choice of the right solvent to use is not an easy one. Several factors must be considered:

1. Is the solvent compatible with the substrate to be cleaned and can it dissolve the contaminant?
2. Are the solvent and equipment safe for the worker?
3. Can the specific solvent and level of solvent be in compliance with local, state, and federal regulations and company policy?
4. Do the solvent, process, and equipment match the required production rate and product flow?
5. Are the solvent and equipment, taken together as a system, within budget?

In order to choose a candidate solvent for a specific process, it is helpful to consider the physical properties. Each of the solvents listed in Table 25.1 is commercially available, is used in a wide range of applications, and is acceptable as a vapor phase solvent. A review of the properties will assist in choosing the right solvent for a specific application. This list is not inclusive by any means. Several solvent suppliers provide excellent solvents that are an azeotrope or blend of the listed solvents for specific purposes. As an example, the HFC (hydrofluorocarbon) and HFE (hydrofluoroether) solvents can be blended with *trans*-1,2-dichloroethylene where additional solvency is required. Some of these solvents can be mixed to form an azeotrope that once mixed stay together in the same ratio throughout the boiling, rinsing, and vapor stage.

The Kauri-butanol (KB) value is a rough measure of solvency power. Generally, higher-KB-value solvents are effective in removing heavy organics such as oils and greases. Cyclohexane is a good example of a high KB solvent, one that excels in removal of rosin flux, oils, and heavy greases. Lower-KB-value solvents are used in critical cleaning where particle removal and light organics are found. Solvents with low KB numbers are important for cleaning applications because it is important to balance the aggressiveness of the solvent for soils of interest with the potential for substrate damage. An extreme example would be removal of epoxy. About the only solvent strong enough to attack epoxy is *N*-methyl pyrrolidone (NMP). If the substrate were steel, NMP would be acceptable but if the substrate were a printed wiring assembly, NMP would be too aggressive.

Accurate KB determinations cannot be made for oxygenated solvents like acetone and IPA. Acetone can be effective in adhesive removal. IPA is used extensively in critical cleaning for electronics, medical implants, aerospace, disk drive, or where a spot-free hydrophilic surface is required.

The boiling point is an important consideration when the contaminant or the substrate is temperature sensitive. An example would be removal of wax with Perchloroethylene (PCE). PCE has a boiling point

TABLE 25.1 Physical Properties of Cleaning Solvents

	Kauri-Butanol Value	Boiling Point (°C)	Vapor Density	Surface Tension (dynes/cm)	Vapor Pressure at 25°C (mm Hg)	Heat of Vaporization (cal/g)
Trichloroethylene	129	87	4.53	28.7	70	56.4
Perchloroethylene	90	121	5.76	32.3	20	50.1
Methylene chloride	136	39.8	2.93	27.2	350	78.7
n-Propyl bromide	125	71	4.25	25.3	111	58.8
HCFC (AK-225 AES)	41	52	7	16.8	291	40.6
HFC (Vertrel XP)	9.4	52	7.86	15.1	253	tbd
HFE-71IPA	10	54.8	7.51	14.5	207	39.5
Acetone	NA	56	2	22.7	229	134.7
Cyclohexane	58	80.7	2.9	24.9	95	85
Isopropyl alcohol	NA	82	2.1	21.7	40	166.1
N-methyl pyrrolidone (NMP)	350	204.3	3.4	40.7	0.24	127.3
D-Limonene	67	154	4.73	25	2	NA
Trans-1,2-dichoroethylene	117	47.8	3.34	27.5	330	72

of 121°C. Most waxes melt and degrease nicely at that temperature. However, some substrates cannot tolerate that temperature. It is also important to consider the purity of the vapor blanket in the cleaning machine. At higher temperatures, the possibility that organics other than the solvent will be present in the vapor phase increases; these organics can recontaminate the substrate during vapor phase drying.

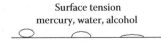

FIGURE 25.2 Effect of surface tension on drop shape.

Vapor density is a measure of the weight of the vapor blanket where air = 1. All of the selected solvents are heavier than air. That is good. It helps keep the solvent in the machine. Emissive solvent loss is a function of the boiling point and the vapor density. Proper equipment design is also important to minimize emissive loss.

Surface tension, the storage of energy at the surface of liquids, impacts the effectiveness of soil removal from product with a complex configuration. Surface tension tries to minimize surface. Imagine a droplet of water on a flat surface. The high surface tension of water (78) causes the water to form a bead. This force makes it difficult for water to penetrate tight crevices. The surface tension of the selected solvents is much lower. A similar-sized droplet of any of the solvents would spread over the surface rather than bead up (Figure 25.2).

The solvent is able to creep into tight spaces, dissolve contaminate, and then be flushed out with fresh solvent. The lower the surface tension, the better.

As illustrated in Figure 25.3, contaminant is trapped in a small hole. Water, due to high surface tension, cannot penetrate. Surfactant could be added to assist, but could be left behind during water rinsing. Solvent, with low surface tension, can penetrate a small hole and completely remove the contaminant.

The vapor pressure of a given solvent is the pressure the vapor exerts over the liquid at a given temperature at dynamic equilibrium. This physical characteristic determines what solvents can be used as the sole solvent in vapor phase cleaners. Note that NMP and d-limonene have low vapor pressures. They also have high boiling points and cannot be easily converted from liquid to vapor. Both are excellent solvents, but should only be considered in co-solvent systems where a second solvent such as isopropyl alcohol can be used as the rinsing and vapor phase portion of the cleaning cycle.

Heat of vaporization is the measure of energy necessary to convert liquid to vapor at the boiling point. Note that most solvents used for vapor degreasing have a much lower heat of vaporization than the energy required for water (539 cal/g). Compared with water, this translates to quicker drying times and lower energy costs.

Safety and Environmental Concerns

Once candidate solvents have been chosen for solvency and compatibility, the next step is to evaluate the solvent and equipment for worker safety and regulatory compliance. There are many methods of

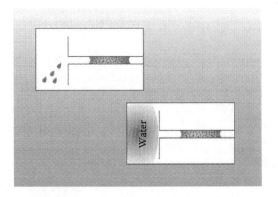

FIGURE 25.3 The small hole problem.

evaluation and each user must determine the degree of importance of various safety issues and also must comply with local requirements as well as company dictated requirements.

HAPs Title III of the 1990 Clean Air Act Amendments lists 187 compounds that are classified as hazardous air pollutants (HAPs). Their emission to the atmosphere must be tightly controlled. At the very least, the HAPs solvents must be used in a cleaning machine that complies with the National Emission Standard for Hazardous Air Pollutants (NESHAP). Due to their known toxicity and potential damage to the environment, it is the opinion of this writer that HAPs solvents should only be used in airtight machines. An airtight system is completely sealed and does not allow solvent vapor to escape into the operator area. This is accomplished by sealing the outside door to the load lock chamber and purging the load lock chamber before and after the cleaning cycle [1].

Many companies have made the decision to move to alternate solvents to avoid the complications of dealing with HAPs solvents.

TLV: Threshold limit values are guidelines established by the American Conference of Governmental Industrial Hygienists (ACGIH) to assist industrial hygienists in making decisions regarding safe levels of exposure to various hazards found in the workplace. TLVs reflect the level of exposure to a vapor or gas that a worker can experience without an unreasonable risk of disease or injury. A low TLV number could suggest the solvent be used in an airtight machine and/or that the workplace and worker be carefully monitored so that the exposure level does not exceed the TLV.

Flash point: Any solvent listed with a low flash point should only be used in equipment properly designed for low flashpoint or combustible solvents. There are many advantages to using low flashpoint solvents in an appropriately designed system. Solvents such as IPA, cyclohexane, and acetone are excellent solvents relative to their physical properties and low toxicity. They clean well and produce a spot-free finish. They are easy to dispose of and are inexpensive when compared with other alternatives. The disadvantage is that since they are flammable, they require more expensive equipment to safely operate.

Total hazard value: There are many methods of evaluating the safety of solvents. TLV, VOC, HAPs, PEL, AEL, ODP to name a few. The list goes on and on. If you have an environmental health and safety staff, they are no doubt experts in this field. If you need help, you might want to consider a rating system called the Indiana Relative Chemical Hazard Score (IIRCHS) developed by the Clean Manufacturing Technology and Safe Materials Institute (CMTI) located at Purdue University [2].

The formula for the assignment of the total hazard value covers many health and environmental concerns. An excellent reference is an article entitled "Solvent by the numbers" by Charlie Simpson published in the January 2002 *CleanTech* magazine [3].

Process Flow and Production Capacity

Now that solvent compatibility and safety have been addressed, the next step is to consider process flow and production capacity. Product offerings start with manual open top designs (Figure 25.4) that provide single station cell cleaning.

For higher volume, or difficult-to-dry parts, a two-sump cleaning system with superheated drying would be a better choice (Figure 25.5).

The vapor is passed through a hot water heated radiator where the temperature is raised up to 10°C–50°F. The superheated vapor then quickly removes liquid solvent from the part.

High-volume manufacturing may require a fully automated system. Figure 25.6 is an example of an automated system that can completely wash, rinse, and dry a basket of parts every 2 min.

The solvent of choice will also impact the equipment decision. Some of the solvents are acceptable to use in open top manual machines, while other solvents, because of their environmental and worker safety concerns, are best used in airtight or vacuum machines (Figure 25.7).

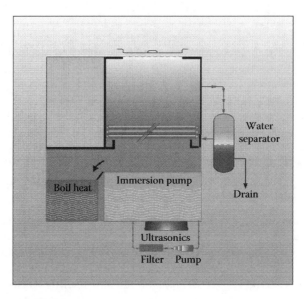

FIGURE 25.4 Single station cell cleaner.

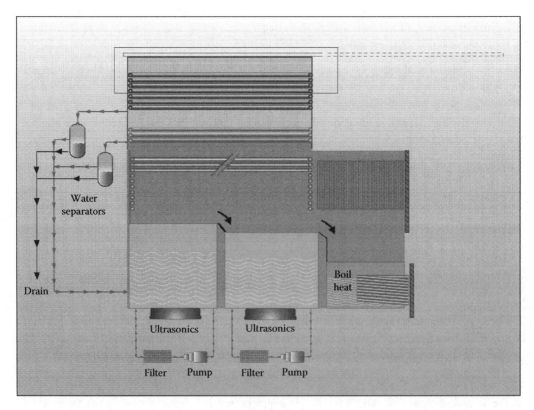

FIGURE 25.5 Two-sump cleaner.

FIGURE 25.6 Fully automated cleaner.

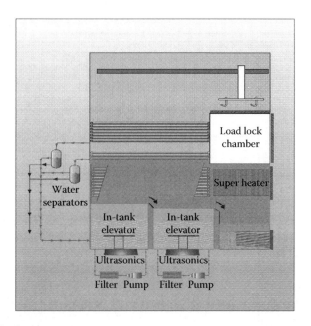

FIGURE 25.7 Airtight cleaner.

Low Flashpoint Systems

Low flashpoint solvents require equipment built specifically to address flammability issues. Such systems can be used with low flashpoint solvents such as acetone, cyclohexane, IPA, NMP, D-Limonene, as well as the other solvents that do not exhibit a flash point. They can be either closed systems or open top systems. These systems are designed to meet applicable NFPA (National Fire Protection Association) codes 30, 34, 70, and 79. They are wired to prevent migration of vapor, have heat or gas detection systems in place, and come with CO_2 suppression. Low flashpoint systems are more expensive than standard solvent systems; however, the higher equipment cost is easily offset by the fact that many flammable solvents are relatively low in cost.

Bi-Solvent and Co-Solvent Systems

The co-solvent process, not to be confused with the bi-solvent process, was introduced in 1991. Co-solvent processes use two solvents, for example an aliphatic hydrocarbon and HFE or HFC, that are mixed in the boil sump. After initial cleaning in the solvent mixture, there is a separate rinse step often with HFE or HFC. The stronger aliphatic hydrocarbon dissolves the contaminant in the boil sump and the HFE or HFC rinses or dries the parts. A co-solvent system works well with light soil loading. Where heavy contamination exists, the rinse sump quickly gets contaminated, organics are left on the part, and there is no method to purge the contamination except by distillation where some of the expensive rinse solvent ends up in the waste. In addition, many of the solvents are VOCs; emissions are often regulated in areas of poor air quality.

In the continued march toward more environmentally friendly cleaning, a new process called bi-solvent has been introduced. This process is very appealing where VOCs are heavily regulated or discouraged by company policy. The two-solvent process uses two SCAQMD (South Coast Air Quality Management District)-certified clean air solvents to remove difficult-to-clean contaminants like oils, heavy waxes, and pitch. The solvents are methyl soyate, an inexpensive biodegradable soybean solvent and 3M™ Novec™ 7200, an HFE. The system operation is depicted in Figure 25.8. Because SCAQMD is located in an area of very poor air quality in Southern California, VOCs and air toxics are strongly

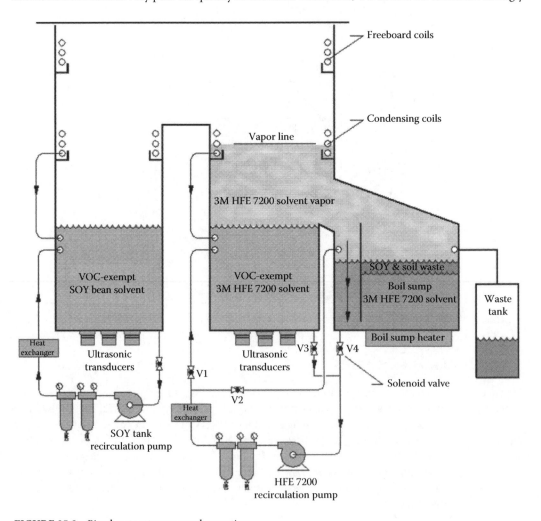

FIGURE 25.8   Bi-solvent system: normal operation.

discouraged. In response, SCAQMD has set up a Certified Clean Air Solvents program [4] to indicate cleaning agents that the agency considers to be environmentally preferable, based on their test protocol. This author is not aware of any required permitting in the United States for this process.

The first production system was installed in an optics manufacturing facility. Optical lens contaminated with pitch or wax is first lowered into the left sump filled with methyl soyate. The contamination is dissolved by the solvent and the process is accelerated by the use of high-frequency ultrasonics. Solid particles are removed by filtration. The parts are then transferred to the second sump on the right filled with HFE. This HFE solvent dissolves the methyl soyate film. This rinse stage is also ultrasonically agitated and filtered. The final cleaning stage is the vapor phase where the parts are finally rinsed and dried in HFE. The lenses exit this stage spot free and dry [5].

Eventually, typically after months of production cleaning, the rinse sump becomes contaminated with methyl soyate. This process is one of the few cleaning systems that can actually clean itself. The reclamation process, as shown in Figure 25.9, works in the following manner. The cleaning process is halted and the boil sump is chilled. Upon cooling, the methyl soyate and the contaminants float to the top. The contamination is then forced to the waste tank. The HFE solvent is saved. The reclamation process takes less than an hour after which the system is returned to production mode.

The final step is the all-important budget for the project. A solvent cost comparison appears in Table 25.2.

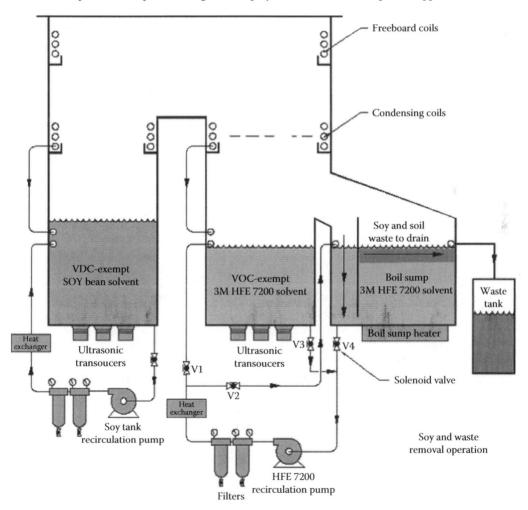

FIGURE 25.9 Bi-solvent system: reclamation mode.

TABLE 25.2   Approximate Relative Solvent Cost

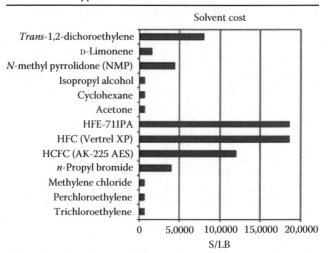

NMP

It appears that NMP will be considered for a lower TLV rating per recent CAL-OSHA PEL activities [6]. The current TLV is 10 according to AIHA. In my personal experience, NMP odor reaches an objectionable level as it reaches 95°F. One method to clean with NMP at higher temperature level and still maintain a safe operating environment is to inert the NMP liquid sump with an IPA vapor blanket in a dual-solvent ultrasonic vapor degreaser. The IPA vapor traps the NMP vapor and does not allow escape to the air. The IPA vapor is condensed by a water-chilled condensing coil. Liquid IPA in a second sump then dissolves the NMP residue and produces a clean and dry part. Part drying is accomplished by dwelling in the IPA vapor blanket until the part reaches vapor temperature. When the part exits the IPA vapor, the heated part flashes dry.

Conclusions

Equipment cost varies with the size, materials of construction, cleaning cycle, and degree of automation. Prices can range from less than $10,000 to over $1,000,000.

Vapor phase solvent cleaning is an important and valuable tool. This cleaning process offers complete washing, rinsing, and drying in a small footprint while minimizing energy, floor space, and process time. With today's efficient equipment and safe solvents, the process is expected to flourish.

References

1. Apparatus and method for precision cleaning and drying systems. Inventors: Wayne Mouser, Randy Honeck, Matthew Bartell. U.S. Patent 6231684.
2. Indiana Relative Chemical hazard Score (IRCHS). www.engineering.purdue.edu/cmti/
3. C. Simpson. Solvent by the numbers, January 2002. *CleanTech Magazine*. https://engineering.purdue.edu/CMTI/IRCHS/Solvents%20by%20the%20Numbers.htm
4. SCAQMD certified clean air solvents. http://www.aqmd.gov/rules/cas/prolist.html
5. Method, apparatus, and systems for bi-solvent based cleaning. Inventors: Wayne Mouser, Russell Manchester, William Barrett, Fredrick Bergman. U.S. Patent 7604702.
6. B. Kanegsberg. Cal/OSHA PEL activities: NMP. http://www.bfksolutions.com/Newsletter%20Archives/V4-Issue%204/PELActivitiesNMP.html

26

Overview to "Nonchemical" Cleaning

Ed Kanegsberg
BFK Solutions

What Is "Nonchemical" Cleaning?

Five chapters in this book are in the category termed "nonchemical cleaning." "Nonchemical cleaning" or "dry-chemical" cleaning are terms used to describe those processes that use neither liquid water nor liquid or vapor organic solvent as the cleaning agent; and both terms are to some extent misnomers. Several of the processes that are called nonchemical are inherently chemical in nature, involving changes in the chemical or molecular constituency of the material being removed. In addition, some of the processes are not completely "dry"; they involve liquid phases during part or all of the process.

Why Is "Nonchemical" Cleaning Utilized?

These processes are becoming increasingly important because, for the most part, they are considered to be "green." They have either a perception or reality of being nonhazardous to workers, the environment or the product, with little or no waste disposal issues. In many cases, they also are less resource intensive than traditional liquid processes, conserving energy, water, and/or petroleum reserves. With increasing societal pressures of regulatory and economic conditions, there are many applications where these processes become the best choice.

Applications for nonchemical processes cover the entire spectrum from gross cleaning of paint or grime to delicate processing of semiconductor wafers and precision optics. Applications to clean objects as large as entire building exteriors or to clean particles as small as 10 nm [1] are accomplished using nonchemical processes.

Categories of Nonchemical Processes*

Dry Media

Removing soil from a surface requires energy to disrupt the adherence of soil to substrate. Most nonchemical processes utilize the energy associated with the impact momentum of the cleaning media onto the surface. This impact momentum cleaning is essentially line-of-sight and, therefore, is frequently not applicable to cleaning complex components with small crevices or blind holes.

Dry media cleaning is usually by impact momentum. Momentum is created either by a forced spray (blast) or in an agitated container. The media includes hard pellets and powders including steel shot, sand, and silicon carbide. Softer media, with less potential for surface damage, but also with less energy for soil removal, include baking soda, nut shells, and talc. These techniques are frequently used as "pre-cleaning" or deburring steps and are not always recognized as a cleaning process. Micro-abrasive micro blasting (see Chapter 27, Swan) allows directed media blast and may be useful for miniature components.

Non-Dry Media

Such solid media as water ice and dry ice (CO_2 pellets or CO_2 snow) are usually considered dry media and indeed do have a similar impact momentum cleaning action. However, both ice and dry ice can liquefy due to the high pressures associated with the momentum of impact, giving these media an added chemical solvent mechanism. CO_2 snow is treated in more depth in Chapter 28 (Sherman).

Water vapor, as high-pressure steam, is also a blasting media. It has both gross and fine cleaning applications (see Chapter 31, Friedheim and Gonzalez). The potential for water condensation can be a problem if water-sensitive substrates are involved.

Liquid and supercritical CO_2 (see Chapter 29, Nelson), because they are neither aqueous nor organic solvents, are frequently lumped into the category of "nonchemical" cleaning agents. However, CO_2 is a very effective chemical solvent in its liquid or near-liquid state.

Dry Chemical Cleaning

UV/Ozone and Plasma cleaning are considered "nonchemical," even though their cleaning action is inherently chemical. Both use high energies to produce either highly reactive atoms (ozone) or reactive ions (plasma) that react with organic contaminants, converting them into simpler, readily removed compounds. The techniques are best utilized in critical final cleaning of relatively uncontaminated substrates. They are not line-of-sight. Because of the high energies of the atomic reactants, there is the potential for substrate modification. Plasma at low pressures is treated in Chapter 30 (Sautter and Moffat). A brief description of atmospheric pressure plasma and UV/Ozone techniques is given later in this chapter.

Cleaning with Light

Two techniques that employ light as the cleaning agent are UV/ozone and laser cleaning. Both of these techniques are described later in this chapter.

---

* This section is adapted from an article in *Controlled Environments Magazine* [2].

TABLE 26.1 Summary of Nonchemical Cleaning Technologies

Technique	Industrial or Critical	Soil Level	Potential for Substrate Damage
Hard media	I	High	High
Soft media	I, C	Low to high	Low
Solid CO_2	C	Low	Low
Liquid, supercritical CO_2	C	Low to moderate	Low
Ice	I	High	Moderate
Steam	I, C	Low to high	Low to moderate
UV/Ozone	C	Low	Low to high
Plasma	C	Low	Low to high
Laser	I, C	Low to high	Low to high
Fluidized dry media	I	High	Low to moderate

Source: Adapted from Kanegsberg, B. and Kanegsberg, E., *Controlled Environment Magazine*, 2007.

Limitations of Nonchemical Cleaning

As with traditional aqueous and organic solvent processes, there is no magic-bullet nonchemical approach. Each process has its own advantages and limitations.

Even when the process agents themselves leave no hazardous residues, care must be taken to be sure that the soils that have been removed by the process are properly disposed. Also, even if chemical exposure is not an issue, worker protection from dust, heat (or cold), pressure, or radiation may be required. "Nonchemical" does not inherently mean safe.

In addition to safety for the workers and the environment, the effect on the surface of the product being cleaned is of great importance. Table 26.1 summarizes the various nonchemical technologies along with the potential to create substrate damage.

Summary of "Nonchemical" Chapters in *Handbook for Critical Cleaning*

The nonchemical chapters in this book treat the spectrum of applications. Solid CO_2 (Chapter 28, Sherman), Supercritical CO_2, $SCCO_2$ (Chapter 29, Nelson) and low pressure plasma (Chapter 30, Sautter and Moffat), and steam (Chapter 31, Friedheim and Gonzalez) were included in the first edition. They have been rewritten or updated for the second edition, sometimes with the inclusion of coauthors (Chapter 30, Sautter and Chapter 31, Gonzalez) and in one case (Chapter 28, Sherman) by a new author. A fifth chapter, Cleaning with Micro Sandblasters (Chapter 27, Swan), has been added for the second edition.

Solid CO_2, $SCCO_2$, and low pressure plasma are most frequently employed for critical cleaning of surfaces where there is little residue to be removed. Steam and abrasive micro-blasting can remove larger deposits of soils but also have applications where small amounts of soil need removal.

For $SCCO_2$, which requires high pressures and low-pressure plasma, which needs a near vacuum, a sealed enclosure is required. Solid CO_2, steam, and abrasive micro-blasting do not inherently require a confined enclosure for operation. However, for confinement of removed material and worker protection, an enclosed chamber may be desirable.

Other Nonchemical Approaches

There are several other technologies that bear mention in this book. They all have applications in cleaning but are not as well known or widespread. With increasing society pressure to go "green," it is expected that these technologies will find increasing use in the future.

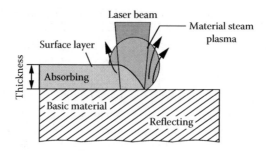

FIGURE 26.1 Schematic of cleaning by laser ablation. (Courtesy of Adept Laser Systems, Kansas City, MO.)

Laser

Laser cleaning has been available for many years. Applications include cleaning centuries of accumulated grime from sculptures or buildings [3], labeling by selective removal (cleaning) of coatings, removal of oxide layers on metals, and removal of absorbed water from semiconductor wafers [4].

In most laser cleaning applications, infrared laser radiation is absorbed by the soil. The soil is heated to the point of vaporization so it essentially boils off (Figure 26.1). In many cases, such as with metals, the laser wavelength is not absorbed by the underlying substrate, so when the soil has been removed, the laser heating ceases. This minimizes any damage to the substrate itself.

In some cases, a liquid might be used in conjunction with laser cleaning [1], to penetrate the contaminant layer. When the liquid is vaporized by the laser pulse, the rapid expansion of the vapor causes the overlying contaminants to be blasted away.

Care must always be taken to avoid worker exposure to the direct laser radiation. In addition, the material being removed can be hazardous and a containment method may need to be employed.

Atmospheric Plasma

Compared with low-pressure plasmas, atmospheric plasmas require higher energies for plasma creation. However, as the name implies, they do not require the confinement chamber to create the near vacuum required for low-pressure plasma. The mechanism of cleaning with atmospheric plasma is similar to that for low-pressure plasma. The plasma, an electrically neutral "gas" of charged particles, interacts both physically and chemically with the surface.

In an atmospheric plasma system, the plasma is generated by a high-voltage discharge and directed by pressurized air as a jet to a nearby surface [5]. Therefore, it can be aimed at a small area, which makes it useful for such applications as mask removal. The air jet also helps with removal of dislodged particles from the surface. The fact that the plasma is charged but is electrically neutral reduces static electrical effects. This makes removal of particles easier.

Plasmas are very chemically active. Many applications for plasmas, both high and low pressure, are for surface modification as well as cleaning. Activation of surfaces, including plastics and composites, can aid in the adhesion of subsequent coatings (Figure 26.2). Many of these processes, although not termed "cleaning," have a cleaning effect.

The plasma itself is electrically neutral and is confined to a small area near the ejection point. In spite of the high energy of the plasma atoms and ions, generally there is little increase in temperature of the substrate. Many applications include treatment of plastics or materials that are incompatible with high temperatures or chemical solvents.

As in other cleaning technologies, the user needs to be cognizant of any hazards associated with removed material and provide proper exhaust and filtering as required.

FIGURE 26.2 Atmospheric plasma as a pretreatment before painting. (Courtesy of Plasma Treat, Elgin, IL.)

UV/Ozone

Ultraviolet (UV) light has long been known to cause chemical changes. Fabric color fading and skin tanning are commonly observed consequences of UV exposure. UV lamps have been used for sterilization purposes.

UV causes depolymerization of hydrocarbons, breaking them down to simple gas molecules CO_2 and H_2O. In this way, UV is a cleaning agent. In some early tests [6], wafers contaminated by human skin oil were clean after about 1 h of exposure to 253.7 nm UV light.

Ozone (O_3) is a strong oxidizing agent and also can decompose hydrocarbons. In the same series of tests [6], ozone cleaned the contaminated wafers in about 10 h.

When UV and ozone are combined, there is a powerful synergy. The cleaning time was reduced to 90 s when the wafers were simultaneously exposed to both 253.7 nm UV and O_3. Addition of 184.9 nm UV further reduced the cleaning time to 20 s. This latter wavelength is absorbed by oxygen (O_2), and helps generate ozone, while the 253.7 nm UV dissociates both hydrocarbons and ozone. Thus, with the combination, ozone is continuously being created and destroyed. In the process, atomic oxygen (O^-) is generated and is believed to be responsible for the accelerated cleaning action. A summary of the wafer tests is presented in Table 26.2.

UV/ozone cleaning is effective only for very light soil loadings. The wafers in the tests described above were pre-cleaned to remove gross contamination, but, by contact angle or auger testing, still showed the presence of a film of contamination prior to UV or UV/ozone exposure.

Applications for UV/ozone have primarily been in semiconductor wafer area. Other applications are those with accessible surfaces, such as precision optics. UV or UV/ozone are line-of-sight processes, so are not effective on complex shapes or with holes that are not accessible to the light.

TABLE 26.2 Exposure Types versus Cleaning Times

Exposure Type	Time to Reach Clean Condition
"Black light" (>300 nm)	No cleaning
O_3, no UV	10 h
253.7 nm, no O_3	10 h
253.7 nm + O_3	90 s
253.7 nm + 184.9 nm + O_3	20 s

Source: Vig, J., *Handbook of Semiconductor Wafer Cleaning Technology,* W. Kern, Ed., Noyes Publications, Park Ridge, NJ, 1993.

UV/ozone has also found utility in cleaning and sterilization of water. For surfaces that can oxidize, such as silver, exposure times need to be limited. Once the contamination layer has been removed, the substrate can oxidize.

UV exposure will keep the surface clean. Where recontamination during storage is to be avoided, UV exposure after cleaning has been shown to be an effective method.

UV/ozone cleaning is usually performed inside a controlled environmental chamber, so exposure is generally not a problem. Care must be taken to avoid exposure either to UV light or ozone. Most plastics and glass effectively shield UV. OSHA has set a limit of 0.1 ppm for ozone exposure.

Fluidized Dry Media

When someone mentions the word "fluid," the word "liquid" is generally thought to be synonymous. However, it is possible to have "dry" fluids. When air is bubbled through a powder, the powder may resemble a liquid by flowing through and around objects placed in it. It is still not a liquid, since it does not have the surface tension associated with molecular fluids, which make them "wet" a surface. Also, because the material is a solid, this fluid does not boil when heated to temperatures in excess of the boiling point of liquids. So it is possible to have a very hot fluid.

One application of a hot fluidized solid is for cleaning. If parts containing organic contaminants such as polymers or oils are placed in a hot fluidized bath, the fluid very effectively transmits the heat to the surface. Pyrolysis, dissociation of the organic molecules, occurs. Pyrolysis is different than burning since it can occur in the absence of oxygen. The surface contaminants are reduced to simpler compounds such as CO_2 and elemental carbon.

One material used for fluidized dry baths is aluminum oxide [7]. The bath can be heated to up to 600°C. This is an effective technique for removing gross contamination levels from metals or ceramics such as residual plastic on molds or paint coatings. The substrate must be capable of withstanding the high temperatures of the bath.

The fluidized bath treatment has disadvantages in that the cleaning action is limited and because the process leaves considerable residue. The powder does not have much of an abrasive quality, so it does not really scrub the surface. At the same time, therefore, it normally will have minimal effect on the underlying substrate. Residual contaminants can be removed by tapping or brushing after removal from the bath. In some cases, a subsequent liquid cleaning technique may be desired.

Figure 26.3 shows an example of a component cleaned by a fluidized alumina oxide system. Temperature is about 500°C with about 25–40 min cleaning time.

Fluidized baths are very hot. So appropriate controls and safeguards must be used in handling components and equipment. In normal operation, the bath is allowed to cool until the parts can be safely removed.

FIGURE 26.3 Metal component cleaned with fluidized alumina oxide bath. (Courtesy of Techne, Inc., Burlington, NJ.)

Summary

Nonchemical cleaning techniques have been the orphans of the cleaning community, recognized as being present, but not getting near the attention that either aqueous or solvent cleaning has. As the number of options for solvent cleaning decreases and given the energy demands and floor space required for most aqueous applications, it is expected that the alternatives presented by some of the nonchemical approaches will be utilized to a greater extent. It is not predicted that either solvent or aqueous cleaning will disappear; both are likely to continue to dominate the market. However, as more and more users recognize the other options available, a better match can be made between the requirements of the cleaning process and the methods to achieve these requirements.

References

1. D. Bauerle et al., Laser cleaning and surface modifications: Applications in nano- and biotechnology, In *Laser Cleaning II*, D. Kane, Ed., World Scientific Publishing, Ltd., London, U.K., 2006.
2. B. Kanegsberg and E. Kanegsberg, Contamination control in and out of the cleanroom: Non-chemical cleaning, *Controlled Environment Magazine*, September 2007.
3. Laser cleaning of the Nickerson House exterior, The Richard H. Driehaus Museum, http://www.driehausmuseum.org/preservation/exterior.php
4. A. Laymon, DPSS Lasers Inc., Personal communication.
5. I. Melamies, Sa(v)fe with plasma, *Process Cleaning Magazine*, May–June 2009.
6. J. Vig, Ultraviolet-ozone cleaning of semiconductor surfaces, in *Handbook of Semiconductor Wafer Cleaning Technology*, W. Kern, Ed., Noyes Publications, Park Ridge, NJ, 1993.
7. D. Sager, Techne Inc., Personal communication.

27

Cleaning with Micro Sandblasters

Jawn Swan
Crystal Mark, Inc.

Overview

Sometimes you need to get things dirty to get them clean. That is the case with micro sandblasting. The micro sandblaster has many applications and cleaning is one of them. Micro sandblasters are tools, and the secret to success with most tools is knowing how to use them.

To understand the process of cleaning with a micro sandblaster, we need to understand the underlying basics, including the components of a micro sandblasting system and the primary variables.

A micro sandblasting system is comprised of

- Micro sandblaster
- Work chamber (to contain the spent abrasive)
- Dust collector (to collect the spent abrasive for disposal)
- Compressed air or gas
- Air dryer (to dry the compressed gas or air)

In micro sandblasting, very small abrasive particles are mixed with a compressed gas, such as air. The mixture is transported through a hose and the flow is then focused through a small nozzle to the area to be sandblasted. The ability to adjust the amount of abrasive introduced into the gas stream and regulate the gas pressure adds another level of control to the equipment.

History of the Micro Sandblaster

In the 1940s, teeth were drilled with what is called a low-speed dental drill. This was a belt-driven affair powered by an electric motor connected by belts and pulleys to a handpiece. The vibration created by this mechanism was not appreciated by patients.

The modern micro sandblaster was invented in the 1940s as an alternative to the "low-speed" dental drill. The inventor was a dentist from Corpus Christy, Texas, named Dr. Robert Black. He called the technique "air-abrasion" [1,2].

Dr. Black found that he could cut tooth structure and clean teeth with fine abrasive particles propelled by a compressed air or gas. He found that sodium bicarbonate was very good for cleaning teeth. The sodium bicarbonate was harder than the tarter but softer than the tooth enamel.

The air-abrasion technology was accepted by the dental community and for a short time hailed as "state-of-the-art technology." It was not to last. The restoration materials at the time were either gold or mercury silver amalgam. It required a high level of training and skill to create the required geometry using air abrasion in teeth in order to use these restoration materials.

The modern air turbine engine (also called "the high-speed handpiece") dental drill was invented in 1953 and quickly became accepted as the best tool for preparing a tooth for restoration. It was faster than air abrasion and the learning curve was not as steep.

The victory of the high-speed handpiece in the technology war resulted in a surplus of air-abrasion units. A booming post-war economy presented new challenges, especially in the aerospace and electronics industries, along with new opportunities for micro sandblasting. Engineers found new uses for these surplus tools. Applications were developed so rapidly that the surplus was depleted. In the 1950s, new machines were designed and built specifically for industrial use.

Variables

Variables affecting the process include the abrasive itself, air pressure, nozzle configuration, the distance from and angle to the substrate, and the abrasive to air mixture.

Exploring the Variables

Abrasive

With any micro sandblasting process, if you do not control the abrasive you do not control the process.

Typically, a micro sandblaster does not recycle the abrasive. Abrasive is only used once. This is because on impact, the abrasive breaks down into smaller particles and becomes contaminated.

Hardness

The hardness of the abrasive relative to the cleaning task at hand is an important factor. Many cleaning applications require the abrasive to be softer than substrate. While there are exceptions to the rule, it is a good guide to start with softer abrasives and work up to harder materials.

Size

A larger abrasive particle delivers more kinetic energy to the substrate than a smaller particle. The typical abrasive range for micro sandblasting is 300 μm or less. Most micro sandblasting applications use abrasives within the 10–160 μm range. Larger and smaller particles can be used, but it is outside the norm.

Shape

Abrasive particles that have sharp and pointed edges cut or strip away surface material upon impact faster than rounded or spherically shaped particles.

The hardness and shape of commonly used abrasives are indicated in Table 27.1.

Most abrasive powders sold for micro sandblasting are not hazardous. However, it is important to obtain a material safety data sheet (MSDS) for the abrasive you are using.

TABLE 27.1 Hardness and Shape of Common Abrasive Powders

Abrasive Powders	Hardness	Shape
Silicon carbide	Knoop 2500, Mohs 9.0+	Sharp and blocky
Aluminum oxide	Knoop 2000, Mohs 9.0	Sharp and blocky
Aluminum oxide (white)	Knoop 2000, Mohs 9.0	Sharp and blocky
Crushed glass	Mohs 5.5	Sharp and blocky
Glass beads	Mohs 5.5	Spherical
Dolomite	Mohs 3.5 to 4	Blocky
Walnut shell	Mohs 3 to 4	Blocky
Plastic blast	Mohs 3.5	Blocky
Soda bicarbonate	Mohs 2.5	Crystalline and soft
Carbo-blast (wheat starch)	80 ± 10 shore, D scale	Crystalline and soft

Air Pressure

The most common gas used in micro sandblasting is compressed air. Whatever compressed gas is used, it must be dry.

Pressure creates speed. Higher pressure increases the abrasive speed and will therefore deliver more kinetic energy to the substrate.

Normal operational pressures range from 5 to 135 psi. However, micro sandblasters can be made to operate lower than 5 psi and at pressures above 200 psi.

Nozzles

Nozzles range between 0.005″ and 0.125″ diameter and rectangular-shaped nozzles range from 0.006″ × 0.012″ up to nozzles 0.016″ × 0.190″. The nozzle size is the limiting factor of how much abrasive and compressed gas is used. It is important to remember when the diameter of the nozzle doubles, the result is four times the flow.

Materials

Tungsten carbide is the material of choice for most micro sandblaster nozzles. Tungsten carbide is very hard and shock resistant. Another choice for round nozzles is sapphire. Sapphire is harder than tungsten carbide and has exceptional life in many applications. In many instances, the longer life can justify the additional cost.

Shapes

Typical nozzle tip shapes are either round or rectangular. While straight nozzles are most commonly used, both tungsten carbide and sapphire tips can be mounted into nozzle bodies so the tips are straight, 45° or 90°. Table 27.2 summarizes nozzle sizes and shapes.

Special Nozzles

Special nozzles are made if the application requires them. For example, placing a hole in the side of a tungsten carbide tube and plugging the end results in a nozzle that is commonly used to get deep into small diameter holes or tubes.

Nozzle Distance

The stream of abrasive and gas emerges from the nozzle in a cylindrical shape for a short distance, and then diverges into a cone-shaped spray with an included angle of about 7°–9°. Holding the nozzle

TABLE 27.2   Common Nozzle Sizes and Shapes

Typical Nozzle Diameters	Typical Rectangular Nozzles	Typical Sapphire Nozzles Diameters
0.005″	0.006″ × 0.012″	0.015″
0.007″	0.006″ × 0.020″	0.018″
0.011″	0.006″ × 0.040″	0.026″
0.014″	0.006″ × 0.060″	0.035″
0.018″	0.006″ × 0.100″	0.045″
0.021″		
0.026″	0.008″ × 0.040″	
0.032″	0.008″ × 0.080″	
0.040″	0.008″ × 0.100″	
0.045″	0.008″ × 0.125″	
0.060″	0.008″ × 0.150″	
0.070″		
0.086″	0.013″ × 0.150″	
0.108″		
0.125″		

close to the substrate will result in a more defined pattern. However, if the nozzle is too close, it will interfere with the expansion of the air and therefore impede the velocity of the abrasive being ejected from the nozzle.

The further the nozzle is from the substrate, the larger the abrading pattern becomes. The speed of the abrasive slows as the distance from the nozzle increases.

In most cleaning applications for nozzle diameters in the 0.018″–.045″ diameter range, the nozzle distance ranges from ½″ to 2″.

Nozzle Angle to Substrate

While the nozzle is typically held perpendicular to the substrate in a cutting or drilling applications, for most cleaning applications, it is desirable to hold the nozzle at an angle in relation to the substrate. In most cases, the operator works the nozzle spray pattern in the direction to be cleaned so that the overspray will be in the right direction.

An additional benefit of blasting with the nozzle at an angle is it allows the abrasive to get under the material to be removed from the substrate.

Abrasive to Air Mixture

In both cutting and cleaning, the concentration of abrasive particles in the air stream has a direct effect on the substrate. Too little abrasive to air and the operation is not productive. Too much abrasive to air and there is not enough room for the air to expand and impart velocity to the abrasive. This becomes more important when working with small diameter nozzles at low pressures.

Work Chambers

The *work chamber* purpose is to

1. Contain the spent abrasive particles ejected from the micro sandblaster.
2. Make the object to be cleaned visible.
3. Protect the user from inhaling possible harmful particles.

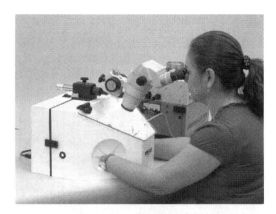

FIGURE 27.1 Work chamber for small parts. (Photo courtesy of Jawn Swan, Crystal Mark, Inc.)

Many times, the object to be cleaned is very small and must be done using a microscope for precise work (Figure 27.1). A smaller work chamber does not require as much air flow as a large one to contain spent abrasive and keep the work area clean.

Work chambers can vary in size depending upon the need. Figure 27.2 is a work chamber designed for large parts, up to 350 lb. The part is placed on the trolley (center left) and rolled into the work chamber. A Lazy Susan mounted on the trolley enables the operator rotate the part for ease of cleaning. This large work chamber connects to a dust collector using a flexible duct. The more volume a work chamber has, the more air flow is required by the dust collector to keep the area clean.

Sometimes, the object to be cleaned is so large, it cannot be put into a chamber; the operator should be protected (Figure 27.3) with safety glasses and respirator.

In another large object application (shown in Figure 27.4), a small bottomless box is held against the mosaic in one hand by the operator while the micro sandblaster nozzle is held with the other. The small box is connected to an industrial vacuum type of dust collector. Note this operator is wearing eye protection and has a breathing mask.

FIGURE 27.2 Work chamber for large parts. (Photo courtesy of Jawn Swan—Crystal Mark, Inc.)

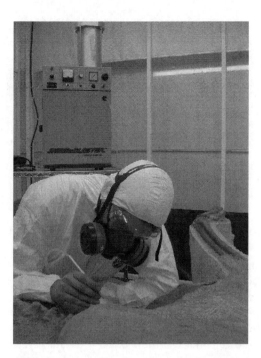

FIGURE 27.3   Personnel protection for large objects. (Photo courtesy of Nate Murphy, Judith River Dinosaur Institute.)

FIGURE 27.4   Cleaning prior to protective coat. (Photo courtesy of Dusan Stulik, @ J. Paul Getty Trust; http://www.getty.edu/conservation/field_projects/vitus/vitus_images.html)

Dust Collectors

Shop vac or commercial shop vacuums are not recommended for micro sandblasting applications because the paper filter fills with abrasive and inhibits the volume of air required to evacuate the chamber. The build up of abrasive on the filter increases the vacuum pressure and often causes a collapse of the paper filter cartridge, and leaks.

Industrial dust collectors are recommended for micro sandblasting applications because they are designed for moving large volumes of air and collecting micron-sized particles. Industrial dust collectors come in many sizes from small to huge and will capture the abrasive particles for disposal and return filtered air back into the room.

The most common filter types for industrial dust collectors are cloth bag and cartridge.

Cloth bag filter types are generally lower cost, smaller, and have manual shakers to dislodge dust particles into a drawer or bin for disposal. The filters must be shaken daily or more often for best performance.

Cartridge filter type dust collectors are generally higher cost, larger, and clean the filters by means of timed compressed air pulses. Most cartridge filter dust collectors are built for continuous operation and have bins or drums for holding the spent abrasive and debris.

If the object to be cleaned is made of or contains hazardous material, then the dust collector should be fitted with high efficiency particulate air (HEPA) filters. "The HEPA filter shall exhibit a minimum efficiency of 99.97% when tested at an aerosol of 0.3 micrometers diameter" [3].

The most common abrasives listed above are not listed as hazardous materials. However, it is important to control the spent abrasive and capture it for proper disposal. It should also be noted that OSHA considers particles $10\,\mu m$ and smaller to be a health hazard.

Cleaning Examples

Art Conservation

The conservation of a famous medieval mosaic provides a good example of the use of a micro sandblaster for cleaning an irreplaceable work of art. The project was a collaboration between the Getty Conservation Institute and the Office of the President of the Czech Republic (4).

The mosaic is on the exterior of St. Vitus Cathedral in Prague and is called *The Last Judgment*. It is considered the most important and the oldest monumental exterior medieval mosaic north of the Alps. The mosaic covers $904\,ft^2$ of the cathedral's south face. The mosaic was comissioned by the Holy Roman Emperor Charles IV in 1370 and finished a year later. For most of its existence, the brilliant colors of *The Last Judgment* mosaic have been rendered invisible, covered over by a layer of corrosion that has repeatedly formed after each cleaning.

The potassium-based composition of the mosaic's glass exposed to rain, wind, snow, temperature variations, and environmental pollutants, cause it to corrode. Repeated attempts at conservation were unable to prevent the recurrence of the grayish layer of corrosion that obscures the mosaic. Many of the failed restoration attempts only added to the cleaning problem.

This difficult and historical problem was tackled with years of research, in collaboration with Material Science Department of UCLA to determine a system to prevent further corrosion of the glass after cleaning. A multilayer protective coating system was developed for this purpose. The cleaning method also required testing and research to avoid further damage to the medieval glass tesserae.

Tests were made on several types of glass including the original tesserae to determine the best abrasive, air pressure, and nozzle distance to be used.

The test found that:

1. Crushed glass from Crystal Mark, Inc. removed the corrosion layer and old restoration materials without damaging the tesserae.
2. Holding the nozzle at an angle worked best.

FIGURE 27.5 The central panel of The *Last Judgment* mosaic before cleaning. It is located on the exterior of St. Vitus Cathedral in the heart of Prague Castle. (Photo courtesy of Dusan Stulik, © J. Paul Getty Trust [5].)

3. Adjusting air pressure independent of abrasive volume was beneficial.
4. A small handheld work chamber could contain the spent abrasive and debris.

Another difficulty was the need to work in situ. Scaffolding was erected and the micro sandblasters and dust collectors were placed upon it along with the conservators.

The corrosion covering the mosaic is cleaned (Figure 27.4) prior to application of the protective coating system. After extensive studies and testing, the cleaning method selected was micro sandblasting using crushed glass. The use of such particles, which are harder than the corrosion layer but softer than the mosaic tesserae, thoroughly removed the corrosion layer without affecting the original glass surface [6] (Figures 27.5 and 27.6).

Over 2000 lb of crushed glass were used to clean the mosaic. It was a slow process, cleaning nearly a million small glass and stone tiles individually by the hands of a team of conservators. All done with four micro sandblasters, with the largest nozzle being 0.032″ in diameter.

Cleaning Fossils

Fossils are the remains of dead organisms preserved in sedimentary stone after water seepage replaces the atoms of the once living tissue with minerals. The remains either represent part of the organism itself or some imprint made by the organism's body. After laying underground for thousands and millions of years, the fossilized organism is usually incrusted with dirt and rock. Paleontologist and those who prepare fossils call this entrustment "fossil matrix."

FIGURE 27.6 The central panel of *The Last Judgment* mosaic after cleaning, regilding, and surface protective coating application. (Photo courtesy of Dusan Stulik, © J. Paul Getty Trust [5].)

Fossils come in all sizes from microscopic to really very large. Smaller specimens are cleaned in small work chambers. Larger specimens may require specially built tables and fixtures (see Figure 27.3).

The fossil matrix can be anything from soft chalk to very hard limestone, and the density can vary within a single fossil. Micro sandblasters are used only after major portions of the matrix have been removed with air scribes or other hand tools. They are ideal for very fine detail work when preparing fossils, especially small fossils.

It is very important for the preparer to learn the hardness properties of the abrasives available. Using an abrasive that is too hard can and will damage a fossil. A good preparer has various hardness of abrasives on hand. Common abrasives for preparing fossil are aluminum oxide, dolomite, sodium bicarbonate, glass beads, and plastic.

It takes much skill and observation to learn how to use a micro sandblaster to its fullest. A new user quickly learns that the stream of abrasive can remove the fossil as well as the matrix. The preparer must observe the process directly, with the help of some type of magnification if possible. Moving the nozzle back and forth around the area and changing the angle of attack will let the preparer see the matrix removal process and see when the fossil is revealed.

Radiaspis

Preparing a fossil is as much an art as it is science. The graceful and elegant trilobite Radiaspis (Figure 27.7) was prepared by Jeff Hammer, Hammer & Hammer Paleotek, Inc. Jeff stated it was the most difficult trilobite he as ever done. Radiaspis measures approximately 2 ½″ × 1 ¼″ × ½″ and took 3 months to prepare with a SWAM-Blaster® micro sandblaster using a variety of abrasives and of nozzles measuring 0.007″ to 0.032″ in diameter at air pressures as low as 2 psi.

FIGURE 27.7   Radiaspis prepared by Jeff Hammer. (Photo courtesy of Jeff Hammer.)

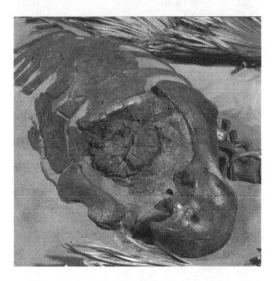

FIGURE 27.8   Thescelosaurus. (Photo by Jim Page, North Carolina Museum of Natural Sciences.)

Thescelosaurus

It is very rare when soft tissue becomes fossilized. Another amazing fossil is the world's first dinosaur specimen with a fossilized heart (Figure 27.8). The dinosaur was given the name "Willo" by the discoverer Michael Hammer [7].

Michael Hammer said, "He as able to keep intact the fine mud impressions of the horny bill structure of the lower jaw as well as the seven scale impressions over the cheek. Had I not been under a scope (microscope) using the micro sandblaster, these structures would not be known about today nor would the most important of all, the 'heart'."

Conformal Coating Removal

The following seven paragraphs are from a NASA Evaluation [8].

Background

Conformal coatings are required on printed wiring assemblies (PWAs) used for space flight applications to provide protection and to extend the life of the assemblies in harsh environments.

There are four major types of coating materials, each developed to suit specific applications. These are: acrylics, urethanes, silicones and parylenes. Depending upon the type of coating material and product requirements, conformal coating may be applied by dipping, brushing, spraying, dispensing or chemical vapor deposition.

In order to rework or repair parts on PWAs, the conformal coating must be removed, either entirely or in specific areas. The most commonly used methods for removal involve thermal, chemical, mechanical and micro abrasive processes. Recent environmental regulations such as the Montreal Protocol and Clean Air Act have had a significant impact on solvent based conformal coating removal processes, particularly with regard to control of volatile organic compounds (VOCs) and ozone depleting chemicals (ODCs). Equipment suppliers for coating removal systems have responded by developing environmentally acceptable methods.

The micro abrasive blasting process, a precise mixture of dry air or an inert gas and an abrasive media is propelled through a tiny nozzle attached to a stylus which is either handheld or mounted on an automated system. This allows the mixture to be pinpointed at the target area of the conformal coating to be removed. A vacuum system continuously removes the used materials and channels them through a filtration system for disposal. The process is conducted within an enclosed anti-static chamber and features grounding devices to dissipate electrostatic potential.

In the micro abrasive blasting technique offers a fast, cost-effective, easy to control, and environmentally friendly non-solvent based method to remove conformal coatings. The system can remove conformal coatings from a single test node, an axial leaded component, a through-hole integrated circuit (IC), a surface mount component (SMC) or an entire printed circuit board (PCB).

Micro abrasive systems inherently generate static electricity as the high velocity particles impinge on the surfaces. The voltage generated at the area of impact can cause electrostatic discharge (ESD) damage to the parts and electrical circuits on a PWA.

The current trend in the electronics industry towards higher speed, greater packaging density and lower power consumption has increased static sensitivity so that new techniques emerging in the field of coating technology need more stringent attention than ever before. An ideal blast media in a coating removal process would be the one which will not generate voltage greater than the ESD susceptibility of a particular device on a PWA or damage the board assembly, and is environmentally acceptable [8].

Controlling ESD

Figure 27.9 shows the view inside an ionized work chamber with the point ionizer nozzle assembly. Note the operator is wearing a ground strap. As stated in the background of the NASA report, micro

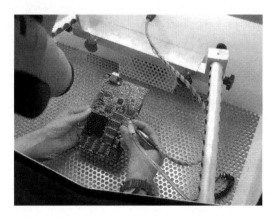

FIGURE 27.9 Inside an ionized work chamber. (Photo courtesy of Jawn Swan—Crystal Mark, Inc.)

FIGURE 27.10   Selectively removing coating with a point ionizer.

sandblast "systems inherently generate static electricity as high-velocity particles impinge on the surfaces."

The ESD generated can be controlled by

Point Ionizer at the Nozzle

The point ionizer is very effective in reducing ESD. The ionization point is located adjacent to the nozzle (Figure 27.10). The abrasive flows through a plume of ionization and the static charges are removed.

Ionization of the Work Chamber

Once particles are ejected from the nozzle, they bounce around creating static charges. Grounded surfaces and placing ionization in chamber prevents ESD from building up.

Static Dissipative Hoses on Micro Sandblaster

Friction of the abrasive inside the hose generates static electricity. Static dissipative hose prevents a place where a static charge can build, inside or outside of the hose.

Low ESD Generation Abrasive

Such an abrasive was developed by Crystal Mark, Inc. for ESD-sensitive applications called by its trade name, "#25A Carbo-blast." It is a wheat starch-based material, water soluble, biodegradable, and considered environmentally friendly. NASA tests have shown the combined overall average ESD surface voltage to be under 30 V when used in conjunction with point ionizer and ionized work chamber.

An additional benefit to using the wheat starch material is most types of coatings can be removed selectively relatively quickly without any visible damage or penetration of the PCB surface and it does not remove gold plating.

Compared to chemical methods for removing conformal coatings, the cost of micro sandblasting is a bargain. The following paragraph (and Table 27.3) are from the *Joint Service Pollution Prevention Opportunity Handbook* Rev 5/03.

Micro-abrasive sand blasting operations generate less hazardous waste than chemical stripping since solvents are not used. The decrease in hazardous waste helps facilities meet the requirements of waste reduction under RCRA, 40 CFR 262; the Pollution Prevention Act (42 USC 13101-13109); and Executive Order (EO) 13148, *Greening the Government Through Leadership in Environmental Management*; and may also help facilities reduce their generator status and lessen the amount of regulations (i.e., recordkeeping, reporting, inspections, transportation, accumulation time,

TABLE 27.3   Removal of Polyurethane Coating from Printed Circuit Boards: Traditional Chemical Removal versus Micro-Abrasive Blast System

	Traditional Chemical Removal	Micro-Abrasive Blast System
Capital costs ($)	0	8,500
Material costs ($)	1,875/year (solvent)	2,670/year (blast material)
Labor costs ($)		
Removal ($)	135,000	2,247.75
Handling ($)	1,124.55	1,124.55
Cleaning ($)	3,373.65	562.27
Waste disposal ($)	12,000	661
Total ($)	153,373.20	7,265.57

emergency prevention and preparedness, emergency response) they are required to comply with under RCRA, 40 CFR 262. In addition, the decrease in the amount of solvents on site decreases the possibility that a facility will exceed reporting thresholds of SARA Title III for solvents (40 CFR 300, 355, 370, and 372). Furthermore, when used as a substitute for chemical cleaning processes, micro-abrasive sand blasting will likely decrease VOC emission levels, thereby potentially reducing the need for an air permit under 40 CFR 70 and 71 [9].

Manual Systems

Most conformal coating applications are for limited use, such as rework. A single component needs to be replaced or the coating is where it is not wanted. The simplest micro sandblasting workstation for conformal coating has four elements:

1. Micro sandblaster
2. Ionized work chamber
3. Point ionizer
4. Vacuum filtration to collect the spent abrasive

Figure 27.11 is a photo of a system that contains all four elements plus a microscope.

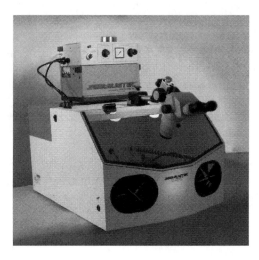

FIGURE 27.11   Basic system for conformal coating removal. (Photo courtesy of Jawn Swan, Crystal Mark, Inc.)

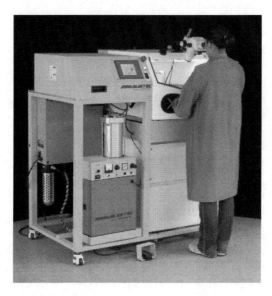

FIGURE 27.12   Automated conformal removal equipment. (Photo courtesy of Jawn Swan, Crystal Mark, Inc.)

Automated Systems

Micro sandblast equipment has been developed into automated conformal removal machines (Figure 27.12). The point ionizer and nozzle are mounted on an XY stage and the PCB is mounted on a turn table. Using a touch screen, the operator inputs the area to be cleaned, starts the machine, and walks away to do other tasks. The system can also be used as a manual workstation.

Cleaning Metal on Ceramic

Metal is often selectively put onto ceramic. In some manufacturing processes, it is desirable to metalize the entire part or surface, then selectively remove the metal to create the desired features. One such case is the making of some types of piezo devices.

Ceramic can be coated several ways. For this discussion, we will use only two: electroless nickel plating and thick film-silver dipping. The metal coating in both processes will cover the entire surface of the ceramic.

A ceramic-like barium titanate is coated entirely with electroless nickel. For it to work as a piezo electrical device, it must be polarized. It cannot be polarized unless there is a complete conductive disconnect. Electroless nickel on barium titanate can be readily removed using micro sandblasting. A fine 50 μm crushed glass will remove the electroless nickel with only minor effect to the barium titanate.

When the ceramic barium titanate has been coated by means of thick film-silver dipping and dried, the coating can be removed using 44 μm glass bead before firing. After firing, it can be removed with crushed glass with minimal effects to the barium titanate.

Usually, the metallization removal is done using some sort of fixture or tooling. This can be as simple as protecting the metallization to remain with a mask and moving a handheld nozzle back and forth over the area to be cleaned. In some situations, the best results are obtained with a fixture holding the nozzle constant distance. That is to say a nozzle held at 0.050″ from the substrate will cut "clean" path. The nozzle moving at a constant rate will remove a constant amount of metallization.

Summary: Cleaning with Micro Sandblasters

When trying to solve a cleaning problem with micro sandblasting, the first step is to determine what abrasive to use. Start your cleaning experiments with softer abrasives first. If that doesn't work, try the next hardest. (Refer to Table 27.1 for abrasive hardness.)

Choose a nozzle to fit the application. If the application is very small, start with a small nozzle; if it is a large area to be cleaned, start with a larger nozzle. (Refer to Table 27.2 for nozzle sizes.)

As not to damage the part to be cleaned, start with a low air pressure. You can always turn it up.

Too much abrasive to air will slow down the particles being ejected from the nozzle. Generally, if you can see a stream of abrasive coming out of the nozzle, it is too much.

For most cleaning applications, the nozzle distance is ½–2 in. This is not a rule; it is a starting point and guideline.

Use a work chamber and dust collector to fit the application. Be safe. Know the materials you are working with. Ask your abrasive supplier for MSDSs. Inquire if the object to be cleaned is made of or contains hazardous material.

If your cleaning application requires control, start with the abrasive powder.

As a rule, don't reuse the abrasive powder, because once used, it is out of control. With any micro sandblasting process, if you do not control the abrasive, you do not control the process.

References

1. Black, R.B. Airbrasive: Some fundamentals. *J. Am. Dent. Assoc.* 1950, 41:701–710.
2. Black, R.B. Application and reevaluation of air abrasive technique. *J. Am. Dent. Assoc.* 1955, 50:408–414.
3. http://www.hss.energy.gov/csa/csp/hepa/docs/std3020.pdf
4. http://www.getty.edu/news/press/conserv/prague.html
5. http://www.getty.edu/conservation/field_projects/vitus/vitus_images.html
6. http://www.getty.edu/conservation/publications/newsletters/9_3/prague.html
7. http://www.naturalsciences.org/microsites/dinoheart/mediakit/index.html
8. http://misspiggy.gsfc.nasa.gov/tva/harrydoc/ESD_Effects.pdf
9. http://205.153.241.230/P2_Opportunity_Handbook/2_II_5.html

28

Cleaning with Carbon Dioxide Snow

Robert Sherman
*Applied Surface
Technologies*

Introduction

Manufacturers and researchers must clean items effectively, quickly, and affordably without damage, and at the same time minimize environmental and safety risks. This is never easy when many cleaning methods exist, and in the past 25 years, CO_2 cleaning methods have added new alternatives.

There are four major CO_2 cleaning methods—pellets, snow, liquid, and supercritical. In dry ice pellet cleaning, macroscopic dry ice impacts a surface, and cleaning is done via a thermo-mechanical shock. CO_2 snow cleaning relies upon smaller and less dense dry ice to remove particles by momentum transfer and organics by a solvent process. Both pellets and snow are inline processes; one piece at a time is cleaned. For liquid and supercritical CO_2 ($SCCO_2$), the cleaning is a batch process relying upon CO_2 solvent properties.

In today's "greener" world, cleaning with CO_2 is a better choice. All the CO_2 cleaning methods satisfy the ever more stringent industrial, research, and environmental demands. They are nontoxic, nonflammable, and non-ozone depleting, and human exposure poses few risks except for CO_2 buildup, frostbite (if applied directly to the skin), and oxygen displacement. The cleaning process can be residue-free and nondestructive. The contaminants can be carried away for venting or disposal.

First, we will discuss the CO_2 phase diagram and thermodynamics.

Thermodynamic Properties

The CO_2 phase diagram in Figure 28.1 shows the three familiar phases solid, liquid, and gas. The initial state for pellets is a solid at atmospheric pressure, an obviously nonequilibrium state at room temperature and atmospheric pressure. For snow, the initial condition is at room temperature and on

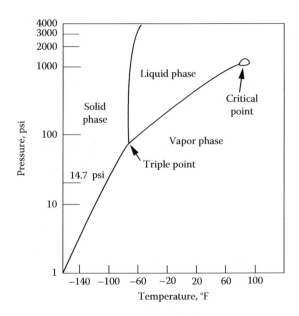

FIGURE 28.1 The carbon dioxide phase diagram—axes not to scale.

the liquid–gas line at 800 psi. After cleaning with pellets or snow, the CO_2 is easy to dispose of, either as a gas (for snow) or solid slowly sublimating to gas (for pellets).

Snow or pellet formation starts from the liquid or gas phase under pressure. Expansion through an orifice leads to a drop in pressure and the formation of a less dense solid phase (snow). Compression results in the formation of fully dense pellets. Liquid CO_2 cleaning is usually between 300 and 800 psi. For $SCCO_2$ the pressure must exceed 1070 psi and the temperature must be above 30.6°C (87°F).

CO_2 Snow Cleaning

CO_2 snow cleaning is a straightforward surface cleaning process in which a stream of small dry ice particles strike and clean a surface via physical and solvent interactions. CO_2 snow cleaning is simple to do; all you need is a source of CO_2, a hose to get the CO_2 to an on/off valve, and a nozzle. There are no chemical reactions or abrasive processes. These interactions remove particles of all sizes, from visible down to 0.03 μm, and also remove low levels of organic residues as effectively as solvents. The large range of particle removal along with the organic removal makes CO_2 snow cleaning unique in its potential.

CO_2 snow cleaning is a nonabrasive precision cleaning process. It is aimed at removing particles and thin organic residues. Thick oily layers are best cleaned by other methods. CO_2 snow cleaning is a line-of-sight method and surface impingement is required, especially for organic removal. Before discussing applications, we will discuss snow formation and cleaning mechanisms.

Snow Formation

The phase diagram tells us little regarding snow formation. Instead, one needs to look at the CO_2 pressure–enthalpy diagram shown in Figure 28.2. In addition to the three phases, the pressure–enthalpy diagram indicates the regions of pressure and enthalpy where the phases coexist. Ideally, CO_2 expansion through an orifice is a constant enthalpy process. This means that the pressure drops along a constant enthalpy line. Only adiabatic nozzles hold this condition. It should be noted that, contrary to a common misconception, CO_2 snow formation does not go through the triple point. The initial conditions for

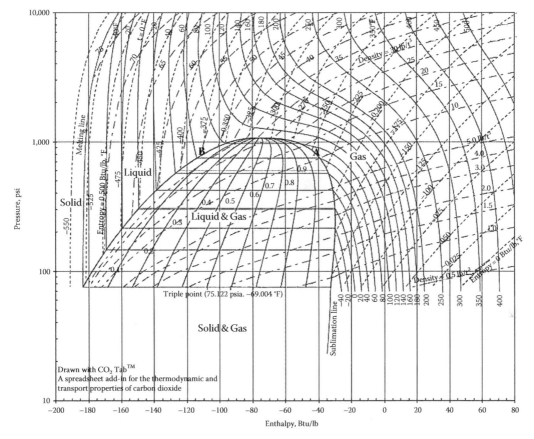

FIGURE 28.2 Pressure–enthalpy diagram for CO_2. Note starting points for CO_2 gas feed at point A and CO_2 liquid feed at point B.

liquid CO_2 or gas are at about 800 psi and are at the liquid–liquid vapor boundary for liquid CO_2 feed and at the vapor–liquid vapor boundary for CO_2 gas feed. Point "A" is for CO_2 gas, Point "B" is for liquid CO_2. With adiabatic nozzles, the pressure drops are vertical lines here. With non-adiabatic nozzles, the enthalpy increases, which leads to more CO_2 gas formation.

Cleaning Mechanisms

Overall, CO_2 snow cleaning removes particles physically bound to surfaces and also organic residues. It is a non-abrasive physical cleaning process, unlike sand blasting. CO_2 snow cleaning cannot replace sand blasting, acid etching, or any other gross removal process.

Whitlock [1] first discussed the two primary mechanisms for CO_2 snow cleaning, namely particle removal by momentum transfer and organic removal by the solvent action of the transient liquid CO_2 phase. These two processes are well-accepted, but they are not the only cleaning mechanisms.

Particle Removal

Particle removal results from the combination of momentum transfer and an aerodynamic drag force. The aerodynamic drag exerts a force on a particle that is proportional to its cross-sectional area. Gas flow alone cannot generate sufficient forces to remove micron and submicron particles bound by physical means, such as van der Waals forces.

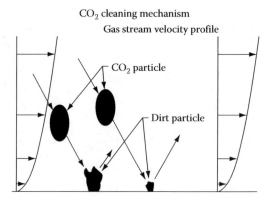

FIGURE 28.3 Cleaning mechanism for particle removal showing momentum transfer and the aerodynamic drag force.

While inside the near surface "dead zone" of no flow, where the aerodynamic drag force is zero, CO_2 snow cleaning can improve the likelihood of particle removal. CO_2 cleaning introduces mass as snow particles. The snow striking the particle is shown in Figure 28.3. This collision transfers momentum from the snow to the surface contaminant. This momentum transfer has the potential to overcome the surface adhesive forces and liberate the surface particulate. Once free, the particle is easily carried away with the flowing gas. Momentum transfer enables submicron particle removal.

Organic Removal

Hydrocarbons can be removed by CO_2 cleaning because liquid CO_2 is an excellent nonpolar solvent for hydrocarbons. During snow impact, stresses increase at the snow-surface interface and the pressure can easily exceed the dry ice yield stress and triple point pressure [1]. The dry ice particle liquefies and acts as a solvent while in contact with the surface (Figure 28.4). When the snow particle starts to rebound off the surface, the interfacial pressures decrease and the dry ice particle re-solidifies, carrying the contamination away.

This mechanism for the removal of organics is supported by the results of M. M. Hills [2]. She determined that organics that are easily absorbed in liquid CO_2 are removed; organics insoluble in

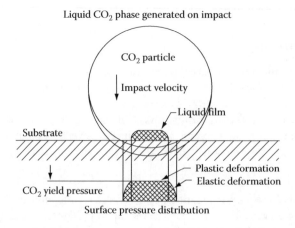

FIGURE 28.4 Cleaning mechanism for organic removal showing the liquid phase at the surface–dry ice interface.

liquid CO_2 are at best, slowly removed. For organics not soluble in CO_2, a slower freeze-fracture removal process was proposed in which the snow freezes the deposit and breaks it off of the surface.

In other instances, the CO_2 snow can be tuned to strike a surface with lower velocity and larger snow. In this case, the particle removal may be non-contact and relies on a combination of shear stresses generated by sublimating large snowflakes and thermophoresis [3]. In this case, the snowflake does not contact the surface, but glides above a cold gas layer generated by sublimating snow. Hill argues that this shear force removes particles at a greater rate than normal aerodynamic gas flows. The freed particle is then "attracted and held to the snow flake by thermophoretic forces—a force arising from thermal gradients."

Examples

For any cleaning process, controlled experiments are required to demonstrate quantitatively the cleaning effectiveness and efficiency.

Hoenig [4], who introduced CO_2 snow cleaning, presented qualitative data showing particle removal. Whitlock [1], and then later Sherman and Whitlock [5], performed experiments quantifying particle and hydrocarbon removal from surfaces. For particulates, Whitlock dispersed and then counted and sized micron and submicron particles on a wafer. Next, he cleaned the same area with CO_2 snow, and counted and sized again. A removal ratio of over 99.9% was achieved for particles larger than $0.1\,\mu m$.

To test cleaning for organics, the surface chemistry of new silicon wafers, contaminated wafers, and then CO_2 snow-cleaned wafers were all measured at the same areas to ensure proper comparisons. Cleaning with CO_2 snow removed all visible signs of the contamination, and XPS measurements showed hydrocarbon levels that were lower than the new wafers. These results demonstrate that CO_2 snow cleaning can remove hydrocarbon contamination and can also reduce the native hydrocarbon contamination.

Sherman et al. [6] used optical microscopy to illustrate the removal of silicon dust particles and organics from a specific area. Figure 28.5a shows a scribed and contaminated silicon wafer with many particles near the scratch. After CO_2 snow cleaning, no particles are visible at the same area, as shown in Figure 28.5b. These data are typical of particle removal from many different substrates.

Optical micrographs comparing the removal of a facial grease residue from given area before and after cleaning illustrate the removal of an organic film. A pair of micrographs showing the organic contaminated and then CO_2 snow-cleaned areas are illustrated in Figure 28.6. This removal of visible organic contamination by CO_2 snow cleaning is typical of many thin oily deposits and fingerprints on many different surfaces.

Equipment

CO_2 snow cleaning is accomplished with simple straightforward high-pressure spray systems, consisting of a CO_2 source, hose for CO_2, valve, and nozzle. A typical system, shown in Figure 28.7, includes a cylinder

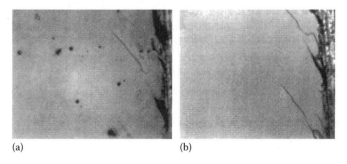

(a)   (b)

FIGURE 28.5 Micrographs at 1000× magnification showing a scribed region of a wafer (a) before and (b) after CO_2 snow cleaning.

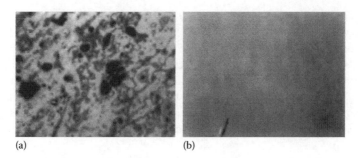

(a)   (b)

FIGURE 28.6   Optical micrographs at 1000× magnification of a scratched Si wafer that was (a) contaminated with facial grease and then (b) CO_2 snow cleaned.

fitting, tubing, on/off gun or valve, filter, and nozzle. The available on/off controls include solenoid, pneumatic, manual valves, and handguns. For the most part, CO_2 snow cleaning equipment is inexpensive; costs range from about $1800 to about $4000 for most systems.

In the following sections, we will discuss CO_2 cleaning equipment and methods in greater detail, including cleaning issues and process parameters such as recontamination, moisture condensation, and static charge.

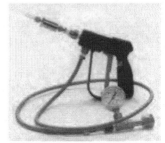

FIGURE 28.7   A typical manual based CO_2 snow cleaning unit with a hose, handgun, filter, and a nozzle.

Nozzles

The nozzle design is the most important part of the CO_2 snow cleaning system. Hoenig's [4] original instrument used a simple tube nozzle surrounded by a large expansion tube. This configuration allowed for the agglomeration and growth of snow, but at the expense of velocity. Smaller orifices and expansion zones are now common. Such non-adiabatic nozzles, for liquid CO_2 only, provide higher velocities and smaller snow sizes than the original Hoenig nozzle. The increased velocity is expected to improve the efficacy of organic removal.

Whitlock et al. [7] introduced a venturi (adiabatic) double expansion nozzle for CO_2 snow cleaning. This nozzle works with either liquid or gaseous CO_2. Sherman [6] simplified this to a single expansion venturi nozzle and obtained similar results. Studies using a single expansion venturi nozzle (Jacobs [8] and van der Hocke [9]) indicate that particle removal down to 0.03 μm can be achieved.

Nozzle design is vital to successful CO_2 snow cleaning. The most efficient nozzles incorporate a continuous venturi converging-diverging design. Other nozzle designs are available, including small diameter orifice tubes, metering leak valves, and others. However, with these nozzles, sudden expansions occur that violate the constant enthalpy condition and these nozzles do not work as well with a CO_2 gas source.

Different cleaning abilities can result from different designs. Nozzle designs can result in snow velocities from 5 m/s to over 100 m/s, depending upon orifice diameter and design, input and exit geometry, additional accelerants, and expansions.

Many nozzle concepts have been developed. Swain [10] introduced a nitrogen boost nozzle. This nozzle has a chamber where the CO_2 liquid droplets nucleate as snow. This agglomeration causes the snow to lose velocity, potentially compromising both particle and organic removal efficiencies. To counter the lost velocity, Swain introduced a vortex nozzle to mix snow with high-pressure (up to 100 psi) nitrogen to increase velocity. Swain indicates better cleaning results using the vortex nozzle, in comparison with simply mixing with nitrogen. Jackson [11] further developed Swain's concept by mixing heated and deionized nitrogen with CO_2. Goenka et al. [12] explored designs aimed at more aggressive cleaning. These designs used air or nitrogen to boost velocity and impact force.

The nozzles described earlier are round sources and clean less than 1 cm diameter spot. Layden [13] developed a continuous large area nozzle. It offers a double expansion nozzle with an exit can ranging in width from 1 to 10 cm. The Layden nozzle operates with either gas or liquid CO_2. Krone-Schmidt [14] developed a multiple orifice nozzle made from numerous individual nozzles.

Moisture, Static Charge, and Recontamination Issues

Moisture condensation, static charge, and recontamination are potential problems that occur during CO_2 snow cleaning. These potential risks must be addressed in setting up CO_2 snow cleaning. The CO_2 stream cools the surface. This can lead to moisture or ice condensation that can potentially interfere with cleaning or cause recontamination. Therefore, the user must minimize or eliminate moisture condensation. Simple methods include using room temperature or heated nitrogen purges, heating with a hot plate, infrared lamps, or a hot air gun. Others have used dry boxes. Most good dry boxes operate at very low dew points (–20°C to –50°C) [15,16]. Dry boxes have load locks for sample loading and unloading, thus maintaining a moisture-free cleaning chamber.

Static charges exist during CO_2 snow cleaning. Antistatic devices can be used during cleaning. Grounding a sample is a good start, but insulators need additional protection. Common antistatic discharge sources can be placed near the sample or even placed on the gun to assist in static control. Generally, antistatic discharge bars are placed in dry boxes to assist in these dry environments. Common antistatic sources and guns are good accessories to any CO_2 snow cleaning system.

Once a contaminant is removed, it must not redeposit on the substrate surface. Redeposition can result from improper workspace design as well as from poorly chosen equipment and techniques. Cleaning should not be performed in small or crowded areas. The stream should be allowed to escape from the cleaned surface with no walls or equipment near the sample to obstruct the rebounding stream. The turbulent stream can remove particles from the walls [3], or allow backflow to the cleaned sample. Hence, the user must make sure that there is adequate space around the sample. Also, for dry boxes, HEPA filters have been used to capture the released particles.

The cleaning procedure is also critical. Cleaning must proceed in a systematic method from a cleaned area to a dirty area. Cleaning must never be performed toward previously cleaned areas and the stream must always be aimed at the unclean areas and toward a vent.

New contamination sources can come include the cleaning equipment, the CO_2, and the cleaning process. The equipment can shed particles that can enter the CO_2 feed stream. Therefore, care is required in choosing material and surface finishes for all materials, valves, and hoses. For critical cleaning applications, choose electropolished stainless steel valves along with parts cleaned to industry standards. Point-of-use filtration is strongly recommended. This means placing a high-quality filter just before the nozzle.

Whitlock [1] and Sherman [6] investigated the sources and consequences of impurities in CO_2 feeds. Whitlock, using many different CO_2 grades, identified heavy hydrocarbons as a source of submicron particulate residue. Under best conditions, the residue was less than one particle per square centimeter. Sherman tested different CO_2 sources to clean a "clean" wafer surface. He identified the SFC (supercritical fluid chromatography) grades as the best CO_2 source for precision cleaning. SFC grades tend to have the lowest levels of heavy hydrocarbon impurities. For common CO_2 feeds, he found that hydrocarbons were a common contaminant.

Surface Damage

A common question is whether the incident snow can damage a surface. In almost every case, the answer is no. There are limitations, primarily related to severe thermal shock, mechanical strength of the substrate, and other factors. If a sample is properly supported, it can be cleaned, and even small flexible items such as the ends of individual fiber optics connectors have been successfully cleaned. In fact, for the vast majority of samples, no damage or changes can be found unless unique thermal stress issues are present. Problematic materials include KBr and LiF. While patterned wafers can be cleaned, MEMS (microelectromechanical systems) devices pose a challenge. Methods have been developed to clean CMOS (complementary metal oxide semiconductor) chips with delicate wire bonds.

Overall, changes may be limited to the molecular scale. Optical and electron microscopy have failed to detect damage on most surfaces even at high magnifications. AFM (atomic force microscopy) has shown the extent of expected surface changes. Examples of detected surface modification include atomic or molecular displacements, and, on soft surfaces, a smoothing effect only seen by AFM.

Applications

CO_2 snow cleaning is versatile. Cleaning applications range from simple laboratory applications to production contamination control. Remember, it is a line-of-sight process intended only for particles and thin oil layer. Applications include cleaning many types of materials, optics, vacuum components, hard disk drive components, parts, wafers, and more. The technique has gained wider acceptance and has entered many new fields because patents and published and unpublished reports of cleaning applications developed by manufacturers and users. We will discuss a few applications to indicate the wide range.

Materials

In the early 1990s, Sherman [17,18] demonstrated particle and organic contaminant removal from a wide variety of materials, including wafers, metals, ceramics, glass, optics, and polymers.

Many different wafer substrates have been cleaned. In all cases, cleaning removed visible contamination, removed particles, and reduced the hydrocarbon background on the surface. Wafer substrates included Si, Ge, InP, GaAs, patterned Si chips, hybrids, photoresist, Si_3N_4, and diamond. Automated systems have cleaned wafer-carrying cassettes [19], wafer handling equipment, chips, and chip modules.

Similar reductions in surface particle and hydrocarbon contamination have been noted for optical substrates, including soda lime glass, quartz, and for an array of coated glass substrates (InSnO, ZnSe, and SnO, for example) [17,20]. In these cases, the cleaning process removed particles and organics without altering, removing, or abrading the coating. Carbon dioxide snow cleaning has also been used to clean optical components; including mirrors, gratings, and filters.

Sherman [17] explored polymer cleaning and documented the removal of silicones, amide wax, and general hydrocarbons from different substrates. While polymers can be successfully cleaned, moisture control and sample support are critical issues.

The removal of gross organic-based contamination from a non-conducting gold contact has been achieved. Initial surface analysis, prior to CO_2 snow cleaning, indicated less than 1 atomic percent gold and over 80 atomic percent carbon. Silicon was also detected. After CO_2 snow cleaning, the gold surface concentration increased to over 24 atomic percent; the carbon concentration decreased to below 64 atomic percent; and silicone was not detected. Another way of viewing this data is to compare the carbon to gold surface compositions. Before cleaning, this ratio was 281; after cleaning, this ratio is 2.6. This represents a decrease of over 100 in the relative extent of surface contamination. Further, the gold contact became conductive after cleaning.

CO_2 snow cleaning has many uses within analytical chemistry. Examples include cleaning the equipment, standards, samples, and sample holders. Applications include surface analysis, scanning electron microscopy, optical spectroscopy, and AFM studies [8]. CO_2 snow cleaning removed particles and organics and led to better images and cleaner surface chemistries. For AFM, as first noted by Morris [21], CO_2 snow cleaning removes the "nanoscum" that causes poor imaging. Since then, many AFM users have used CO_2 snow cleaning as part of their sample preparation techniques.

Jacobs [8] presented an example of the efficacy of CO_2 snow cleaning in surface preparation for polymer adhesion to quartz surfaces. Initial solvent cleaning and wiping resulted in polymer decohesion. After CO_2 snow cleaning, the polymer thin films adhered to the quartz. Further analysis using AFM showed an array of numerous small particles on the initial surface (Figure 28.8). After

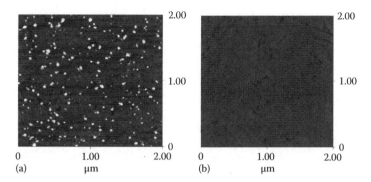

FIGURE 28.8 AFM image comparison of (a) before and (b) after cleaning a quartz surface.

CO_2 snow cleaning, the particles were removed. The images shown in Figure 28.8 are $2\,\mu m$ across, so the smallest particles imaged are about $0.03–0.04\,\mu m$.

Vacuum Technologies

Applications in vacuum technologies include cleaning of electron, ion and x-ray optics systems along with samples, substrates, components, and mass spectrometers. Cleaning can also be used during system fabrication [20]. Snow cleaning can remove machining oils, particle residues, and other debris from stainless steel surfaces and weld rods before welding. This process has led to less staining near the weld zones and fewer weld defects. CO_2 snow cleaning has also been used to assist in the final cleaning of vacuum parts after initial gross soil removal.

A study by Layden and Wadlow [22] compared solvent cleaning to CO_2 snow cleaning of a residual gas analyzer (RGA). Solvent and ultrasonic cleaning required total disassembly of the system, which was time-consuming. Repeated solvent cleaning led to an improved, though still slow pump downtime for the reassembled system. With CO_2 cleaning, only the filament and beam aperture had to be removed, and pump downtimes were further decreased to acceptable times. RGA analysis after solvent cleaning indicated hydrocarbon and alkaline contamination. After CO_2 snow cleaning, RGA analysis indicated that these contamination was reduced or eliminated. This study indicates the potential utility of CO_2 snow to simplify and improve the efficacy of cleaning of complex vacuum equipment.

In another application, Bailey and Mitavaine [23] successfully developed equipment for CO_2 snow cleaning of the electron guns for television cathode ray tubes.

Optics

Cleaning glass substrates before coating is a major application of CO_2 snow cleaning. CO_2 snow can be either the initial or final cleaning step of uncoated and coated glass, lenses, ceramic, or semiconductor substrates. CO_2 snow cleaning can also be used for individual optical components during production or assembly.

One example concerns a complex gyroscope subassembly that was cleaned in a dry environment. The removal of particles and residual organic contamination was critical for the end application; CO_2 snow cleaning fulfilled the requirement [24]. Examples of related optical cleaning applications include analytical standards, satellite optics, precision optics, and large telescopes. The key to proper optical cleaning is to ensure a dry, static-free environment.

Brandt [25], using an AFM to compare the relative flatness of a surface before and after CO_2 snow cleaning demonstrated that CO_2 snow cleaning could produce a flatter surface on a gold-coated quartz mirror. The enhanced smoothness is consistent with damage parameters for materials with low yield stresses. Zito [26] designed nozzles and methods for cleaning large optics. Other telescope users have built their own large-scale systems for cleaning the optics. The general observation is that snow cleaning can restore the reflectivity of a telescope mirrors to the initial values. For large telescope mirrors, the high elevation above sea level and cold climate conditions provide the dryness necessary for cleaning. At lower elevations, a dry environment or thermal input may be needed to counter moisture condensation.

Hard Disks

Cleaning hard disk drive subassemblies and parts has always presented challenges to contamination control engineers and CO_2 snow cleaning has met these needs. Almost all parts of a typical hard disk drive assembly have been successfully cleaned, including sliders, heads, head stack assemblies, aluminum base castings, head gimbals, and even coated hard disks. The key concern for this industry in cleaning parts is to ensure that there is no recontamination of the parts. Semiautomated and automated disk cleaning units are available.

Clean Rooms, Process Equipment, and Tooling

CO_2 snow can be used to clean clean rooms and semiconductor processing equipment. Contrary to expectations, the key to this process is to get the particles airborne, so that the room's filters can capture them. One must clean counters and the exterior of process equipment during maintenance, when critical surfaces are covered and the equipment is closed. Unpublished verification tests have indicated the efficacy of cleaning [27].

For process equipment, the chamber is opened during routine maintenance; and snow cleaning is added to the normal cleaning process. As with cleaning walls and the exterior of process equipment, the HEPA filters remove the airborne particles. In an unpublished study, a series of tests were performed on a silicide deposition system in a clean room during maintenance when critical surfaces were covered [27]. After monitoring the new snow cleaning process for several months, it was found that the particle populations on test wafers were reduced by a factor of 5. The reduction in particle population was associated with increased yields. Similar results have been obtained in non-clean-room environments and also mini-environments such as dry boxes or portable clean chambers placed over a cleaning station.

Other Applications

In a patent application, Kolb et al. [28] shows that CO_2 snow cleaning can mimic, and possibly even clean, as well as CO_2 pellets. They describe CO_2 snow cleaning to remove paint from a surface. A flashlamp or radiant energy source initiates weakening of the paint, and then CO_2 snow removes the loosened paint. An integral aspect of this method is a confined space for the lamp and the snow cleaning process.

Another interesting application is the cleaning of chips with attached wire bonds. The key steps are to adjust the snow velocity so as not to damage the delicate wire bonds [29].

Overall, a multitude of applications for CO_2 snow cleaning exist. Most manufacturers can assist and advise you as to what can or cannot work.

Carbon Dioxide Pellet Systems

The second of the four CO_2 cleaning methods is a solid phase method, CO_2 pellets. The technique is best described as "sand blasting without sand." It is a simple technique where macroscopic dry ice pellets are propelled by compressed gas to forcefully strike and clean a surface. Impact and thermomechanical forces do the cleaning. As with CO_2 snow, CO_2 Pellet cleaning is a line-of-sight process; and the technique uses neither chemical reactions nor direct mechanical abrasion. Pellets striking the surface are adjusted to remove gross contamination and overlayers without damaging the base material. After cleaning, the pellets sublimate to the gas phase, so all the user has to do is remove the debris on the floor. There is no separation of the cleaning media from the debris.

CO_2 pellet cleaning is considered a gross cleaning process in that it is aimed at removing unwanted overlayers such as grime, paint, casting, and mold residues. It is now a common and acceptable cleaning method. Keep in mind that it is not a gentle process.

Pellet Formation

The key factors in CO_2 pellet blasting are obtaining and then accelerating the pellets. In the phase diagram (Figure 28.1), we see that pellets are not in equilibrium at atmospheric conditions. Therefore, the pellet source must be either large dry ice blocks, smaller pellets stored in a freezer, or liquid CO_2 for conversion into pellets. Generally, the pellets are small, about 2–3 mm in diameter and twice as long; however, a range of sizes has been used. A number of methods such as shaving or milling have been developed to convert large blocks of dry ice into pellets of small, reproducible size. Once reduced in size, the pellets are fed into a high-pressure stream. One problem is that pellets can seize and agglomerate, blocking the flow. Shakers have been used to ensure a consistent flow of pellets in feed hoppers.

Onsite pellet production can be appropriate for large-scale operations. Production requires liquid CO_2, snow making nozzles or horns, and snow compression devices such as extruders. There are patents relating to this process. For example, a patent assigned to Westinghouse Electric Corp. [30] and others describes a contained pellet making and blasting system.

Cleaning Mechanisms

The cleaning method is fairly simple. A cold hard object hits the surface and the contamination is knocked off. CO_2 pellet cleaning shares many concepts with plastic bead, sand, and soda blasting, with one major difference. When the cold pellet hits the object, there is instant cooling; and the pellet starts to self-destruct, releasing large volumes of gas. The combination of impact energy, thermal shock, and forces generated by the pellet's sudden phase change to gas (impact forces accelerate the sublimation) results in surface cleaning via thermomechanical forces. The cleaning has been compared to microexplosions of gas on the surface along with severe thermal shock and abrasion. The phase change from solid to gas expands the volume over 700 times; this can result in the delamination of overlayers once they are broken or cracked [31]. However, well adhered overlayers with similar thermal expansion to the substrate will likely not be removed. The relative softness of the pellets compared to sand or other grit may make it possible to clean softer metals such as copper or aluminum without unacceptable surface damage [32].

Examples

Pellet cleaning applications range from paint and grime removal to decontamination of nuclear residues in buildings. Most of these applications lie outside the scope of this book on critical cleaning. Because pellet cleaning has not drawn as much academic interest as CO_2 snow or $SCCO_2$, the most expeditious approach is probably via an online web search.

Equipment

Equipment requirements are much more complex for pellets than for CO_2 snow. In addition to a pellet source, a large air compressor for pellet acceleration, a mixing chamber, hose(s), and a nozzle are required.

Both one- and two-hose systems are commercially available. In a single-hose system, the dry ice pellets and compressed air are fed to the nozzle via the same hose. This setup can allow for higher pressures, and hence higher velocities and impact energies. The one-hose systems require more complex airlocks and feed systems. Double-hose systems have separate hoses for dry ice and compressed air. Though easier and less costly to manufacture, these systems tend to have lower velocities and impact energies.

Cleaning Issues

Pellet systems are geared to gross cleaning. Therefore, issues such as recontamination and source purity that are critical to success with CO_2 snow are not as relevant for pellet cleaning.

One concern is the potential for product damage due to thermal shock. Dry ice pellets have been used for cleaning heated foundry molds and plastic injection molds. In both cases, the molds are hot and the potential concern could be raised as to whether the cold dry ice pellets might result in thermal shock and cracking. An analysis by Linder [31] suggests otherwise, even when cold pellets impact a heated mold. Thermal shock is one of the most important mechanisms in removing dissimilar overlayers. However, the thermal changes are limited to near the surface. With certain exceptions, thermal shock does not reach the yield stress. Therefore, Linder concluded that cleaning steel and most other metal molds is safe.

Abrasion is not an issue in most cases. The hardness of dry ice pellets is lower than that of other common grit materials. Therefore, gross abrasion tends not to be an issue for heavy industrial applications. However, soft substrate materials and fragile items have to be approached carefully. CO_2 pellets can cause shot peening damage on soft materials [32].

Applications

Numerous applications exist for CO_2 pellets with the proviso that it is a line-of-sight method and, in contrast with CO_2 snow, is aimed at gross cleaning applications. Some benefits include cost reduction, elimination of aqueous and solvent cleaning agents, a generally favorable worker safety and environmental profile, and speed.

Examples of the array of applications include mold and die cleaning in foundries and casting industry; ovens, conveyers, hoppers, tanks in the food industry; paint removal in the aircraft industry; printing presses; nuclear and biological decontamination; general cleaning of industrial grime (paper mills, electrical generators and machinery, cement); historical restoration (except for paintings or delicate items); disaster recovery; and restoring wooden decks.

Paint removal was the first well-known application for CO_2 pellets. This application may require optical or thermal radiation input to weaken the paint bonds. Many U.S. patents discuss this in detail [33]. The U.S. Military and suppliers have invested time and effort in developing the process. The approach has been used for difficult applications such as removal of sealant residues from aircraft sections.

Bjornard et al. [34] used pellets to clean large glass substrates immediately prior to thin film coating. The group addressed problems related to moisture condensation. They patented the use of dry boxes for cleaning, along with load locks before and after cleaning. The process was automated. Another application involves reworking of used semiconductor fab parts. Some evaporated metals and oxides can be removed from shields in certain cases, though this process cannot remove certain metal overlayers.

Gillis et al. [35] patented a mobile decontamination system using CO_2 pellets. They teach approaches to field decontamination using CO_2 pellets. The major components, including the compressor, CO_2 tank, air dryers, load locks, and cleaning chamber, are on skids. While the initial application was aimed at cleaning up nuclear waste in a room, the approach could be extrapolated to similar applications.

Liquid and Supercritical CO_2 Systems

Because liquid CO_2 has excellent solubility for hydrocarbons, it has been adopted as a solvent in special situations. Since liquid CO_2 has almost zero environmental impact, the main limitation is that it does not exist as a stable compound at atmospheric pressure and temperature. At room temperature, the equilibrium pressure is 800 psi. Therefore, the main limitation is the need for high-pressure equipment. Absent this limitation, it would be reasonable to speculate that the solvent-based process would have gained wider acceptance.

Though initially designed for metal parts cleaning, liquid-based CO_2 systems' major impact has been, and may well be in garment cleaning, as a replacement for perchloroethylene in dry cleaning. Capital equipment costs are relatively high. Pressure vessels, CO_2 circulation systems, and filtering systems are required. Systems developed for dry cleaning are now being used for consumer-based cleaning and other applications [36]. This is batch cleaning, and issues of cleaning delicate or breakable parts have to be addressed.

Advancements were made by Micell Industries to extend the use of liquid CO_2 beyond the removal of hydrocarbons to a range of organic compounds [37]. The approach by Micell and others has been to add surfactants, enzymes, and other cleaning compounds to the liquid CO_2, and also to introduce agitation, ultrasonics, and centrifugation. These enhancements may allow cleaning of complex and heavily soiled manufactured items. While dry cleaning for garments using liquid CO_2 is catching on, the costs remain high compared to conventional methods.

The $SCCO_2$ is a higher-pressure version of liquid CO_2 cleaning. $SCCO_2$ has the best efficacy of hydrocarbon removal of the four methods. By using pressures and temperatures above the critical point, a new phase, called the superfluid phase, is formed. In this phase, there is no difference between liquid and gas properties. The CO_2 has the solvency properties of the liquid and the penetration power of a gas. $SCCO_2$ is discussed more fully in Chapter 29 (Nelson) [38].

Conclusions

Four approaches to CO_2 cleaning are discussed and compared:

Snow
Pellets
Liquid
Supercritical

The focus of the discussion is on cleaning with CO_2 snow in which a stream of snow is created and used for line-of-sight cleaning. The thermophysical properties of CO_2 allow the removal of particles in a range of sizes as well as thin-film organic residues. The carbon dioxide snow cleaning process can address the contamination control problems, ranging from laboratory and clean room applications to a wide variety of industrial applications. Examples include:

- Optics—UV, IR and visible lenses, laser filters, substrate cleaning, and mirrors
- Vacuum technologies—cleaning vacuum system and components, deposition systems
- Microelectronic—substrate preparation, wire bond pads, ICs, hybrids, and tooling
- Contamination control—cleaning clean rooms, process equipment, and tooling
- Substrate preparation—metals, wafers, ceramics, polymers, and glass

By optimizing the CO_2 snow cleaning process, it is possible to remove particulates and low levels of organic contaminants from a wide range of materials while avoiding moisture and inadvertent recontamination. It is also possible to automate the cleaning process. Overall, the work done in the past 15–20 years has taken CO_2 snow cleaning from a laboratory niche cleaning method to its present state as an accepted solvent-free, dry cleaning process.

References

1. Whitlock, W. 1989. Dry surface learning with CO_2 snow. In: *Proceedings of the 20th Annual Meeting of the Fine Particle Society*, Boston, MA.
2. Hills, M. M. 1995. *J. Vac. Sci. Technol. A*, 13A, 30–34.
3. Hill, E. February 1994. *Precis. Clean. Mag.*, 2, 36.
4. Hoenig, S. A. August 1986. *Compressed Air Mag.*, pp. 22–25.

5. Sherman, R. and W. Whitlock. 1990. *J. Vac. Sci. Technol. B*, B8, 30–34.

6. Sherman, R., D. Hirt, and R. Vane. 2002. *J. Vac. Sci. Technol. A*, 12A, 1876–1881.

7. Whitlock, W. H., W. R. Weltmer, and J. D. Clark. 1989. U.S. Patent 4,806,171, February 21, 1989.

8. Jacobs, K. 1997. Personal communication. See discussion of AFM data in the quartz section at www.co2clean.com/afm.htm.

9. van der Donck Jacques, C. J. The 2004 MSF Conference, unpublished.

10. Swain, E. A., S. R. Carter, and S. A. Hoenig. 1992. U.S. Patent 5,125,979, June 30, 1992.

11. Jackson, D. P. 1998. U.S. Patent 5,725,154, March 10, 1998.

12. Goenka, L. N. 1995. U.S. Patent 5,390,450 February 21; U.S. Patent 5,405,283, April 11; U.S. Patent 5,514,024, May 7; U.S. Patent 5,944,581, August 1999.

13. Layden, L. L. 1990. U.S. Patent 4,962,891, October 16, 1990.

14. Krone-Schmidt, W. 2001.U.S. Patent 6,173,916, January 16, 2001.

15. Krone-Schmidt, W. and J. R. Markle. 1994. U.S. Patent 5,316,560, May 31, 1994.

16. Drexler, D. and J. Yurko. Personal communication.

17. Sherman, R., J. Grob, and W. Whitlock. 1991. *J. Vac. Sci. Technol. B*, B9, 1970–1977.

18. Sherman, R. and P. Adams. 1995. In: *Proceedings of CleanTech*, Rosemont, Illinois, Witter Publishing, Flemington, NJ, p. 271.

19. Bowers, C. W. 1998. U.S. Patent 5,766,061, June 16, 1998.

20. Sherman, R. unpublished results.

21. Morris, W. 1992. Personal communication.

22. Layden, L. and D. Wadlow. 1990. *J. Vac. Sci. Technol. A*, A8, 3881–3883.

23. Bailey, R. A. and J. C. Midavaine. 1997. U.S. Patent 5,605,484, February 25, 1997.

24. Hubner, E. 1994. Personnel communication.

25. Brandt, E. S. and B. A. Simpson. 1998. U.S. Patent 5,765,578, June 16, 1998.

26. Zito, R. 1997. U.S. Patent 5,775,127, May 23, 1997.

27. Layden, L. Personal communication.

28. Kolb, A. C., L. W. Braverman, C. J. Silberman, R. R. Hamm, and C. Michael. 1997. U.S. Patent 5,613,509, March 25, 1997.

29. Sherman, R. and A. Lee. Unpublished work.

30. Palmer, C. E. 1996. U.S. Patent 5525093, June 11.

31. Linder, D. R. In: *Proceedings of Precision Cleaning'97*, Cincinnati, Ohio, Witter Publishing, Flemington, NJ, p. 242.

32. Uhlmann, E. and M. Krieg. 2005. Available at www.shotpeener.com/library/pdf/2005072.pdf

33. Cates, M. C., R. R. Hamm, M. W. Lewis and W. N. Schmitz. 1994. U.S. Patent 5328 517, July 12; U.S. Patent 5782253, July 21, 1998; Llyod, D. L. 1996. U.S. Patent 5571335, November 5; and others.

34. Bjornard, E., E. W. Kurman, D. A. Shogren, and J. J. Hoffman. 1997. U.S. Patent 5651723, July 29.

35. Gillis, P. J., B. Sklar, A. J. Kral, and M. C. Randolph. 1992. U.S. Patent 5123207, June 30.

36. Dewees, T. G., F. M. Knafel, J. D. Mitchell, R. G. Taylor, R. J. Iliff, D. T. Carty, J. R. Latham, and T. M. Lipton. U.S. Patent 5267455, December 7, 1993; U.S. Patent 5279615, January 18, 1994; Jackson, D. P. and O. F. Buck. U.S. Patent 5013366, May 7, 1991; Chao, S. C., T. B. Stanford, E. M. Purer, and A. Y. Wilkerson. U.S. Patent 5467492, November 21, 1995.

37. McClain, J. B., T. J. Romack, and J. P. DeYoung. US Patent 6030663, February 29, 2000.

38. W. Nelson. Cleaning with dense-phase CO_2: Liquid CO_2, supercritical CO_2, and CO_2-snow, Chapter 8, *Critical Cleaning Processes*, Taylor & Francis, Boca Raton, FL.

29

Cleaning with Dense-Phase CO₂: Liquid CO₂, Supercritical CO₂, and CO₂ Snow

William M. Nelson
Obiter Research, LLC

Solvents for Cleaning

In terms of cleaning, a solvent is a substance, single or multicomponent, capable of dissolving other substances to form a homogeneous system (solution). Once the contaminant is dissolved in the solvent, both can be removed from the substrate. The criteria for what determines a good solvent depend on the use and ultimate level of cleanliness sought during the process. A solvent may be defined in rough terms as any liquid that serves as a carrier for another substance or as a means of extracting or separating other substances. New criteria for solvent selection include the necessity of leading to minimal environmental impact during and after their use.

In practice, many solvents are mixtures rather than pure compounds. The most readily available solvent is water, but next in importance comes a group of organic liquids and their mixtures. Presently, as mentioned earlier, there is a concern for the use of environmentally benign solvents, and there is increasing interest in alternatives to the more traditional solvents. For example, there remains a challenge to clean effectively manufactured metal parts so that they are suitable for their intended use, and any residual waste streams can be efficiently, effectively, and economically handled. Traditional cleaning solvents will see continued use, albeit in more restricted and regulated manner. This creates a demand for new classes of solvents, or as in the case of CO_2, further extension of solvent applications that meet all the above criteria.

Solvents play an essential role in many industrial coating and cleaning applications. However, choosing the appropriate solvent or solvent blend for a particular application is often not easy. Many variables must be considered, and the choice is system dependent. To use a new solvent, a product typically needs to be reformulated to maintain desired performance attributes.

Today, efforts are being guided by the principles of Green Chemistry, in particular the efficient use of materials and energy, development of renewable resources, and design for reduced hazard, including the design and use of alternative solvents [1]. With an increasing emphasis for all businesses to reduce their environmental footprint, there is considerable interest in new technologies aimed at reducing current waste streams. Cleaning technologies (preparing a material's surface for subsequent steps in an industrial process) can have a significant environmental impact, and are therefore candidates for new technologies.

The available solvents have given individuals in the cleaning industry a great set of tools. Now, however, judicious preparation and careful reasoning must guide the selection of cleaning technologies. Insofar as cleaning creates a novel waste stream, the movement toward zero discharge of pollutants into the air, water, and soil will limit future choices. With ozone-depleting chemicals (ODC) being phased out, many manufacturers are struggling to find efficient and effective replacement solvents and cleaning agents. The ODC phaseout and a host of other environmental and safety concerns have prompted the development of alternative cleaning agents. The first decade in the twenty-first century has seen rapid improvements in solvents that may address this challenge.

The quality and suitability of the cleaning process is heavily dependent on the quality of the solvent utilized, and this is especially true for applications demanding high levels of cleanliness. The solvent is either an active agent in the process or is the stage on which the process occurs [2]. Solvent use in cleaning will continue to be pervasive. The challenge to the cleaning industry will be to adopt the most environmentally benign and efficacious technology.

The literature reveals that use of CO_2 has permeated almost all facets of the chemical industry and that careful application of CO_2 technology can result in products (and processes) that are cleaner, less expensive, and of higher quality [3]. Cleaning with CO_2 is advantageous in that after cleaning, the only waste streams generated are the isolated contaminants that were removed from the part that was cleaned. There are no large, liquid solvent waste streams to treat (as there are with aqueous cleaning) or air streams to treat (as is the case with some solvent cleaning solutions). The challenge will be to harness CO_2 to accomplish the cleaning tasks.

Dense-phase CO_2 is a true solvent, and it can serve as the dissolving media for specific cleaning applications. In this chapter we briefly examine (1) the solubility character of CO_2 in its liquid and supercritical (SC) phases, (2) the use of co-solvents and entrainers to enhance solubility, (3) the equipment and engineering necessary to implement this cleaning technology and explore some areas where this technology has its largest current use.

Properties of Carbon Dioxide

As Christine Cline wrote in 1996, "CO_2 is four cleaning agents in one. Its supercritical state ($SCCO_2$) provides chemical extraction and solvent cleaning for organic compounds; its liquid state CO_2 (LCO_2) provides degreasing by quickly removing oils and greases; its solid/gaseous state removes submicron particulates and light organic material and its solid state removes gross contaminants from a wide variety of surfaces and substrates. CO_2's cleaning performance is enhanced by its low viscosity, high density, solubility parameter, compatibility and low latent heat of vaporization" [4].

The four practical ways of using carbon dioxide for cleaning are (1) dry ice pellets, (2) "snow" particles, (3) LCO_2 washing systems, and (4) $SCCO_2$. In this particular chapter, we will be concerned with the use of CO_2 as a solvent (numbers 2, 3, and 4). The cleaning process will depend on the LCO_2 solvent properties. These properties are augmented or altered by the addition of co-solvents or surfactants. The liquid-based CO_2 washing systems rely on the liquid phase solvent properties. When $SCCO_2$ is used, the systems rely exclusively on CO_2's unique SC fluid properties.

Carbon Dioxide

CO_2 is a colorless gas. Industrial applications of CO_2 are in solid (dry ice), liquid, and gaseous forms; these can be found in a variety of such applications as beverage carbonation, welding, chemicals manufacture, and cleaning. It occurs in the products of combustion of all carbonaceous fuels and can be recovered from them in a variety of ways. CO_2 is also a product of animal metabolism and is important in the life cycles of both animals and plants. It is present in the atmosphere in small quantities (0.03%, by volume).

Growing concern over the impact on global climate change of the buildup of greenhouse gases (GHGs) in the atmosphere has resulted in proposals to capture CO_2 at large point sources and store it in geologic formations, such as oil and gas reservoirs, unmineable coal seams, and saline formations, referred to as carbon capture and storage (CCS) [5]. There will be ample quantities of this gas for use as a solvent, and the issue will be to ensure its purity.

CO_2 is not very reactive at normal temperatures. However, it does form carbonic acid, H_2CO_3, in aqueous solution. This will undergo the typical reactions of a weak acid to form salts and esters. A solid hydrate, $CO_2 \cdot 8H_2O$, separates from aqueous solutions of CO_2 that are chilled at elevated pressures. It is very stable at normal temperatures but forms CO and O_2 when heated above 1700°C.

CO_2 has several advantages: environmental acceptability, nonflammability, and noncorrosivity. Additionally, carbon dioxide has no ozone depletion potential, and, while it does have some global warming potential, its use in cleaning operations would contribute insignificantly to global warming as it will be previously captured CO_2 that is released into the atmosphere.

Phases of CO_2

As a chemical, the physical form and chemical properties of CO_2 are dependent on temperature and pressure. A pressure–temperature (*P–T*) phase diagram, shown in Figure 29.1, illustrates the phase changes of CO_2, where the three phases of solid, gas, and liquid are indicated.

This chart indicates the temperature and pressure conditions that must be used if CO_2 is to be used for cleaning. The substance remains a liquid, as long as the temperature and pressure fall within the $CO_2(l)$ region. CO_2 is supercritical when its pressure and temperature are beyond the critical point. Notice that

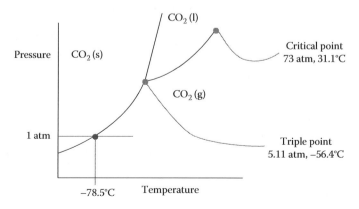

FIGURE 29.1 CO_2 phase diagram.

the triple point of CO_2 is well above 1 atm. Notice also that at 1 atm CO_2 can only be the solid or the gas. LCO_2 does not exist at 1 atm. Dry ice (solid CO_2) has a temperature of $-78.5°C$ at room pressure which is why a person can get a serious burn (actually frostbite) from holding it in their hands.

Liquid and Supercritical Carbon Dioxide and CO_2 Snow

Liquid CO_2

Although LCO_2 does not exist at normal room pressures, it does exist at slightly elevated pressure. A laboratory cylinder of CO_2 will contain LCO_2 at a pressure of about 75 psi at room temperature. On a particularly hot day (above 88°F), the LCO_2 will pass though its critical point and the contents of the cylinder will exist as the SC fluid.

The LCO_2 cleaning technology used alone under these conditions is a solvent much like room temperature 1,1,1-trichloroethane (TCA). As such, it will remove many but not all types of contaminants. Contaminants that are not soluble in LCO_2 alone can be solubilized or otherwise separated by employing proprietary additives, modifiers, or mechanical adjuncts in the process [6].

Liquefied CO_2 can be used as a solvent for cleaning. Process temperatures generally range between 50°F and 70°F, and process pressures range from 750 to 1200 psi. The area in the CO_2 phase diagram wherein the substance is liquid is shown in Figure 29.1.

Early studies showed that LCO_2 and $SCCO_2$ can be used as a solvent substitute in the polymerization of hydrocarbon monomers. However, because it was found that CO_2 alone is a poor solvent, further efforts focused on the development of additives to enhance the solvent capability of CO_2. This resulted in the development of a class of CO_2-compatible surfactants that may make the technology a viable and nonpolluting surface-cleaning option [7].

The LCO_2 immersion cleaning process can meet a variety of cleanliness requirements, ranging from visually clean to more rigorous quality standards requiring such sophisticated test methods as nonvolatile residue analysis, infrared spectroscopy, or scanning electron microscopy. Functional testing, such as that measuring weld joint porosity and adhesive strength, has also been conducted to evaluate the technology.

Supercritical CO_2

CO_2 becomes "supercritical" when it is heated above its critical temperature—the point beyond which it cannot be liquified and compressed (see Figure 29.1). CO_2 becomes supercritical at temperatures above 87.8°F (31.1°C) and pressures above 1072 psi (73.8 bar) [8]. $SCCO_2$ applications typically operate at temperatures between 90°F and 120°F (32°C and 49°C) and pressures between 1070 and 3500 psi.

There are two unique points on this phase diagram: the "triple point" and the critical point. The "triple point" is the lower point and is the unique combination of temperature and pressure at which all three phases (gas, liquid, and solid) exist simultaneously.

The second unique point is the critical point. It can be shown experimentally that for every liquid there is a point along the boiling point curve where the line between the liquid and gaseous phases disappears. This is called the critical point. At temperatures higher than this point, we can no longer think of there being two phases; there is only a single phase. This single phase is a very dense gas, or frequently called a critical fluid. Another way of thinking about this is to remember that at the critical temperature or above, we can no longer compress the material to a liquid, no matter what the applied pressure is. These critical fluids have unique properties and are used for many commercial processes.

One use of $SCCO_2$ is the remediation of soils contaminated with petroleum hydrocarbons, polyaromatic hydrocarbons, polychlorinated biphenyls, dioxins, heavy metals, and other organic and inorganic compounds that have resulted from industrial activities, accidental spills, and improper waste disposal practices. SC fluid extraction is a remediation technology for contaminated soils. It is a simple, fast, and selective solvent extraction process that uses a SC fluid as the solvent. In SC fluid extraction, the extracted contaminants first dissolve into the SC solvent and then these contaminants are separated from the SC solvent via a simple change in pressure and temperature conditions or by using a separation process [9]. The success of SC fluid extraction as a method for removing these contaminants from soils is highlighted and some of the future research needed to develop it as a commercial-scale economic remediation technology as well as a surface cleaning technology is discussed.

CO_2 Snow

Taking advantage of the thermophysical properties of CO_2, an effective cleaning stream containing solid CO_2 snow can be created. The process can efficiently remove particles of all sizes and organic residues. The parameters necessary to perform cleaning include purity, filtration, static charge control, moisture control, and feed pressure. By paying attention to these details, the CO_2 cleaning process can be automated and particles as well as organics can be removed from a wide range of materials [10].

CO_2 snow cleaning technology has demonstrated the ability to clean a variety of contamination types in a non-damaging, environmentally friendly manner. CO_2 snow cleaning methods have been used to clean both particles and light organics and currently can be found in a variety of high-tech industries to include, semiconductor (Si and GaAs), flat panel display, disk head and media manufacturing, and fiber optics [15].

Engineering Condensed-Phase CO_2 Cleaning

In order to use condensed phase CO_2 in industry, the CO_2 gas needs to be pressurized and heated (see Figure 29.1). One current application places parts to be cleaned into a pressure vessel into which CO_2 gas is introduced. The temperature and pressure are raised until the SC state is reached.

A basic system for this process then consists of six components:

- Compressor
- Heat exchanger (heating)
- Extraction vessel (pressure vessel)
- Pressure control valve (expansion)
- Heat exchanger (cooling)
- Separation vessel

Figure 29.2 shows the basic components that comprise a condensed-phase CO_2 cleaning system. CO_2, which may be stored as a gas or in liquid form, is compressed above its liquid or critical pressure by a pump.

The compressed CO_2 is then heated to its liquid phase or to above its critical temperature in a heater, or sometimes in the cleaning chamber. Any parts in the cleaning chamber are cleaned by being surrounded by the liquid or fluid. Typically, the cleaning chamber will include an impeller to promote mixing.

Condensed-phase CO_2 (containing dissolved contaminants) is then bled off to a separator vessel, where the liquid/fluid is decompressed and returned to a gaseous state. The contaminants remain in liquid form and are collected out the bottom of the separator, while the gaseous CO_2 is sent through a chiller to return it to a liquid form for storage to be reused again. This closed-loop recycling of the CO_2 means that only a small portion of the cleaning solution has to be replaced over time due to system leakage. The now clean parts can be removed from the chamber and are usually immediately ready for the next step in the manufacturing process, since no drying or rinsing is required to remove residual cleaning solution. With some plastics, which can absorb CO_2, a bakeout may be needed.

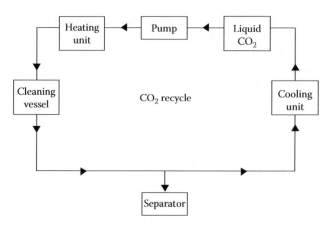

FIGURE 29.2 Basic schematic of a supercritical carbon dioxide cleaning system.

Process temperatures may range from 95°F to 149°F (35°C–65°C). Pressures vary from about 1070 to 4000 psi. Nonmetallic materials must be tested for compatibility with the entire process. The process works well for removing trace fluids. Some suppliers also claim effective removal of particle contamination, but this will depend on physical conditions and added entrainers. It may be possible to fine-tune the operating pressure and temperature to match the soil being removed. Parts that cannot be subjected to elevated atmospheric pressures cannot be cleaned with SC fluids. This process is used in the precision cleaning industry, and may become more broadly applicable.

Details of Condensed Phases

Solubility

An experimental determination of the solubilities of solid mixtures in SC fluids was made [11]. Examples of the solubility measurements are becoming more common in the scientific literature (Table 29.1) and [12].

Dense-phase fluids have both liquid- and gas-like properties. This property will confer highly desirable solubility characteristics. This allows the liquid or fluid to penetrate very small gaps and complex assemblies, which will enhance the potential range of substrate geometries (Table 29.2).

Co-Solvents

If condensed-phase CO_2 is going to be implemented on a commercial scale to remove contaminants from material surfaces, the costs associated with this process must be reduced and the range of solubilities must be broadened. These goals can be made more attainable with the use of an entrainer (or co-solvent)

TABLE 29.1 Examples of Solubility Measurements in $SCCO_2$

Issue	Result	Reference
Improving polymer solubility	Surfactants' improved solubility	[35]
Low solubility of organophosphorous compounds	All compounds' solubility improved	[36]
Solubilities of amorphous polymers	Describes conditions for solubility	[37]
Separation of vegetable cuticular waxes	Enhanced separation	[38]
Effect of co-solvents	Measured effects	[39]
Determine effects of methanol as co-solvent	Measured effects	[40]
Determine effects of *n*-pentanol as co-solvent	Measured effects	[41]

TABLE 29.2 Examples of Substrates for $SCCO_2$ Cleaning

Missile gyroscopes
Accelerometers
Thermal switches
Nuclear valve seals
Electromechanical assemblies
Polymeric containers
Special camera lenses
Laser optics components
Porous ceramics
Metal parts

to increase the solubility of the solute in the SC phase, reducing the size of the required extractor and/or lowering the pressure needed to effect the desired extraction.

Dense CO_2 extraction is an emerging technology in cleaning and remediation. The successful design and implementation of processes based on this technology require the accurate determination of solute diffusivities in the systems [13]. An example of this comes from the recovery of a coke catalyst through $SCCO_2$ cleaning. A recovered intermediate selectivity was achieved by the $SCCO_2$-cleaned catalyst, suggesting an effective cleaning of Pd sites and a possible reconstruction of the residual coke layers onto the surface caused by the solvent nature of $SCCO_2$. Scanning electron microscopic (SEM) pictures show that the cleanup of the used catalyst with $SCCO_2$ removes most of the coke and oily materials from the surface [14].

Substrate Compatibility

Case studies and research results of dense CO_2 cleaning and drying are helpful. For example, some work on drying demonstrates the benefits of CO_2. Several CO_2-soluble surfactants with different molecular architectures were investigated as possible agents to remove rinse water from aqueous-based photoresists utilizing $SCCO_2$ drying process. Hydrocarbon and fluorocarbon surfactants having a short chain length and polymeric surfactants based on block or random copolymers were selected for this study. SEM images confirmed that the positive resist patterns were preserved without any deformation or damage by rinsing with deionized water followed by surfactants-aided $SCCO_2$ drying [16]. The lessons learned from true, long-term implementation of a technology are always the most valuable to others considering applying the technology themselves (Table 29.3).

While test results show the technology works quite effectively in many cases, there are some published case studies from industry that resulted in decisions not to pursue the technology for full-scale application.

In one instance, tests of $SCCO_2$ cleaning were performed for a manufacturer on metal discs contaminated with oil-type residues in a 1 L system at 180°F (82°C) and 2000 psi [17]. $SCCO_2$ cleaning was not pursued due to the predicted long payback time (10 years) of the investment, and the high operating pressure of the system (which was viewed as a safety hazard).

TABLE 29.3 Examples of Compatibility Tests

Issue	Result	References
Concerns over $SCCO_2$ cleaning polymers	$SCCO_2$ cleaning can be adjusted to have no detrimental effect on crystalline polymers	[42]
Oil removal from rings, washers, and plates	$SCCO_2$ cleaning removed 97%–99.95% of the oil	[43]
Different substrates, including aluminum, glass, copper, brass, stainless steel, and epoxy boards	$SCCO_2$ good for water-sensitive or high-temperature sensitive parts	[44]
Soils and other solid materials, containing residual pesticides	Supercritical fluids can be applied to these substrates	[45]
Various metals, plastics, and epoxies needed to be precision cleaned, using CFC-113	$SCCO_2$ was used; excellent payback and >90% reduction in ODS	[46]
Vapor degreaser was used to clean gyroscope parts contaminated with machining coolants, silicone oils, and damping fluids	$SCCO_2$ works well for parts with complex shapes	[47]
High cleanliness standards are required for precision cleaning	$SCCO_2$ was shown to work more efficiently for contaminated plastic than cleaning with CFC-113	[21,27,47,49–51]
Replace PERC as a dry-cleaning solvent	$SCCO_2$ is equally effective	[48]

In another study, AT&T researchers tested a number of different cleaning alternatives to replace 1,1,1 trichloroethane vapor degreasing. The requirement for additional additives and co-solvents to obtain acceptable cleaning results led the researchers to eliminate $SCCO_2$ from consideration [18].

$SCCO_2$ cleaning appears to be effective in a number of instances from a technological standpoint, but some earlier technical limitations combined with the economies of the process have resulted in a slow rate of implementation in the private industrial sector.

Effect of Mixing on $SCCO_2$ Cleaning

Test results indicate that increasing the internal agitation and temperature in the cleaning chamber reduces the time needed to clean the metal parts. The SC cleaning process, when run at optimum conditions, appears to use less energy than conventional vapor degreasing operations. Furthermore, cleaning results attained with $SCCO_2$ plus mixing compare favorably with conventional solvent cleaning [19].

Two studies that discuss the effect of mixing on $SCCO_2$ cleaning resulted in somewhat different conclusions. As mentioned earlier, a manufacturer of metal discs was not able to determine that mixing had any effect on cleaning efficiency, but that a larger sample size may have given a more definitive result [17].

A more in-depth study specifically designed to examine the effect of fluid turbulence on $SCCO_2$ cleaning concluded that mixing has an effect. Researchers at Pacific Northwest National Laboratory completed the study. The researchers recommended that agitation be used whenever possible for $SCCO_2$ cleaning applications to help maximize cleaning efficiency, and that mixing rates can be optimized to minimize power costs [20].

Studies have demonstrated that $SCCO_2$ is an excellent solvent for oils such as hydrocarbons, esters, silicones, perfluoropolyether, halocarbon-substituted triazines, and organosilicones with various reactive functionalities [21]. The ability to dissolve a particular oil or polymer at any given pressure will greatly depend on the molecular weight and structure of the material. The ability of SC fluids to dissolve many types of oils and organic materials, coupled with the ability to penetrate minuscule pores and interstices of metal, ceramic, and composite parts, suggests that these fluids could partially replace chlorofluorocarbon (CFC).

In conclusion, dense CO_2 is the most promising of the alternative technologies due to its low cost, low toxicity, nonflammability, and environmental acceptability [21].

Examples of Industrial and Chemical Applications of Dense CO_2 Cleaning

A cleaner, by definition, removes dirt or other extraneous material from a surface. An effective cleaner is able to perform its task because of its ability to wet the surface, penetrate the soil, lift and remove the soil, and hold soil in suspension (so a surface can be wiped or rinsed).

The degree of surface cleanliness must meet the following two criteria: (1) it must be sufficient for subsequent processing, and (2) it must be sufficient to ensure the future reliability of the device or system. Beyond this, there are numerous factors that affect the quality of the cleanliness. Three drivers, or forces, have contributed to the recent attention given to these solvents (Table 29.4).

TABLE 29.4 Drivers Influencing Adoption of Condensed-Phase CO_2 Cleaning

Driver	Description
Environmental	Environmental problems associated with common industrial solvents (mostly chlorinated hydrocarbons)
Economic	Increasing cost of regulated solvent use
Technological	Inability of traditional techniques to provide the necessary separations needed for emerging new industries (microelectronics, biotechnology, etc.)

The availability of inexpensive, nontoxic solvents such as liquid or $SCCO_2$ and their attractive properties has renewed interest in the applicability of these solvents, especially in the area of cleaning. In the following section, we present several industrial applications of dense CO_2 cleaning and rinsing with references.

Substrate Drying

Condensed CO_2, when regarded as a solvent, may benefit from the enormous data of the science of solvents [22]. After a surface was cleaned with an aqueous agent, residues, water, and surfactant were displaced from the cleaning vessel with CO_2 assisted by gravity to reduce the rinse time from well over 10 min to approximately 1 min. The low interfacial tension produced by surfactant and $SCCO_2$ is beneficial in the removal of the cleaning solution [23].

Surfactants

$SCCO_2$, like any cleaning technology, works better on certain classes of soils. $SCCO_2$ has been employed to successfully remove oils such as hydrocarbons, esters, silicones, perfluoropolyethers, halocarbon substituted triazines, etc. It is an elegant cleaning method with many environmental benefits. Early studies showed that $SCCO_2$ can be used as a solvent substitute in the polymerization of hydrocarbon monomers. However, because it was found that CO_2 alone is a poor solvent, further efforts focused on the development of additives to enhance the solvent capability. This resulted in the development of a class of CO_2-compatible surfactants that may make the technology a viable and nonpolluting surface-cleaning option [7].

It has been shown that high-pressure $SCCO_2$ shows potential as a cleaning medium for removing hydrocarbon machine coolants from metal substrates [27]. In addition, $SCCO_2$ by itself can be tuned so as to remove the contaminants listed in Table 29.5.

Use with Photoresists

An extensive discussion of SC media is to be found in Savage's review [24]. SC fluids are effective cleaning agents because of their ability to penetrate substrates rapidly and penetrate small interstitial spaces. After dissolving any contaminants, the critical fluid is removed easily and completely because it lacks surface tension. Solvent properties can be adjusted by small changes in temperature and pressure, and this allows CO_2 to dissolve a range of organic compounds. A clear example of this is found in recent research with photoresist polymers. Photoresist polymers with high aspect ratios are presently cleaned with aqueous solutions. For high aspect ratios, the pattern collapses during the drying step. The origin of resist pattern collapse is the surface tension of the rinse liquid. The use of CO_2 as a rinse liquid should help prevent the resist pattern collapse problem [25].

Cleaning Rollers for Printing and Packaging Industries

$SCCO_2$ actually has physical properties somewhere between those of a liquid and a gas. SC fluids are able to spread out along a surface more easily than a true liquid because they have lower surface tensions than liquids. At the same time, a SC fluid maintains a liquid's ability to dissolve substances that are soluble in the compound, which a gas cannot do. In the case of $SCCO_2$, this means oil and other organic contaminants can be removed from a surface even if it has an intricate geometry or includes cracks and crevices.

Printing or film coupling processes use engraved rollers whose surface is formed by microscopic cells that carry inks or adhesives on the film.

TABLE 29.5 Contaminants Removed by $SCCO_2$

Silicone oils
Flux residues
Petroleum oils
Machining oils
Dielectric oils
Lubricants
Adhesive residues
Plasticizers
Fats and waxes

During their use, cells are progressively filled up of residual dry ink and/ or adhesive that reduces their efficiency. Cleaning of cells is very complex due to their microscopic dimensions. A new technique based on the use of SC mixtures (CO_2 and organic solvents) to clean engraved rollers was developed. An almost complete cleaning was obtained in 40 and 60 min operating with a $SCCO_2$ plus 80% w/w of N-methyl pyrrolidone [26].

In general, this process cannot remove contaminants which do not dissolve in CO_2 (Table 29.6).

TABLE 29.6 Contaminants Not Removed by $SCCO_2$
Rust
Scale
Lint or dust
Ionic species
Metal salts
Many (but not all) fluxes

CO_2 Snow Cleaning

Indium tin oxide (ITO) glass cleaning is LCD and other flat panel display industry's key technologies. At present, the usual wet cleaning technology consumes a large amount of water and chemicals, and produces a large amount of contaminant venting. CO_2 snow jet spray cleaning has been successfully applied to cleaning the surface of semiconductor chip, vacuum devices, and space telescopes. The surface cleaning of ITO film was carried out with CO_2 snow jet treatment. Experimental data show that the CO_2 snow jet treatment effectively removes particulate and hydrocarbon on ITO surface [28].

Cryogenic Aerosol Cleaning of Photomasks

A dry cleaning technology using CO_2 cryogenic aerosols has been reported. The cleaning mechanism relies on aerosol particles to overcome the force of adhesion of the contaminant particles on the surface. Particle removal is possible without degradation or etching of underlying film or the need for drying with isopropyl alcohol as in wet cleaning. Experimental results with silicon wafers show that the removal of submicron particles is 10% higher than larger particles. Inorganic contaminants such as ammonium sulphate, commonly known as "haze," is removed by cryogenic aerosol cleaning with 99% efficiency, as seen using optical inspection tool. The results show that over 16 cleaning cycles, the change in transmission is 0.04% and the change in phase is 0.37°. Thus, a noninvasive cleaning for submicron particles from photomasks is possible with CO_2 cryogenic aerosols [29].

Dry Cleaning of Fabrics

The removal of particulate soils from textile in dry cleaning with CO_2 is insufficient compared to perchloroethylene (PERC). Especially the removal of relatively small particles poses a problem. Various anionic and amine-based surfactants have been investigated to enhance the removal of particulate. The use of these surfactants, however, does not bring particle removal up to the level of PERC. The removal of non-particulate soils in CO_2 with water, surfactant, and co-solvent is better than in PERC. The use of a co-solvent (iso-propanol) had a positive effect on the removal of particulate and non-particulate soils when using amines. However, when anionic surfactants were used, the addition of a co-solvent had a pronounced negative effect on particulate soil removal [30].

Economics

Until recently, the economics behind condensed-phase CO_2 use in cleaning has been prohibitive when applied in the cleaning industry. Equipment for SC fluids cleaning tends to be costly; process development is very application specific [31]. With the current shift in environmental awareness, this is changing.

Researchers at Pacific Northwest National Laboratory during a $SCCO_2$ cleaning market assessment completed in 1994 identified some of the reasons for lack of demand in the private sector for $SCCO_2$ cleaning (Table 29.7) [32].

TABLE 29.7 Potential Barriers to Acceptance of SCCO₂

Higher capital costs for SCCO₂ systems relative to other cleaning technologies
Lack of awareness of SCCO₂ cleaning technology
Substrate to be cleaned lacks compatibility with SCCO₂ or with high pressures
A perception that SCCO₂ cleaning does not remove particulates effectively
The requirement for a continuous process (SCCO₂ cleaning is a batch process)
The existence of established aqueous cleaning technologies to replace solvent vapor degreasers

The initial cost of a condensed-phase CO_2 system can become acceptable when the environmental benefits (health and safety and regulatory) and the energy savings—overall the return on investment—are considered. The high capital cost of $SCCO_2$ cleaning systems can be attributed, in part, to the high-pressure cleaning chamber and the valves and instrumentation required for the system. The increasing mass production of CO_2 will allow vendors to realize the economies of scale. The time required for cleaning and stripping processes can be reduced by as much as 80%–90%.

Published investigations into the economics of $SCCO_2$ cleaning show that operational costs of $SCCO_2$ cleaning are quite reasonable and often lower than solvent vapor degreasers or aqueous cleaning systems. CO_2 is relatively inexpensive and can be reused in most $SCCO_2$ cleaning systems. Waste treatment due to cleaning is minimal, as no waste stream is generated besides the actual contaminants removed from the parts being cleaned. Unfortunately, the initial capital costs for $SCCO_2$ systems are usually higher than other alternatives—sometimes by a significant amount.

Findings on economics include the following:

- Utility costs for the unit will be relatively insignificant when compared to operations and maintenance labor costs [33].
- A much higher volume application would be required to justify the $SCCO_2$ system [34].

Conclusions

In terms of the drivers affecting the adoption of any cleaning technology (technological, regulatory, and economic), dense-phase CO_2 is favorably positioned in all areas. As the global economics of this cleaning technology become better, cleaning with LCO_2, $SCCO_2$, and CO_2 snow will become more prevalent.

References

1. Beach, E. S., Cui, Z., Anastas, P. T. (2009) Green chemistry: A design framework for sustainability. *Energy & Environmental Science* **2**(10), 1038–1049.
2. Nelson, W. M., Art in science: utility of solvents in green chemistry, in *Green Chemistry: Frontiers in Benign Chemical Syntheses and Processes*, Anastas, P. T., Williamson, T. C., Eds. Oxford University Press: Oxford, U.K., 1998, p. 200.
3. Beckman, E. J. (2004) Supercritical and near-critical CO2 in green chemical synthesis and processing. *Journal of Supercritical Fluids* **28**(2–3), 121–191.
4. Cline, C. M. (1996) Emerging technology: Emerging markets. *Precision Cleaning* **4**(10), 11–19.
5. Plasynski, S. I., Litynski, J. T., McIlvried, H. G., Srivastava, R. D. (2009) Progress and new developments in carbon capture and storage. *Critical Reviews in Plant Sciences* **28**(3), 123–138.
6. Jackson, D., Carver, B. (1999) Liquid CO₂ immersion cleaning. *Parts Cleaning* 32.
7. Darvin, C. H., Lienhart, R. B. (1998) Surfactant solutions advance liquid CO₂ cleaning potentials. *Precision Cleaning* **6**(2), 28–29.
8. Smith, J. M., Van Ness, H. C. *Introduction to Chemical Engineering Thermodynamics*. 4th edn. McGraw-Hill: New York, 1975, 54 pp.

9. Saldana, M. D. A., Nagpal, V., Guigard, S. E. (2005) Remediation of contaminated soils using super-critical fluid extraction: A review (1994–2004) *Environmental Technology* **26**(9), 1013–1032.

10. Sherman R., Depalma, P. (1998) CO_2 process parameters and automation potentials. *Precision Cleaning* **6**(6), 22–27.

11. Liu, G. T., Nagahama, K. (1996) Solubility of organic solid mixture in supercritical fluids. *Journal of Supercritical Fluids* **9**, 152.

12. Anitescu, G., Tavlarides, L. L. (2006) Supercritical extraction of contaminants from soils and sediments. *Journal of Supercritical Fluids* **38**(2), 167–180.

13. Fu, H., Coelho, L. A. F., Matthews, M. A. (2000) Diffusion coefficients of model contaminants in dense CO_2. *Journal of Supercritical Fluids* **18**(2), 141–155.

14. Trabelsi, F., Stuber, F., Abaroudi, K., Larrayoz, M. A., Recasens, F., Sueiras, J. E. (2000) Coking and ex situ catalyst reactivation using supercritical CO_2: A preliminary study. *Industrial and Engineering Chemistry Research* **39**(10), 3666–3670.

15. Brandt, W. V. (2001) Cleaning of photomask substrates using CO_2 snow. *Proceedings of SPIE—The International Society for Optical Engineering* **4562**(II), 600–608.

16. Lee, M. Y., Do, K. M., Ganapathy, H. S., Lo, Y. S., Kim, J. J., Choi, S. J., Lim, K. T. (2007) Surfactant-aided supercritical carbon dioxide drying for photoresists to prevent pattern collapse. *Journal of Supercritical Fluids* **42**(1), 150–156.

17. *Supercritical Fluid Extraction Cleaner Application: Texas Instruments Incorporated*. Toxics Use Reduction Institute: Lowell, MA, 1994.

18. Gillum, W. O. Replacement of chlorinated solvents for metal parts cleaning. *Precision Cleaning'94*. Witter Publishing, Flemington, NJ, 1994.

19. Silva, L. J. (1995) Supercritical fluid for cleaning metal parts. *The Hazardous Waste Consultant* **13**, 1.25.

20. Phelps, M. R. *Waste Reduction Using Carbon Dioxide: A Solvent Substitute for Precision Cleaning Applications*. Pacific Northwest National Laboratory: Richland, WA, 1994.

21. Gallagher, P. M., Krukonis, V. J. Precision parts cleaning with supercritical carbon dioxide, in *Solvent Substitution for Pollution Prevention*, U.S. Department of Energy and U.S. Air Force, Editors. Noyes Data Corporation: Park Ridge, NJ, 1993, p. 76.

22. Connors, K. A. *Chemical Kinetics: The Study of Reaction Rates in Solution*. VCH Publishers, Inc.: New York, 1990, 480 pp.

23. Keagy, J. A., Zhang, X., Johnston, K. P., Busch, E., Weber, F., Wolf, P. J., Rhoad, T. (2006) Cleaning of patterned porous low-k dielectrics with water, carbon dioxide and ambidextrous surfactants. *Journal of Supercritical Fluids* **39**(2), 277–285.

24. Savage, P. E., Gopalan, S., Mizan, T. I., Martino, C. J., Brock, E. E. (1995) Reactions at supercritical conditions—Applications and fundamentals. *AICHE Journal* **41**, 1723.

25. Jincao, Y., Matthews, M. A., Darvin, C. H. (2001) Prevention of photoresist pattern collapse by using liquid carbon dioxide. *Industrial and Engineering Chemistry Research* **40**(24), 5858–5860.

26. Della, P. G., Volpe, M. C., Reverchon, E. (2006) Supercritical cleaning of rollers for printing and packaging industry. *Journal of Supercritical Fluids* **37**(3), 409–416.

27. Salerno, R. F. High pressure supercritical carbon dioxide efficiency in removing hydrocarbon machine coolants from metal coupons and components parts, in *Solvent Substitution for Pollution Prevention*, U.S. Department of Energy and U.S. Air Force, Editors. Noyes Data Corporation: Park Ridge, NJ, 1993, p. 98.

28. Li, J.-J., Qi, T., Li, S.-L., Zhao, G. (2007) Cleaning of ITO glass with carbon dioxide snow jet spray. *Proceedings of SPIE—The International Society for Optical Engineering*, 6722, art. no. 67224I.

29. Banerjee, S., Lin, C. C., Su, S., Chung, H. F., Brandt, W., Tang, K. (2005) Cryogenic aerosol cleaning of photomasks. *Proceedings of SPIE—The International Society for Optical Engineering* 5853 PART I, art. no. 12, 90–99.

30. Van Roosmalen, M. J. E., Woerlee, G. F., Witkamp, G. J. (2004) Surfactants for particulate soil removal in dry-cleaning with high-pressure carbon dioxide. *Journal of Supercritical Fluids* **30**(1), 97–109.

31. Kanegsberg, B. (1996) Precision cleaning without ozone depleting chemicals. *Chemistry & Industry*, 787.

32. Pacific Northwest National Laboratory: Richland, WA, 1994.

33. Barton, J. C., The Los Alamos Super Scrub: Supercritical Carbon Dioxide System Utilities and Consumables Study, Los Alamos National Laboratory: Los Alamos, NM, 1994.

34. Licis, I. J., *Pollution Prevention Possibilities for Small and Medium-Sized Industries—Results of the WRITE Projects*. U. S. Environmental Protection Agency, Washington, DC, May, 1995.

35. McClain, J. B., Betts D. E., Canelas, D. A. (1996) Design of nonionic surfactants for supercritical carbon dioxide. *Science* **274**, 2049.

36. Meguro, Y., Iso, S., Sasaki T. (1998) Solubility of organophosphorus metal extractants in supercritical carbon dioxide. *Analytical Chemistry* **70**, 774.

37. O'Neill, M. L., Cao, Q., Fang, R., Johnston, K. P., Wilkinson, S. P., Smith, C. D., Kerschner, J. L., Jureller, S. H. (1998) Solubility of homopolymers and copolymers in carbon dioxide. *Industrial & Engineering Chemistry Research* **37**, 3067.

38. Stassi, A., Schiraldi, A. (1994) Solubility of vegetable cuticular waxes in supercritical CO₂ isothermal calorimetry investigations. *Thermochimica Acta* **246**, 417.

39. Anitescu, G., Tavlarides, L. L. (1997) Solubilities of solids in supercritical fluids 2. Polycyclic aromatic hydrocarbons (Pahs) plus CO₂/cosolvent. *Journal of Supercritical Fluids* **11**, 37.

40. Souvignet, I., Olesik, S. V. (1995) Solvent–solvent and solute–solvent interactions in liquid methanol/carbon dioxide mixtures. *Journal of Physical Chemistry* **99**, 16800.

41. McFann, G. J., Johnston, K. P., Howdle, S. M. (1994) Solubilization in nonionic reverse micelles in carbon dioxide. *AICHE Journal* **40**, 543.

42. Sawan, S. P. *Evaluation of the Interactions between Supercritical Carbon Dioxide and Polymeric Materials*. Los Alamos National Laboratory: Los Alamos, NM, 1994.

43. Novak, R. A. Cleaning of precision components with supercritical carbon dioxide, in *International CFC and Halon Alternatives Conference*. U. S. Environmental Protection Agency, Washington, DC, 1993.

44. Williams, S. B. *Elimination of Solvents and Waste by using Supercritical Carbon Dioxide in Precision Cleaning*, LA-UR-94–3313. Los Alamos National Laboratory: Los Alamos, NM, 1994.

45. Knez, Z., Riznerhras, A., Kokot, K., Bauman, D. (1998) Solubility of some solid triazine herbicides in supercritical carbon dioxide. *Fluid Phase Equilibria* **152**, 95.

46. Hunt, D. How one of the largest air force users is getting out of CFCs, in *Proceedings of the 1992 International CFC and Halon Alternatives Conference*, 1992.

47. Weber, D. C., McGovern, W. E., Moses, J. (1995) Precision surface cleaning with supercritical carbon dioxide: Issues, experience, and prospects. *Metal Finishing* **93**, 22.

48. McCoy, M. (1999) Industry intrigued by CO₂ as solvent: 'Green' processes based on supercritical carbon dioxide are moving out of the lab. *Chemical & Engineering News* 11.

49. McGovern, W. E., Moses, J. M., Weber, D. C. The use of supercritical carbon dioxide as an alternative for chlorofluorocarbon (CFC) solvents in precision parts cleaning applications, in *Proceedings Air Pollution Control Association Annual Meeting*, 1994.

50. Purtrell, R., Rothman, L., Eldridge, B., Chess, C. (1993) Precision parts cleaning using supercritical fluids. *Journal of Vacuum Science and Technology* **11**, 1696.

51. Silva, L. J. (1995) Supercritical fluid for cleaning metal parts. *The Hazardous Waste Consultant* **13**, 1.25.

30

Gas Plasma: A Dry Process for Cleaning and Surface Treatment

Kenneth Sautter
*Yield Engineering
Systems, Inc.*

William Moffat
*Yield Engineering
Systems, Inc.*

Introduction

The importance of plasma processing to current manufacturing practices cannot be overstated. Over the past 50 years, applications have been developed to remove material in a controlled fashion, deposit material layer by layer, and modify surfaces such that subsequent depositions adhere well. Not only are plasma processes versatile, but they are also efficient and present chemistry at its most fundamental (one might say most elemental) state. The effluents resulting from the cleaning process require minimal processing.

Though laboratory examples of plasma phenomena were known from the 1800s, the first practical applications were developed in the 1940s. Forensic analysis found the controlled ashing of animal tissue to be useful for spectrographic analysis. In the 1960s, plasma ashing proved useful for removing photoresist from some of the earliest wafers. Applications soon spread to deposition, etch, and surface activation as the chemistry and physics involved became better understood.

Technology

What is plasma? Simply stated, plasma is partially ionized gas, an evenly balanced mix of positive ions and negative electrons. As energy is added to molecules, matter is transformed from solid to liquid to gas. As more energy is added to a gas, the gas can be excited to the point where electrons are moving freely instead of being bound in a molecule. The atoms can break apart into ions and electrons. Plasmas are electrically conductive because the electrons are free floating. Plasma characteristics are different enough that it is considered a unique fourth state of matter. There are many examples in nature. On a grand scale, the sun is huge mass of plasma. Closer to home, there are the Aurora Borealis and on a more local scale are lightning bolts. How do we harness these forces to make them useful in a manufacturing environment?

When one thinks of plasma phenomena in nature, it is usually associated with extreme energy and high temperatures. The sun's corona is an example of high-temperature plasma where the gas is completely ionized and both the gases and the electrons are at very high temperatures. However, for industrial applications, we discuss weakly ionized plasma. This is low-temperature plasma in which a significant but low fraction of the gas is ionized. In these nonthermal plasmas, even though the electrons may be excited to an energy of 1 eV (or approximately 10,000 K), the temperature of the ions and neutral species is only a few degrees above ambient. A neon light is an example of this kind of plasma. Creating a pretty glow is one thing, but we are looking to create a force to modify materials.

There are two phases to the phenomena with which we are working. First is the creation of the plasma, which consists of a pure gas, or mixture of gases. Once the plasma species have been created, the target surfaces will be exposed to the ions and radicals to be modified. Some systems separate the area of plasma creation and the reaction chamber. This is known as downstream plasma and will be discussed later. The reaction phase of the operation also consists of two competing phenomena. The first is physical erosion or sputtering. The ions are accelerated to the surface and knock loosely bound materials off the surface. Inert gases such as argon or helium do not react with the surface; they simply knock off material such as aluminum oxide. The other phenomenon is chemical reaction during which ions and radicals react with materials on the surface. For instance, oxygen ions react with carbon containing organic films to form carbon monoxide and carbon dioxide.

In order to form the plasma, an electric field is added to the gas in a chamber, and plasma is formed between the anode and cathode. In order to keep the current flowing, the electric field is switched at high frequency, usually in one of three bands: low frequency at 30–50 kHz, high frequency at 13.54 MHz, and microwave at 2.54 GHz. Generally, if a gas is excited at atmospheric pressure, it will quickly return to its ground state unless the amount of voltage is extremely high. This is due to the number of collisions between gas molecules. In order to sustain the plasma and do work on a surface, the number of collisions between molecules must be limited. However, if the pressure is too low, the mean free path between molecules is too far to sustain plasma. Most industrial plasma processing takes place under vacuum between 1 and 150 Pa of pressure (0.05–1 T). Although there are a number of publications that describe atmospheric processes, most of these are surface activation or biological applications. It can be argued that the chemistry of the plasma is the same and is independent of the power source. However, each method of dissociation has its merits and disadvantages. For example, a low-frequency field allows for a longer forward time between switching that would tend to accentuate more physical impact than would a high-frequency field.

Applications

What are the uses of plasma in industry, how are engineers making use of them, and more importantly, how does one go about setting up a process? As alluded to earlier, there are a wide variety of current uses of plasma technology. Cleaning applications include removal of organic residues such as photoresist, epoxy bleed-out, machine oil residues, excess solder flux, skin oils, mold release agents, and even slag from laser cutting operations. Plasma is useful for removal of inorganic contaminants as well, including metal oxide and metal etch. In the biomedical field, plasma systems have been found to be extremely efficient in decontamination applications. There is widespread use of plasma in assembly operations for surface activation and adhesion promotion prior to epoxy bonding, wire bond, molding, encapsulation, and inking.

Setting Up a Process

With all the different applications and equipment designs, how does one approach selecting the right process? First, one must decide on the goals of the process, and then tailor the process to emphasize the characteristics most appropriate. Ask a series of questions such as

1. Is the material to be removed chemically reactive? Examples include organic residues of photoresist, oils, or adhesives.
2. Is the substrate sensitive to oxidation in an oxygen-rich environment?
3. Is the film thin loosely bonded? One example is native aluminum oxide.
4. Is the objective to make the surface uniform for epoxy bonding?

These questions address the fundamental operating parameters of a plasma process: gas selection, requirements of physical removal versus chemical reaction, as well as electrical sensitivity of the parts being cleaned. Table 30.1 lists some of the most common gases used in plasma cleaning applications and a guide to applications.

Carefully selecting the operating conditions results in a highly repeatable manufacturing process. There are several modes that plasma systems may run. As shown in Figure 30.1, the plasma species are

TABLE 30.1 Process Gas Selection Guide

Gas	Chemical Activity	Physical Mass	Uses and Comments
Argon	Inert	High	Physical ablation (surface roughening and activation prior to bonding)
Oxygen	Oxidation	High	Removal of organic contaminants creating CO and CO_2 Similar in mass to Argon
Helium	Inert	Low	Cleaning delicate substrates such as very thin oxide insulators or critical optical components that may be damaged by Argon mass
Hydrogen	Reducing	Low	Metal oxide removal Usually used mixed with argon or nitrogen
Ammonia	Reducing	High	Surface activation and oxide removal
Forming gas	Reducing	High	Cleaning halogen contaminants Metal oxide removal Mixture of hydrogen with nitrogen
CF4		High	Generally at 5%–20% concentrations with oxygen for enhanced organic removal Higher fluorine concentrations used for etching Si In a pure form, used to form polymer coatings
Air	Oxidation	High	Noncritical applications Organic contaminant removal Surface activation

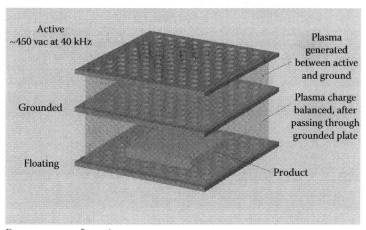

Downstream configuration

FIGURE 30.1 Shelf configuration in a parallel plate capacitive plasma system.

generated between a power source and ground. Shown in the figure is an example of Downstream Mode in which the samples sit outside the plasma generation area. Electrons produced as process gases are excited and grounded, shielding electrically sensitive substrates from high-energy electrons. This mode enhances chemical activity and provides the mildest physical abrasion.

Substrates can also be placed on a grounded shelf within the plasma generation area for a more aggressive etch. In this mode, all the generated species would be able to react with the substrate, providing a kinetically enhanced physical etch. Placing parts on the Power or Active shelf provides the most aggressive etch as the gas species are accelerated to the plate, increasing the sputtering activity.

In addition to selecting the process gas and operating mode, the most important parameters are the plasma time and energy.

Cleaning Effects

Judging Cleaning

Once the cleaning process is complete, how can one tell that the surface is clean? One of the simplest and most powerful techniques is through the use of water contact angle. Different materials have characteristic surface energies. A droplet of Deionized water forms a bead that can be measured (Figure 30.2).

As the surface is exposed to plasma, it becomes more hydrophilic, and the droplet spreads out. Precise measurements require an instrument called a goniometer, but the before-and-after effect can be observed qualitatively by placing a small drop of water on the surface (Figure 30.3).

Diagnostics

One of the most common questions is, "How can I tell that the plasma is correct and that the system did its job?" There is a two-part answer to this question. First, if the system has a view port, the best method of telling if the system is running properly is simply by observing the plasma. If the plasma is flickering

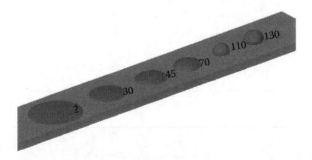

FIGURE 30.2 A surface showing various degrees of surface energy. Left to right the surface goes from hydrophilic to hydrophobic.

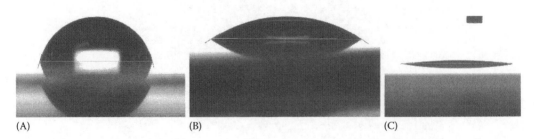

FIGURE 30.3 A plasma-treated surface as seen through goniometer optics: (A) silicon untreated, (B) 30 s argon plasma, and (C) 2 min argon plasma.

or flashing, there is likely arcing in the system due to charge buildup and discharge. If the plasma is steady, look at the color of the plasma. Each gas has a unique color plasma.

While argon produces a deep blue or purple color, helium produces a vivid green color. Oxygen yields a light blue to gray color. If the color of the plasma changes, an air leak is the likely source. Air produces a yellow or pink plasma (pink is from the contribution of nitrogen.)

Case Studies

Pre-Wire Bonding

ESCA (electron spectroscopy for chemical analysis, also known as x-ray photoemission spectroscopy, XPS) has shown that the presence of carbon on a surface limits the quality of the wire bond.[1] Using ESCA, the relative level of carbon contamination on a copper surface, for example, can be determined from the ratio of the area of the carbon (C) and copper (Cu) peaks. Since water drop measurements are much easier to perform, it is good to know that the measured ESCA C/Cu ratios correlate extremely well with water drop contact angles measured on the same surface.

Bonding pads on the substrate are also subject to various and inconsistent levels of contamination. One source of contamination on the die surface is fluorine ion, most likely left as a residue on the wafer or die during the fab process.[2] While the presence of this contamination may not show wire bond degradation immediately, there is evidence showing correlation of the presence of Fl element and bond failure due to the migration of fluorine that results in embrittlement of the wire.[3] Removal of this contaminate may require the use of argon bombardment. This is an important application of plasma, that is, the removal of trace amount of impurities such as inorganic ions by the use of Argon.

Marking

Particularly with aqueous-based inks, printing or marking on many different types of surfaces is improved by the use of plasma. The replacement of organic solvent-based inks with aqueous-based inks does not provide the same consistency of adhesion. In our experience, heat or hydrogen flame treatment is inconsistent: plasma treatment results in better surfaces. Some situations show better results than others because of the nature of the ink, the encapsulant, and the marking process. Plasma treatment ensures uniformity of the process.

Die Bonding

Plasma cleaning or treatment of substrates improves adhesion of the epoxy and provides a better bond between the die and the substrate. This bond provides better heat dissipation. Studies have also shown that there is less delamination at the die when samples have been plasma treated.[4]

Cleaning after Laser Cutting

Materials such as Parylene may be used to provide a chemically resistant barrier layer for semiconductor devices. However, in order to make electrical contact with the device, holes must be cut in the Parylene. Laser cutting often leaves a slag or residue of partially melted material that must be removed before further processing can continue. One effective method for removing the laser slag is to plasma clean the material with an inert gas such as argon (Figure 30.4).[5]

Medical/Dental

In the medical industry, plasma processing is being used for applications as diverse as instrument cleaning, sterilization, and tooth bleaching. In a recent study, it was shown that a helium plasma with just 0.2% oxygen was most effective at sterilizing *Escherichia coli* and *Bacillus subtilis* endospores.[6]

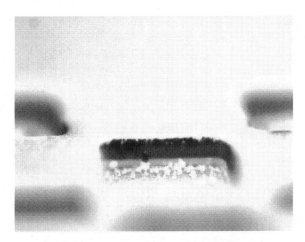

FIGURE 30.4   Laser-cut openings in a polymer film. An argon plasma was used to remove the laser slag.

Encapsulation

In this process, we are asking the molding compound to provide, in addition to all of the other required properties, good adhesion to a number of different surfaces. Depending on the type of package, the molding compound must adhere to the substrate material, solder mask, die, and the metal bond pads. There are many applications that involve the bonding of one material to another, for example, plastic to plastic or metal to plastic. As indicated in the study by Herard,[7] plasma improves the quality of the bond (Figure 30.5).

Fluxless Soldering

A patented plasma process has been shown to give good welding results with typical lead and gold solder formulations, without the use of flux, and hence without the requirement of a flux removal step.[8]

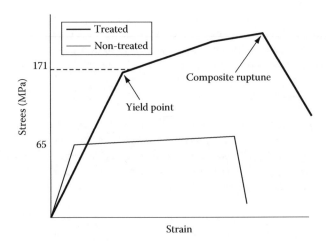

FIGURE 30.5   Pull testing indicating improved adhesion after plasma treatment.

Summary

Plasma processing has been shown to be robust, cost-effective, and environmentally friendly. The number of applications is growing rapidly and is limited only by the imagination of the creative process engineer.

References

1. Djennas, F. et al. Investigations of plasma effects on plastic packages delamination and cracking. *Electronic Components and Technology Conference*, Orlando, FL, June 1–4, 1993.
2. Goodman, J. et al. Fluoride contamination from fluoropolymers in semiconductor manufacture. *Solid State Technology*, July 1990, 65–68.
3. Gore, S. Degradation of thick film gold bondability following argon plasma cleaning. *ISHM Proceedings*, 1992, pp. 737–742.
4. Oren, K. A case study of plastic part delamination. *Semiconductor International*, April 1996, 109.
5. Fisher, J. ITRI reports on micro-vias, *Printed Circuit Fabrication*, 58–60.
6. Hong, Y. F. et al. Sterilization effect of atmospheric plasma on *Escherichia coli* and *Bacillus subtilis* endospores. *Letters in Applied Microbiology*, January 2009.
7. Herard, L. Surface treatment for plastic ball grid array assembly and its effect on package reliability. *Proceedings of the ISHM Workshop on Flip Chip and Ball Grid Array*, Berlin, Germany, November 11–15, 1995.
8. Koopman, N. et al. Solder flip chip developments at MCNC, *Presented at ITAP*, 1996.

31

Superheated, High-Pressure Steam Vapor Cleaning

Max Friedheim
PDQ Mini-Max

Jose Gonzalez
PDQ Mini-Max

Introduction

Steam, a natural phenomenon used in cooking and medication, is known to all of us. Steam drives ships, blows whistles, and, in general, is part of our lives. We take it for granted. Heat up some water above 100°C (212°F) and we get steam. Traditional steam cleaning has been used successfully for a number of years for janitorial and food service applications. In this chapter, however, we discuss a patented cleaning technique based on superheated, high-pressure steam vapor.

The basis of steam vapor cleaning or aqueous/waterless cleaning is readily explained by analogy with steam from a tea kettle. Almost everyone at one time or another has put his hand over the tea kettle into the flow of steam vapor, felt the warmth, and marveled in the delight. But how many have thought to ask, "How come we are not scalded by boiling water?" Simply put, the steam vapor is composed of single molecules of water, not water droplets. Water droplets have a high specific heat and can store much heat energy, which can potentially cause the burns or scalds when the common old-fashioned steam jenny or steamer is used.

Operation

This new technology sometimes enables the use of water without an additive package as the sole primary cleaning technique. In the steam vapor cleaner, distilled, deionized, or any potable filtered water is drawn from a reservoir. A metered amount of liquid is injected into the patented chamber and instantaneously is converted to steam vapor under pressure. The equipment is activated via a hand or foot switch; and the equipment discharges high-pressure, dry steam through a wand or handle. Efficacy of cleaning is based on

- Heat
- Pressure
- Vapor phase water

Heat, when applied to the contaminant such as oil or grease, helps to liquefy the soil so that it is more readily removed. In addition, the pressures used in the range of 200–300 PSI mechanically assist in dislodging and removing contamination from the surface. For many benchtop-type applications, steam vapor cleaning offers advantages over traditional hand-wipe cleaning. There is no wipe cloth, swab, or solid applicator. Instead, the combination of hot steam and pressurized blast enable cleaning in tight spaces, complex geometries, blind holes, and under close-spaced components.

The equipment can be small, and highly portable. As such, the technique can be used for small-scale benchtop cleaning or even in field repair, where the wand is handheld. The duration of the steam blast is controlled by the operator. The operator typically activates generation of steam vapor by pressing an activation switch for a second or two. The resulting flow of high-pressure steam vapor continues for approximately 10–20 s, allowing the part to be cleaned. In the smaller models, a short recovery time is required for more steam to be generated. After the chamber has purged itself of all liquid, the user again activates the system. With some practice, the operator is able to time the bursts to provide a nearly continuous flow of steam. In addition, the technology can be adapted for automated, continuous generation of steam vapor for large-scale operations.

Technology Development

In the decade following the first edition of *Handbook for Critical Cleaning*, going green became the password to successful manufacturing. Chemicals and solvents were replaced by aqueous processes. In most areas, people turned to water-based chemical cleaning. Problems were associated with conversion to aqueous processes. Examples include protecting the operator from hot water burns, minimizing high energy consumption, contending with wastewater streams, and an inability to control flow into areas that are more inaccessible. These problems resulted in increased labor and expense.

The technology of steam to clean has taken on a more prominent roll, because it is water based. The steam cleaners that evolved new configurations were developed.

Examples include dishwasher type configurations, water heated and mixed with high-pressure water pumps, as well as bigger, more powerful steam cleaners. As the superheated steam vapor technology has become more wildly known, there has been additional evaluation of the technology to resolve existing problems.

Additives

In some cases, cleaning and drying can be accomplished with water alone. With heavily soiled surfaces, cleaning can be enhanced by application of surfactant solution directly to the part to be cleaned. Steam vapor is then used to heat and remove the cleaning agent and contaminant.

Avoiding corrosion during cleaning of carbon steel is another potential problem. In such cases, an appropriately designed rust inhibitor can be added for applications where potential corrosion of the part is an issue.

Drying

While many consider liquid aqueous cleaning to be environmentally preferable in terms of pollution prevention, one potential drawback to liquid aqueous cleaning is the need for a drying step. Drying may add to the cost of the process in terms of

Equipment
Drying time
Drying temperature
Cooldown time
Energy
Holdup in production flow

In the case of steam vapor cleaning, because cleaning is accomplished with heated, pressurized, water vapor, the cleaned parts are typically dry and ready for the next step in the manufacturing process, be it plating, painting, or further fabrication.

Safety

Because the steam vapor is generated almost instantaneously, no steam is stored under pressure. This provides safer operation for the worker. Because steam vapor cleaning is free of water droplets, the operator gains a measure of safety with steam vapor cleaning, even at the high pressures employed. However, gloves may be considered depending on the specific application, and goggles are recommended to protect the eyes from debris.

In addition, the equipment may be used in a waste management cabinet to minimize worker exposure to debris from the part being cleaned, as well as entrapment of the residue being removed.

Waste Stream Management

Waste stream management is a concern in any cleaning process. It is generally understood that organic solvents must be handled as hazardous waste. However, liquid aqueous systems must also be managed as waste streams. The additive package itself may be unsuitable for discharge to a sewer line. In addition, even where the surfactant package is said to be biodegradable, soils and trace metals may result in need to treat, wash, and rinse water as hazardous waste. Even with filtration and evaporative techniques, this can add significant costs to the process.

In contrast, because the temperature at which the steam vapor is being produced is 500°F, the vapor evaporates leaving only the residue of contaminants for disposal. The residue is typically collected on rags or absorbency pads, resulting in a relatively concentrated, manageable waste.

Energy and Water Usage

Compared with many other cleaning techniques, steam vapor technology is very low in energy consumption because only a small mass of water is being heated. Many cleaning systems require constant heating of a fairly large mass bath. In steam vapor cleaning, electrical current is drawn only when the chambers are heating. In addition, we cannot take the availability of water for granted. In many areas, water is a costly commodity. In steam vapor cleaning, the average water consumption per 8 h shift is relatively low.

Steam Vapor Cleaning: Estimated Water Consumption

Size of Steam Vapor Cleaner	Gallons of Water (Average) per 8 h
Small	1
Medium	2
Large, continuous steam capability	5

General Applications

Solvent-based cleaning, for all the safety and environmental problems involved, has advantages of rapid solubilization of a wide range of soils. With liquid aqueous-based cleaning, it is sometimes necessary to use additional cleaning approaches to achieve the desired cleanliness. Steam vapor cleaning has utility not just on its own but also as an adjunctive technique to complement solvent or liquid aqueous approaches.

In some applications, a chemical alternative or solvent must be used because the contaminant is such that only a chemical agent can provide the needed solubility or "softening" power to allow it to be readily removed. Examples include burnt on carbon deposit, paint, and heavy industrial grease. In such cases, steam vapor can be used as a final rinsing, drying, or detailing process. In such cases, steam vapor is used to remove the final traces of cleaning agent residue, soil, or surface oxidation, while eliminating the wastewater stream.

In addition, while a large-scale cleaning operation may be adaptable to most of the parts being cleaned, there may be parts that do not lend themselves to the standard cleaning process, or must be cleaned rapidly, on short notice. Steam vapor cleaning can be used to add flexibility to the general cleaning system.

No cleaning system is perfect. With many cleaning systems in use today, a certain proportion of parts are not cleaned acceptably. This in turn necessitates costly time-consuming hand, probing-type detailing. Adding steam vapor technology as a final detailing tool aids in final inspection and acceptance of parts, and may allow manufacturers to do their jobs more efficiently and economically.

Other cleaning systems may rely on applying abrasion or force to remove contaminants. Pressure washers use water blasting, ultrasonic cleaning uses implosions to loosen dirt, and enclosed-type cleaners use spray and chemical under high pressure to dislodge soils. Steam vapor technology has essentially no abrasive quality as the vapor consists of individual water molecules, so it may be preferable for fragile delicate applications, not affecting the integrity of any surface.

Cost-Saving Estimates

A U.S. Navy report[1] by their Fleet Activity Support Technology Transfer (FASTT—P-2) that using this technology as a viable alternative to solvent cleaning and degreasing of weapons, automotive parts, electronics, printed circuit boards, ground support equipment, and other items estimates a capital cost of approximately $8,300, an annual saving of nearly $400,000 with a payback period of under 1 year, actually in less than 10 days!

Case Studies

Electronics Assembly, General Cleaning

Steam vapor cleaning has been evaluated for the removal of flux and other contaminants for surface-mounted assemblies. Tests were performed for the U.S. Navy at Crane, Indiana. Ten mother boards and 26 interface cards were cleaned with steam vapor technology. No damage due to heat or electrostatic discharge was detected. The U.S. Navy has authorized the use of the steam vapor technology with avionics and other applications.

Final Surface Preparation Prior to Laser Welding, Biomedical Application

A manufacturer of stainless steel needles for biomedical applications used steam vapor cleaning to improve surface cleanliness prior to laser welding.

The overall process of unwinding the roll of stainless steel strips, bending, shaping, and laser welding the product requires 4.5 h. Prior to laser welding, the original process called for the strip to be run through solvent, and then wiped dry between paper towels. It is important to remove all traces of solvent; any residual solvent interferes with laser welding. Unfortunately, residual solvent produced welding "misses," resulting in an unacceptable reject rate. An automated steam vapor cleaner was implemented prior to laser welding. The reject rate was reduced to negligible levels, production was increased by over 30%, and solvent usage was eliminated.

Detail Cleaning of Refrigeration Equipment

B/E Aerospace, Galley Products Group, in Anaheim, California, produces some 90% of the airline galley refrigeration equipment in use worldwide. B/E also repairs refrigeration in-house and specifics options for field repair. After some process optimization, B/E has introduced steam vapor cleaning to replace some mineral spirits cleaning of segments of refrigeration tubing.[2]

Initially, assemblers, accustomed to cleaning with mineral spirits, were unfamiliar with the new technology. By making the equipment available in the shop, operators found a number of applications for steam vapor cleaning. It is currently in regular use in assorted applications as a final cleaning technique.

Printing Equipment

The Los Angeles Daily News is subject to stringent requirements for solvent elimination mandated by the South Coast Air Quality Management District (SCAQMD).

The Los Angeles Daily News implemented steam vapor cleaning technology for an assortment of printing-related cleaning applications. As a result, solvent tank cleaning at the facility was eliminated. Emissions of volatile organic compounds were reduced from 20 to 2.2 ton annually.

Rigid Tube Extenders

A new patent pending process utilizing rigid tube extenders has greatly enhanced the utility of steam cleaning. The tube extenders are attached to a flexible wand assembly that then carries the superheated steam vapor from the generators into the dirty tube assemblies. The rigid extenders are designed to eject the steam vapor as they are passed in and out of the tube, thus heating the interior surfaces. Cleaning can be accomplished in minutes using water with no additives or, if necessary, with added rust inhibitor.

The rigid extender tubes can be attached to scrubber attachments that may consist of abrasive material, metal wire brushes, etc. This provides the option of enhancing the cleaning action of steam by adding significant cleaning force. The result is increased efficacy of soil removal with reduced cleaning time.

Newer Cleaning Applications

Enhanced Tube Cleaning

Steam cleaning with rigid tube extenders has proved useful in tube cleaning. Tubes, be they barrels of weapons, hydraulic lines, or oil lines, have always posed a cleaning problem. To successfully remove soils, the cleaning process must be designed so that solution enters the tube, removes items such as burnt on carbon and "fowling," and then leaves the tube dry and free of residue. Of course, this all has to be accomplished in a short time.

In one comparison study, military mortar tubes (60 or 81 mm) were cleaned. Prior to utilization of steam technology, these weapons could not be properly cleaned without a total disassembly followed by hand-scrubbing with brushes and chemicals. The area of the firing pin at the base of the weapon was particularly difficult to clean. The time expended for this task was three or more hours. Further, cleaning was not totally successful. The life expectancy of the weapon was reduced and there was no safe, secure way to inspect for internal stress cracks. In addition, the resultant chemical waste hazards had to be contended with.

With the steam cleaning technology with special brush attachment, the task can be completed in 15–20 min, not hours; and the interior and firing pin are clean and dry. Wastewater disposal is reduced considerably (about a quarter glass full including the fouling inside and out).

Medical Instruments

Steam cleaning with the rigid tube extender technology has found utility in the hospital environment where biohazard reduction prior to sterilization is imperative; instruments used in central services and operating rooms must be properly cleaned to remove assorted debris, including protein and blood.

A mechanical cleaning technique, even in the hands of experienced technicians, is a time-consuming task, one that is difficult to accomplish in a consistent manner without irreversible damage to the instrument.

For example, a flexible reamer used in surgery is very difficult and time-consuming to clean. The outer layer of flexible stainless steel braid becomes imbedded with surgical residue that must be removed prior to sterilization. Because it is flexible, it is readily deformed during cleaning to the extent that reuse is impossible. Replacements cost hundreds of dollars.

An alternative process utilizing steam cleaning a rigid tube extender was tested. The high-pressure steam vapor is emitted forward and sideways. This enables the medical reamer to be thoroughly cleaned inside and through the steel braided external walls. This process is completed in seconds and has been adopted by medical facilities nation wide. Patents are applied for the technology just described.

Autoclaves

Steam cleaning is, of course, not a replacement for terminal sterilization. However, autoclaves must themselves be cleaned. In the hospital/medical environment, cleaning the large autoclaves is normally a 3–4 h event. With superheated, high-pressure steam vapor, this procedure is accomplished in approximately 30 min. The cleaning agent is solely water converted to steam vapor.

Note: The steam equipment is registered as an FDA Class II registered device (DOC # E197753).

Conclusions

Steam vapor cleaning has found wide application in such diverse areas as electronics, aircraft, ground equipment, plant maintenance, and biomedical. Because of the diversity and relatively low capital investment, it has been adopted by the U.S. military, fortune five hundred companies, and small- to medium-sized manufacturers. Still, there are many facets to using steam vapor technology, some of which are yet to be discovered. An in-depth evaluation[3] by the Naval Air Warfare Center Aircraft Intermediate Maintenance Facility at Coronado, California, concludes that there are so many potential applications for steam vapor cleaning that "we haven't even scratched the surface yet." To date, the manufacturer of this equipment has observed no injury to products and no personnel injury. Further, plastics such as conformal coating on electronics assemblies are not damaged or removed when steam vapor is applied.

Steam cleaning technology has been used to resolve problems in the pharmaceutical, electronic, automotive, chemical, military, and many other industries. This environmentally friendly cleaning technique also reduces energy consumption, eliminates the need for costly installations, and impacts the bottom line in a positive manner.

The MiniMax steam cleaner is listed under the USEPA SNAP program as an acceptable substitute for ozone depleting compounds in metal cleaning, electronics cleaning, precision cleaning, aerosols, and an array of other applications (*Federal Register*, September 28, 2006).

Note: Technology and applications of steam vapor cleaning refer to various models of the PDQ Mini-Max Steam Vapor Cleaner.

References

1. U.S. Navy report by Fleet Activity Support Technology Transfer (FASTT—P-2). http://205.153.241.230/P2_Opportunity_Handbook/8_I_11.html
2. Petrulio, R. and B. Kanegsberg. Practical solutions to cleaning and flushing problems, *Presentation and Proceedings*, *CleanTech '98*, Rosemont, IL.
3. Naval Air Warfare Center Aircraft Intermediate Maintenance Facility at Coronado, CA. http://205.153.241.230/P2_Opportunity_Handbook/8_I_6.html

32

Making Decisions about Water and Wastewater Processes

John F. Russo
Separation Technologists

Introduction

Water is one of the most important liquids used in cleaning processes. Users should be concerned about water purity at each stage of a process to make the best decisions about water and wastewater and the solid waste generated by a parts cleaning process. As new technologies are introduced, users have more options in source (tap) water and wastewater treatment than ever before. This usually adds to the complexity of decision making, especially if the most effective, least-cost solution is the objective. In this chapter, water treatment terms are defined and various water processes are explained and compared. It is designed as an introduction for those new to the water treatment field, as well as a reference for experienced users.

Water and Wastewater Operational Factors

The following list of factors begins with the most important a new user should consider first. For example, discharge regulations are the first factor because if no wastewater can be discharged or the facility is near a stream of water, lake, or other sensitive discharge area, it will be much more difficult and expensive for management. The next most important factor is the wastewater volume. If the volume is small, the wastewater can be hauled at a low cost except in a remote location. If the volume is large, hauling or evaporation, unless solar, can be very expensive. Source (tap) water availability can be a serious concern—especially future changes in availability. Also, it is very difficult to treat water such as brackish water with very high total dissolved solids (TDS). A new plant location can be greatly affected by these two factors alone.

To achieve the optimal washing system at the lowest cost, users must consider source (tap) water, wastewater treatment, and solid waste disposal. By using the list below, the cost consequences of a wrong decision decrease substantially. Potential factors include the following, which correspond to major sections in this chapter:

- Discharge regulations
- Wastewater volume
- Water purity specification
- Source (tap) water treatment
- Wastewater treatment
- Wastewater treatment for new processes
- Overcapacity of the current wastewater treatment system

Discharge Regulations

Usually, regulations are discussed after wastewater treatment designs are presented. Since regulations can substantially affect the optimal design of a wastewater treatment system for a parts washing operation, they should be considered early in the design process.

Water and Wastewater

For water treatment specialists, the word "water" generally refers to water obtained from a variety of sources, while wastewater refers to any water used in a process that is reused or discharged. For example, water used simply as cooling water through a heat exchanger without any treatment other than transferring heat through a metal tube can be referred to as either water or wastewater in the industry. There is no absolute convention about this. All wastewater and solid waste, including sludge from processes, must comply with federal, state, and local regulations.

As regulations increase, it becomes more difficult to discharge wastewater without treatment. Throughout the United States, discharge of wastewater is controlled by three categories of regulations: POTW, ground water (septic system or well), and surface water (streams, rivers, ponds, and lakes). (POTW is a publicly owned treatment works or municipal sewer district.) At a minimum, state and local agencies must comply with federal regulations. State regulations might be more stringent than federal regulations, and local regulations might be more stringent than state regulations. For discharge to surface water, a NPDES (National Pollutant Discharge Elimination System) permit must be obtained from both federal and state agencies. Sometimes a local community might require that no industrial wastewater be discharged, even if the water or wastewater meets federal drinking water standards.

In many places in the United States, a permit is required to discharge wastewater from an industrial process even for a small batch-type cleaner (like a household dishwasher). Even if a user has tested his wastewater and it is in compliance with regulations, a local agency will usually still require a permit. A user is strongly advised to notify local regulatory agencies to avoid future issues and fines after the fact.

In the past, testing samples were usually taken at the end of the sewer pipe from the building. However, in increasingly more states, wastewater is tested as it comes from the equipment within a user's facility. This makes compliance more difficult. The old adage, "The solution to pollution is dilution," which is the illegal action of diluting wastewater from a process with source (tap) water to meet discharge regulations, has been declining for decades.

User concerns include

- Can I discharge any wastewater at all to the POTW, ground water, or surface water?
- Is a permit required? Are there regulations?
- What wastewater is acceptable for discharge? What is the concentration of contaminants?
- What permits might be required and air pollutants regulated for an evaporator?

Solid Waste

Federal regulations define solid waste as a solid material, liquid, or sludge, for example, the sludge from an evaporator. Users generating solid waste with toxic metals are liable for any of their solid waste hauler's or regenerator's violations of environmental laws from the time the wastewater or solid waste leaves the generator's facility (toxic metal generated by the original company's manufacturing process) to the ultimate disposal site. The expression, "cradle to grave," is commonly used to describe this concept. Any entity between the generator and the end user, knowingly or unknowingly, can be liable for damages or a sensitive user's losses.

Sensitive users and generators of toxic metals who rent or own deionizer (DI) tanks that are sent back and forth to a regeneration vendor to treat or replace their spent granular media (activated carbon and resin) must be greatly concerned. The typical design of most DI operations is to have one activated carbon tank followed by two or more deionized (DI) tanks. During this operation, there are two sources of toxic metal cross contamination. The first is the used (spent) activated carbon. In the regenerator's facility, after the carbon is removed from the tank(s), there is always residual carbon with toxic metal contamination inside crevices in the tanks. Also, once the carbon is removed, there is a possibility that the new carbon is put into the wrong tank that contains residual toxic metals in tank crevices and shipped to a sensitive user.

Sensitive users include kidney dialysis, biotech, hospital, residential, medical research, or other applications. Such toxic metal cross contamination can have serious and even catastrophic effects, not affects on humans, major medical research studies, or manufacturing processes.

The second source of contamination is from ion exchange resin tanks used to make high-purity deionized water. The vendor regenerates the resin tank(s) so the resin regains its ability to deionize water. There are two methods to treat the resin inside the same tanks or have it bulk regenerated. In bulk regeneration, spent resins from one user or many users are blended together and chemically treated.

There are three sources of toxic metal contamination during regeneration (chemical) of the spent resin. The first and second are the same as described with activated carbon. The third occurs when toxic metal resin from bulk regeneration is combined in error with a sensitive user's resin and then put into the sensitive user's tanks. Also, even though most of the toxic metals are removed during the regeneration process, they are *never* completely removed.

The best way to eliminate the potentially catastrophic threat of toxic metal cross-contamination is to use a vendor that only treats toxic metal wastewater and does not treat source (tap) water for sensitive users using rental DI tank services. The next best possibility is doing what a nationwide vendor does in providing rental DI services to both toxic metal and sensitive users. It has one facility just for toxic metals because of concern about liability. The vendor also takes other precautions, for example, having tank connections with reverse fittings to make sure that a casual user does not make an error by using tanks that should only be used for sensitive users.

There are four characteristics (federally mandated) of solid waste that a user must comply with: ignitability, reactivity, corrosivity, and toxicity, to determine if it is hazardous or not before disposal. (See EPA "Code of Federal Regulations," "Protection of the Environment," Section 40.) The two most frequently used characteristics are corrosivity and toxicity. The pH of the waste determines the corrosivity, while TCLP (toxicity characteristic leaching procedure) measures the amount of toxic metal. Sometimes, the wash chemical has a high pH and the washing process might clean parts containing toxic metals. The TCLP test is the minimum compliance requirement throughout the United States. Certain states might have a more stringent TCLP procedure.

The TCLP test identifies eight toxic heavy metals—cadmium, barium, silver, chromium, lead, arsenic, mercury, selenium (my acronym is "CBS CLAMS")—with allowable concentrations. For example, in the washing process for electronic assemblies (see "Closed-Loop Design"), the parts are soldered with solder that usually contains lead and silver. In this washing process, all of the consumable parts of the process, such as filters, activated carbon, ion exchange resin, and other solid waste, must be tested to determine whether they are toxic or not before disposal. The solid waste and the wastewater from this process must be compliant with two different metal regulations (lead and silver) and the solid waste used in the process must also meet the minimum TCLP requirements. Hauling hazardous waste usually costs up to about several hundred dollars per drum, while hauling nonhazardous waste usually costs up to about $50 per drum.

Occasionally, an entire process generating solid waste is defined as hazardous even though parts of it are not, according to the TCLP test. For example, a user might have an electroplating process with one of the eight toxic metals. Even though the ion exchange resin (solid waste) used on the rinse water is not hazardous according to the TCLP test, the ion exchange resin is still considered hazardous because the entire process has been designated by the EPA to be hazardous. In the electronic assembly example (see "Closed-Loop Design"), the process could have as many as two toxic metals (lead and silver). The entire washing process is not designated as hazardous waste; however, the resin and carbon must be tested to ensure that the concentration of both toxic metals does not exceed the TCLP test limits. Whenever toxic metals are present, the user must prove that the solid waste does not exceed regulatory limits before disposal.

Wastewater Volume

Determining the volume of wastewater from a cleaning process is very important because it can have a major influence on handling the water, wastewater, and solid waste generated by the process. For small volumes, cleaning processes generating less than about 25–75 gal/week of wastewater, if hazardous, are likely best hauled by a licensed carrier. For larger volumes, recycling all wastewater or discharging to the POTW, ground water, or surface water supply might be used.

Water Purity Specification

Determining the water purity specification is not easy in many cases and some investigation is often necessary. Information from trade associations, competitors, or related processes is very helpful. If these sources are inadequate, a user might have to experiment on a small scale or make the determination during the actual production process. The latter situation has a downside risk of too many part failures.

Measuring Water Purity

In many applications, a user must be concerned about measuring those characteristics of source water (tap water or raw water) from a lake, river, well, or wastewater that affect the quality of the parts being cleaned. In many applications, two characteristics are measured—dissolved and undissolved contaminants. In addition, another contaminant, colloids (see "Reverse Osmosis"), has characteristics of both dissolved and undissolved contaminants (see Table 32.1).

Dissolved Contaminants

Dissolved contaminants include ionic compounds such as sodium chloride, calcium carbonate, and many others that form ions in water. They are measured by a total dissolved solids (TDS), conductivity, or resistivity meter. Other dissolved contaminants such as sugar, starches, and other water-soluble organic compounds are not ionic and cannot be detected by these three types of meters.

Low-purity water is usually expressed in conductivity or TDS, while high-purity water is usually expressed in resistivity. Conductivity or TDS measurements are often used to determine the capacity of ion exchange resins or the rejection capability of membrane systems using nanofilters (NF) and reverse osmosis (RO).

TABLE 32.1 Resistivity, Conductivity, and TDS Conversion Chart

Resistivity (Ohm/cm) at 25°C	Conductivity (μS/cm) at 25°C	Dissolved Solids (ppm)	Approximate (GPG) as $CaCO_3$
18,000,000	0.056	0.0277	0.00164
15,000,000	0.067	0.0333	0.00193
12,000,000	0.084	0.0417	0.00240
10,000,000	0.100	0.0500	0.00292
5,000,000	0.200	0.100	0.00585
2,000,000	0.500	0.250	0.0146
1,000,000	1.00	0.500	0.0292
500,000	2.00	1.00	0.0585
300,000	3.33	1.67	0.0971
200,000	5.0	2.50	0.146
100,000	10.0	5.00	0.292
50,000	20.0	10.0	0.585
30,000	33.3	16.7	0.971
20,000	50.0	25.0	1.46
10,000	100.0	50.0	2.92
5,000	200	100	5.85
3,000	333	167	9.71
2,000	500	250	14.6
1,000	1,000	500	29.2
500	2,000	1,000	58.5
300	3,300	1,670	97.1
200	5,000	2,500	146
100	10,000	5,000	292

Source: Owens, D.L., *Practical Principles of Ion Exchange Water Treatment*, Tall Oaks Publishing, Littleton, CO, 1995. With permission.

Note: Approximate grains/gallon calculated by dividing ppm column by 17.1.

Most meter manufacturers use different algorithms to convert the conductivity electrical measurement to a TDS scale reading. The conversion factor from several sources can vary from about 0.4 to 0.7.[1-6] Thus, it is likely that different meters will give different results. The reason for the differences between meters is that source (tap) water throughout the United States at different locations is different in composition. Then, add the additional complexity of wastewater that has a far greater number of compositions than source (tap) water. Even if the TDS of the two different water samples were the same, they could still have a different effect on resin capacities because of the different weights and charges of the ions in the water or wastewater samples. If TDS is not accurate enough, a complete water analysis can be performed.

For simplicity, all TDS readings used in this chapter are determined by multiplying the conductivity readings by a conversion factor of 0.50. A simple correlation to remember is that a conductivity reading of one microsiemens (1 μS/cm) = resistivity of 1 megohm cm = approximately 0.50 ppm TDS (see Table 32.1). Since wastewater might contain ions that differ substantially from natural water supplies, this conversion will likely be even less accurate.

Higher dissolved ionic content (higher TDS) means higher conductivity or lower resistivity of the water or wastewater. A conductivity meter is best for water approximately 1 μS/cm or higher. For lower conductivity (lower TDS), resistivity is the preferred measurement. In both examples, accuracy and readability of the meter are improved.

Another important consideration is that readings from approximately 1–18.2 megohm cm are best made with a resistivity meter with its cell inserted in a flowing stream of water. (18.2 megohm cm is the highest water purity possible.) At 10 megohm cm or higher, this is the only way to make an accurate measurement. Stagnant water absorbs ionic impurities from the plastic or metal pipe and reduces the water purity reading.

Undissolved Contaminants

Undissolved contaminants do not affect the electrical properties of source (tap) water and wastewater, and are measured by different methods. There are far more different kinds of contaminants in wastewater than in source (tap) water. Some of the most common contaminants are measured by:

- *Total suspended solids* (TSS) measures the weight of all particles that do not pass through a 0.45 μm absolute rated membrane filter. This helps a user determine the particle loading of the water or wastewater to design a filtration system for a washing application.
- *Fat, oil, and grease* (FOG) measures any compound (vegetable or animal fats, petroleum and synthetic oils, lubricants, and some sulfur compounds) extracted by a fat-soluble solvent. It is used to determine whether a user complies with the discharge regulations of a POTW.
- *Biological oxygen demand* (BOD) is a measure of the amount of oxygen required by aerobic microorganisms to decompose organic matter in a given period of time. This is usually used for discharge compliance.
- *Chemical oxygen demand* (COD) is a measure of the amount of oxygen required to oxidize compounds such as sulfides, salts of metals, and organic compounds with potassium permanganate. This is used for discharge compliance, but much less often than BOD.
- *Total organic carbon* (TOC) is usually a measure of the total amount of oxidizable organic matter in high-purity water.

pH

pH is a measure of the acidity, neutrality, or basicity of water and is expressed as the negative log of the hydrogen ion concentration, or $-\log [H^{+1}]$. A pH reading below 7 is an acid condition, 7 is a neutral condition, and above 7 is a basic condition. In certain applications, the pH of the source water for wash chemicals and rinse water can be very important.

Source (Tap) Water Treatment Options

All water processes require an initial charge of source (tap) water from a well, river, lake, or a transported supply of water (bottled or from a tanker truck). Many operations might require a continuous supply. Two options for source (tap) water are no treatment and dissolved solids (ions) removal. Other options such as mechanical filtration, water softening, and others (see "Other Treatment Options").

IMPORTANT: Materials of Construction Used in Water and Wastewater Systems

It is best to use corrosion-resistant materials for wetted surfaces in any system with water or wastewater with very low or very high TDS. If low TDS water (resistivity of about 100,000 ohm cm and higher or conductivity of 10 μS/cm and lower) is in contact with any material that is not plastic, or 304 stainless steel or better, corrosion might result. At 10 megohm cm and higher, corrosion will result; fluorocarbons (Teflon® and others) and 316 stainless steel are superior. Materials like Teflon are superior to 316 stainless steel. High TDS water like seawater (about 30,000 ppm) is well known to be corrosive to many different metals.

No Treatment

Sometimes the source (tap) water is of sufficient purity that no treatment is necessary. Laboratory scale or pilot testing can help provide guidance for this determination. If a pilot scale test is not practical for some reason, it might be necessary to go to a full production scale with a backup plan to treat the source water as quickly as possible if this option is insufficient.

Dissolved Solids (Ions) Removal

Industrial water treatment commonly uses RO, deionization (DI), and much less often, electrodeionization (EDI) to reduce dissolved solids (sometimes called minerals) and ions from water or wastewater. RO removes far more dissolved solids than a nanofiltration (NF) membrane process. It is essentially identical in design to RO except for the membranes and other incidental differences. NF is used for water softening, industrial wastewater treatment processes, and other specialty applications where RO water is not necessary.

Reverse Osmosis

RO is a membrane process that removes all particles, molecules, colloids (ranging in size from about 0.1–0.001 μm),[7] ions and other species down to a size of about 200 molecular weight and larger from water or wastewater. Even though RO removes all particles, it is used primarily for dissolved solids, not particle removal. It is not designed to remove significant amounts of microorganisms or colloids.

RO is a process in which a pump is used to force water or wastewater through a permeable membrane barrier, as shown in Figure 32.1. The key component of an RO unit is the membrane, which is made from a thin film of specially treated plastic, most often constructed in the form of a spiral. Membranes vary in size from about 2–12 in. in diameter to several feet long. Water is forced through the membrane at pressures as high as about 1000 lb/in.²

RO separates the source (tap) water or wastewater into a permeate stream (product water for use in a process) and a wastewater stream (wastewater to a POTW). There are two key performance characteristics describing an RO system—% recovery and % rejection. The first is the percentage of lower ionic content water recovered (permeate) of the total amount treated by the RO. The second is the percentage

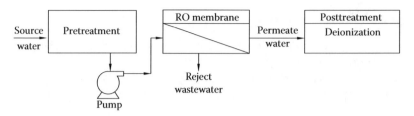

FIGURE 32.1 Single-pass RO system.

of much higher ionic content water rejected by the membrane. Essentially, an RO removes ions from water, concentrates them, and then discharges them to a POTW.

For the first performance characteristic (% recovery), an RO system usually recovers from 15% to 75% of all water or wastewater the RO treats. (15% is the typical recovery of a household RO drinking water system.) Therefore, from 85% to 25% goes to a POTW. As a membrane ages, its ability to reject dissolved solids decreases; the practical life of a membrane is from 3 to 5 years. In wastewater applications, the life could be substantially reduced. In the second performance characteristic, an RO usually rejects from 85% to 99% of all ions. Ionic concentration is very often the most important factor in distinguishing the different grades of water purity used in industrial applications. Ions have a wide variety of weights, shapes, and charges that determine the percentage rejection. For example, a higher charged ion like sulfate (SO^{-2}) has about a 95% rejection percentage, while a lower charged ion like sodium (Na^{+1}) has about a 90% rejection rate.[8]

Colloid reduction is very important when treating water and wastewater with RO because the colloids will likely irreversibly foul membranes, even sometimes in small quantities. These contaminants have a size between particles and ions, but are not truly soluble in water, as salt is. Typically, they do not settle as fast as larger particles do, but under the right conditions, they agglomerate like colloidal iron in source (tap) water. If there are substantial amounts of colloids, more pretreatment is required to prevent possible irreversible fouling of the membranes, often requiring special cleaning or replacement. Membranes can be cleaned, depending on the cost effectiveness of the procedure—additional equipment, labor, number of membranes, ability of getting adequate additional life, and other factors.

A typical RO system might consist of a pretreatment stage using one or more of the following: mechanical filters (multimedia filter, filter cartridges, or other), adsorptive media (activated carbon), anti-scaling chemical, pH treatment, and water softening. This is followed by a high-pressure pump, membrane(s), and storage tank. The posttreatment might include an ultraviolet light, repressurization pump, and, if higher purity water is required, DI (deionization) or EDI (electrodeionization) could follow. Sometimes cartridge membrane filters down to 0.2 μm could follow. The selection of these processes depends on a source water analysis and the water purity objective.

Another RO design is a double-pass RO, which is an RO followed by another RO in a single unit. This process can be followed by either DI or EDI. Usually, the sensitivity of an RO to increased costs from treating water or wastewater is much less than for an ion exchange system. This is why RO is often used as pretreatment before an ion exchange system when dissolved solids increase substantially. For example, the cost to treat 100 or 1000 gal with RO is much lower in terms of the cost of membrane life as compared to ion exchange resin whose operating cost would be about 10 times greater.

RO rejects essentially the same percentage of ions whether the incoming water to the RO has hundreds of parts per million TDS or thousands of parts per million of TDS. For example, the incoming water to an RO might have a TDS of 200 ppm with 95% rejection. In another example, the incoming water might have a TDS of 2000 ppm with the same rejection. The amount of minerals remaining in the product water (reject stream) from the 2000 ppm incoming water will be about 10 times greater than in the 200 ppm water.

Deionization

Deionization (DI) is a process using ion exchange resin to remove dissolved solids (ions) from water and wastewater.

> Ions are charged atoms in water. For example, when table salt, sodium chloride (NaCl), is dissolved in water, the ions form loosely held pairs, Na^+ and Cl^-. This occurs with many other compounds (mostly salts) that dissolve in water.

Occasionally, DI is referred to as demineralization, an older term used less frequently today. DI water is any water treated by a deionizer from which dissolved solids are removed, water resistivity increases, and TDS and conductivity decrease. There is no specific water purity measurement that defines the term, deionized water. Even though RO removes dissolved solids similar to DI, it is not referred to as deionized water, but RO water. Usually deionization does not require a similar amount of pretreatment as RO does. DI cannot remove the fine particles that RO can.

There are three basic deionizer designs:

- Two-bed weak base
- Two-bed strong base
- Mixed bed

Each deionizer design uses cation and anion ion exchange resin. There is weak acid and strong acid cation resin and weak base and strong base anion resin. Weak acid cation resin is used in specialty applications and will not be discussed. The lowest purity deionized water is produced by a strong acid cation and a weak base anion resin in series; a strong acid cation and strong base anion resin produces the next higher purity; and a mixed bed (composed of a mixture of strong acid and strong base resin) produces the highest purity water. These last three resin combinations are most often used (Table 32.2).

Both weak and strong base anion resins are the same except that the molecules have different functional groups, which give each a unique ionic removal property. Strong base anion, most commonly used, removes dissolved carbon dioxide and silica in addition to what weak base anion resin removes.

Strong acid cation and strong base anion resins are made of polystyrene and are cross-linked with divinylbenzene. Cation resin has a great number of spherical beads with a diameter of about 0.025 in. Each resin sphere is composed of an even far greater number of polystyrene molecules with negatively charged sulfonic acid functional groups with a positively charged hydrogen (H^{+1}) ion attached. Strong base anion resin is very similar, except that it has positively charged quaternary ammonium functional groups with a negatively charged hydroxyl (OH^{-1}) ion attached.

DI water is produced in a two bed deionizer in two steps. Cations in a source (tap) water or wastewater are removed and replaced with hydrogen ions, while all of the anions in the water or wastewater remain the same. This produces very low pH water from the acids created during the exchange process.

TABLE 32.2 Deionizer Designs vs. Water Characteristics

Water Characteristics	Two-Bed Weak Base Deionizer	Two-Bed Strong Base Deionizer	Mixed Bed Deionizer
Purity (megohm cm)[a]	0.02–0.6	0.1–0.9	1.0–18.2
pH	6 or lower	8.0+	Close to 7
Carbon dioxide and silica removal	No	Yes	Yes
BOD and COD reduction	Essentially none	Essentially none	Essentially none

Source: Otten, G., *American Laboratory*, July 1972.

[a] These are the typical ranges of water purity that are dependent upon feedwater characteristics.

For example, the cation resin removes the sodium (Na^+) ion from sodium chloride (NaCl) and replaces it with a hydrogen ion (H^{+1}) to become hydrochloric acid (HCl). (This is the reason the word "exchange" is used to describe this process.) There are cations such as potassium (K^{+1}), calcium (Ca^{+2}), ferrous iron (Fe^{+2}), and many others. In the second step, the acid enters the anion tank and the resin removes the chloride ion (Cl^{-1}) from the hydrochloric acid and replaces it with a hydroxyl ion (OH^{-1}). One H^{+1} ion from the cation resin and one OH^{-1} ion from the anion resin make water (H_2O). There are anions such as sulfate (SO_4^{-2}), nitrate (NO_3^{-2}), silica (in a variety of compositions and charges), and many others. The above chemical reactions occur in one tank in a mixed bed deionizer containing both cation and anion resin.

A *two-bed weak base* deionizer produces the lowest resistivity (higher conductivity and TDS) water. The resin capacity—number of gallons of water treated per ft^3 of ion exchange resin—depends on the capacity of the cation and anion resin. Both have a nominal capacity of about 30,000–40,000 grains/ ft^3. (A range is given for these capacities because it better represents a typical user's experience.) At the endpoint, the water purity drops off somewhat sharply. For wastewater, it is not unusual for the decline to be much more gradual. This is also characteristic of the two bed strong base and mixed bed designs, which follow.

A *two-bed strong base* deionizer uses the same cation resin as a two bed weak base deionizer. However, the strong base anion resin increases the purity of the water more than a two bed weak base. The higher water purity is achieved by removing dissolved carbon dioxide (carbonic acid) and silica ions, causing the resistivity of the water to increase. The higher removal of ions by the anion ion exchange resin reduces the nominal capacity to about 20,000–30,000 grains/ft^3.

A *mixed bed* deionizer contains exactly the same resin used for a two bed strong base design, except that cation and anion resins are intimately mixed in one tank. It produces the same effect as multiple two bed strong base deionizers in series. It is commonly used when water purity above one megohm cm is required and has a nominal capacity of about 10,000–13,000 grains/ft^3.

Even though TDS is an approximation of the capacity of a deionizer, a water analysis is better when designing a large system over about 10 gpm. For a two bed deionizer, the higher the flow rate, the more important a water analysis is in optimizing system design. Regardless of these factors, if the TDS is too high at about 500–1000 ppm, the cost of replacing or regenerating the resin becomes prohibitive unless the amount of wastewater treated is small and infrequent. So, RO alone or as pretreatment to the ion exchange resin makes the process much more economical.

Ion exchange resins have specific capacities, that is, the ability to remove ions from a given number of gallons of water. The capacity is inversely proportional to the TDS of the water. For example, if the TDS goes from 100 to 300 ppm, the capacity of the resin is decreased to one-third of its original capacity. For example, for 30,000 grains/ft^3 resin, the new capacity would be 10,000 grains/ft^3. The capacity of resin mostly depends on, in no specific order, concentration of different ions, concentration of foulants (e.g., colloids), flow rate, volume, depth and cross-sectional area of the resin in a tank, water purity, amount of chemical used to regenerate the resin, and temperature of water. In the preceding discussions, the highest resin capacity numbers stated are theoretical and are typical for several ion exchange resin manufacturers under ideal conditions.

For higher water purity, some of these resin capacity factors reduce the full utilization of the theoretical capacity of the resin and have an increased adverse effect on the resin's ability to produce high-purity water. This is commonly observed when treating lake water supplies in the Northeastern part of the United States. Universal colloids, as they are sometimes called, are very long chain molecules up to about one million molecular weight that prevent and interfere by blocking the accessibility of ions to resin sites, thereby reducing capacity. However, the mechanism for the reduction of water purity is different. In this case, colloids most often have ions bound on the molecules that are not easily removed by standard resins. These long-chained molecules with few ionic charges on them are very susceptible to removal because of the low ionic charge per unit weight of the molecule. When the molecules are released from the resin, the resistivity meter can detect the ionic charges on the molecules as the water passes through the cell of the resistivity meter.

Calculating Ion Exchange Resin Capacity

A TDS meter (see "Dissolved Contaminants") is most often used. It is much simpler than getting a complete water analysis, which is far more accurate, but much more expensive and time consuming. To be useful, all readings from a TDS meter have to be divided by 17.1 as calcium carbonate.[9] Therefore, 1 grain/gal = 17.1 ppm/grain (as calcium carbonate). For example, a water sample has a TDS of 100 ppm (as calcium carbonate). (Grain is an ancient unit of weight measurement referring to average size grains of dry wheat. 7000 grains of wheat equals one pound.)[9] To convert the ions to grains/gal, the following calculation is made: 100 ppm/gal ÷ 17.1 ppm/grain = 5.8 grains/gal. Therefore, in 1 gal of water there are 5.8 grains of TDS (total dissolved solids). Thus, 30,000 grains/ft³ (resin capacity) ÷ 5.8 grains/gal = 5,172 gal/ft³ of DI water produced before the resin is exhausted, requiring replacement or regeneration. This calculation applies to two bed strong base, mixed bed, and water softening processes.

Deionizer Resin Operating Options

An important variable that affects the capacity of ion exchange resin is whether it is used to treat source (tap) water or wastewater. Source (tap) water resin can be regenerated many times and still have good capacity for about 10 years for water softeners and 5 years for deionizers before replacement with virgin (new) resin. However, resin capacity varies greatly for wastewater because it usually has to treat a far greater concentration of foulants and oxidizers. This resin might not be regenerated economically or produce high-purity water, or both; thus requiring frequent replacement of virgin resin as often as each time it is used. Operating experience is the best way to determine whether to use regenerated resin or virgin (new) resin each time. Resin options include

- Off-site regenerated
- On-site regenerable
- Disposable

Regeneration is a process of chemically treating spent ion exchange resin to restore its capacity to remove dissolved solids. There are two ways that resin can be regenerated—off-site and on-site.

Off-site regenerated resin is a method in which a vendor takes the exhausted tanks to his facility and regenerates the resins with acid and caustic chemicals. The vendor then returns the tanks to the customer.

On-site regenerable resin is typically used for large volume users with more than about 20 gpm (gallons per minute). The resins are regenerated inside tanks at the user's facility with the same chemicals as used for off-site regeneration. The user might have to treat the wastewater produced by the regeneration process for pH. If any toxic metals are present, the user treats the toxic metals to meet federal, state, and local regulations.

Disposable resin has no waste stream to treat at a user's facility since the contaminants are held on the resin. After the resin is used once, it is discarded to a nonregulated or hazardous landfill after it is tested according to federal, state, and local regulations, as discussed earlier.

RO or DI or Both

Generally, RO is preferred when

- Requiring lower operating costs for reduction of TDS from source (tap) water containing 200 ppm or more (the higher, the more economical)
- Eliminating use of strong acid and caustic chemicals used to regenerate DI resins
- Requiring much lower concentration of small, nonionic substances

Generally, DI is preferred when

- Requiring no wastewater to be discharged
- Existing wastewater treatment system is at over-capacity
- Waiting for a wastewater discharge permit for an RO

- Requiring higher water purity that RO alone cannot produce
- Requiring a simple, flexible pilot system
- Producing similar water purity even when input water quality varies greatly
- Requiring flexibility in supplying water purity at a range of flow rates with the same equipment
- Requiring elaborate pretreatment for an RO system to operate cost effectively
- Requiring a simple system for low flow rate application

Generally, both RO and DI are preferred when

- Reducing TDS using an RO to make DI more economical

Case Histories

- *RO-only application*: In 2002, a customer with an RO-only application used well water with a conductivity of about 476 µS/cm and hardness of about 210 ppm. The untreated water compromised the chemical bath performance and caused spotting that would affect further surface treatment of the metal parts. In the final RO design, the hardness was too high, so a chemical anti-scalant was used to prevent fouling the RO membrane with hardness (calcium and magnesium carbonate scale). Otherwise, the RO would not produce the proper amount of treated water. The final design included a source (tap) water storage tank, and then an RO that produced permeate (product water) directly to the operations without a storage tank (called a direct feed design). The final water purity produced was about 15–25 µS/cm. System cost was $15,200.
- *DI-only application*: Our customers used closed-loop DI for electronics assembly operations. (See the first case history in "Closed-Loop Design.") System costs range from several thousand to $70,000.
- *Both RO and DI application*: In 2001, a customer who worked with automotive glass specified an RO and DI equipment to meet a water purity specification of 10–15 megohm cm at 5–7 gpm. The design used an RO to pretreat the water into mixed bed DI tanks and then into a storage tank. (The storage tank provided for periodic peak flow demands during cleaning operations.) The water continuously recirculated in a loop from the storage tank back to the storage tank to maintain high purity. The loop had another mixed bed, a UV light, and a 0.2 µm absolute membrane cartridge filter. It might have been possible to achieve the water purity specification with only DI. However, if lake water were used, colloids could make it much more difficult for DI to achieve the water purity. System cost was $32,100.
- *RO or DI application*: A washer manufacturer's customer wanted to remove lubricants from strip steel. The washer manufacturer tried high-purity DI water, which was a typical solution for all of their washing applications. However, even though DI water produced a spot free rinse, it caused another condition called "flash" rusting. This happens when steel parts are rinsed with higher water purity than necessary to achieve spot-free rinsing. The problem was eliminated when either a two bed weak base deionizer or an RO was used to produce low water purity, followed by an air blower.

Wastewater Treatment Options

The following decision making is the most complicated part of the selection process. Each of the options has to be analyzed to determine its cost effectiveness in meeting a user's objectives. It is very important in this evaluation that the entire process be reviewed initially to decide the critical success factors and margin of safety. After these are listed, another review must be made of the next most important variables. This method ensures that all key variables are considered; otherwise, much time and resources can be lost because they were not identified earlier in the evaluation. Typically, a closed-loop system uses the lowest amount of makeup water, while a cleaning process without any water reuse requires the largest amount of water.

For example, consider including a provision for future products that might require cleaning parts at a higher level of cleanliness than is customarily used in the original design. In another example, a user

Washer equipment

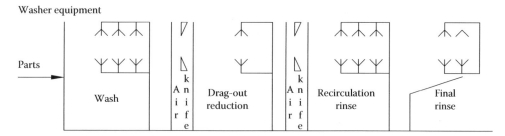

FIGURE 32.2 Conveyorized washer design. Note: This schematic represents a conveyorized washer. It can also be visualized as a multistage cabinet washer where all of the parts remain stationary and are subjected to each cleaning step. The parts are moved manually or automatically from one cleaning step to another in a dip tank cleaning process.

might inquire about future regulations for the contaminants being complied with and make accommodations for them. Regulations very seldom become less stringent.

Cleaning operations can use the sequence of operations as shown in Figure 32.2 for a conveyorized washer. This schematic shows the same processing steps that would be used for a series of dip tanks or a multistage cabinet (like a dishwasher). These other two designs look quite different from the conveyorized design.

No-Discharge Wastewater Option

The key to any no-discharge wastewater option is the ability to reuse wastewater and generate the least amount of solid waste. Sometimes wastewater from one application can be considered acceptable source water for another process. Cascade counter-flow rinses are very often used; they are a good example of wastewater reuse in the same process. With this method, the highest-purity water is used at the end to rinse the parts in the process. The wastewater flows in the opposite direction to the parts being cleaned as it cascades to the previous step in the process. Each time the wastewater is reused, the overall cost of water for the process and the amount of wastewater generated decreases as compared with using the highest purity water for each rinse stage and dumping it to drain.

A user might decide not to discharge wastewater because

- Return on a no-discharge system's capital equipment investment and operating cost are very favorable.
- Federal, state, or local regulatory agencies prohibit discharge of industrial wastewater.
- High cost of wastewater treatment.
- Restriction in volume of wastewater discharge.
- Uncertainty of water availability.
- Liability for ground or surface water contamination.
- High monitoring cost for discharging wastewater.

Closed-Loop Design

The design in Figure 32.3 allows no wastewater discharge to a POTW because it is all recycled to the same process. A closed loop is not easily attained, but for some processes, it is the most cost-effective, ideal solution.

Closed-loop design benefits include

- Recovery of high-cost DI water
- No wastewater discharge to POTW
- No wastewater tests, permits, inspections, and reports
- Wastewater converted to hot, DI water

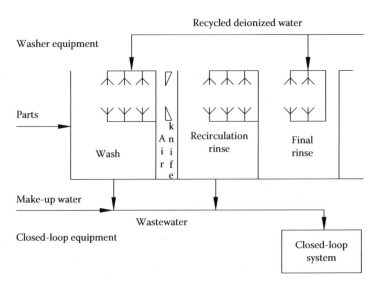

FIGURE 32.3   Washer and closed-loop wastewater system design.

- Wastewater recycled nearly indefinitely
- Reduction of energy and water usage by at least 90%
- Minimal amount of new source (tap) water
- Water purity ranging from low to high
- Solid waste is not hazardous in standard applications

Case Histories

- From the mid-1980s to today, over 450 electronic assembly customers worldwide have greatly benefited from our closed-loop water recycling system designs. All wastewater is recycled back to the washing process after it is treated. In an electronics assembly application, a manufacturer takes printed circuit boards, inserts numerous electronic devices on the boards, fluxes the boards, and then solders the devices. The flux is sometimes removed with DI water, and the wastewater from the washing process is processed by a closed-loop system. Today it is the design standard in this industry.[10] However, for many nonelectronic-assembly applications, the capital cost might be similar, but the operating cost for such a closed-loop system is often prohibitive because of the high dissolved solids (ions) in the wastewater. This causes a high operating cost because of the consumption of activated carbon and ion exchange resin.

This closed-loop design results in substantial savings. The return on investment is achieved within 6–12 months, after which the cash flow is positive.[11] That is, the cost to operate will be less than for a system that discharges all of the wastewater to a POTW. For example, suppose the TDS of the source (tap) water is about 300. In a once-through system, the source (tap) water with the same TDS is continuously fed to the parts washer, and then to waste treatment and a POTW or just to a POTW. However, in a closed-loop system, the water returns to the closed-loop system at 20–50 ppm instead. (The range depends on the soldering chemistries used in the process.) The saving is the difference between the cost of deionizing water at 300 and at 20–50 ppm. Also, there are savings from the energy recovered from recycling heated water used in the washing process, which ordinarily goes to the drain in a once-through system and there is minimal need for new source (tap) water. However, as much as 200 gal/day of source (tap) water is required because of the DI water that evaporates from board drying while operating 8 h/day. Some of this water-laden air can be economically recovered in some cases. The higher the TDS of the source (tap) water, the greater the return on a user's investment.

This closed-loop design uses a combination of particle, organic, and ionic removal media, allowing the water to be completely reused. Depending on the level of reliability required of an electronic assembly, the media is replaced after the resistivity drops below the water purity specification. The life of the ionic removal media depends on the soldering materials used in the process. Operating experience has demonstrated that soldering fluxes alone can cause the operating costs to vary as much as 50%.[12] Depending on the amount of certain soldering materials, such as water soluble tape, temporary mask, defoamers, dissopads, or other materials used, the operating costs can be so high that it is prohibitive to operate. A thorough knowledge of the soldering materials' impact on the operating cost of the system is crucial in achieving the lowest operating cost.

The solid waste (particle-removal filters, activated carbon and ionic removal media) generated must be tested (TCLP test) to determine compliance with federal, state, and local regulations prior to disposal. The key toxic contaminants generated by this process are lead and, sometimes, silver, both of which are regulated. From numerous installed operating systems, it has been found that the first particle-removal filter can be hazardous waste because of the solder balls being washed off the boards during the soldering process. Usually, the organic removal media and the ionic removal media are considered nonhazardous.

- In 1999, a customer required a recycling system to remove dissolved chrome from the rinse water using ion exchange resin. This process eliminates wastewater treatment, does not discharge wastewater, and substantially reduces water consumption, while providing chrome-free rinse water. System cost was $5100.
- In 2002, a nozzle manufacturing customer tested nozzles at multiple test stations to ensure they allowed water to flow within an acceptable range. The specification required continuous warm water in a narrow range, no odor from the water, water purity in the proper range to produce a spot free rinse, yet not too high to cause corrosion, and a specific air saturation of the water. The final design consisted of mixed bed DI tanks, deaerator, ozonation, ultraviolet radiation, and water storage tank. Our customer was very concerned about odors from the water, so both an ozonator and UV radiation were used. Most often one or the other is sufficient. System cost was $151,700. (Note: Detailed specifications are not stated to maintain confidentiality.)
- In 2009, a fuel cell customer was using an ink matrix in their process that required the wastewater to be hauled. The company wanted to substantially reduce the amount of waste to haul. The objective was achieved by using a UF membrane and DI. The UF membrane system produced two streams, the reject stream that contained the concentrated solid contaminants and the product water that was recovered from the wastewater. This product water was deionized and used again in the process. The solid waste generated by the UF was a small fraction of the amount of wastewater previously hauled. System cost was $75,300.

Zero-Wastewater Discharge Design

In the previous section, an electronics assembly case history was discussed using a closed-loop design. Many other parts are also washed, such as automotive, computer, and others, which might have oil and other contaminants. In these applications, the amount of dissolved solids (ions) or TDS in a wash tank are far greater than in electronics assembly applications. This makes the capital and operating cost of a closed-loop design prohibitive unless substantial design changes are made. Even with such changes, the costs are higher than for a closed-loop system.

Figure 32.4 shows a zero-wastewater discharge design that represents the typical stages of a variety of washer designs. Parts can be washed in a series of dip tanks, multi-stage cabinet design (dishwasher style), or conveyorized design. This design, like the electronics assembly closed-loop design discussed in "Closed-Loop Design," does not allow wastewater discharge to the POTW. In the electronics application, the solid waste had the toxic metals, lead and silver, but passed the TCLP test, making it non-hazardous. In nonelectronics applications, the solid waste is usually an alkaline wash chemical, which might be hazardous, depending on the pH, oil, or other characteristics.

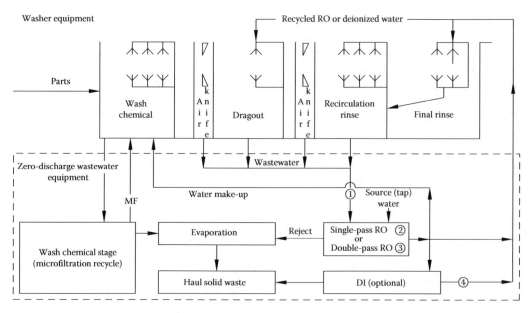

FIGURE 32.4   Washer and zero-discharge wastewater system design.

The first section is the wash chemical stage, which is designed to remove oil and other contaminants. If there is emulsified oil, recycling through a microfiltration membrane system (mechanical process) removes it. (Emulsified oil is an intimate mixture of oil and alkaline wash chemistry that does not separate.) It will separate into two components if mechanical or chemical methods are used. Free oil can be removed by numerous mechanical devices that allow the oil to accumulate at the top and be removed. In low-volume applications, it is most cost effective to haul away the wastewater.

The second section is the dragout stage, which is designed to reduce the amount of dissolved solids (ions) with the least amount of rinse water. One or more mechanical techniques are used, including air knives, rinse water spray, misting, and letting the excess wash chemistry drip off the parts. (An air knife is a long, narrow, slotted device that spans the width of a conveyorized washer. High-pressure air is forced through the slot to more effectively remove water from blind holes and other areas of a part.)

The third section is the final rinse water stage, which is used to rinse the parts with a recirculation rinse and final rinse with the highest purity water. Usually the final rinse cascades back to the previous recirculation rinse section.

Wash Chemical Stage

There are three general methods of handling wash chemical wastewater without discharging to a POTW—hauling, evaporation, recycling, or a combination.

The best method depends on four factors:

- Volume of wastewater
- Cost of hauling wastewater
- Cost of evaporation
- Cost of recycling

The cost of hauling, evaporation, and recycling are dependent upon local conditions. The discussions that follow provide guidance on what to consider. In addition, the stringency of discharge regulations and whether or not the wastewater and solid waste are hazardous can be significant factors.[13]

Volume of Wastewater

The volume of wastewater is the key factor for the next three methods for disposing of the unusable wash chemical. Estimates have to be made of the volume of wastewater to determine which is most cost effective.

Hauling

Hauling is most economical for low-volume applications because it eliminates the need for any capital equipment for treatment. A common carrier is allowed to haul waste as long as the wastewater is not hazardous. Otherwise, the solid waste must be manifested and hauled by a licensed trucking company to an authorized facility. In addition, even though federal, state, and local regulatory agencies might not classify the solid waste as hazardous, a local landfill owner might have his own regulation disallowing this solid waste. In some cases, for a new process, hauling might be used as a temporary measure until a final decision is made.

Evaporation

Evaporation is a method of separating a liquid from solids, typically by heating the wastewater with gas, electricity, solar energy, or vacuum distillation. When the amount of wastewater is much larger, hauling is not cost effective unless evaporation is used to reduce the volume, making it cost effective. It can greatly reduce the amount of wastewater to be disposed of by about 70%–95%. This is usually an energy-intensive process and the cost of the energy must be considered. If there are other processes in a plant producing excess heat that can be used, or if solar energy is available, evaporation can be very economical.

After evaporating the volatiles, the remaining contaminants might be solid waste with toxic metals or have a very high pH (regulations vary throughout the United States) that makes it a hazardous waste. The water vapor from evaporators (electric or gas) is like distilled water and has a water purity similar to medium purity DI or RO that can be reused in the washing process. An analysis of the cost of condensing the water vapor and the usefulness of the higher purity water can be made to determine its cost-effectiveness. The cost of evaporation and hauling is not significantly affected by the concentration of the toxic metals in the wastewater. Therefore, concentrating the liquid might be best. Depending on the regulations in some states or local areas, the vapors can be highly regulated and an air discharge permit required. These areas are sensitive air pollution regions of the United States.

Recycling

Recycling can be accomplished by using either a microfilter or an ultrafilter membrane.[14,15] (See Table 32.4 for particle size removal ranges.) These filters are made of polymeric, ceramic, or stainless steel materials. There are advantages and disadvantages of each of these materials. The ultrafilter is made only of polymeric materials in order to achieve removal of shorter chain molecules that the microfilter cannot remove very effectively. The membrane separates the alkaline wash into two streams: (permeate) alkaline chemistry and (reject) concentrated emulsified oil and free oil (small amounts). Depending on the molecular size of the components of the wash chemical, either a microfilter or ultrafilter can be very effective. In most cases, a microfilter at a nominal rating of 0.2 µm is very effective for higher molecular weight contaminants and it operates at a much higher flow rate than an ultrafilter. It is best to remove the free floating unemulsified oil before either microfiltration or ultrafiltration (UF) processing; otherwise, the flow rate through either could be substantially reduced and could even irreversibly foul the membranes.

During either microfiltration or UF (especially), some of the larger molecules (key ingredients) of the alkaline wash chemistry are removed. The amount of passage depends on the size of the molecules in the wash chemistry components and the membrane porosity. Some chemical suppliers offer an add-back package to rebalance the alkaline wash chemistry. Experience from operating such systems has shown that the life of a wash chemical is extended from three to ten times.

Recycling design benefits include

- Maintain a consistent wash chemistry
- Reduce average levels of emulsified oil on parts
- Reduce maintenance cost by not replacing the wash chemistry as frequently
- Reduce chemical consumption by extending the life of wash chemical
- Reduce water consumption from less frequent wash chemical replacement
- Reduce solid waste and wastewater hauling costs

A user's ability to achieve these benefits depends on a careful evaluation of the parts washer, cleaning chemistry, membrane unit, oil-based contaminants, and other factors. First, review the entire process, user's objectives, and potential benefits. Next, demonstrate the recyclability of the wash chemistry on a laboratory scale. A pilot test is performed at the user's facility to corroborate the benefits of recycling on a larger scale, and for the customer's understanding of the operating requirements of the membrane unit. Other chemicals and membranes might be tested to achieve optimum results. After this procedure is followed, the user can have complete confidence in the successful operation of a full-scale production washing process.

Case History

In 1998, a manufacturer of metal computer parts had difficulty removing water soluble oils and suspended particles that coated metal parts before electroplating. Even though the wash chemical was being replaced frequently, the failure rate of the next process step (electroplating) was too high. Neither an oil skimmer nor a coalescer reduced the failure rate because they could not remove the emulsified oil that caused the problem. Replacing the wash chemical even more frequently was too expensive. A pilot system was shipped to the manufacturer to ensure the equipment would achieve its purpose and the equipment operation was understood by the user. Even though the wash chemical bath was at 170°F, the temperature could have been even higher. The microfilter was very successful in maintaining both the emulsified oil concentration at a continuous low level. Other benefits such as reduced chemical costs, maintenance downtime, and others mentioned above contributed to making this a very cost-effective solution. The life of the chemical bath was extended by more than five times. System cost was $27,600.

Dragout Stage

The dragout stage uses mechanical methods to reduce the amount of wastewater coming from the wash chemistry tank before it gets to the next stage. The greater the volume of water produced, the larger the RO pretreatment system, and the more costly the zero-wastewater discharge equipment. Therefore, an effort should be made to achieve the highest concentration of contaminants in a given amount of dragout water.

The dragout method for a series of dip tanks could consist of orienting parts with blind holes to drain, holding wet parts over the wash tank for a period of time, air spraying parts, and other techniques. All actions are performed over the wash tank to reduce the amount of wastewater generated that has to be treated and chemicals required to replenish the wash tank. The parts can also be briefly sprayed with rinse water over an empty tank just before immersing in a rinse tank. For a multi-stage cabinet design (dishwasher style) orient parts with blind holes to drain and hold wet parts before going to the next stage to allow better drainage. For small parts in large quantities, the orientation of the blind holes to drain might not be practical. A brief rinse will be very effective.

For a conveyorized washer, the dragout stage is illustrated in Figure 32.4: an air knife, a low flow rinse spray, and another air knife. An old conveyorized washer without these features might be modified. As the washed parts enter the dragout stage, the air knives remove excess wash water from the parts and return it to the wash tank for reuse. They simultaneously reduce the amount of additional chemistry required to replenish the wash tank. A brief rinse removes more of the wash chemistry to reduce the TDS, allowing the RO to attain higher water purity. The final air knife removes more of the rinse water off the parts. Most of the rinse wastewater from any of these three washing system designs (dip tanks, multi-stage cabinet, or conveyorized), can be used as makeup water to the wash tank. Conserving water and chemicals is the best way to achieve the lowest operating cost system.

Rinse Water Stage

TDS reduction process options include

- Single-pass RO
- Double-pass RO
- Single-pass or double-pass RO followed by DI

Table 32.3 lists ways to produce low, medium, and high water purity as predicted by different treatment methods using incoming wastewater at 5000 ppm TDS (1). 5000 ppm is an arbitrary TDS of the wastewater in Figure 32.4 coming from air knife, dragout, air knife and recirculation rinse going to a single-pass or double-pass RO. It allows sample calculations to be made for illustrative purpose.

The final rinse water treatment method depends on the amount of TDS dragged out of the wastewater. For example, if the dragout stage is very effective, a single-pass RO might be adequate; otherwise, use a double-pass RO. With an installed washer, a user might have to consider the cost of modifying the washer's dragout section, if possible. The additional benefit of a double-pass RO is that a user might be able to achieve the specified water purity without requiring a DI closed loop. It is better to use a double-pass RO only, rather than a single-pass RO followed with a DI unit; otherwise it will be more costly. In all cases the lower the TDS of the wastewater to the TDS reduction process, the less extensive the treatment equipment and the lower the operating cost.

Calculating Water Purity Specifications

The following sample calculations allow a user to make an evaluation of different technologies, designs, and capability. These three water purity specifications are arbitrary and are not industry standard:

- *Low purity* (2): Depending on the wastewater TDS, a single-pass RO might achieve this specification. As an example, the wastewater has 5000 ppm TDS and the RO rejection rate for dissolved solids is 95%. Therefore, 5000 ppm × 5% (100% − 95% = 5%) = 250 ppm of the dissolved solids remaining in the product water. The remaining contaminants become the waste stream. Using the conversion factor of 0.5 (see "Dissolved Contaminants"), 250 ppm ÷ 0.5 ppm/µS/cm = 500 µS/cm (2000 ohm cm).
- *Medium purity* (3): If low-purity water is not acceptable, a double-pass RO raises the resistivity from the single-pass RO to about 40,000 ohm cm. Take the 250 ppm from the above low-purity option and recalculate: 250 ppm × 5% = 12.5 ppm ÷ 0.5 ppm/µS/cm = 25 µS/cm (40,000 ohm cm).

TABLE 32.3 Zero-Discharge Wastewater to Produce Low-, Medium-, or High-Purity Rinse Water

	Washer Stages		
Water Purity	Wash Chemical (1) (ppm)	Dragout (ppm)	Recirculation and Final Rinse Water Stage (Ohm cm)
Low purity (2)	5,000	Depends on process	Single-pass RO 2,000 (250 ppm)
Medium purity (3)	5,000	Depends on process	Double-pass RO 40,000 (12.5 ppm)
High purity (4)	5,000	Depends on process	Double-pass RO plus DI 1,000,000+ (less than 0.5 ppm)

- *High purity* (4): Adding a final closed-loop DI (conventional mixed bed DI or EDI) to either a single-pass RO (see low water purity) or a double-pass RO (see medium water purity) will achieve this specification. The water purity will be above 1,000,000 ohm cm (1 megohm cm).

Both single-pass and double-pass RO designs have a reject stream that goes to an evaporator. This eliminates the final wastewater stream from the process, thereby attaining a zero-discharge wastewater design. The solid waste produced is not likely to be hazardous unless it contains one or more of the eight toxic metals mentioned in "Solid Waste." If they are present, the waste must be subjected to a TCLP test to determine the concentration and whether any of the toxic metals exceed the regulation limit. If even one metal exceeds the regulation, the solid waste must be hauled by a licensed carrier as hazardous solid waste. Also, too low or too high pH is a regulatory concern.

Discharge Wastewater Option

A user might discharge wastewater because

- Wastewater meets discharge regulations directly without any treatment
- No-discharge wastewater option does not meet corporate return on investment objective
- Ability to treat wastewater for discharge at low cost
- Availability of source (tap) water at low cost
- Hauling costs for wastewater are too high
- Lack of interest in pioneering water conservation, reducing the amount of contaminated wastewater or solid waste

The following sections describe appropriate wastewater discharge options for various contaminants and conditions.

Fat, Oil, and Grease

Some washing applications might have any of these three kinds of contaminants, fat, oil, and grease (FOG), coming off the parts being cleaned. Usually, the contaminants produce free-floating oil and some emulsified oil. Free-floating oil forms on the surface of the wash chemical, while emulsified oil forms throughout the wash chemical. Usually emulsified oil appears milky white in larger quantities (dyes will change the color); in smaller quantities, it might be cloudy or not visible.

The ability to remove oil, fat, or grease contaminants as free floating, dispersed, and/or emulsified forms with the proper method depends upon a few key factors: amount and type of contaminants, temperature, and wash chemical (emulsifiers and surfactants). Some factors in selecting equipment include available space, maintenance, capital and maintenance costs, operating skills required, and other operating conditions.

The following is a summary of the most common technologies and techniques used to remove free and emulsified oil from water with an indication of the capabilities of each:

	Free Oil	Emulsified Oils
Decanter	Yes	No
Oil skimmer	Yes	No
Coalescer	Yes	No
Thin-film oil separation	Yes	No
Centrifugation	Yes	No
Activated carbon	Yes (very limited)	Yes (very limited)
Chemical precipitation	Yes	Yes
Dissolved air flotation	Yes (no chemical)	Yes (with chemical)
Microfiltration	Low amounts	Yes
Ultrafiltration	Low amounts	Yes

Free Oil

Free oil or unemulsified oil is always considered free-floating oil. There are many proven, long-used methods and even a newer technology, thin film oil control, which can be used:

- *Decanter* (gravity separator) allows the free oil to rise to the surface, separating it from water, and spilling over a weir into a container.
- *Oil skimmer* includes an oil-attracting material in the form of a belt, disk, or similar device to remove free oil from the surface of the wastewater.
- *Coalescer* is a device constructed of materials, usually polymeric media that allows the adherence of very small droplets of oil that grow in size and are released to the surface of the water when large enough, and then removed by a decanter.
- *Thin-film oil separation* equipment is a newer method that utilizes Bernoulli's principle. In the three step process, the oil is concentrated, and then allowed to flow over a weir for disposal. There are no moving parts or consumables.
- *Centrifugation* spins the wastewater at a high velocity forcing the densest particles and organic compounds to separate from other particles and compounds.
- *Activated carbon* is available in a granular or cartridge design and is effective for very low amounts of free oil; otherwise it would be prohibitive in cost.

Chemical precipitation, dissolved air floatation and microfiltration and UF will be discussed in the next section.

Emulsified Oil

Here are three methods that use chemicals and membranes to separate emulsified oil from water:

- *Chemicals* are used as a general purpose method for small or large amounts of emulsified oils.
- *Dissolved air flotation* (DAF) is a completely different design used most effectively for large operations. This design uses air that attaches to free or dispersed oil and facilitates its rise to the wastewater surface for easy removal. Chemicals are used to allow removal of the emulsified oil.
- *Membrane filtration,* such as microfilters and ultrafilters, are ceramic, metal, or polymeric, and are very effective in meeting local discharge regulations. A microfilter can process much larger volumes of wastewater than an ultrafilter on an equal basis of surface area of the membrane. Therefore, it is better suited for large flow rate applications. However, an ultrafilter has a smaller pore size, so it is more effective in removing low-molecular-weight organic molecules. Both of these technologies are increasingly competing with chemical precipitation and even DAF for large flow applications. Membranes can only process small amounts of free oil before the flow rate through the membranes is reduced. However, a free oil removal device eliminates this restriction. There are several important variables that must be considered when making a decision about whether a membrane or DAF is best.

pH

Sometimes the wash tank of a cleaning operation contains alkaline wash chemistry with a pH higher than the local discharge limit. This condition can be corrected by using an acid pH chemical control unit. Less frequently, the use of an alkaline chemical control is necessary for applications with a low pH.

Biological Oxygen Demand and Chemical Oxygen Demand

Often state and local regulatory agencies have discharge limits for BOD (biological oxygen demand) and, to a much lesser extent, COD (chemical oxygen demand). The BOD test method determines the amount of oxygen consumed by microorganisms to decompose organic matter in water or wastewater.

The COD test method uses a chemical oxidant to determine indirectly the amount of oxygen required to oxidize both organic and inorganic contaminants in water.

COD contamination sometimes requires chemical oxidant, carbon adsorption, ultraviolet oxidation, or ozonation to achieve further reduction. A membrane process such as UF or NF can be used, but they are more often used in a recycling process where the permeate would be reused; otherwise the waste stream must be hauled or evaporated and then hauled.

Toxic Metals

The following methods are widely used to remove different toxic metals to meet discharge limits. In many cases, there is not one single form of a metal; rather, there are two or more forms simultaneously. It is not unusual that as many as three of these technologies are used together:

- Mechanical filtration (particulate)
- Chemical precipitation (particulate, colloidal, and chelated)
- Membrane filtration (particulate and others)
- Ion exchange resin (ionic)

Mechanical filters are useful for particles of about 1 μm and larger, at low concentration levels, and small to moderate volumes of wastewater treated. Filter replacement cost is the most important factor. This is a low capital cost option with the smallest space requirement.

Chemical precipitation is most often used for large volumes because it is economical. It is designed to remove chelated, non-chelated, and colloidal toxic metals. A chelate is a metal ion that is chemically bound by other ions, making it difficult to separate. With a few exceptions, this process can be used to remove chelated toxic metals too. This is the next highest capital cost option, which occupies a larger space than mechanical filters.

Membrane filtration is very effective, especially for moderate to large applications for all forms of these contaminants. There are three types of membranes that can be used, depending on the state of the toxic metal. RO can be used for dissolved metals, but the user must be careful of the chelated forms of metals, which might be colloidal, and which could irreversibly foul the membranes (see "Reverse Osmosis"). Neither UF nor MF can remove ionic forms of metals. UF can remove suspended, colloidal, and chelated forms of metals. MF's effectiveness is limited in removing colloidal and chelated forms of metals. Membranes are very flexible in their removal capability, but are usually the higher capital cost systems in this grouping. In large applications, chemical precipitation and membrane filtration must be studied carefully to determine the most cost-effective approach.

Ion exchange is a very simple method used primarily for two reasons: The ionic metal concentration cannot be removed by any of the three other methods and other forms of the metal (colloidal and chelated) do not exceed the discharge regulations. Neither chelated nor colloidal forms are effectively removed by resin. Under these conditions, a UF membrane will eliminate the problem. If suspended solids contain metals, they can be removed by mechanical filters before the particles get to the ion exchange process.

There are three types of resins used: standard deionizing (cation and anion or mixed bed), specific heavy metal, and cation or anion with other combinations of resin. There are two criteria in making the decision about which to choose—cost and capacity. The standard resins are less expensive, but have a lower capacity; while the specific heavy-metal resin is more expensive, it has a higher capacity. The key difference between these two is that the deionizing resin removes ions that do not have to be removed, while the specific heavy-metal resin does not. The key question is, "Does the higher capacity resin justify its cost in terms of the amount of metal it removes?" For low-volume applications the standard resin is probably best. But at higher volumes, since the cost becomes significant, the best action is to either do pilot testing or just try the different resins in the actual production operation. The last type, cation or anion with other combinations of resin, is too specific to the conditions, usually with several solutions, and is too complicated to discuss here. In all of these decisions, pH, TDS, competing ions, and other factors, each or all of them together, could have a major impact upon the effectiveness of the resins used.

The final selection of the best method or combination depends on the type of metal form (particulate, colloidal, ionic, chelated, or a combination), flow rate, total flow per day, concentration of contaminants, and other factors. In some membrane applications, the wastewater processed could be reused. The waste stream would contain the concentrated metal and the oils would be hauled as hazardous waste.

Case Histories

- In 1998, a customer was washing beryllium copper strips that had water-soluble stamping oils and metal particles on the surfaces. An oil skimmer and coalescer were used to remove the free-floating oils. However, there were bare spots causing poor adhesion, resulting in later post cleaning plating part failures. After reviewing the entire cleaning process, the chemical cleaner was changed along with adding a chemical recycling system that could remove the emulsified oil that could not be removed previously. Reductions in chemistry use, physical labor, waste disposal, and product rejects led to an annual cost saving of almost $120,000 per year. System cost was $29,800.
- In 2005, a customer decided to discharge wastewater to a POTW because it was not economical to recycle the wastewater. The wastewater stream flowing at 4 gal/min had a zinc concentration that was above the local regulatory discharge limit. Zinc-selective ion exchange resin was the best solution to remove the heavy metal. Two systems cost $6400 each.
- 1n 2008, a large military aerospace R&D center for advanced electronics had toxic metals in their process that would violate local discharge to a POTW if not treated. Four of these toxic metals had to meet these regulations. The technologies that were pilot tested proved that an ultrafilter and ion exchange would solve the problem. The ultrafilter was necessary because of the small particle size and ion exchange was used for the ionic forms of some of the metals. System cost was $44,500.

Other Treatment Options

Other treatment options can be used individually or with the treatment methods discussed previously. The following sections describe some of the most common.

Distilled Water

Distilled water is a common, low-TDS water, similar to what RO and ion exchange can produce. It is produced by simply heating water to boiling and condensing the vapor, which then becomes distilled water. It is used most often for pharmaceutical and laboratory applications where bacteria-free water is necessary. It is usually not economical for industrial applications as compared with the other options described previously. It is mentioned here because it is used in many small applications and readily available in small quantities.

Water Softening

There are some applications in which softened water is sufficient to achieve a cleanliness specification. This process might be used alone to treat the source (tap) water for washing parts or, more often, as pretreatment to RO.

Water softening is a process of removing mostly hardness minerals (ions), calcium and magnesium, and to a much lesser extent, dissolved iron or manganese and replacing them with sodium cations. The key component of a water softener, as in a deionizer, is the ion exchange resin inside a tank (Figure 32.5). The cation ion exchange resin is the same as used for deionization, except that it is in a different form— sodium (Na^{+1}) instead of hydrogen (H^{+1}). Softening the source (tap) water has essentially no effect on the TDS, conductivity, or resistivity of the water.

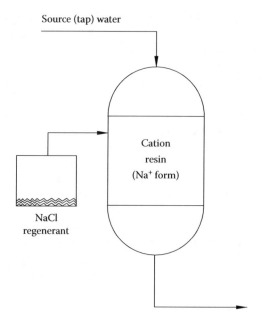

FIGURE 32.5 Water softener.

After all of the ions are exchanged, the ion exchange resin becomes spent. The resin is regenerated (reversing the process) by flowing concentrated sodium or potassium chloride through the resin during a multistage process, performed manually or automatically. During the regeneration process, the sodium (from sodium chloride salt) or potassium (from potassium chloride salt) replaces the hardness cations that the softener removed. This process is very inexpensive compared to regenerating a deionizer. There are factors such as regenerant concentration, iron fouling of the resin, and others that can significantly influence the actual capacity of the resin. The minerals remaining after softening do not cause a hard scale or soap curds (bath ring) to form on the sides of the tank containing the softened water. The amount of soap, detergent or alkaline cleaner required to wash parts is greatly reduced.

Soft water can achieve visually spot-free parts as long as minerals are lower than about 50 ppm and depending on other factors, especially the surface shininess of the parts. A drop of source (tap) water or wastewater allowed to evaporate on a shiny test metal panel will usually leave a more observable residue than with softened water. The appearance of any residue is highly dependent on the level of polish of the part. The more highly polished, the more observable any residue or spots are. The great saving as compared with DI water makes this technology worthwhile to evaluate.

The ion exchange resin capacity calculation for a water softener is the same as for a deionizer (see "Deionization").

Electrodeionization

Electrodeionization (EDI) is an ion exchange process that produces high-purity water greater than 1 megohm cm resistivity (1 μS/cm) without the use of chemical regenerants—hydrochloric acid and sodium hydroxide. However, pretreatment is far more important than for a conventional deionization system. Also, it is most often used as a posttreatment to an RO membrane process. It is used far less often in industrial processes instead of conventional DI, but is increasingly gaining popularity because it does not use hazardous chemicals.

Mechanical Filters

Mechanical filters are physical barriers for removing particles from water. Mechanical filtration is the most common method used to remove particles from water and wastewater in washing processes. The particles can vary from very coarse to a sub-micrometer (sub-micron) level, that is, less than 1 μm in size. There can be a very large overlap in the removal capabilities of the various technologies that follow. See Table 32.4 for a chart of the different types of contaminants and the separation technologies used.

There is neither an industry-wide micrometer nor filtration efficiency rating that defines all of these filters. For example, a 10 μm rated bag filter might have an efficiency rating of 70%, that is, 70% of all 10 μm particles are removed, while a 10 μm cartridge filter (non-membrane type) might remove 90%. Beta ratios are sometimes used to rate the effectiveness and efficiency of filters. If the beta ratio is 10, the filter has an efficiency rating of 90%, while a filter with a beta ratio of 100 has a 99% efficiency rating. There is far more reliability of filter micrometer ratings and efficiencies for a given manufacturer. Also, there are certain industries that use standard criteria accepted only within that industry.

As a general rule, the rank order from high to low flows of the four most common filter designs is granular media, bag, and the two cartridge types. However, depending on the application, it is possible to operate any of the following four designs at flow rates as low as a few gallons per minute:

- *Granular media filters* are composed of a single media or multimedia with various grades of sand and other minerals. They are used primarily to remove suspended particles from 20 to 40 μm in size and larger from source (tap) water. They can remove much finer particles, but not as effectively. As a reference point, a grain of table salt is about 125 μm.

TABLE 32.4   Particle Size Removal Range by Filtration

These sizes of well known objects and particulates illustrate the size of the micrometer (or micron).

Human hair 75 μm — Protozoan cyst 4 μm — Smallest bacteria 0.2 μm — Not to scale

Synthetic dye	Rosin smoke	Fly ash
	Oil smokes	Coal dust
	Tobaco smoke	
	Metallurgical dusts and fumes	
		Ammonium chloride fume — Cement dust
Aqueous salt		Sulfuric concentrator mist — Silican sand
	Carbon black	Pollen
Metal ion		
	Zinc oxide fume — Paint pigments	Insecticide dusts — Mist
Sugar molecule	Colloidal silica	Ground talc
		Spray dried milk
	Albumin protein molecule — Alkali fume	
Atomic radi		Milled flour — Pin point
	Atmospheric dust	
	Asbestos	Protozoan cysts
	Endotoxin/pryogen	Red blood cell diameter — Human hair
	Viruses	Bacteria
		A.C. fine dust
Reverse osmosis	Ultrafiltration	Particle filtration
Nanofiltration	Micro filtration	

Partical diameter angstrom units, Å

1   10   100   1,000   10,000   100,000   1,000,000

0.0001 0.0002 0.0005 0.0008 0.001 0.002   0.005 0.008 0.01   0.02   0.06 0.08 0.1   0.2   0.6 0.8 1   2   6   8 10   20   50 80 100

Particle diameter micrometers, μm

Source: Courtesy of Water Quality Association (WQA), Lisle, IL.

- *Bag filters* are manufactured from felt materials, both woven and nonwoven, and typically have a higher contaminant loading and a lower cost per pound of contamination removal than cartridge filters. They are the most commonly used filters in industry. They are primarily designed for higher flow and higher solids loading applications. Usually, they are not as effective and reliable as cartridge filters for particles in the lower micrometer sizes.
- *Cartridge filters* are commonly used filters made from a wide variety of plastic and natural fibers, such as polypropylene and cotton, in a large variety of designs such as molded, fiber wound, and pleated paper. They are most often used for lower flow rates and higher efficiency, low micrometer removal applications. Efficiency is defined as the ability of a filter to remove particles at a stated micrometer (micron) rating. For high-flow-rate and high-volume applications, granular or bag filters are most often used first and sometimes followed by cartridge filters, if needed.
- *Cartridge membrane filters* are manufactured from a variety of plastic and inorganic materials with different shapes (flat sheets, tubes, spiral wound tubes, and other forms). They are designed to remove very small particles and organic molecules from a liquid stream. Microfilters (MF) are rated at about 0.05–1.0 μm. Ultrafilters (UF) are rated to remove essentially all particles and molecules from about 10,000 to 1,000,000 molecular weight from water.

Sometimes a microfilter from one manufacturer is called an ultrafilter by another manufacturer. To compare one membrane with another, a user must determine the test method the manufacturer used for the rating. This rating problem can be extreme, for example, a membrane manufactured from a plastic material, such as polysulfone, polypropylene, or nylon, rated at 0.2 μm might reject 99.999% (beta ratio of 10,000) of all bacteria, whereas a ceramic membrane with the same rating will have dramatically lower removal efficiency and be unable to significantly remove any bacteria.

Membranes are available in many shapes, sizes, and with varying removal effectiveness. They can be in a disk, cartridge (5–30 in. long), or in large microfiltration systems with one or several membranes and many other configurations. This allows a user the ability to test a filtration application on a small scale at a very low cost.

None of these membranes remove dissolved solids (ions) from water. However, for example, an ultrafilter will effectively remove colloids and other high-molecular-weight substances, such as surfactants, while a microfilter membrane will not do so effectively. However, both types can be used to recycle wash chemicals (alkaline cleaners). Ultrafilter membranes are much more effective than microfilters in removing large organic molecules (macromolecules and lower-molecular-weight petroleum products). This is desirable for discharge to a POTW, but not for recycling a wash chemical. This will result in more add-back chemical additions. Membranes are also mentioned in "Recycling" in the "Wash Chemical Stage" section and in "Emulsified Oil" in the "Fat, Oil, and Grease" section.

Adsorption (Activated Carbon)

Adsorptive technology uses large surface areas to remove contamination. Activated carbon is one of several adsorptive materials used. It is a granular media made by heating carbon-containing materials, such as coal, coconut shells, and similar substances, in the absence of air, producing a porous material with a very large surface area. It is usually used as a pretreatment method to remove chlorine and long-chained organic molecules (primarily decayed vegetation in reservoirs and lakes) prior to ion exchange resins and some RO systems. Activated carbon has a catalytic corrosion effect on stainless steel tanks used to hold the carbon. The carbon can cause pinhole leaks; lining the tanks with plastic solves this problem.

Oxidation

Oxidation is a process, with or without a chemical, that can oxidize chemicals, dissolved species, deactivate microorganisms (bacteria, viruses, and others), and reduce the amount of organics and many other species and substances in water or wastewater. Even air alone can be used effectively, for example, oxidization of dissolved iron and manganese to form particles that are removed by mechanical filtration prior to an RO. Oxidation is usually achieved by a chemical or air.

Ozonation is a nonchemical oxidation process that uses ozone gas, a powerful oxidant, to greatly reduce the concentration of microorganisms, organic compounds, and other chemical compounds in water. This method can eliminate the need for chemical dosing used for chlorine, potassium permanganate, and others.

Ultraviolet Radiation

Ultraviolet radiation (UV) sanitizes (not sterilizes) water or wastewater to control microorganisms. (In the medical, pharmaceutical, and similar fields, sterilization means a 100% kill rate of all microorganisms.) UV units are too often referred to as "sterilizers," implying that they have the ability to produce microorganism-free water. However, UV sanitizers do greatly reduce the numbers of a wide range of microorganisms with a wide range of degree of effectiveness. There are two types of lamps, 254 nm wavelength, the most common, and 184 nm. The latter produce a slight amount of ozone in addition to their sanitizing ability, thus increasing their effectiveness against microorganisms. If unfiltered water or wastewater is used, it might drastically reduce the effectiveness, because the particles or contamination block the UV radiation and reduce the ozone's effectiveness before it reaches the microorganisms. Prefiltration is important to ensure continuous optimum operation. It can be used alone or as a post-treatment to a DI water system, pre- or post-RO, or alone.

Wastewater Treatment for New Processes

First, it is advisable to have a water treatment specialist involved in the early stages of process decisions. Second, to reduce the uncertainty of wastewater treatment decisions, a user should determine the local source water conditions, similar processes in the industry (competitors), availability of hauling, allowance of temporary discharge to define the process, and piloting the process, all of which greatly prevent overdesign costs. The less that is known about a process, the greater the margin of safety that is usually necessary to ensure a treatment system that will meet a user's requirements. A user should attain maximum flexibility before buying a permanent system. A user should be concerned about the following three areas.

Source Water Treatment

If a water sample is available, it is best to have it analyzed, especially if high-purity water is necessary. It is best to wait for the results of the analysis before renting a long-term system or buying a permanent system, unless the uncertainty of the treatment process is minimal. Any water can be treated to meet a user's need; it is a matter of cost.

No-Discharge Wastewater

It is difficult to achieve an economical wastewater treatment system for a no-discharge wastewater design because of unknowns, for example, type of wash chemical, specific contamination generated by the process, surface quality of the parts, and other factors. For small volume applications, the entire wash tank and rinse water could be hauled. For large volumes of wastewater, where hauling might be a problem and the user is on a POTW, it might be possible to discharge it with minimal treatment with a waiver from the local regulatory agency until the final design is achieved. If on a septic system, river, or other body of water, hauling is the only practical way. Another alternative for any of the above could be a temporary treatment system alone or along with hauling until enough data are gathered to define the final, permanent treatment system.

Wastewater Discharge

If a user has decided to discharge to a POTW, it is necessary to obtain the discharge regulations to determine the wastewater conditions that must be met to obtain a permit as soon as possible. It is easier to prepare for this application than for a zero-discharge design because there are far fewer conditions affecting the final design. For example, for most alkaline cleaning applications, pH and oil are the two key concerns. For pH adjustment, equipment is usually easily obtainable on relatively short notice. The amount of oil in the wastewater is more difficult to assess and could lead to a large, unnecessary initial expenditure if a large margin of safety is required, such as considering a UF membrane or chemical treatment system. In such cases, a discharge waiver from a POTW would be of great value until the final effluent is tested.

Overcapacity of the Current Wastewater Treatment System

Usually in such applications, treatment at the source of the discharge can be a primary solution. The cost of expanding the entire wastewater treatment system is usually much more than trying to reduce the amount of wastewater going to the central treatment system. A careful evaluation of all discharge sources should be made to determine the most viable option from a cost standpoint. For temporary overcapacity applications, hauling might be most economical.

Conclusion

Selecting the best source water and wastewater treatment processes for cleaning applications requires a methodical approach. It is best to make a quick review of the entire process while moving forward toward the final design. There might be a critical interdependency of some parts of the cleaning process and wastewater treatment system. A user does not want to miss a critical item in the analysis, especially early in the analyzing process. Sometimes a simple change can alter the entire economic equation, changing a previously uneconomical solution into an economical one or vice versa. The general trend is toward increasing stringency of discharge regulations. Continual vigilance by users in maintaining their knowledge of current water and wastewater treatment practices will ensure the most effective, low-cost design in the future. Most important of all is to make selecting the best wastewater treatment system a team effort; there will be fewer errors and recriminations later.

References

1. Osmonics Corporation, *Pure Water Handbook*, Osmonics Inc., Minnetonka, MN, 1991, p. 10
2. Owens, D., *Practical Principles of Ion Exchange Water Treatment*, Tall Oaks Publishing, Inc., Littleton, CO, 1985, p. 160.
3. Myron L Company, TDS/conductivity conversion chart, P/N SSB AB 01-08 with graph of 442 natural water standard and TDS/conductivity conversion chart with graphs of DS meter curve, NaCl, and DS meter curve, natural water (442), 2008.
4. Mettler Toledo Thornton, *200CR Conductivity/Resistivity Instrument, Instructional Manual*, P/N 84295 Rev. K, February 2008, p.13.
5. McPherson, L., Correlating conductivity to PPM of total dissolved solids, *Water/Engineering & Management*, August 1995, pp. 31–33.
6. Chemtrol, A Division of Santa Barbara Control Systems, *Conductivity and TDS*, available at http://www.sbcontrol.com/tds.htm.

7. Parehk, B.S., ed., *Reverse Osmosis Technology*, Marcel Decker, Inc., New York, 1988, p. 169.

8. Byrne, W., *Reverse Osmosis: Practical Guide for Industrial Users*, Tall Oaks Publishing Co., Littleton, CO, 1995, p. 5.

9. Owens, D., *Practical Principles of Ion Exchange Water Treatment*, Tall Oaks Publishing, Inc., Littleton, CO, 1985, p. 14.

10. The Association Connecting Electronics Industries (IPC), *Aqueous Post Solder Cleaning Handbook* IPC-AC-62A, January 2006, pp. 11–14, 31–35.

11. Russo, J.F. and Fischer, M., Operating cost analysis of PWB aqueous cleaner systems: Zero-discharge water recycling system vs. once-through, presented at *Third Int. SAMPE Electronics Conference*, June 20–22, 1989.

12. Borek, K., *A Clean Break from CFCs, Circuits Assembly*, Miller Freeman Inc., San Francisco, CA, February 1992, p. 43.

13. Kieper, T. and Russo, J.F., Closed-loop alkaline recycling proves an award winning application, *Parts Cleaning*, May 1999.

14. Rajagopalan, N., Lindsey, T., and Sparks, J., Recycling aqueous cleaning solution, *Products Finishing*, July 1999.

15. Quitmeyer, J.A., Sifting through filtration options, *Precision Cleaning*, December 1997, 16–23.

16. Otter, G., American Laboratory, 1972.

Resources

Water Quality Association, *WQA Glossary of Terms*, 1993.

Kunin, R, *Ion-Exchange Resins*, Robert E. Krieger, 1985.

American Water Works Association, *Water Quality and Treatment*, McGraw-Hill, New York, 1990.

Byrne, W., *Reverse Osmosis: Practical Guide for Industrial Users*, Tall Oaks Publishing, Littleton, CO, 1995.

33

Overview of Drying: Drying after Solvent Cleaning and Fixturing

Barbara Kanegsberg
BFK Solutions

Introduction

Drying is the third step in the cleaning process, and it is one that is too often neglected.[1] Drying is removing residual rinsing agent and, in some cases, residual washing agent, from the part without redepositing soil on the part and without damaging the part itself. The drying process has to be chosen carefully with consideration of the product, process flow, and ultimate use of the product. In planning a cost-effective cleaning process, budget some money for an effective drying system. Particularly with aqueous cleaning, many manufacturers have found it to be the most time-consuming step in the cleaning process. This is because drying requires that the product be heated and agitated. If the product cannot tolerate a high temperature, for physical drying, the drying step must be longer and slower. In addition, if a fairly large component must be heated to a high temperature, it may be too hot to handle for longer than you may find tolerable to achieve efficient process flow.

This chapter will consider drying after solvent cleaning, the importance of fixturing, the critical human factor, and attempts to substitute drying for washing and rinsing.

Do You Need Drying?

Dry critically. If residual cleaning agent or rinsing agent is not an issue for product performance, if residual cleaning or rinsing agents can be tolerated in the next step, perhaps drying is not an issue. More often than not, however, drying is required.

Do not operate under the delusion that a carefully designed wash and rinse system are a good start and that a drying system will be upgraded under next-years budget.

We see too many critical applications, even in clean rooms, where product is meticulously washed and rinsed, then left out to air dry. This is a recipe for product failure, because clean rooms are not particle-free. One savvy technician showed us critical components drying on a clean room bench and quipped "we call this the contaminant redeposition process." As it turned out, he was right.

Are you asking too much of the drying process? Drying is not a substitute for washing and rinsing. We have observed drying steps of several days duration introduced in an attempt to correct inadequate soil removal. The results were semi-successful; more effective washing and rinsing were required. We frequently hear that cleaning is not necessary, because there is a heat-treat step. While heat-treat carbonizes some soils, it can also serve to chemically modify soils and make them exceedingly adherent. This can compromise surface quality. We have pragmatically observed that the performance of critical aerospace product is markedly improved by introducing effective cleaning processes prior to heat treat.

Solvent Drying

The two chapters in this section are concerned with drying in aqueous processes, and this is indeed appropriate because water is, well, wet. Water has a much higher boiling point and higher surface tension than many solvents, so it is more difficult to get rid of. Perhaps you are thinking: I am purchasing a highly sophisticated solvent system. Drying will not be a problem.
Think again!

Those considering solvent systems also need to plan carefully to achieve adequate drying in an efficient, cost-effective manner. Some of the principles (although not all of the specifics) discussed in the chapters on removal of water are very applicable to designing a good solvent drying system. I suggest you pay particular attention to discussions of vacuum drying and the need to avoid recontamination of the part during the drying step.

In general, all drying systems, solvent and aqueous, should be properly vented to minimize employee exposure to undesirable vapors. Considering that our understanding of what is undesirable is still limited, and considering that many different solvent residues may be present, it is best to err on the side of caution even when removing seemingly benign solvents or aqueous additives. At the same time, solvent traps may be needed to assure environmental compliance and neighborhood safety.

Those of us experienced in cleaning regularly observe problems associated with solvent drying. Some are equipment related; others are related to employee education and inappropriate handling of components. Inadequate and improper drying in solvent cleaning processes can result in

- Product damage
- Production slow-down
- Solvent loss

With aqueous systems, drying is often optional. In solvent systems, the drying portion of the process is may be so integrated a part of the cleaning process as to be indistinguishable as a separate set of features. It is important that you identify those features relating to drying and then determine which features will be most important in your application. Some features associated with solvent drying include

- Hoists
- Superheated vapor zones
- Freeboard
- Sample rotation
- Fixture design
- Vacuum-assisted removal of solvent

Solvent drying is often thought of as automatic. However, drying is relative. One needs to consider not only "How clean is clean enough?" but also "How dry is dry enough?" The concept of adequate drying is relative to process requirements. Perhaps it is not necessary to dry. In industrial applications, a very light coating of mineral spirits or of a light oil can protect the part from corrosion. In contrast, with other components, particularly those containing plastics and those having complex geometries, you need to consider the consequences of not removing solvent. Residual solvent can interfere with subsequent process steps such as coating and physical assembly.

Consider the consequences of outgassing.[2] Outgassing can be thought of as a subtle form of solvent incompatibility or cleaning agent residue. Minute crevices in components can entrap solvent. Plastics and epoxies can adsorb solvent. The extent of solvent adsorption depends on the material, the chemical, and the cleaning process conditions (such as temperature, pressure, duration of cleaning). At ambient temperature, the solvent can be released over a period of days, weeks, or even years. It is particularly critical to avoid outgassing in sealed systems, which are expected to last for decades. In a sealed gyroscope, released solvent may chemically react with the flotation fluid, producing a medium that corrodes delicate coil windings, resulting in unexpected, catastrophic product failure.[3] One can think of other sealed, critical systems, such as pacemakers where neither the doctor nor the recipient wants to deal with catastrophic product failure.[4,5] Even in nonsealed systems, for biomedical implantables and other critical applications, avoiding solvent outgassing would seem to be a reasonable policy on general principles. Drying is particularly critical for medical devices. Where the product comes in contact with tissues or introduces materials into the bloodstream, residue must be minimized; and that includes not only thin film and particulate residue but also vapor phase contaminants. One also might want to avoid outgassing where the part is sealed or bagged. If metal parts with residual water are placed in plastic bags for storage or shipment, they can corrode or discolor. Similarly, solvents can gradually escape or outgass from parts in plastic bags, perhaps interacting with the plastic, perhaps recondensing on the parts—in general, making a mess.

How do you know when there is outgassing? There are analytical techniques such as residual gas analysis (RGA) and head-space gas chromatography (GC). In both cases, a protocol is established to speed up the outgassing process, systematically collect the vapors, and then analyze them. However, only you and your coworkers can make the decision as to where outgassing is an important factor. If so, you need to dry. As with removal of water, drying to avoid outgassing can be accomplished by both chemical and physical methods. You may decide to use both methods.

Removal of solvent can be accomplished by chemical methods. Isopropyl alcohol (IPA) is sometimes used to displace water. IPA can also be used as a second rinsing and drying step with high boiling solvents such as those based on mineral spirits, ester blends, long chain alcohols, and terpenes.[4] However, for some applications, with very complex components, particularly those containing complex composites and plastics, IPA may leave the component wet, for the purpose of the process under consideration.[6,7] In such cases, the IPA would classically have been displaced by a perfluorinated material (PFC). Today, because of concerns about using materials with a long atmospheric lifetime, hydrofluorocarbons (HFCs) or hydrofluoroethers (HFEs) are preferred over PFCs in such applications. The chapter in this book by Mouser describes a bi-solvent system in which a high-boiling-point, bio-based solvent is followed in sequence by an HFE.

Solvent drying (i.e., removal of excess solvent) can also be accomplished by physical methods, including spin-drying and vacuum, and vacuum with heat. In manual systems, as might be used with very low-volume, high-precision processes, vacuum bakeout in a small oven can be useful. In larger airless or airtight cleaning operations, the vacuum drying system may be an integral part of the cleaning system.[8] Appropriate vacuum drying can complete the process of cleaning under vacuum. In all cases, the process must be carefully adjusted to the product in question and to the total volume of parts to be processed. Where low flashpoint solvents are involved, the drying system must have appropriate engineering controls to prevent fires or explosions.

In liquid/vapor phase cleaning or degreasing systems (classic open-top degreasers), the components must be held in the vapor zone and then above the vapor zone to achieve adequate final cleaning, rinsing, and drying. It is critical to handle solvent properly to avoid solvent loss. In addition to technical considerations, there are both regulatory and economic factors related to solvent drying. For certain solvents, notably in the Federal Halogenated Solvents NESHAP,[9] the dwell time and rate of removal are spelled out. Local regulations may also call out certain requirements to avoid emissions of air toxics (HAPs) and volatile organic compounds (VOCs). The reader is reminded to explore the fascinating, often contradictory and Byzantine world of solvent control as it applies to the regulatory climate in his or her location.

Even for relatively unregulated solvents, it is important to remember that cleaning agents are not free. Some of them cost over $15/pound. Particularly with large-scale operations, containing the solvent can be an important economic consideration even for relatively inexpensive solvents. Even in well-designed, automated systems, assuring complete drying of the parts is a critical part of solvent containment.

Fixturing

Appropriate fixturing and positioning of the components cannot be overemphasized in achieving adequate drying. This applies to aqueous, semiaqueous, and solvent systems. If the parts are not adequately exposed to the drying agent or the drying environment, there will be problems. In aqueous systems, if the parts are not adequately exposed, water can recondense on the parts, and drying may not be achieved within a reasonable time.

Designing the fixturing system is very process-specific, and the difficulties are not limited to aqueous processes. Time and again, I have observed that in the first trial run of even very well-contained, sealed solvent systems with vacuum drying, after processing the first batch of product, visible solvent is present. More pointedly, when I remove a clean part for inspection, a puddle of solvent may land on my shoe. This does not mean that the system is performing poorly. Often, it means that the parts have to be better positioned so that the solvent will flow out of the part. Part rotation can be helpful in this regard.

Some of the factors may be summarized as follows.

Fixturing Factor	Positive Impacts on Drying	Provisos
Basket design, large proportion of mesh or screen	Adequate exposure of parts to heat, forced air, or drying solvent. Avoid trapping water or drying agent	Large, solid, non-mesh, metal areas of fixtures may trap water or drying agent
Materials of construction	Able to withstand heat. Must have long-term compatibility with chemical drying agent	Exercise care in adapting existing baskets to new drying process
Size, strength relative to product load	Adequate size allows flexible processing	Overloading baskets can mechanically damage fixturing, produce inadequate drying
Part rotation	Achieve thorough drying. Decrease drying time	Parts can be mechanically damaged during rotation, particularly when not immersed in solvent
Sample positioning	Decrease drying time	Improper positioning can cause water or solvent to remain trapped

The Human Factor

Employee education and the input and feedback of production personnel, particularly the technicians, is critical in optimizing the drying systems. This factor is often neglected in solvent drying process. In manual systems, there is the tendency for the operator speed up the degreasing process by dunking the parts in the solvent, yanking the basket out, holding it briefly in the now destroyed vapor zone, and then racing across the floor while rapidly shaking the parts. This phenomenon is popularly referred to as the "dunky-do" approach. To counter this, employee education and, more important, employee training are crucial.[10]

Certainly, good equipment design and systems automation are important. The increasing emphasis on cost-containment and rapid manufacturing tends to counteract even the most thoughtfully designed solvent cleaning system. All too often, this author has observed that a relatively sophisticated cleaning system will be purchased and cleaning and drying times set appropriately. The employees are then "trained," and manuals outlining complex procedures are installed. Then, within a few weeks, there are complaints of solvent loss, solvent odor, and inadequately cleaned parts with pockets of solvent. An investigation usually reveals that someone (typically someone on that infamous third shift—the equivalent of "the butler" in mystery stories) has taken shortcuts.

In aqueous processes, employees may remove the parts from the drier sooner than would be desirable, or even skip the drying step completely. They may contaminate the parts by attempting to speed up drying using air hoses, or they may damage parts by increasing drying temperature.

Typical undesirable shortcuts in both aqueous and solvent systems include

- Speeding up the drying cycle
- Shortening the entire process
- Overloading baskets of parts

The importance of employee buy-in cannot be overemphasized. You may think you have purchased a fail-safe, totally automated system. In my experience, the creative and determined employee can override a fail-safe system and destroy process control in almost as short a time as it takes the average 4-year-old to remove a child-proof medicine cap.

Usually, employees have an immediate reason for shortcutting the system, and it is typically related to total processing time and immediate perceived profits. An employee may intellectually understand that a new aqueous or solvent process will take, for example, 20 min for complete cleaning and drying. If the old process only took 10 min, or if there has been a sudden increase in production volume, that employee may independently or even under direct orders from the supervisor decide to cut back on the process time. More often than not, this means cutting back on the drying time on the grounds that drying is an extra nonessential part of cleaning. Often, in cutting back on drying, they have compromised the most important step. It is critical to ask: are we unwittingly rewarding supposed efficiency at the expense of adequate process control? The only way to combat the problem of shortcutting the drying step is to reward good process control to at least the same extent as we reward rapid production rates.

Conclusion

In summary, drying is a critical, time-consuming portion of the cleaning process for aqueous, semi-aqueous, and solvent cleaning. It is also a part of the process that is overlooked and underfunded on a regular basis. Once the question of how dry is dry enough has been answered, it is important to follow-through with appropriate choices in drying method and fixturing, and then with appropriate on-going employee training and education.

References

1. B. Kanegsberg. Washing, rinsing, and drying: Items to consider for the optimization of your cleaning process. *Metal Finishing Magazine*, September 2005. http://www.metalfinishing.com/view/4194/washing-rinsing-and-drying-items-to-consider-for-the-optimization-of-your-cleaning-process/
2. B. Kanegsberg and E. Kanegsberg. Outgassing. *A2C2 Magazine* (Now *Controlled Environments Magazine*), June 2005.
3. B. Kanegsberg, B. Abbink, K.T. Dishart, W.G. Kenyon, and C.W. Knapp. Development and implementation of non-ozone depleting, non-aqueous high precision cleaning protocols for inertial navigation subassemblies. *Microcontamination'93 Proceedings*, San Jose, CA, 1993.
4. B. Kangsberg, H. Mallela, H. Dominguez, and W.G. Kenyon. Integrating precision de-oiling and defluxing processes in high volume manufacturing systems. *IPC Presentation and Proceedings*, San Diego, CA, 1995.
5. B. Kanegsberg. Cleaning for biomedical applications. *Presentation and Proceedings, Precision Cleaning '97*, Cincinnati, OH, 1997.

6. B. Kanegsberg. Cleaning systems for low-flashpoint solvents. *Precision Cleaning Magazine*, 3(3), March 1995, 21–28.

7. M. Carter, E. Andersen, S.-C. Chang, P.J. Sanders, and B. Kanegsberg. Cleaning high precision inertial navigation systems. A case study and panel discussion. *Presentation and Proceedings, Clean Tech' 99*, Rosemont, IL, 1999.

8. M. Ohkubo. An airtight argument: Vacuum solvent cleaning systems work. *Precision Cleaning Magazine*, 7(1), January 1999, 24–29.

9. National emission standards for hazardous air pollutants [*Federal Register*: December 2, 1994] ENVIRONMENTAL PROTECTION AGENCY 40 CFR Parts 9 and 63 [AD-FRL-5111–3] RIN 2060-AC31. http://www.epa.gov/fedrgstr/EPA-AIR/1994/December/Day-02/pr-184.html. Note: NESHAP updated May 3, 2007, see www.epa.gov/fedrgstr/EPA-AIR/2007/May/Day-03/a7668.htm

10. R. Petrulio and B. Kanegsberg. Back to basics: The care and feeding of a vapor degreaser with new solvents. *Presentation and Proceedings, Nepcon West '98*, Anaheim, CA, 1998.

34

Drying

Daniel J. VanderPyl
Sonic Air Systems, Inc.

A Historical Perspective on the Drying Process

The continuing evolution of manufacturing technology has led to a wide range of purpose-built machinery for automated and semiautomated processes in parts cleaning and drying. Prior to today's prominence of aqueous cleaning technology, those using solvent/chemical cleaning and degreasing processes were unconcerned about drying.

Often, the solvent-based process resulted in cleaning and subsequent evaporative drying simultaneously. The process of chemical cleaning resulted in a clean, spot-free, generally high-quality part, although maintaining product cleanliness was complicated by a continuous need to filter and replenish chemicals. Little was known about the health or environmental effects of such chemical use.

The 1987 Montreal Protocol initiated the phaseout of ozone-depleting chemical cleaning processes, and its signees began a movement taking aqueous cleaning from a cleaning option to a cleaning standard. Drying was an afterthought in the early aqueous cleaning systems. As such, approaches to drying consisted of a wide variety of methods that were often inefficient or ill-suited to the application. As the impact of various drying methods on product cleanliness became apparent, the focus on drying quickly sharpened.

The printed circuit (p.c.) board industry led the way. It pioneered new aqueous cleaning equipment and effective chemical replacement technologies. Throughout the 1980s, product quality in the p.c. board industry was the driving force while manufacturing efficiency was a much lower priority. Today, this scenario has reversed itself in the p.c. board industry along with most other high-volume production environments; process costs and throughput rates are now scrutinized just as much as quality.

Currently, aqueous cleaning systems are effective and able to match the cleanliness of most of the original chemical-based cleaning systems. Customer knowledge of cleaning/drying processes has increased tremendously. Customers have learned that cleaning and drying now go hand-in-hand; you cannot have one without the other. Today, there is more specific drying required than ever before. As always, the specifications of the drying system are intimately connected to the needs of the subsequent manufacturing steps whether they are parts inspection, component assembly, painting, marking and coding, or packaging.

Equipment manufacturers have developed machinery and technology to match these increasingly stringent cleanliness requirements. The U.S. military specifications (mil-specs) were a driving force in many of these refinements, which then spurred equipment manufacturers to identify drying as the weak link to total product cleanliness.

With the new focus on total product cleanliness, process costs soared. Drying systems were cumbersome, operating costs were high and throughput capacity was poor. As the world economy continued to evolve and prices in the technology sector (i.e., computers and communications equipment) dropped, competitive forces dictated that improvements in manufacturing efficiency were an integral part of tomorrow's profits. Thus evolved a menu of drying options. The secret of their best use lies in matching the appropriate technology to the cleaning method to attain the desired levels of cleanliness and productivity at the least cost.

Environmental Stewardship Can Be Both Green and Lean

One hundred and ninety-five countries signed the 1987 Montreal Protocol and agreed on methods to reduce and/or eliminate chemicals, which could harm the Earth's atmospheric ozone layer. So began the worldwide realization that we must minimize man's impact on our planet and the use of the finite resources that we all share.

The reduction in each country's CO_2 emissions, particularly those resulting from fossil fuel power plants, is another means of protecting our environment. In developing the most efficient parts drying methods, we can be ecologically responsible by reducing the total energy demand while simultaneously reducing the per part processing costs.

Today the challenge is how to remove water-based residuals from the surface of parts with an endless array of shapes, sizes, and throughput rates. Because compressed plant air is a utility in most factories, using it to blow water off parts seems a natural thing to do. Likewise, electric, natural gas-fired, and stream-fed heaters are also viewed as logical ways to dry water remaining on parts. However, selecting the method that is both the best use of your capital equipment investment while yielding the lowest energy and labor costs per part being dried requires an understanding of drying options as well as an understanding of which options can best fit your process requirements. So now let us explore the options.

The Definition of Drying for Aqueous Parts Cleaning

There are two primary categories of parts drying technologies for aqueous cleaning systems. The first is chemical displacement drying where typically organic solvents are used to displace water from a component. The second is physical drying, which is the topic of this chapter.

For the purposes of this chapter, we will focus on the most widely used methods of physical drying, i.e., those that use an exchange of air throughout the component by various means to blast, strip, or evaporate moisture from the exterior and interior of a component.

Air is the primary element interacting with the moisture on the surface of the component. We therefore define air drying as any mechanical action to remove water/moisture from the surface that does not rely solely on natural evaporative drying. In the following pages, we will delineate several types of air drying systems with the related applications and examine the impact of component design on dryness standards, production rates, operating costs and worker safety.

Charting the Best Drying Method

In order to select the most suitable drying technique for various types of parts, a chart is provided as a general drying guideline (Figure 34.1). Since it is very difficult to factor in all of the product variables, such as complexity of part geometries, surface finish or material of construction, the chart is based on simple smooth metal parts. With the Y-axis representing the total surface area (L × W × H) of each part and the X-axis being the quantity of the parts being dried per hour, the letter codes at each intersect point represent the most commonly used drying method.

Please keep in mind that some products (not represented in the drying graph) such as flat sheet material, continuous strip or web products and the large internal surface area of mixing tanks and other vessels, may have aqueous wash cycles followed by a drying step. One or more of the following drying techniques may be employed, but again, the total surface area to be dried and units per hour will dictate the drying methods that can be used.

The letter codes for the various drying techniques are:

SD, spin drying
DD, desiccant drying
FA, forced air/no heat
FH, forced air/with heat
HC, high-velocity compressed air
HB, high-velocity blower air

Individual part size – total surface area (IN2)													
5000	HC	LV	LV	HC	RH	HB	HB	HB	HB	HB	HB	HB	HB
2000	HC	LV	LV	LV	RH	HB	HB	HB	HB	HB	HB	HB	HB
1000	HC	LV	FH	FA	RH	RH	HB	HB	HB	HB	HB	HB	HB
500	HC	FH	FH	FA	RH	RH	HB	HB	HB	HB	HB	HB	HB
200	HC	FH	FH	FH	RH	RH	HB	HC	HC	HB	HB	HB	HB
100	HC	HC	FH	FH	RH	RH	HC	HC	HC	HB	HB	HB	HB
50	HC	HC	HC	RH	RH	RH	HC	HC	HC	HC	HB	HB	HB
20	CA	HC	HC	RH	RH	RH	FH	HC	HC	HC	HB	HB	HC
10	CA	CA	HC	HC	RH	RH	FH	FH	FH	HC	HC	HC	HC
5	CA	CA	CA	HC	HC	HC	FH	FH	FH	FH	HC	HC	HC
2.5	CA	CA	CA	CA	FH	FH	FH	FH	FH	FH	SD	SD	SD
1	CA	CA	CA	CA	CA	FH	FH	FH	FH	SD	SD	SD	SD
	1	10	25	50	100	200	500	1000	5000	10000	15000	20000	25000

Drying volume – parts/hour

The above chart represents the most common drying technique used and assumes part geometry to be moderate complexity with no delicate handling of the parts required.

Legend:		
CA - Compressed air spot drying	HB - High-velocity blower air	RH - Radiant heat
FA - Forced air no heat	HC - High-velocity compressed air	
FH - Forced air with heat	LV - Low-velocity air	

FIGURE 34.1 General guidelines for drying technique selection.

LV, low-velocity air
RH, radiant heat
CA, compressed air spot drying
VC, vacuum chamber

Types of Physical Drying Techniques for Aqueous Cleaning Technology

The following drying systems will be covered in this chapter:

- Centrifugal spin drying
- Desiccant bulk drying
- Forced air drying/without heat
- Forced air drying/with heat
- Forced air drying/without heat
- High-velocity air blow-off (compressed air and blowers)
- Low-velocity air drying
- Radiant heat
- Spot drying (vacuum hoses and compressed air blow-off nozzles)
- Vacuum chamber drying
- Drying mixing tanks, vessels, and blenders

There are many good options to achieve efficient drying. Making the optimal choice is difficult. Just as there is no ideal, one-size-fits-all cleaning technique, the drying system must be suited to the components and to the overall manufacturing processes. Therefore, we are indicating the pros and cons; as well as applications and misapplications.

Centrifugal Spin Drying

Centrifugal spin drying involves high-speed centrifugal spinning of up to hundreds of batched components to remove moisture from component surfaces following cleaning, cooling, and plating. Commercially available spin dryers are rated by loading capacity and range from one to several hundred pounds per basket. Components are transported in a basket and either manually or automatically hoisted from the final rinse process to the centrifugal spin dryer. Cycle time can last from 2 to 20 minutes, depending on component complexity. At speeds up to 1500 rpm, care must be taken not to damage precision or delicate components.

Ideal Application

Centrifugal drying is best suited to

- Components of less than one cubic inch in size
- Simple geometry
- Few if any blind holes or crevices
- Not requiring critical drying

A typical example is parts on plating lines.

Misapplication

The technique is not well suited for drying complex machined components requiring absolutely dry and spot-free surfaces.

Advantages

Advantages include low operating cost and a small footprint for batch applications with adequate dwell time at the drying step. In addition, the noise level is low.

Disadvantages

On the other hand, because baskets must be constantly loaded and unloaded, the method is relatively labor intensive. With heavily loaded baskets, there is the potential for worker injury. As production rates increase, the tendency is to load baskets more and add more spin drying units, rather than upgrade drying technology.

Impact of Component Design

Centrifugal drying is more likely to become ineffective for parts with complex shapes or with densely packed batches. It can only be used for durably designed components.

Drying Effectiveness

Component geometry influences drying effectiveness in centrifugal spin dryers more than most other drying methods, and moisture retention can be a problem. Forced air heating may be used to supplement drying capacity.

Drying Impact on Cleanliness

The potential for component damage makes spin drying unsuitable for precision components. Clean components wearing against one another can result in surface damage, deformation, or dislodging of particles, thus defeating the cleanliness requirement of more critical applications.

Equipment Menu

The primary equipment required is a spin dryer (0.5–7.5 HP). Larger dryers equipped with heated-air circulation are also available. Ancillary equipment includes an automated or semiautomated hoist.

Associated Energy and Labor Requirements

Loading and unloading time can make labor costs significant. The number of components per batch impacts the energy cost per component dried. As with all forms of drying, there is an optimum point of effectiveness with the spin dryer; overloading the spin basket leads to diminishing drying effectiveness and increased per part drying costs.

Safety and Environmental Issues

The greatest potential issue for employees is the physical handling of baskets of components, both during loading and unloading of the chamber and removing dry parts.

Desiccant Bulk Drying

Desiccant bulk drying is one step better than nature's own evaporative process. Desiccant material draws moisture from ambient air and is often used as both a curing and a drying process. Storage in a large chamber or room with desiccant material would be the final drying step to achieve as close to zero moisture content as possible. Frequently, desiccants are used for removing moisture from compressed air sources and in large batch drying of agricultural products. It is the least often-used method to dry manufactured parts.

Ideal Application

Desiccant drying is ideally suited for products where a moisture content of 1%–5% as a portion of total weight is needed for maximum part stability or shelf life and where heat or other aggressive drying methods may be detrimental to the parts. For example, any assembly with a latex component may benefit from desiccant drying.

Misapplication

Desiccant drying is not appropriate for continuous manufacturing processes where drying cycle time must be kept to a minimum. It is not suitable for applications where absorbed moisture is not the issue.

Advantages

It is much easier to control the uniformity of dryness among all the components with desiccant drying than with radiant heat or forced air drying. Radiant heat and forced air create different rates of drying within the same batch or process.

Disadvantages

The drying chamber typically has a significant footprint, the cycle time is long, and labor costs associated with sample handling are high.

Impact of Component Design

Desiccant drying performance is impacted by the porosity of the product. The more porous, the longer it will take to draw out the moisture. Typically, the goal is to reduce the moisture content of a given part to between 1% and 5% of moisture as a percentage of the component's weight.

Drying Effectiveness

The level of drying using this method is measured by the humidity in the chamber. Desiccant drying in large bulk processes makes it easy to measure the level of retained moisture.

Drying Impact on Cleanliness

Outside elements that could compromise the level of product cleanliness are not usually a factor. The cleanliness of porous products has more to do with the forming of the product, and water may be part of that process. For example, in molded parts such as latex, water is an integral part of product molding and additional cleaning is not a required process.

Equipment Menu

The primary equipment consists of a sizeable container or room, from 1 to 1000 ft^3 in volume. Ancillary equipment includes a fan for air circulation and a dehumidifying unit.

Associated Energy and Labor Requirements

Throughput is lowest with desiccant drying. Labor rates are among the highest because product must be manually moved in and out of the chamber or room.

Safety and Environmental Issues

Desiccant drying itself poses relatively few worker hazards. Hazards associated with product composition are a potential problem.

Forced Air Drying (without Heat)

Forced air drying (without heat) is based on the exchange of large volumes of air through an enclosed zone in order to extract ambient moisture. It is an accelerated evaporation process that can also use heated air. A dehumidification unit can shorten the drying cycle.

Ideal Application

The most effective use is with simply designed components that have surface and/or core temperatures higher than ambient. By drawing away air with forced air drying, humidity, and moisture are removed from all but the surface encouraging more evaporation. Without air circulation, such a heated product would make the whole chamber humid.

For example, in the tire industry, using a Banberry rubber extrusion process, a continuous ribbon of rubber product exits at 350°F. It is then immersed in a cold water quench. Forced air accelerating the evaporative process can help remove the water on the rubber strip.

Misapplication

Forced air drying is not well suited to batch drying applications where part configuration and part stacking result in numerous water pockets within the batch. In such situations, the drying time may increase.

Advantages

Initial capital outlay is minimal as no specialized equipment is needed. Noise levels are low and environmental impact is minimal.

Disadvantages

The throughput rate for forced air drying without heat is low; and the drying zone may be large, resulting in a large footprint. There can be problems in process control. For one thing, there can be difficulty in controlling drying effectiveness where ambient temperatures may fluctuate. In addition, product quality may be adversely impacted due to the potential for part recontamination from unclean air.

Impact of Component Design

Typically, forced air drying is used only for simply designed components at low production rates.

Drying Effectiveness

Forced air drying is effective for industrial drying. For example, it can dry rubber products well enough to eliminate potential voids in the stamping or molding phase of production. However, forced air drying without heat is ineffective for parts with complex surfaces requiring a high drying standard.

Drying Impact on Cleanliness

Forced air drying can contaminate the component if the air source is unclean. However, cleanliness is a function of the total manufacturing process, and forced air drying is not used in applications involving critical cleaning.

Equipment Menu

The primary equipment consists of a low-pressure fan assembly adjacent to the product, most commonly an axial or box fan simply blowing large volumes of air across the surface. Ancillary equipment includes a dehumidification unit to decrease drying time.

Associated Energy and Labor Requirements

Where the process is conveyorized (as in most applications of this method), labor costs are low. However, forced air drying may also be used for batch cleaning, where labor costs may be significant. Energy costs are low since blowers use much less energy than heaters.

Safety and Environmental Issues

Worker safety and environmental concerns are minimal, since humid air can be expelled with a roof mounted exhaust fan. Noise is typically not a problem.

Forced Air Drying (with Heat)

In such systems, components are indexed into a drying zone using a circulating fan that warms air up to 200°F above ambient temperature. The heat source is most often electric, but natural gas, heat, and waste steam are also used. Overhead conveyor systems also use forced air drying with heat prior to electrostatic powder paint zones.

Ideal Application

Forced air drying with heat is the method commonly used in batch ultrasonic cleaning systems. Using sliding beam systems, baskets of components are transferred into drying chambers after the final rinse. The forced air with heat provides an accelerated evaporative drying process.

Misapplication

Misapplications of forced air drying are related to the thermal stability of the component as well as to required process time, process flow and part handling/worker training problems. In a continuous conveyorized process, forced air drying with heat can be either a benefit or detriment depending on what happens after the drying process. The time taken to stabilize the parts at room temperature is the price paid to dry those parts with heated forced air. Attempts to speed up the process can cause problems. For example, in a batch-type paint process of cleaning/drying/painting, components are often manually handled in inspection or assembly, and high-temperature parts may burn workers. In addition, increased drying temperatures cause thermal expansion of the component, pushing a precision machined part out of inspection tolerances.

Advantages

Forced air drying with heat is versatile and can be used over a wide range of part and component sizes. Because the air is mixed evenly, the method results in uniform drying. Compared with desiccant drying, it can dry relatively large batches of parts more effectively in a much shorter period of time. In some applications, the waste heat from other processes can be used to create steam for forced heat drying.

Disadvantages

The drying cycle time is directly proportional to variations in batch size and component orientation. If an elevated core temperature is required, drying time and cost will increase. Parts may become too hot to handle in the next step of production. Heat may expand components beyond inspection tolerances. Heat may bake on contaminants depending on the contaminant level in the final rinse stage.

Impact of Component Design

Particularly in ultrasonic batch-cleaning systems, consistency of batch size, component geometry, and orientation in each basket must be consistent for consistent results. Complex, ornate components with blind holes present the most difficult challenge in maintaining drying/cleanliness consistency.

Drying Effectiveness

Drying quality will be affected for both batch and inline continuous processes if there are any thermal constraints related to the product and/or variations in component design, volume of parts per batch or drying rate.

Drying Impact on Cleanliness

Forced air drying with heat results in accelerated evaporation of moisture from components. Any particulate suspended in the surface liquid will ultimately be baked onto the surface of the component as it is dried. Therefore, the water quality of the final rinse zone in terms of dissolved contaminants and the filtration needed to maintain water quality and eliminate particulates is integral to final product cleanliness.

Equipment Menu

The primary equipment required includes a squirrel cage fan/blower (high volume, low pressure) and an inline duct heater. The equipment is predominantly electric. However, natural gas may be used in a continuous conveyor system. HEPA filters are commonly used for precision cleaning applications. For batch processes, the drying zone is typically an enclosed chamber placed subsequent to the ultrasonic cleaning and rinsing tanks.

Associated Energy and Labor Requirements

Process costs depend on the drying cycle required for each component. Process costs for batch processes can be considerable. For example, running 10 batches a day over an 8 hours day requires a fan and heat source. The lowest temperature likely to result in effective drying would be approximately 200°F. Let us assume that production levels were to increase to 20 batches a day. In order to achieve the same level of drying, either the temperature would have to increase, or the number of drying chambers would have to be doubled. While the energy consumed per component dried might not increase, total energy usage would be likely to increase considerably.

In addition, for batch processes, labor costs are incurred for loading and unloading. Lag time for cooling may add extra handling and labor costs. Inline applications generally reduce labor costs. However, conveyor length and lag time in the production flow may involve added costs. The throughput rate can be anything desired within the limits of the temperature constraints of the parts. The utilizable temperature becomes the limiting factor.

Safety and Environmental Issues

Workers must be protected against burns from the heat chamber and from heated components. Environmental impact can vary according to the application.

High-Velocity Air Blow-Off (Compressed Air and Blowers)

A high-velocity air blow-off is any air stream directed at the surface of a product to create a sheer force, which strips liquid from the product. It is one of the most widely used drying methods. The air source can be generated from high-pressure plant air systems, compressed nitrogen or self-contained blower systems. High-velocity air streams can generate static charge. While the moisture being removed from the part counteracts this tendency, static build-up in parts drying applications is of concern.

Compressed Air

Compressed air systems are defined as drying systems using a compressed air generator with a minimum of 10 psi (pounds per square inch) and supplying air through air knives and nozzles.

Ideal Application

Compressed air is best for the small-scale drying of parts. The approach is practical for parts of less than 6 in.^{2} of cross-sectional area and travelling single file at less than 5 ft/min.

Misapplication

Compressed air, when used to dry components greater than 36 in.^{2} of surface area on any one side, will generally result in very high operating costs. Also, components with critical cleanliness requirements must be protected from oil and condensation produced by the compressed air system.

Advantages

Compressed air systems of adequate capacity are often already in place. Equipment is available from a range of suppliers. The systems consume relatively little space. Air knives and nozzles are compact; piping systems are generally less than one inch in diameter.

Disadvantages

Energy consumption can be up to 75% more than with blowers. Because most compressors are oil-lubricated, filtration of the air stream is required to prevent recontamination of the parts. In addition, compressors condense liquid from the air stream, resulting in part contamination. Finally, compressed air produces a low-frequency noise that is audible at considerable distances from the blow-off zone. Workers may find the noise uncomfortable or unacceptable.

Impact of Component Design

Component design and fixturing considerations during cleaning and drying impact the choice of compressed air versus a blower system.

As a rule of thumb, compressed air can be effectively used for drying parts measuring less than 6×6 in. of cross-sectional area with blind holes and crevices that measure less than ½ in. diameter with hole depth a minimum 5× hole diameter. Components with large, smooth surface areas and simple geometry are poor candidates for compressed air, as operating costs will skyrocket.

Drying Effectiveness

The ability of air to strip moisture from complex or critical surfaces relies on the air having a straight path to the respective surface or area. A part can be easy to dry, but if parts are stacked 10 on a rack, air flow and effectiveness are restricted. Moisture condensation is a potential problem.

Drying Impact on Cleanliness

Compressed air may carry particles of oil and dirt, which would compromise precision cleaning. Also, condensation from the high-pressure air can recontaminate the parts with moisture. Therefore, compressed air is typically used for industrial processes, not for precision cleaning.

Equipment Menu

The primary equipment consists of an electrically operated dryer, an oil separator, a piping system, a compressed air knife or nozzles, available plant air, and a receiver tank, which is placed as close to the blow-off point as possible. If the drying process is not continuous, this tank can be used as a reservoir for compressed air. Ancillary equipment includes an enclosure, tunnel or chamber where blow-off can take place to reduce ambient noise and provide an opportunity to exhaust moisture-laden air as the product is dried.

Associated Energy and Labor Requirements: Additional Cost Considerations

A wide range of compressors and accessory items, nozzles, knives, and other blow-off devices are available, but the end user is generally left to decide how many of what configuration is required.

It is important to note that compressor costs are commonly higher than for a blower system and use approximately 75% more energy. The decision on whether to use compressors or blowers must be made on a trade-off basis considering existing resources and return on investment (e.g., is sufficient compressed air already available?). In general, if you have neither blowers nor compressors, a blower will generally cost less in terms of initial cost and operating cost.

In addition, for more exacting processes, while compressed air is available with very stringent air drying capabilities, in most cases the cost of filters, separators, oil filters, and energy use is prohibitive.

Finally, for larger parts or precision cleaning, where workers must carefully and individually clean parts, added labor costs must be considered.

Safety and Environmental Issues

Noise is a problem with compressed air. It generally operates at a lower frequency range than high-velocity air blowers, and this allows sound levels to travel greater distances than blower system air sounds. For example, at 100 ft, a blower system might read in the high 70 db (decibel) range. A compressed air nozzle at 90 db, located 100 ft away, would still generate 85 db. The high-frequency whistling noise that occurs as air is blown off surfaces causes random spikes of impact noise that often exceed 100 db. The cost of the materials capable of absorbing lower-frequency sound levels is generally very high.

Small bits of debris blown from blind holes can be a worker safety problem, and workers should wear protective goggles. Sound levels often require hearing protection. Contaminants may be blown into the environment, and condensation can blow into associated work areas causing slippery floors or fouling of work sites. If the systems cannot be adequately enclosed or ventilated, operators must also be protected from breathing atomized contaminants.

Blowers

High-velocity air blow-off can be achieved using a dedicated blower producing between 0.5 and 5.0 psi, with the air directed through air nozzles or air knives. Blower systems together with infrared lamps are the most widely applied method of drying components in conveyorized cleaning and plating systems. In addition, high-velocity air is often used for critical cleaning applications such as electronics assembly and medical devices.

Ideal Application

High-velocity air blow-off with air blowers is most effective with any component size greater than a 6×6 in.^{2} surface area, having simple to moderate surface complexity, and for production rates greater than 100/h.

Misapplication

Blowers are not suitable for extremely small components, complex geometries and blind holes, low production rates, and short cycle times. Process equipment design may restrict nozzle or air knife access, necessitating small piping and unacceptably high pressure. However, adding heat can increase drying capability.

Advantages

Blowers can be sized to accommodate a wide variety of components and production rates. High-velocity blowers provide the most energy-efficient of blow-off air methods. The heat of compression assists in the drying. Air can be supplied oil- and moisture-free, and self-contained air delivery without air pressure fluctuations is achievable.

Disadvantages

Blowers cannot remove liquid from complex parts with unexposed surfaces. Exhaust air requirements are much higher than for compressed air. Large 2 and 3 in. diameter feed and connecting piping for air knives and nozzles take up additional space.

Impact of Component Design

Small intricate components and those with blind holes more than 3× the depth of the hole diameter are not dried effectively.

Drying Effectiveness

Compared to compressed air, the greater range of air volume that can be used in blowers allows greater versatility.

Drying Impact on Cleanliness

Sealed bearing blowers can deliver oil-free air. However, any oil-lubricated blower is vulnerable to seal failure resulting in the contamination of the entire cleaning system. On a positive note, because the high-velocity strips away all moisture, the system does not create enough pressure to produce condensation that might re-contaminate the part. In addition, use of high-impact air effectively eliminates baking on of particulates entrained in surface liquids.

Equipment Menu

The primary equipment consists of a self-contained motor assembly blower, an air filter/silencer assembly (to filter out ambient air and dirt particles) and air knives. Ancillary equipment includes electric inline heaters, a recirculating blower with piping and filters and exhaust fans connected to drying chambers.

The option of a recirculating air blower should be considered to cut process costs and to avoid contamination. Air introduced into a drying chamber must be properly exhausted to avoid introducing heat, moisture, and contaminants into the surrounding work environment. The heat introduced can be comparable to leaving a door open in a factory all day long.

Associated Energy and Labor Requirements

High-velocity air blowers are the most efficient means of forced air blow-off. While little labor is associated with the drying process, complex parts may require spot blow-off following exit from the drying chamber.

Throughput rates for blower systems are generally high and are based on matching blower size to throughput demands. Assistance of an experienced equipment vendor can be very helpful in giving specific application advice in designing the system and evaluating the appropriate size and/or number of blowers to support demand.

Safety and Environmental Issues

Noise with high-velocity blowers can be significant. The blower system operates at the higher end of the audible frequency range, i.e., 90 db. But at 100 ft, that same blower system might read in the high 70 db range (every point is a multiple of 10, therefore, there is a tremendous difference between 85 and 80 db). Proper enclosure can reduce sound impact.

In-line heated air or recirculated blower air can increase surface temperature of the components being dried. Subsequent handling by operators must be considered.

If air from the blower is not recirculated, an exhaust system must be installed. Controls depend on specific regulatory requirements.

Low-Velocity Air Drying

Low-velocity air knife drying systems involve large volumes of air introduced through an air knife plenum exiting at velocities no greater than 10,000 fpm. Although initial velocities are low, because of the wider air path and greater volume of air, the drop in pressure is lower compared with other systems. This effectively results in a longer air path.

Ideal Application

The ideal application for low-velocity air drying is conveyorized cleaning of components with simple geometries and with low throughput rates.

In the early days of aqueous cleaning, parts cleaners frequently used low-velocity air drying systems. They worked well because the component geometry was simpler and throughput rates and conveyor speeds were slower. The low velocity was effective for many applications, as the dwell time within the drying zone was correspondingly much greater than with the higher production speeds of today. Today, higher impact velocity will compensate for higher throughput rates by creating greater sheer force for more effective drying despite shorter dwell time.

Misapplication

Despite advances in the technique, low-velocity drying is not readily adaptable to parts with complex geometries. Low-velocity air drying is unsuitable for high-speed throughput of anything but the smoothest of surfaces.

Advantages

Advantages include low initial capital outlay, low operating costs, low maintenance, and low noise levels.

Disadvantages

The method is slow and is ineffective with complex parts. Equipment size is large, particularly relative to drying capacity.

Impact of Component Design

Simple component designs are best suited to low-velocity air drying.

Drying Effectiveness

Low-velocity air drying is effective for simple parts at low speeds. It can work effectively at higher throughput rates if critical levels of drying are not required or with supplementary inline heating.

Drying Impact on Cleanliness

The air from both low velocity fans and high velocity blowers is only as clean as the filter which is used. Although 10μ air filters are the norm for the inlets of most fans and blowers, high efficiency particle air (HEPA) filters at 99.97% efficiency must be used for any critical cleanliness application. HEPAs come in many shapes and sizes, but the only ones that "really" work are the inline canister type mounted between the fan/blower and the air knives. The air pressure drops can hurt drying performance, so make sure the HEPA is rated for the air flow and with low pressure loss.

Equipment Menu

The primary equipment is a centrifugal fan assembly (1–10 HP), an air distribution manifold, flex-hose and piping, and an air knife plenum. Ancillary equipment includes a shroud enclosing the air knife assembly with an optional hood to draw moisture-laden air from the air knife zone.

Associated Energy and Labor Requirements

While total energy consumption of low-velocity air knives is low, energy usage per component tends to be comparable to higher velocity systems with higher production rate equipment.

In addition, with supplementary inline heating, low-velocity air drying then becomes an energy efficiency problem when compared to other types of technology and may become an issue.

Safety and Environmental Issues

Even with the low-velocity system, undesirable aerosols must be vented and, if required to meet environmental regulations, appropriately trapped.

Radiant Heat

Radiant heat is the process (typically inline) by which a heat source (generally infrared tube lamps, IR) is used to flash dry very thin layers of moisture on the surface of components. It is often used after high-velocity systems, which remove the bulk of the liquid. In some batch processes, it is used for final drying following forced air drying.

Ideal Application

The ideal application is for final drying of parts with complex geometries, where air knives may not completely dry the parts.

Misapplication

Infrared heat is inappropriate as the primary or sole drying source for products having more than a trace amount of liquid residue.

Advantages

Advantages include relatively low capital outlay and the absence of noise or environmental concerns.

Disadvantages

When used without initial drying to remove gross moisture, operating costs and cycle time can be high. The footprint for exclusive IR drying can be high. Dissolved or suspended contaminants can bake onto the surface, and excessive heating can result in throughput, handling and inspection problems.

Impact of Component Design

Complex shapes and surfaces with close dimensional clearance between components in an assembly, i.e., circuit boards, are best suited to the IR drying method. IR provides better control and greater simplicity.

Drying Effectiveness

An IR heater element, like any heater element, is either on or off. The cycle time of the element determines the amount of radiant heat produced. It can be difficult to maintain consistent heating.

Drying Impact on Cleanliness

When IR is the second stage of drying to an air knife system, most contaminants entrained in the surface liquid are carried away with the air velocity. However, if the percentage of surface moisture is more than 5%, or, the level of entrained contaminants is high, the likelihood of undesirable residue increases.

Equipment Menu

Primary equipment consists of IR heater tubes in a shielded or shrouded chamber with insulation to minimize surface heat on the exterior of the chamber along with a heater control system to ensure even element cycle times. Ancillary equipment consists of initial drying with fans, blowers, and/or air knives.

Associated Energy and Labor Requirements

IR heater elements used as the only drying method are likely to result in excessive electrical demands. But as a second stage, it can be a very energy-efficient drying method for complex components.

Automated processes are available.

Safety and Environmental Issues

If components are raised above 125°F, part handling issues must be considered. If preceded by adequate rinsing, there are no obvious environmental issues related to IR drying.

Spot Drying (Vacuum Hoses and Compressed Air Blow-Off Nozzles)

A wide range of products have drying requirements for specific areas of the product but do not necessarily require a completely dry part. In such cases, capital outlay may not be justified relative to labor costs.

Vacuum Hoses

Localized vacuum drying is a manual process used for very low production rates or for component inspection where only certain portions require drying. A vacuum drying system could be a central system or a dedicated vacuum unit.

Ideal Application

The ideal application is for simple components in industrial grade cleaning.

Misapplication

Examples include parts, complex parts and blind holes where air cannot be exchanged adequately to draw liquid to the vacuum source or critical applications where contact with the vacuum nozzle could compromise cleanliness.

Advantages

Advantages include low equipment and energy costs. The technique may avoid scattering of large amounts of moisture and contaminants. In addition, it may be possible to combine the drying and inspection steps.

Disadvantages

Contact with the vacuum hose and/or bristled pick-up nozzles may result in contamination. Process control is difficult.

Impact of Component Design

Localized vacuum is ineffective with complex components.

Drying Effectiveness

Process control is operator dependent; effectiveness is variable. Localized drying is most commonly used in quality control operations adjunctive to critical inspection of machined components.

Equipment Menu

Equipment consists of vacuum systems or portable vacuum units.

Associated Energy and Labor Requirements

Vacuum units require manual handling of each component. With low production rates, labor costs may be acceptable. The energy consumption for a vacuum is 1–3 HP; energy per component is highly variable. A vacuum source running continuously with sporadic parts spot drying may be more costly to operate than an on-demand compressed air nozzle.

Safety and Environmental Issues

When using a portable vacuum, residual chemicals may cause problems as potential fire hazards.

Spot Drying, Compressed Air Blow-Off Nozzles

Compressed air blow-off is the most common spot drying method due to the availability of compressed air and its ability to blow moisture from confined spaces.

Ideal Application

Spot drying with air blow off is uniquely suitable for blind holes or crevices where moisture is randomly trapped. The technique is used for low-volume, intermittent production and for inspection.

Misapplication

The most common misapplication is where compressed air nozzles are operated by line workers repeatedly for a non-variable product line. These applications are much better suited to high-velocity blow-off systems.

Advantages

The small hand-held nozzles are very maneuverable for working with complex parts.

Disadvantages

The equipment is noisy and can scatter moisture and contaminants. Process control is difficult.

Equipment Menu

Equipment consists of localized blow-off nozzles or custom nozzle and tube assemblies connected to a compressed air line.

Associated Energy and Labor Requirements

The method is labor intensive. On-demand use of compressed air generally minimizes the power requirements of the compressed air system applied to spot drying, unless the blow-off nozzles are operated on a continuous basis.

Safety and Environmental Issues

The potential for debris or moisture to be sprayed throughout the work area poses potential worker exposure issues (eyes, inhalation). Scattered contaminant on floors may produce slipping hazards. The equipment can produce intermittent, unpleasant noises.

Vacuum Chamber Drying

Vacuum drying is the process by which components are placed into a sealed chamber where a vacuum is pulled on the component to lower the water vapor point. The moisture becomes an aerosol in the chamber. A filtration system pumps the moisture from the air, returning dry air to the chamber to repeat the process.

Ideal Application

Vacuum chamber drying is best suited to complex geometries and to porous metal components that must be completely dried, as prior to epoxy resin impregnation. It is also useful for metals that must be completely dried to prevent corrosion.

Misapplication

Where above considerations are not a concern, vacuum drying may not be the most efficient or rapid approach.

Advantages

Because high heat is not required, vacuum drying is less likely to damage temperature-sensitive components. Also, energy costs may be less for critical metal components.

Disadvantages

Equipment costs are relatively high with manual systems, component handling (loading), increased labor costs, and process times.

Drying Effectiveness

With proper equipment maintenance, the technique is very effective for most components.

Drying Impact on Cleanliness

The process does not typically produce any contaminants.

Equipment Menu

Primary equipment is a vacuum chamber of appropriate dimensions. Ancillary equipment includes hoists and semiautomatic systems.

Associated Energy and Labor Requirements

The vacuum pump and the air circulation/moisture extraction system can require from fractional to 15–20 HP depending on chamber volume. Labor costs must be included in considerations of manually loaded chambers.

Safety and Environmental Issues

The chamber itself, even though under high vacuum, generally presents no immediate worker safety issues. Loading and unloading of the chamber must be done to minimize worker injury.

Drying Mixing Tanks, Vessels, and Blenders

The focus of this chapter has been the drying of parts. However, mixing tanks, vessels, and blenders used for both powdered and liquid products can also have critical drying requirements. The process of batching and blending products in food, confectionaries, pharmaceuticals, and cleaning agents is most commonly done in a circuit of stainless steel blenders, mixers, helical conveying tubes, and holding tanks. The use of composite holding tanks and vessels is also more common these days, particularly in pharmaceuticals processing.

In all of these environments where multiple product lines are processed and batched using the same tank and blender system, a CIP (clean-in-place) series of high-pressure wash nozzles is used to clean and disinfect the entire circuit. A cost-effective and rapid method of drying all residual rinse water from the entire blender and tank circuit is the use of heated and pressurized air blowing through the system for 15–30 min. A heated pressure blower produces 1 psi back pressure at 160°F and a volume equal to 1 air exchange of the entire system circuit every 30–60 seconds to completely dry and sanitize the tank or blender.

What Happens When Parts Drying Is Also Parts Washing?

Aqueous washing is used to mechanically remove contaminants that are of dissimilar material to the primary product. However, what if, in the manufacturing of a product, there is debris loosened from the surface of the primary product itself? Then, instead of aqueous washing, the use of "air washing" may greatly simplify the cleaning task.

For those situations where an effective high-velocity air blow-off may be used, possibly together with static neutralizing bars, no aqueous wash is even necessary. In many circumstances, the air wash process is both a push (air blow-off) as well as a pull (air vacuum exhaust) to remove the air-borne particles that have been liberated from the surface to ensure that no recontamination of the primary product can occur. These parts air washing systems can be conducted within a glove box environment or in a semi-enclosed zone of a product conveyor.

Integration of Drying Systems with Cleaning Systems

Integration of drying with the aqueous cleaning process is determined, in part, by the type of system employed: inline versus batch cleaning. In conveyorized inline cleaning, cleaning occurs in one zone, rinsing in a second zone, and drying in the last zone. Batch cleaning may be single zone, where all steps take place sequentially in a single chamber; or multiple zone batch, where cleaning, rinsing, and drying steps occur in separate chambers.

Integration is also dependent on the level of drying required.

- Industrial drying demands only visual cleanliness. If it looks clean, then it must be. This level of drying is associated with a broad range of heavy industry manufacturing requirements.
- Precision cleaning is the standard in a wide array of manufacturing, from metals to electronics, where surface contaminants remaining after cleaning and drying are measured by weight, electrical conductivity, optical scanners, and other methods.
- Critical cleaning has the most stringent cleaning and drying needs. Products made with critical cleanliness standards are generally those of the highest technology sector.

Any of these drying levels can be achieved using any of the aforementioned methods of throughput. A single zone batch-cleaning/drying process can work for automotive machine parts as well as for medical fiber optic components.

The key to effective integration of cleaning and drying is the correct evaluation of user needs. This can best be done by starting at the end result (with cleanliness requirements and throughput) and working backward to develop the proper equipment menu.

With today's fast-paced changes in technology and production capacity, users must continuously revisit the process to validate that the method used today still meets current quality production criteria. Increases in production and throughput, the ability to measure contaminants more accurately and increasing understanding of the effect of contamination levels on the quality of the end product all drive cleaning/drying integration.

Unless a component is cleaned and dried in a bubble, where all elements are controlled to absolute values, the potential for contamination exists the moment the component enters a new environment. Critical cleaning applications are done in clean rooms where the humidity level is controlled, yet humidity can increase from people just breathing, compromising the cleanliness of the part. Handling components through conveyorized processes or having air inadequately filtered from the compressed air or blower source compromises the cleaning method. Every engineer responsible for parts cleaning must ask the question, "Does the drying method maintain the results of the parts cleaning method?" If it does, then the integration of the drying and cleaning processes has been successful.

Conclusions

The appropriate dryer is very application dependent. General considerations include part size, production rate, final rinse temperature, drying standard (quantitative or subjective), next step after drying and current drying method. More specific questions related to blow-off drying are indicated in Figure 34.2.

Application questionnaire: drying/blow-off—parts manufacturing

(1) Please describe part:

Dimensions: Length Width Height Diameter ☐ Inches ☐ cm   Temperature of part ☐°F ☐°C

Smooth surface ☐ Thru holes ☐ Pockets ☐
Rough surface ☐ Blind holes ☐ Crevices ☐
Channels ☐ Protrusions ☐ Ribs ☐
Grooves ☐ Other

(2) What is the liquid or material to be removed?

Tap water ☐ Contaminants ☐ Acids/caustics ☐
D.I. water ☐ Dust ☐ Coatings ☐
Coolants/lubricants ☐ Wash solutions ☐ Other

(3) What surfaces require blow-off or drying?

All surfaces ☐ Left side ☐ Front ☐
Top ☐ Right side ☐ Back ☐
Bottom ☐ Other

(4) What is the method in the cleaning process?

Spray nozzles ☐ Dip tank ☐ Conveyor bath ☐ None yet ☐

(5) What type of conveyor is used?

				Width	Height	Length
Reel to reel ☐ Flat belt ☐ Roller ☐
Automatic hoist ☐ Open mesh belt ☐ Chain ☐ Dimensions:
Overhead monorail ☐ Manual feed ☐ Other ☐ Other

(6) Throughput is:

Continuous ☐ Speed is ____ ☐ Feet/min ☐ Meters/min
Indexed (stops and starts) ☐ Please describe:

(7) What is the next step after blow-off/drying?

Packaging ☐ Printing ☐ Oven/autoclave ☐
Next assembly ☐ Electrical testing ☐ Testing ☐
Manual handling ☐ Other

(8) What method do you use to determine dryness?

Visual/touch ☐ Gross weight measurement ☐ Optical scanner ☐ Dielectric testing ☐
Other ☐

(9) What is your current drying method?

Compressed air nozzles ☐ Fans ☐ Manual labor ☐
Compressed air/air knives ☐ Heater tunnel ☐ None-(new applications) ☐
Blowers/air knives ☐ Evaporative drying ☐ Other

(10) How effective is your existing drying method?

Very effective ☐ Somewhat effective ☐ Not effective ☐ Not applicable (new application) ☐

(11) What are the problems and costs associated with your current method of drying?

Quality-high reject rates ☐ Extra conveyor runs ☐
Decreased production ☐ Excess WIP inventory ☐
Excessive labor costs ☐ Excessive energy costs ☐ Other

FIGURE 34.2 Some questions to pose for appropriate dryer selection.

35

Liquid Displacement Drying Techniques

Phil Dale
Layton Technologies

Robert L. Polhamus
RLP Associates

Introduction

Drying techniques for complex or special parts cleaning fail to attract the same attention as cleaning and rinsing technologies. This is perhaps because drying naturally comes last or because it is viewed as an "easy" step, after the tough job of cleaning has been realized.

As removing soiling has its challenges, so too does the drying stage. New "contaminants," even something as seemingly basic as water, can offer potential problems for drying and satisfactory final acceptance of a parts cleaning operation.

The use of liquid displacement drying processes has tended to focus on high-end engineering and electronics applications. These tasks normally involve a measurable cleaning standard, which is more precisely measured than in most general cleaning applications. There is usually a reason to clean the components that provide value to the product, and if they were not clean they would most likely not perform adequately in the final application.

As users become increasingly definitive and subsequently more demanding in their cleaning requirements, the need for precision drying opportunities will grow. In addition, more products are being designed and fabricated to support the development of small geometry and close tolerances. Once water- or other aqueous-based cleaning or rinsing chemistry is entrapped in these devices, the range of effective drying options becomes limited.

Here, we examine the options for drying using liquids to displace and dry parts. This is a widely used technology for a wide range of applications including, but not limited to, electronics, medical, nanotechnologies, semiconductor, and wafer fabrication operations.

If a simple cleaning chemistry can be used with a simple rinse system, then it would be logical to assume that an equally simple drying technique can be employed. Unfortunately, this is often not the case.

When evaluating a potential drying technology, it must be remembered that the product is accepted as clean prior to entering the dryer. Therefore, the dryer could have the potential to recontaminate the product if cleanliness is not maintained at this stage. It would be impractical to expend any amount of time in cleaning a part beyond the capabilities of the drying system. As a result, it is critical that the drying technology be matched to the application.

Various drying technologies offer advantages while suffering from some disadvantages. Any of the three basic drying techniques—evaporative, mechanical displacement, and liquid displacement—can be optimized to provide the proper balance of technical performance and cost.

As in the evaluation of the cleaning and rinsing chemical, once a thorough investigation of the technical requirements of the application has been determined, several drying processes can be evaluated on their technical merit to meet the needs of the application. Having determined the acceptable drying technology, the equipment to implement the process is fairly self-evident and, subsequently, the cost is appropriate to the need.

Although liquid displacement drying is most closely associated with water removal, it can also be used in areas of solvent cleaning. As with water-based cleaning, a multi-tank solvent system may have one or more chemicals that are good cleaning agents but with low volatility and are therefore difficult to evaporate. These products can undergo a rinsing phase with a more volatile chemical and, eventually, either the rinsing chemical or an additional solvent is evaporated from the surface leaving the product surface dry. We focus in this chapter on the removal of water from the substrate although these same principles can be applied to the solvent cleaning systems.

Two liquid displacement processes are commonly utilized. One process involves chemicals that are soluble with water. The second process uses chemicals that are insoluble with water, and these are characterized as either more dense or less dense than water. These processes are described in the following text.

Water-Soluble Displacement

Water-soluble displacement drying is primarily accomplished through two methods. The predominant technique is vapor phase using isopropyl alcohol (IPA). Another process uses an immersion technology in cold high-purity water slow pull with an alcohol vapor layer in the same process chamber. In either case, it is the molecular interaction between the two chemicals that makes the process successful. IPA and water are perfectly miscible materials. That is, they will dissolve or take each other into solution in a nearly unlimited capacity. After the drying process, the effluent from the system will be a homogeneous mixture of alcohol and water. As a result, the alcohol is considered "consumed" in the process and is inappropriate for further drying unless the IPA can be purified and returned to the system. There are a variety of techniques suitable for this purpose and must be evaluated relative to the potential payback based upon the application.

With minimum exception, all liquid displacement techniques utilize batch processing. All of the following process description will center around the batch concept.

Vapor-Phase Process

The vapor-phase process is the most widely used of the liquid displacement processes. The equipment is simple in nature and the process relies very heavily on the principles of physical chemistry. The equipment configuration consists of a single chamber where the entire process is performed. The main features of the process chamber are heating elements, a vapor containment cooling coil, a liquid effluent capture tray, a freeboard zone, and a robotic lift mechanism (Figure 35.1).

The equipment must be prepared to induce and contain the process of vapor displacement. Liquid IPA is introduced into the process chamber to a preset level. Cooling water is circulated through the containment coils at an appropriate temperature and flow rate. If all conditions are acceptable, heat is applied to the liquid bath.

The IPA is heated until it reaches its boiling point. IPA vapors are generated from the boiling liquid and begin to rise upward in the process chamber. Through a combination of condensation on the side

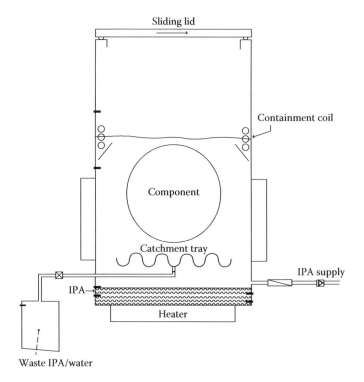

FIGURE 35.1 Vapor-phase process.

walls and a vapor density greater than air, the vapors move slowly upward from the liquid level displacing air in a "plug flow." The vapors will continue to rise upward in the chamber until they contact the containment coil. The containment coil will cause the vapors to condense on its surface, and as the vapors condense, more vapor is attracted to the coil to produce a continuous condensation action. Convection of the vapors toward the coil will produce a limit or ceiling on the height to which the vapors will rise.

As the vapors condense, they are directed back downward along the side walls and returned to the liquid volume. Upon return, the liquid is available to be regenerated into vapor. This process continues indefinitely in the process ready or idle mode until a drying cycle is initiated. Once the vapor zone has been fully established and stabilized, the unit is ready to process work. Prior to the implementation of a vapor-phase process, careful consideration should be given to product orientation and fixture design. Both factors must take into consideration the accessibility of the parts to the vapor, and the ability of the parts to drain freely. The critical nature of these factors will become self-evident from further discussion of the process.

The process contains three zones: the liquid level, vapor zone, and freeboard area. The liquid level is the amount of liquid IPA contained in the process chamber. In most cases, this level is no more than a few inches deep. The second zone is called the vapor zone. This is the area from the top of the liquid to the top of the vapor blanket at the point where it is captured by the containment coil.

The containment coil is located a specified distance from the top opening of the process chamber. The third zone is the area from the top of the vapor zone to the top of the process chamber called the freeboard zone. This dimension should be greater or equal to the height of the vapor zone. This freeboard plays a critical role in the containment of the vapors and the cost-effectiveness of the system.

The work and fixture (load) is placed on the robotic hoist platform or suspended from the robot arm. The temperature of the load must be below the boiling point of IPA, and, in general, the cooler the better. Once loading is complete, the process is initiated and the load is lowered from the home position through the freeboard zone into the vapor zone. This begins the first phase of drying called the displacement phase.

Once the load penetrates the vapor zone, it becomes a preferential condensation site for the vapors versus the containment coil, and the vast majority of the vapor condenses on the load. The rate of condensation on the coil observed during the idle mode will show significant, if not complete, reduction upon insertion of the load. The IPA will condense and mix with the water on the surface of the load. As the IPA liquid accumulates on the surface, it will fall from the surface by gravity taking dissolved water with it. In many cases, the water that is on the surface of the load is high-purity deionized (DI) water, which has a relatively high surface tension. The IPA has a relatively low surface tension, so it will "wet" the surface and also displace water from the load. The liquid condensate and water fall from the product and are collected on a liquid effluent capture tray and directed to waste.

Through the combination of solvation and displacement, all water is removed from the surface. After a period of time, based on water-loading and vapor generation rate, the surface becomes water free. Even though the surface is water free, as long as the load is below the boiling point of the liquid, condensation will continue.

The water layer on the parts has been displaced by a layer of IPA condensate. As the IPA condenses, it transfers energy to the load. Over time, the load temperature begins to rise. As the temperature approaches, the boiling point of IPA, the rate of condensation slows and eventually almost stops. This is the second phase of the process and is referred to as the thermal equilibrium phase. This phase begins when the water is displaced from the surface and ends when the load reaches a near equilibrium with the vapor where equal volumes of condensate and liquid are changing state on the surface. An indication that thermal equilibrium has been reached is when the condensate dripping from the containment coil has reached a rate equal to or close to the pre-cycle idling rate.

At this point in the process, the load is covered with a micro layer of IPA liquid. Although the surface may appear to be dry, this micro-layer exists on the surface in equilibrium with the vapor. In order to "dry" the surface, the load is removed from the vapor zone into the freeboard area. Once in the freeboard area, and even beginning during the transitional phase from the vapor zone into the freeboard area, the micro-layer is allowed to flash dry from the surface.

The IPA vapors that flash from the surface are more dense than air and, if allowed to remain undisturbed, will fall back into the vapor zone. Residence time in the freeboard area will allow these vapors to be recaptured as well as allowing the load to cool prior to removal from the system. After the freeboard residence has timed out, the load is returned to the home position.

The vapor-phase drying system is quite effective, but as can be implied from the process description, works very well on certain product configurations. Since the vapor phase provides little if any mechanical agitation, the process relies strictly on the affinity of IPA for water. If product or fixture configurations trap or hold water, the process is less effective. This is the most significant drawback to vapor phase and must be taken into consideration.

Vapor phase does provide two significant advantages over its competitive liquid displacement techniques. First, the process is extremely simple. It contains limited moving parts and simpler component design. It utilizes the laws of physics to transport the chemical in contact to the load and, in the vapor phase, has the ability to provide equal contact to all surfaces simultaneously.

The second significant advantage is that it can maintain chemical purity levels superior to other techniques. The process generates contaminant free chemistry by the production of the vapor zone. By definition, the vapor that condenses into liquid must contain no nonvolatile constituents since all components of the vapor must be volatile to exist in the vapor zone. The liquid in the sump remains pure since any condensate that will be mixed with water is immediately removed from the process chamber after it has been captured on the liquid effluent capture tray. This process prevents any build up of contamination from the influent alcohol or from the water carried into the dryer.

Dual Vapor Drying Systems

The drive to continually improve drying techniques and the ability to design processes for specific applications is never ending. Recently, a technique has been introduced that uses two immiscible fluids that create a constant boiling blend vapor phase. This technique is referred to as the dual vapor drying system.

The chemicals chosen for this process have the commonality of being relatively high in volatility with similar boiling points and relative vapor pressures. However, in order for this process to be successful, the fluids must be essentially immiscible with each other. By utilizing one chemistry that is soluble with water and one that is not, the amount of chemical that can be returned directly back to the system without contamination is increased, thus reducing overall chemical consumption while still providing effective drying. An additional advantage of this system is that if one chemical is chosen that is nonflammable, it can be used with a flammable solvent with the combined vapor phase being nonflammable.

One such system combines perfluoroether (PFE), where local legislation or regulation permits, with the conventional drying fluid IPA. The process utilized is very similar to conventional IPA vapor phase drying discussed above with minor differences in chemical management (Figure 35.2). From a drying perspective, the parts that can be effectively dried and the time of the process are essentially unchanged relative to conventional IPA drying.

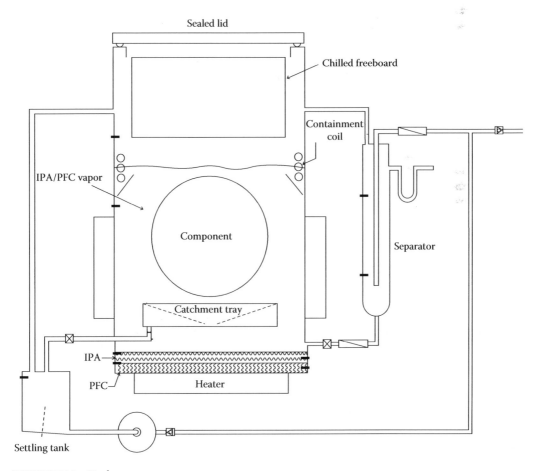

FIGURE 35.2 Dual vapor process.

The significant difference between this process and conventional IPA drying is that less IPA is consumed in the process, the vapor is nonflammable, and gravimetric separation recovers the PFE for reuse in the process.

A volume of liquid that is approximately 50% PFE and 50% IPA (V/V) is placed in the bottom of the process chamber. Since the liquids are immiscible and have a significant difference in density, the PFE will settle to the bottom and the IPA will form a layer on the top of the PFE. Heat is applied to the liquid. As the PFE liquid begins to boil, the PFE vapor must percolate through the IPA layer, heating the IPA, which although not boiling will contribute vapor to the system. Like previously described, the vapor is more dense than air and will displace all air out of the process zone to create a 100% vapor blanket above the liquid to the height of the condensing coil. The relative vapor pressures of each chemical contribute to a mixture of vapor that is approximately 50/50 by volume and is nonflammable.

During idling conditions, the vapor will condense on the coil with the liquid condensate being directed back down the chamber walls to return to the boiling solution. Upon the introduction of a load, the vapor will preferentially condense on the parts to be dried. The liquid condensate will drip off the parts and be collected on a condensate trough described in the conventional IPA process. The condensation of both vapors will contribute heat to the substrate and as the product approaches the temperature of the vapor, the process will eventually cease. Leaving the product in the vapor zone will eventually produce an equilibrium of vapor–condensate on the surface of the parts with its associated microlayer of liquid coating the parts.

The liquid that condenses on the parts is captured by a saucer tray as described in the earlier discussion of IPA vapor phase. However, in this case, the liquid will separate into two phases, a PFE and IPA–water mixture. This liquid effluent is sent to a holding tank where the liquid is allowed to settle and separate into two layers. The more dense PFE will settle to the bottom where it can be extracted and recirculated back into the process chamber. The IPA–water mixture is decanted off for reclaim or disposal. Since the PFE is not chilled, it retains much of its thermal content and is returned to the process chamber at elevated temperatures ready to begin the vapor cycle again. This conserves heat energy by not extracting the thermal value as with the waste IPA–water effluent.

Since IPA is consumed in the process, the level of IPA in the tank will be reduced as a function of process load volume. A special system of liquid-level floats is included in the design to continually make up the IPA volume from a sealed reservoir. By maintaining a relatively consistent ratio of IPA to PFE in the liquid phase, the composition of the vapor will remain constant, providing reproducible results.

In a conventional IPA process, the parts are removed from the vapor zone into the freeboard area above the condensing coil where final flash evaporation takes place. After a relatively short time, the parts are removed from the system. Any residual IPA that is left on the parts or that is distributed into the atmosphere around the unit is of a sufficiently low concentration to avoid any operator exposure or flammability concerns. A significant difference between the conventional and dual vapor process is that the cost of the PFE chemical is such that even minor losses will have cost impact on the process.

The dual vapor system is designed with a chilled freeboard to increase solvent retention. Above the primary condensing coil is a refrigerated plate operating at a temperature below the freezing point of water. This low temperature over a boiling environment creates a temperature inversion producing a dense layer of cold air above the vapor. The parts are raised from the vapor zone into the freeboard, where the vapor pressure of either constituent is essentially zero. As the condensate layer on the parts is flashed off the surface, it is immediately cooled and with its density being significantly higher than air, it will fall back into the vapor zone for recapture. The refrigerated freeboard area works to reduce the amount of the expensive PFE that will exit the system as well as significantly reducing the amount of IPA evolved from the system. Although the cost consideration is not as critical for the IPA, it is a volatile organic compound (VOC) and is tightly regulated by many national and local environmental laws. The design of this system is such as to produce very low solvent emissions. The unit can also be supplied with a lid that seals during the process to lower solvent emissions even further.

In addition to lowering overall vaporous emissions, the dual-phase system reduces the amount of chemical consumed in the process. In the conventional IPA dryer, the entire vapor zone is composed of

IPA and it condenses on the parts which solvates and displaces water from the surface. The volume of liquid IPA generated during the vapor phase drying process is contaminated with water. In some cases, the water content will be a little as 1%–2% or as high as 12%–15%. In either case, the IPA is unusable in this state and must be extracted from the system for disposal or reclaim. The difficulty of removing IPA from the water is exacerbated due to the fact that IPA and water will form an azeotrope which makes simple single-plate distillation inappropriate to provide solution suitable for use. Multi-plate distillation or membrane separation is usually required, often with significant cost impact.

The dual vapor system reduces the amount of water laden IPA waste by simply making less IPA available to be contaminated. Since the vapor phase is roughly 50/50 PFE to IPA, and water is immiscible in PFE, all of the water removed must be contained in the liquid IPA effluent. This automatically doubles the water loading in the IPA by halving the volume condensed.

The major advantage of this process is the ability to efficiently use IPA as the drying fluid, which has established itself as a product of choice. The long history of IPA in this application provides confidence in the performance, whereas this technique eliminates some of the negative aspects. By providing a vapor phase that is nonflammable the safety aspects of using IPA are addressed. Also, since the system has emission control technology targeted at the maximum level of solvent retention, it reduces the potential environmental impact from using IPA in communities sensitive to VOC emissions. Also, the total amount of chemical consumed in the process can be reduced through effective recirculation of the PFE solvent.

One drawback to this process is the potential of increased cost to add the solvent retention technology. Another potential drawback may be in extended process times necessitated by increased residence time in the chilled freeboard to accommodate solvent retention criteria.

Liquid/Vapor Process

A second process that uses a combination of liquid immersion followed by a vapor deposition is called the "Marangoni" process. This process is generally used following a high purity DI water final rinse stage. The principle for displacement differs significantly from the previously described technique (Figure 35.3).

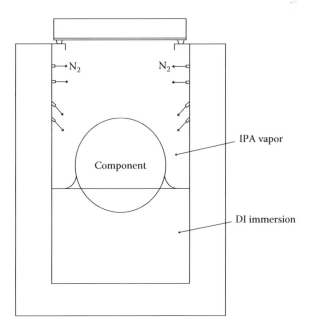

FIGURE 35.3 Marangoni process.

The fundamental principle for displacement of the water relies on the differential surface tension between the DI water and the IPA. Following a high-purity rinse stage, the substrates are immersed in a bath of DI water at ambient or slightly above ambient temperature. The substrates are slowly removed from the water either by gently raising the substrates from the bath or, more generally used in production, slowly draining the water from the tank. At this point, the substrates are generally removed from the carrier through a lift mechanism that suspends them in the liquid bath providing minimal contact points. The slow withdrawal of the DI water creates a sheeting effect on the surface to strip away any residue. A meniscus is formed that trails the liquid withdrawal. As the liquid level is being lowered, IPA vapor is introduced into the system. Some of the vapor coats the exposed surface as some of it is absorbed into the surface layer of the DI water.

At this point, an intriguing phenomenon occurs. A discovery by Lord Kelvin's brother, James Thompson, identified a reaction in fluids of dissimilar surface tensions as related to a teardrop effect witnessed on vessels containing alcoholic beverages. He identified this in a paper published in 1855. Subsequent investigations by a fluid dynamics investigator, Marangoni, provided the theory and the name for the process.

As the IPA is absorbed into the micro-layer of liquid on the surface of the substrate, it lowers the surface tension relative to the bulk liquid in the process tank. The physics of the two dissimilar surface tensions drive the system to equilibrium by forcing the low surface tension liquid toward the liquid of higher surface tension. This produces a flow that essentially strips the water layer from the substrate. The bulk liquid is continually purged with pure DI water to maintain its high surface tension.

Once the substrate has been completely removed from the liquid, a purge of drying atmosphere, usually nitrogen gas, is admitted to the system and any residual IPA that may be on the surface of the substrate is evaporated.

Small amounts of IPA are consumed in the process as it is absorbed into the bulk liquid or evaporated. The DI water must be completely drained from the system and refilled between cycles since it has become contaminated with the IPA.

Fixturing, substrate topography, and fluid motion are all critical to the success of this process. In most commercial systems, the substrate is removed from the carrier during the water removal phase. This is a result of requiring a smooth flow of liquid across the substrate surface which could be impaired or negated by contact points or irregular configurations in the carrier. The product surface must also be extremely smooth in order to guarantee successful water removal. Finally, the motion of the water across the surface must be controlled to eliminate any turbulence. Agitation at the interface of the IPA rich layer and bulk liquid will result in a disruption of the surface tension differential and negate the effects of the Marangoni principle.

The technical requirements stated above make the product configuration much more complex than other dryer designs. More moving parts and the possibility of disturbances in the flow of fluid across the surface make the system less forgiving in its operation. The process also consumes large volumes of DI water since the entire tank volume must be exchanged after each cycle. Even though there is only a small amount of IPA dissolved in the liquid, it is quite difficult to recover in situ and must either be thrown out or recycled.

The major advantage of Marangoni to other principles is that it essentially uses water to dry itself. The small amount of alcohol used is fairly insignificant from a chemical consumption perspective. Also, since the process takes place at room temperature or slightly higher, and the water is removed from the surface without evaporation, the results should be spot free drying and minimal energy consumption.

Liquid Displacement

Both of the previously discussed techniques relied on the solubility of water with the drying chemical (IPA) for success. As was mentioned, the solubility of water in alcohol enhanced the removal process, but rendered the IPA unsuitable for reuse in the system. There exist several chemicals that can successfully remove water from the surface without being soluble. The advantage of these chemicals is that they will reject the water down to part per million levels allowing the drying chemistry to be recycled in the system. Closed-loop recycling of the chemical will significantly reduce the waste stream volume and enhance system economics.

The chemicals that will dry while rejecting the water are also highly volatile and can be recirculated in the system through distillation. Although these various chemicals use the same principles described below, they differ significantly in their behavior with the water once it is removed. Two types of chemicals are used in the liquid displacement technique: one is more dense than water and the other is less dense. Both techniques will be discussed.

Liquid More Dense than Water

Fluids more dense than water have been used for many years in industrial applications. These products were based on chlorofluorocarbons (CFC) chemicals. Two methods dominated the process. One process involved using a CFC with a surfactant, the other blended the CFC with alcohol. The chemical management within the system was slightly different and so was the process.

As a result of the terms of the Montreal Protocol, the CFC chemistries were banned worldwide for use in this application. Several chemical companies have developed replacement chemistries for this process. These chemicals are hydrofluorocarbons (HFC), hydrofluoroethers (HFE), and some brominated hydrocarbons. All of these chemicals are being developed with surfactant additives to assist in the drying process. These products are also being blended with alcohols to replace CFC products in a non-surfactant drying applications.

Displacement Fluid with Surfactant

Unlike the previously discussed techniques that used vapor phase alcohol, the liquid displacement technique requires a minimum of two process tanks. One process tank contains the drying chemical and the surfactant while the second chamber contains pure drying chemical distillate. The chemical management is strikingly similar to the conventional solvent cleaning system, and this makes sense if water is considered to be a contaminant.

In the boil sump, usually situated on the left-hand side of the equipment (Figure 35.4), the drying chemical and surfactant are blended. This mixture is boiled to create a vapor as described in previous techniques. The surfactant is nonvolatile or of extremely low vapor pressure at the boiling point of the drying chemical to contribute little if any vapor. The vapor is condensed on a containment coil and the liquid returned to the right hand immersion sump. This pure distillate stream fills the immersion sump and overflows a weir to return to the boil sump.

The drying process takes place in a manner similar to conventional solvent cleaning. The product is lowered into the unit over the boil sump. As it enters the vapor zone, solvent vapors condense on the part and begin to displace the water on the surface. This liquid falls from the surface into the boil sump. The product continues to be introduced into the equipment until it is fully immersed in the boil sump.

When the product is fully immersed, the combination of the drying chemical and surfactant displaces the water from the surface, including blind holes and other areas where water may be trapped. The water that is displaced rises to the surface of the sump since it is insoluble and less dense than the drying fluid.

The water displaced from the product will accumulate at the surface of the boil sump. To prevent recontamination of the surface of the parts, the surface of the fluid is continually purged by a sparging mechanism to skim the surface layer off the sump and force it into a water separator. The waters separator is a series of baffled chambers that will allow the water to separate from the drying fluid and surfactant mixture for removal from the system. The pure drying fluid–surfactant mixture is returned to the boil sump through the sparging system to continue the process. The water is decanted from one of the baffled chambers and sent to drain or collected for proper disposal.

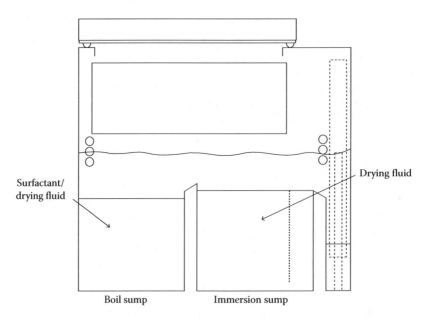

FIGURE 35.4   More dense than water—surfactant/displacement fluid.

After the product has been removed from the boil sump and suspended in the vapor zone to allow bulk liquid to drip off the product, the parts are transferred to the pure distillate immersion sump. The parts will be coated with a micro-layer of drying fluid–surfactant. In the immersion sump, the surfactant is removed from the surface by its solvency in the drying fluid. The mechanism in this sump will be to remove all surfactant from the surface, disperse it throughout the bath, and redeposit it on the surface at the equilibrium concentration of the mixture in the bath. This fluid is continually being returned to the boil sump by overflow of the weir, which returns the surfactant to the boil sump for reuse. Since the surfactant is essentially nonvolatile, the concentration in the system will remain fairly constant across a long period of time and will produce consistent performance. After a specified period of time, the parts are removed from the immersion sump and suspended in the vapor zone above the immersion sump.

The temperature of the immersion sump is below the boiling point of the drying fluid. While the surfactant rinse is occurring, the parts cool down. When they are suspended in the vapor zone after the surfactant removal they will experience a final distillate rinse. Pure distillate from the vapor zone will condense on the parts until they reach the micro-layer of thermal equilibrium, previously described, and the residual layer will be flashed off in the freeboard area for the final "dry" cycle. Water-free and solvent-free parts are removed from the system.

This process offers several advantages over the previously described techniques. The primary advantage is its ability to dry complex geometries by going into full immersion while maintaining chemical purity through the drying fluids ability to reject the water for rapid recycling. Another advantage is that it performs at lower temperatures than vapor phase alcohol dryers. It also has lower chemical consumption than previously described techniques.

The major disadvantage of the system is that the use of a surfactant can raise the suspicion of residue on the product after drying. Although this can usually be addressed by multiple rinse sumps, it could be considered unacceptable in some applications. Other disadvantages are the relatively large footprint of the equipment and the cost of the drying fluid. The drying fluids are quite expensive relative to alcohol and require sophisticated equipment design features to minimize emissions.

Displacement Fluid with Alcohol

The fluids discussed above are also being blended with alcohol to eliminate the need for the surfactant. In this process, alcohol is blended with the drying fluid in the boil sump. The formulation is generally 10% IPA and 90% drying fluid. Although the alcohol is soluble in the drying fluid it does not form an azeotrope at that concentration but will form an azeotrope at a very low percentage as it is boiled and generated into a vapor.

The vapor generated from the boil sump will form an azeotrope with the alcohol, and in a manner described earlier, will be condensed on a cooling coil and sent to an immersion sump (Figure 35.5). The parts to be dried are immersed in the immersion sump. Here, the small amount of alcohol with the high-density drying fluid will combine to remove the water from the surface. The alcohol is preferentially soluble in the water and produces a phase separation creating an insoluble layer of alcohol and water. This liquid layer is less dense than the drying fluid and therefore rises to the top of the sump as previously described in the surfactant system. The surface is sparged to force the insoluble layer into the water separator. The water/alcohol layer is separated from the drying fluid and the drying fluid is returned to the boil sump via the sparging system. Since no surfactants are used and all constituent chemicals are volatile, an immersion rinse after immersion in the drying sump is not required.

The drying cycle is essentially identical to the vapor phase drying cycle described earlier. The chemical composition of the vapor is an azeotrope of the drying fluid/alcohol and will condense on the parts and heat the substrate until condensation ceases. This process will tend to displace any liquid drag out from the water removal step and leave the product water- and liquid-solvent free.

The major advantage of this process is that it is able to use alcohol as a drying fluid without the concern for flammability while also being able to rapidly reject the water from the system. Although this process maintains the advantageous aspect of immersion drying, it produces a waste stream of alcohol and water not generated in the surfactant process. It does eliminate the concern of surfactant residue, but since alcohol is consumed in the process, requires that the alcohol level in the boil sump be monitored and adjusted to produce consistent drying performance. Another aspect of this technique is that

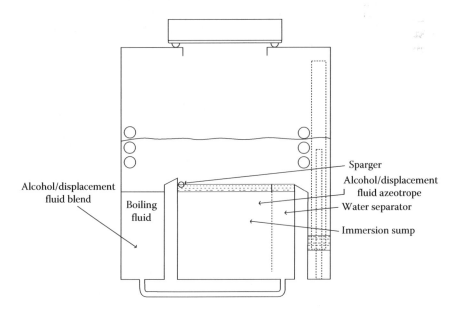

FIGURE 35.5 More dense than water—alcohol/displacement fluid.

the top layer of the immersion must be adequately purged to prevent the redeposition of water from this layer. If this is a critical concern, an additional sump may be used for safety.

Liquid Less Dense than Water

Liquid displacement with fluids less dense than water and insoluble with water have been used for many years. The established techniques usually utilize one of an assortment of hydrocarbons that displace the water from the surface and then are air dried themselves. Many of these techniques are not able to produce residue-free surfaces due to nonvolatile fractions in the hydrocarbon. With water displacing oils, a significant fraction of the solution is nonvolatile with the intended purpose of leaving behind a residual material to protect the surface from oxidation or other aspects of contamination.

These dryers are usually quite simple in design, generally consisting of a single immersion tank (Figure 35.6). The process is also simple. The parts to be dewatered are placed in the immersion bath and usually suspended off the bottom with fixturing or a work rest. Some form of mechanical energy is applied; usually vertical agitation in the fluid, air agitation, or ultrasonics. The water is displaced from the surface, and since it is more dense than the displacement fluid, it sinks to the bottom of the tank. The bottom of the tank below the work rest or below the suspended parts is usually necked down to produce a funnel-shaped section where the water is collected. The water level in the tank can be observed through the use of a sight glass or other methods. When the water accumulates to a proper level, it can be drained from the system off the bottom of the tank. When it is drained, a small layer should be left in the tank to prevent draining of any hydrocarbon into the water effluent.

The advantage of this technology is its simplicity both in process and design. The equipment is very inexpensive as are the displacement fluids. It has a disadvantage in that its ability to totally remove all water from the surface is suspect since it relies strictly on unenhanced chemical incompatibility. Without affecting the substrate affinity for water, it can not prevent redeposition of water should the opportunity arise.

Displacement chemicals that are volatile, less dense than water, and insoluble in water offer an additional drying option. Much of the development work in this area has centered around volatile methyl siloxane (VMS) fluids. These chemicals are insoluble in water and are combined with a surfactant to displace water from the surface of the substrate. The development of this technology is relatively recent and is still being tested to optimize its performance.

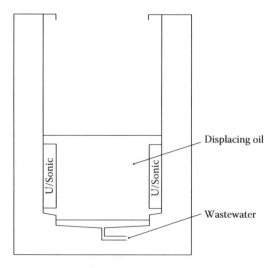

FIGURE 35.6   Less dense than water—oil displacement.

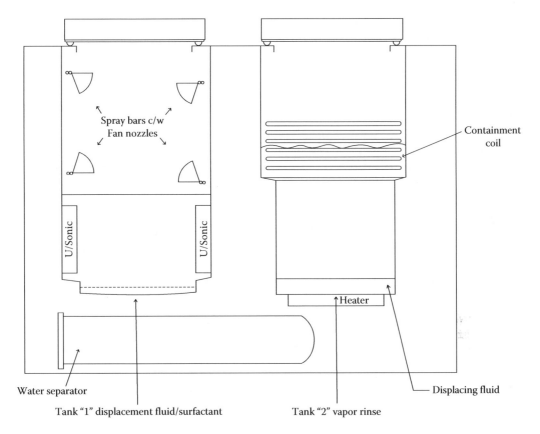

FIGURE 35.7 Less dense than water—Surfactant/displacement fluid.

The process for these chemicals differs from the more dense than water technique. Although the process still uses a vapor phase drying step, the equipment is considerably different than the systems described above (Figure 35.7). The equipment is a two-sump design, but the chambers are isolated in that they do not share a common vapor zone. The primary water displacement takes place in a chamber using room temperature or slightly above ambient drying fluid. The parts are immersed in a bath of displacement fluid and surfactant. The water is removed from the surface, and being more dense than the drying fluid, sinks to the bottom of the liquid bath. After a specified period of time, the parts are removed from the liquid bath and suspended above the fluid.

The parts are coated with the liquid from the immersion sump, which is a combination of displacement chemistry and surfactant. The bulk liquid is allowed to drip off the parts. The parts are then sprayed with recycled displacement fluid which will help to knock off any of the drag out layer coating the parts. After the spray has timed out, the parts are removed from the dewatering chamber and transferred to the drying chamber.

During the dewatering cycle, the liquid immersion sump is continually recycled through a water separation system that strips the water from the solution and returns it to the chamber. This material is also used as the source for the spray following the immersion dewatering.

The drying chamber is a single sump design very similar to the vapor-phase drying system discussed previously, and it essentially performs the same process. In this chamber, the displacement chemical without surfactant is boiled to create a vapor. The vapor is contained with a cooling coil to define a vapor zone, and the cooling coils are located at a depth in the chamber to provide acceptable freeboard. The vapor that condenses on the cooling coil is directed downward along the side of the chamber walls to be returned to the liquid level to complete the distillation cycle.

The displacement fluid wet parts from the dewatering step are introduced into the drying chamber. They are immersed into the vapor zone and the displacement chemical condenses on the surface. This liquid condensate flushes off the displacement fluid and the liquid falls into the liquid level. The vapors continue to condense on the parts until the parts approach the boiling point of the displacement chemical, completing the thermal equilibrium phase of the process. After thermal equilibrium is reached, the parts are elevated into the freeboard area where the residual micro-layer of displacement chemistry is flashed off. The product can be removed from the system water and displacement chemical free.

During the vapor rinse phase, the vapor condenses on the parts and removes the displacement fluid which falls into the sump. This liquid contains displacement fluid and surfactant as well as the pure distillate from the vapor zone. The surfactant is nonvolatile so as the solution boils it does not contribute to the vapor zone, as a result the vapor zone remains pure regardless of how much product is processed.

The introduction of wet parts into the drying chamber, and the collection of the displaced fluid in the liquid level, results in a net increase in liquid volume in the drying chamber. Liquid-level floats monitor the level in the chamber and periodically liquid from the sump is returned to the dewatering chamber. This fluid contains the displacement chemistry as well as surfactant that has accumulated in the drying chamber. This action returns the surfactant to the dewatering chamber to prolong its useful life.

The most significant advantage of this process is that the water leaves the surface and falls to the bottom of the tank. Unlike the more dense than water technique where the water rises, this process eliminates the necessity to remove dewatered product through a layer of liquid that may contain some water. The process is suited to complex geometries because of its immersion step. These chemicals are also more moderately priced than the more dense than water products and have essentially no adverse environmental impact associated with global warming or ozone depletion.

The major drawback of this system is that the chemicals are flammable; however, they are not VOC which may make them acceptable in specific environmental situations. The equipment also has a relatively large footprint.

Summary

The attempt in this limited evaluation is not to make inference that all liquid drying techniques are applicable across the board but merely to show that the availability of drying techniques is almost as diverse as the applications they address. In any case, it is the responsibility of the user to provide the necessary insight into cleanliness standards, evaluation criteria, costs, environmental impact, and other specific considerations relative to his or her particular requirements. He or she will utilize this information in collaboration with the chemical and equipment suppliers to develop his or her appropriate process. As it is with all things, however, change will occur and any decisions made today must be considered tentative. Careful consideration should also be given to future needs and changes in technology so a optimized position can be established that satisfies today's needs while being flexible enough to accommodate tomorrow's.

Glossary of Terms and Acronyms

Note: This is intended as an explanation of some of the more commonly used terms and acronyms related to cleaning chemicals and processes; it is not a complete list of all technical terms or abbreviations used in this book. In general, the included terms are referred to in more than one chapter. These definitions are to be considered to be descriptive rather than necessarily to be formal definitions.

AAMI: Association for the Advancement of Medical Instrumentation, an organization that sets standards to increase the understanding, safety, and efficacy of medical instrumentation.

Abrasive media: Materials used to remove soil via the momentum of impact.

ACGIH: American Conference of Governmental Industrial Hygienists; sets TLVs.

Airless: A description of an enclosed cleaning system that is sealed to contain either full vacuum (~1 mmHg) or a pressure significantly elevated above ambient (~800–10,000 mmHg).

Airtight: A description of an enclosed cleaning system that is sealed to contain a light pressure above ambient, typically about 0.5 psig.

Aqueous cleaner or process: Water based, may contain significant levels of organic and/or inorganic compounds.

ASTM: American Society for Testing and Materials; a group that establishes testing standards.

Atmospheric lifetime: The length of time a chemical may persist in the atmosphere before breaking down to other compounds; a measure of the potential for climate change.

Azeotrope: A solvent blend that, over a limited range of temperatures, maintains the same relative concentrations as the mixture components evaporate.

Benchtop cleaning: Generally referred to as a small-volume, labor-intensive, nonautomated cleaning process performed in the open rather than in specially designed cleaning tanks; examples include overhaul and repair and spot cleaning.

Bio-based cleaning agent: Uses animal or more typically plant materials, favored because ingredients are renewable resources.

Bi-solvent: Patented sequential cleaning agent process.

CAA: Clean Air Act; the U.S. legislation that regulates air quality standards, including the phase out of ODCs.

Cal/OSHA: California Occupational Safety and Health Administration; sets standards that are legally enforceable in California.

CAS: Clean air solvent; cleaning agents that have been analyzed by South Coast Air Quality Management District (SCAQMD) in California and found to meet their stringent environmental requirements for VOCs, ODCs, GWPs, and air toxics.

CAS number: Chemical abstracts service numbers, unique chemical identifiers.

Cavitation: Vacuum "bubbles" created by negative pressures in ultrasonic and megasonic processes.

CFC: Chlorofluorocarbon.

Cold cleaning: A cleaning process in which the cleaning solvent is below its boiling point (as distinguished from vapor degreasing).

Contaminant: Material that has the potential to degrade the appearance or performance of a part, component, or assembly.

Co-solvent: A sequential process using a different solvent for a rinse; or two cleaning agents in the same tank.

D-Limonene: A citrus-derived organic cleaning solvent.

DMSO: Dimethyl sulfoxide; a cleaning solvent.

Dragin: Material (cleaning chemicals and contaminants) brought in from a previous cleaning step.

Dragout: Material (cleaning chemicals and contaminants) carried over to a subsequent cleaning step.

EPA: Environmental Protection Agency; the U.S. government agency responsible for setting and administering air and water standards.

ESCA: Electron spectroscopy chemical analysis; an analytic technique for determining surface contamination; also known as XPS.

Flammable: Used to describe a combustible material that ignites very easily, burns intensely, or has a rapid rate of heat spread.

Flash point: The lowest temperature of a flammable liquid at which vapors are given off to form a flammable mixture with air, near the surface of the liquid or within the container; test methods differ by geographic locality.

Freeboard: A term used in vapor degreasers defined as the distance from the point where the boiling solvent vapor idles to the top of the machine opening.

Freon: A trade name (DuPont) for CFC-113; sometimes applied generically to CFCs.

FTIR: Fourier transform infrared spectroscopy; a surface analytic technique utilizing reflected infrared light to identify types of surface contaminants.

Greenhouse gas: A gas that persists in the stratosphere and acts to trap re-radiated heat from the earth's surface.

GWP: Global warming potential; a relative measure of a material's heat trapping ability as a greenhouse gas.

HAP: Hazardous air pollutant, a U.S. EPA classification of under 200 compounds.

HEPA: High Efficiency Particulate Air, a filter with specific properties used to minimize particulates in cleanrooms or controlled environments as well as to control hazardous emissions.

HFC (or HCFC): Hydrofluorocarbon; a class of chemicals developed as ODC replacements.

HFE: Hydrofluoroether; a class of chemicals developed as ODC replacements.

Hydrophilic: Water soluble.

Hydrophobic: Water insoluble; usually soluble in organic solvents.

IPA: Isopropyl alcohol, a common organic solvent.

IPC: Association connecting electronics industries; a leading source for industry standards in the electronics sector.

ISO standard: Standards adopted by the International Organization for Standardization, a Geneva-based organization that promulgates worldwide proprietary industrial and commercial standards.

KB: Kauri-butanol; a number used to compare the solubility of heavy oils in a particular solvent. It is the volume of solvent required to produce a defined degree of turbidity when added to standard solutions of Kauri resin in *n*-butyl alcohol.

Kyoto protocol: International agreement to limit emissions of greenhouse gases responsible for global warming.

LEL: Lower explosion level; the lowest concentration at which a mixture can explode.

MC (Meth): Methylene chloride.

Megasonics: A cleaning technique utilizing sound waves at frequencies higher than those for ultrasonics, from 500 kHz to 2 MHz.

Montreal protocol: International agreement to limit or eliminate production of ozone-depleting compounds (ODCs).

MSDS: Material safety data sheet.

Neat: A term meaning pure or undiluted.

NESHAP: National Emission Standards for Hazardous Air Pollutants; a series of U.S. federal regulations involving chemicals that can cause air pollution.

NMP: *N*-methyl pyrillodone; a cleaning solvent.

NPB (or *n*PB): *n*-propyl bromide; a cleaning solvent.

NVR: Nonvolatile residue; solid material left behind when a solvent evaporates.

ODC: Ozone-depleting compound, known to persist in the stratosphere and cause depletion of the ozone layer.

ODP: Ozone-depletion potential; a relative cumulative measure of the expected effects on ozone of the emissions of a gas relative to CFC-11.

OSEE: Optically stimulated electron emission; a surface analytic technique that measures the degree (but not the nature) of contamination by using UV light to stimulate the surface to emit electrons.

OSHA: Occupational Safety and Health Agency; the U.S. government agency responsible for setting and administering worker safety standards.

Particulates: Contaminate material with observable length, width, and thickness. In practice, an observable size will be about 0.1 microns or larger.

PCE (Perc): Perchloroethylene.

PEL: Permissible exposure limit. These are exposure guidelines for workers using the given chemical. PELs may be set by EPA or OSHA or by state agencies like Cal/OSHA.

PFC: Perfluorinated compounds containing fluorine and carbon but not chlorine or bromine.

POTW: Publicly owned treatment works; a local water treatment facility.

RCRA: Resource Conservation Recovery Act. Defines hazardous wastes and how to manage them.

REACH: Registration, Evaluation, Authorisation and Restriction of Chemical substances; a new European Community Regulation on chemicals and their safe use.

RO: Reverse osmosis; a filtering mechanism through a semipermeable membrane.

ROHS: Restriction of hazardous substances directive; an EU directive that restricts the use of six hazardous materials in the manufacture of various types of electronic and electrical equipment.

Saponification: The reaction between any organic oil containing reactive fatty acids with free alkalies to form soaps.

SARA: Superfund Amendments and Re-authorization Act. This act requires reporting of inventories and emissions of listed chemicals and groups.

SCAQMD: South Coast Air Quality Management District; the air quality regulating agency in southern California.

SEM: Scanning electron microscopy; a surface analytic technique involving imaging a surface by means of an electron beam.

Semi-aqueous: A sequential process using both organic solvent and water rinse.

SIMS: Secondary ion mass spectroscopy; a surface analytic technique using atoms ejected from a surface to identify contaminants.

SNAP: Significant new alternatives policy; an EPA effort to identify CFC replacement chemicals.

Soil: Matter out of place, contamination.

Solvent: Organic (carbon-containing) liquids; usually distinguished from aqueous.

STEL: Short-term exposure limit; a 15 min TWA exposure that should not be exceeded at any time during the day (see *TWA*).

STOC: Solvent and Adhesives Technical Options Committee; a United Nations UNEP committee that provides a great deal of input into worldwide environmental policy on cleaning.

Stoddard solvent: A common hydrocarbon blend used for cleaning oils.

Stratosphere: The atmospheric layer above the troposphere; considered to be above about 7 mi.

Surfactant: A material added to water or a solvent in order to increase wettability.

TCA: 1,1,1-trichloroethane (also called methyl chloroform).

TCE (TRI): Trichloroethylene.

TDS: Total dissolved solids; a measure of concentration of dissolved contaminants.

TLV: Threshold limit value; a concentration level above which there may be adverse health risks on exposure; usually set by the American Conference of Governmental Industrial Hygienists (ACGIH).

TOC: Total organic carbon; a measure of concentration of organic matter in water.

Troposphere: The lower layer of the atmosphere.

TWA: Time weighted average; an employee's permissible average exposure in any 8 h work shift of a 40 h week (see *STEL*).

UEL: Upper explosion level; the maximum concentration at which a mixture can explode.

Ultrasonics: A cleaning technique utilizing sound waves from 20 kHz to over 100 kHz.

UNEP: United Nations Environment Programme; a United Nations group that includes the STOC.

Vapor degreasing: A cleaning process in which, at least for the final cleaning, the part is suspended above a boiling solvent and is cleaned by the condensate of freshly distilled solvent vapor.

VMS: Volatile methyl siloxane; a silicon-based cleaning solvent.

VOC: Volatile organic compound; responsible for smog formation in the troposphere.

XPS: X-ray photoemission spectroscopy (see *ESCA*).

Index